Handbook of Enology
Volume 2
The Chemistry of Wine Stabilization and Treatments
2nd Edition

Handbook of Enology
Volume 2
The Chemistry of Wine Stabilization and Treatments
2nd Edition

P. Ribéreau-Gayon, Y. Glories
Faculty of Enology
Victor Segalen University of Bordeaux II, France

A. Maujean
Laboratory of Enology
University of Reims-Champagne-Ardennes

D. Dubourdieu
Faculty of Enology
Victor Segalen University of Bordeaux II, France

Original translation by

Aquitrad Traduction, Bordeaux, France

Revision translated by
Christine Rychlewski
Aquitaine Traduction, Bordeaux, France

John Wiley & Sons, Ltd

Copyright © 2006   John Wiley & Sons Ltd, The Atrium, Southern Gate, Chichester,
                   West Sussex PO19 8SQ, England

                   Telephone (+44) 1243 779777

Email (for orders and customer service enquiries): cs-books@wiley.co.uk
Visit our Home Page on www.wiley.com

Reprinted August 2006

All Rights Reserved. No part of this publication may be reproduced, stored in a retrieval system or transmitted in any form or by any means, electronic, mechanical, photocopying, recording, scanning or otherwise, except under the terms of the Copyright, Designs and Patents Act 1988 or under the terms of a licence issued by the Copyright Licensing Agency Ltd, 90 Tottenham Court Road, London W1T 4LP, UK, without the permission in writing of the Publisher. Requests to the Publisher should be addressed to the Permissions Department, John Wiley & Sons Ltd, The Atrium, Southern Gate, Chichester, West Sussex PO19 8SQ, England, or emailed to permreq@wiley.co.uk, or faxed to (+44) 1243 770620.

Designations used by companies to distinguish their products are often claimed as trademarks. All brand names and product names used in this book are trade names, service marks, trademarks or registered trademarks of their respective owners. The Publisher is not associated with any product or vendor mentioned in this book.

This publication is designed to provide accurate and authoritative information in regard to the subject matter covered. It is sold on the understanding that the Publisher is not engaged in rendering professional services. If professional advice or other expert assistance is required, the services of a competent professional should be sought.

Other Wiley Editorial Offices

John Wiley & Sons Inc., 111 River Street, Hoboken, NJ 07030, USA

Jossey-Bass, 989 Market Street, San Francisco, CA 94103-1741, USA

Wiley-VCH Verlag GmbH, Boschstr. 12, D-69469 Weinheim, Germany

John Wiley & Sons Australia Ltd, 42 McDougall Street, Milton, Queensland 4064, Australia

John Wiley & Sons (Asia) Pte Ltd, 2 Clementi Loop #02-01, Jin Xing Distripark, Singapore 129809

John Wiley & Sons Canada Ltd, 22 Worcester Road, Etobicoke, Ontario, Canada M9W 1L1

Wiley also publishes its books in a variety of electronic formats. Some content that appears in print may not be available in electronic books.

Library of Congress Cataloging-in-Publication Data:

Ribéreau-Gayon, Pascal.
   [Traité d'oenologie. English]
   Handbook of enology / Pascal Ribéreau-Gayon, Denis Dubourdieu, Bernard Donèche ; original translation by Jeffrey M. Branco, Jr.—2nd ed. / translation of updates for 2nd ed. [by] Christine Rychlewski.
      v. cm.
   Rev. ed. of: Handbook of enology / Pascal Ribéreau Gayon ... [et al.]. c2000.
   Includes bibliographical references and index.
   Contents: v. 1. The microbiology of wine and vinifications
   ISBN-13: 978-0-470-01037-2 (v. 1 : acid-free paper)
   ISBN-10: 0-470-01037-1 (v. 1 : acid-free paper)
 1. Wine and wine making—Handbooks, manuals, etc. 2.   Wine and wine making—Microbiology—Handbooks, manuals, etc. 3.   Wine and wine making—Chemistry—Handbooks, manuals, etc. I. Dubourdieu, Denis. II. Donèche, Bernard. III. Traité d'oenologie. English. IV. Title.
   TP548.T7613 2005
   663'.2—dc22
                                                  2005013973

British Library Cataloguing in Publication Data

A catalogue record for this book is available from the British Library

ISBN-13: 978-0-470-01037-2 (HB)
ISBN-10: 0-470-01037-1 (HB)

Typeset in 10/12pt Times by Laserwords Private Limited, Chennai, India
Printed and bound in Great Britain by Antony Rowe Ltd, Chippenham, Wiltshire
This book is printed on acid-free paper responsibly manufactured from sustainable forestry in which at least two trees are planted for each one used for paper production.
Cover photograph by Philippe Roy, CIVB Resource Centre.

Contents

Acknowledgments	vii
Part One The Chemistry of Wine	**1**
1 Organic Acids in Wine	3
2 Alcohols and Other Volatile Compounds	51
3 Carbohydrates	65
4 Dry Extract and Minerals	91
5 Nitrogen Compounds	109
6 Phenolic Compounds	141
7 Varietal Aroma	205
Part Two Stabilization and Treatments of Wine	**231**
8 Chemical Nature, Origins and Consequences of the Main Organoleptic Defects	233
9 The Concept of Clarity and Colloidal Phenomena	285
10 Clarification and Stabilization Treatments: Fining Wine	301
11 Clarifying Wine by Filtration and Centrifugation	333
12 Stabilizing Wine by Physical and Physico-chemical Processes	369
13 Aging Red Wines in Vat and Barrel: Phenomena Occurring During Aging	387
Index	**429**

Acknowledgments

The authors would particularly like to thank the following people for their contributions to the new edition of this book:

— Virginie Moine-Ledoux for her work on the use of yeast mannoproteins in preventing tartrate precipitation (Chapter 1), as well as the stabilization processes for protein casse (Chapter 5)

— Takathoshi Tominaga for his elucidation of the role of volatile thiols in wine aromas (Chapter 7)

— Valérie Lavigne-Cruège for her work on the premature aging of white wines (Chapter 8)

— Philippe Darriet for his research into the organoleptic defects of wine made from grapes affected by rot (Chapter 8)

— Cédric Saucier for his elucidation of colloidal phenomena (Chapter 9)

— Michel Serrano for work on clarifying wines by filtration (Chapter 11)

— Martine Mietton-Peuchot for her research into physical processes for stabilizing wine (Chapter 12).

This book benefits from their in-depth knowledge of specialized fields, acquired largely through research carried out in the laboratories of the Bordeaux Faculty of Enology.

The authors are also especially grateful to Blanche Masclef for preparing a large proportion of the manuscript. They would like to thank her, in particular, for her hard work and dedication in coordinating the final version of the texts.

March 17, 2005

Professor Pascal RIBEREAU-GAYON
Corresponding Member of the Institute
Member of the French Academy of Agriculture

PART ONE

The Chemistry of Wine

1

Organic Acids in Wine

1.1 Introduction	3
1.2 The main organic acids	3
1.3 Different types of acidity	8
1.4 The concept of pH and its applications	9
1.5 Tartrate precipitation mechanism and predicting its effects	21
1.6 Tests for predicting wine stability in relation to crystal precipitation and monitoring the effectiveness of artificial cold stabilization treatment	28
1.7 Preventing tartrate precipitation	37

1.1 INTRODUCTION

Organic acids make major contributions to the composition, stability and organoleptic qualities of wines, especially white wines (Ribéreau-Gayon *et al.*, 1982); (Jackson, 1994). Their preservative properties also enhance wines' microbiological and physicochemical stability.

Thus, dry white wines not subjected to malolactic fermentation are more stable in terms of bitartrate (KTH) and tartrate (CaT) precipitation. Young white wines with high acidity generally also have greater aging potential.

Red wines are stable at lower acidity, due to the presence of phenols which enhance acidity and help to maintain stability throughout aging.

1.2 THE MAIN ORGANIC ACIDS

1.2.1 Steric Configuration of Organic Acids

Most organic acids in must and wine have one or more chiral centers. The absolute configuration of the asymmetrical carbons is deduced from that of the sugars from which they are directly

Table 1.1. The main organic acids in grapes

L(+)-Tartaric acid	L(−)-Malic acid	Citric acid
COOH–(HO,H)–(H,OH)–COOH	COOH–CH$_2$–(H,OH)–COOH	CH$_2$COOH–C(OH)(COOH)–CH$_2$COOH

| D-Gluconic acid | 2-keto D-Gluconic acid | Mucic acid |

| Coumaric acid (R$_1$ = R$_2$ = H) / Caffeic acid (R$_1$ = OH; R$_2$ = H) | Coumaryl tartaric acid |

derived. This is especially true of tartaric and malic acids (Table 1.1). The absolute configuration of the asymmetrical carbons is established according to the Prelog rules (1953). Further reference to these rules will be made in the chapter on sugars, which are the reference molecules for stereoisomerism.

1.2.2 Organic Acids in Grapes

The main organic acids in grapes are described (Table 1.1) according to the conventional Fischer system. Besides tartaric acid, grapes also have a stereoisomer in which the absolute configuration of the two asymmetrical carbons is L, but whose optical activity in water, measured on a polarimeter, is d (or +). There is often confusion between these two notions. The first is theoretical and defines the relative positions of the substituents for the asymmetrical carbon, while the second is purely experimental and expresses the direction in which polarized light deviates from a plane when it passes through the acid in a given solvent.

Tartaric acid is one of the most prevalent acids in unripe grapes and must. Indeed, at the end of the vegetative growth phase, concentrations in unripe grapes may be as high as 15 g/l. In musts from northerly vineyards, concentrations are often over 6 g/l whereas, in the south, they may be as low as 2–3 g/l since combustion is more effective when the grape bunches are maintained at high temperatures.

Tartaric acid is not very widespread in nature, but is specific to grapes. For this reason, it is called *Weinsäure* in German, or 'wine acid'. It is a relatively strong acid (see Table 1.3), giving wine a pH on the order of 3.0–3.5.

Tartrates originating from the wine industry are the main source of tartaric acid, widely used in the food and beverage industry (soft drinks, chocolates, cakes, canned foods, etc.). This acid is also used for medical purposes (as a laxative) and in dyeing (for mordanting fabric), as well as for tanning leather. Tartrazine, a diazoic derivative of tartaric acid, is the yellow coloring matter in wool and silk, but is also used as food coloring under the reference number E102.

L(−)-Malic acid is found in all living organisms. It is especially plentiful in green apples, which explains its German name *Äpfelsäure*, or 'apple acid'. It is also present in white and red currants, rhubarb and, of course, grapes. Indeed, the juice of green grapes, just before color change, may contain as much as 25 g/l. In the two weeks following the first signs of color change, the malic acid content drops by half, partly due to dilution as the grapes grow bigger, and also as a result of combustion. At maturity, musts from northerly regions still contain 4–6.5 g/l malic acid, whereas in southerly regions, concentrations are only 1–2 g/l.

Citric acid, a tri-acid, is very widespread in nature (e.g. lemons). Its very important biochemical and metabolic role (Krebs cycle) requires no further demonstration. Citric acid slows yeast growth but does not block it (Kalathenos *et al.*, 1995). It is used as an acidifying agent in the food and beverage industry (lemonade), while sodium (E331), potassium (E332), and calcium (E333) citrate have many uses in fields ranging from pharmaceuticals to photography. Concentrations in must and wine, prior to malolactic fermentation, are between 0.5 and 1 g/l.

In addition to these three acids, which account for the majority of the acidity in grapes, there are also phenol acids in the cinnamic series (e.g. coumaric acid), often esterified with an alcohol function of tartaric acid (e.g. coumaryltartaric acid).

Ascorbic acid (Figure 1.1) should also be mentioned in connection with these oxidizable phenol acids. It is naturally present in lactone form, i.e. a cyclic ester. Ascorbic acid also constitutes a Redox system in fruit juices, protecting the phenols from oxidation. In winemaking it is used as an adjuvant to sulfur dioxide (Volume 1, Section 9.5).

Must and wine from grapes affected by noble and/or gray rot have higher concentrations of acids produced by oxidation of the aldehyde function (e.g. aldose) or the primary alcohol function of carbon 1 of a ketose (e.g. fructose). Thus, gluconic acid, the compound corresponding to glucose, may reach concentrations of several grams per liter in juice from grapes affected by rot. This concentration is used to identify wines made from grapes affected by noble rot, as they contain less gluconic acid than those made from grapes affected by gray rot (Sections 10.6.4, 10.6.5 and 14.2.3). The compound corresponding to fructose is 2-keto gluconic acid (Table 1.1).

The calcium and iron salts of these acids are used in medicine to treat decalcification and hypochrome anemia, respectively.

Calcium gluconate is well known for its insolubility in wine and the turbidity it causes. Mucic acid, derived from galactose by oxidation, both of the aldehyde function of carbon 1 and the primary alcohol function of carbon 6, is just as undesirable. Also known as galactaric acid, it is therefore both

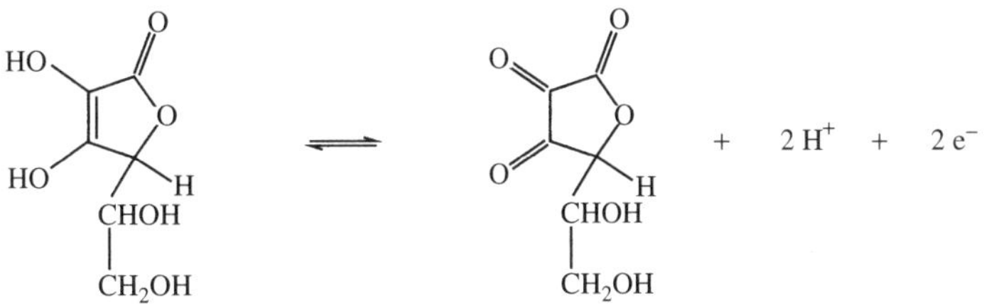

Fig. 1.1. Oxidation–reduction equilibrium of ascorbic acid

an onic and uronic acid. The presence of a plane of symmetry in its structure between carbons 3 and 4 makes it a meso-type stereoisomer. Mucic acid has no optical activity. Its presence has been observed in the crystalline deposits formed throughout the aging of sweet white wines made from grapes with noble rot.

1.2.3 Organic Acids from Fermentation

The main acids produced during fermentation are described in Table 1.2. The first to be described is pyruvic acid, due to its meeting function in the cell metabolism, although concentrations in wine are low, or even non-existent. Following reduction by a hydride H^- ion—from aluminum or sodium borohydride, or a co-enzyme (NADH) from L and D lactate dehydrogenases—pyruvic acid produces two stereoisomers of lactic acid, L and D. The first, 'clockwise', form is mainly of bacterial origin and the second, 'counter-clockwise', mainly originates from yeasts.

The activated, enolic form of the same acid, phosphoenol pyruvate (Figure 1.2), adds a nucleophile to carbon dioxide, producing oxaloacetic acid, a precursor by transamination of aspartic acid.

The enzymic decarboxylation of pyruvic acid, assisted by thiamin pyrophosphate (TPP) or vitamin B1, produces ethanal, which is reduced

Table 1.2. The main acids produced during fermentation

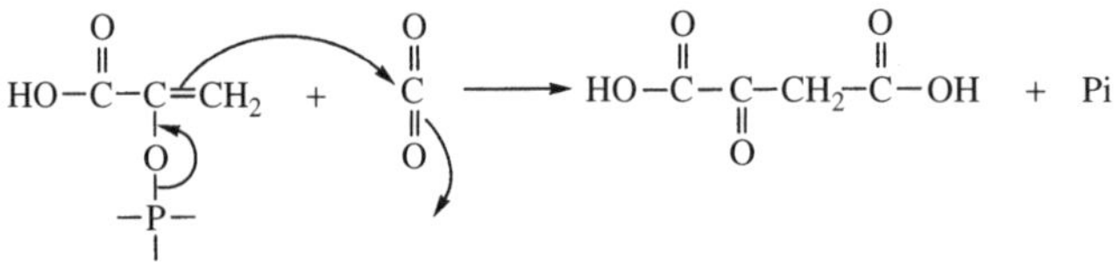

Pyruvic acid	L(+)-Lactic acid	D(−)-Lactic acid
COOH–C(=O)–CH$_3$	COOH–C(HO,H)–CH$_3$	COOH–C(H,OH)–CH$_3$

Succinic acid	Acetic acid	Citramalic acid
COOH–CH$_2$–CH$_2$–COOH	COOH–CH$_3$	CH$_3$–C(OH)–CH$_2$–COOH

Oxaloacetic acid	Fumaric acid
COOH–C(=O)–CH$_2$–COOH	HOOC–CH=CH–COOH

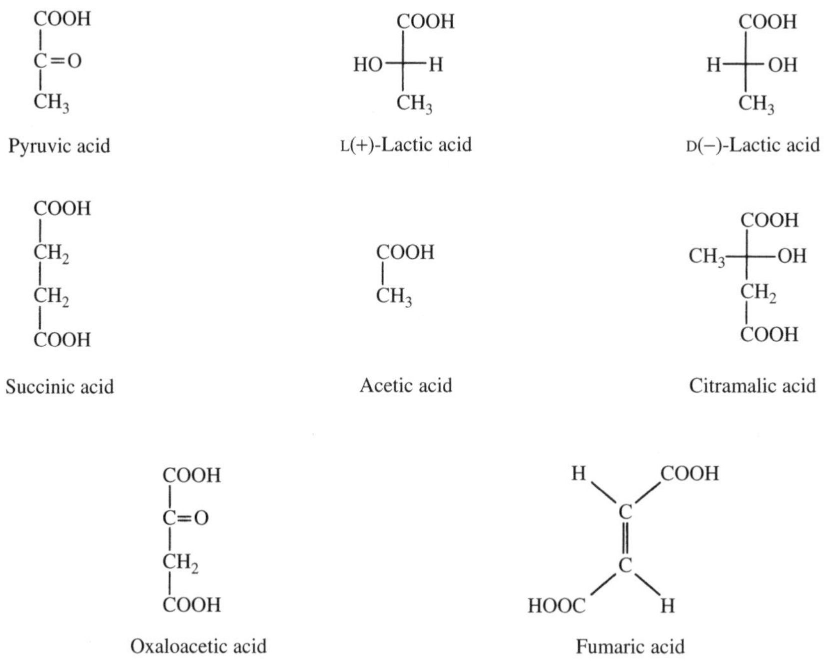

Fig. 1.2. Biosynthesis of oxaloacetic acid from phosphophenolpyruvic acid

Table 1.3. State of salification of the main inorganic and organic acids (Ribéreau-Gayon et al., 1972)

Category	Name	pK_a	Form in wine
Strong inorganic acids	Hydrochloric	Less than 1	Completely dissociated salts
	Sulfuric 1	Approx. 1	
	Sulfuric 2	1.6	
	Sulfurous 1	1.77	Bisulfite acid
	Phosphoric 1	1.96	Phosphate acid
Strongest organic acids	Salicylic	2.97	
	Tartaric 1	3.01	Acid functions partly neutralized and partly free (not highly dissociated)
	Citric 1	3.09	
	Malic 1	3.46	
	Formic	3.69	
	Lactic	3.81	
	Tartaric 2	4.05	
Weakest organic acids	Benzoic	4.16	
	Succinic 1	4.18	
	Citric 2	4.39	
	Acetic	4.73	Free acid functions (very little dissociated)
	Butyric	4.82	
	Propionic	4.85	
	Malic 2	5.05	
	Succinic 2	5.23	
	Citric 3	5.74	
Weak inorganic acids	Phosphoric 2	6.70	
	Carbonic 1	6.52	Free acid functions (almost entirely non-dissociated)
	Sulfurous 2	7.00	
	Hydrogen sulfide 1	7.24	
	Carbonic 2	10.22	
	Phosphoric 3	12.44	
Phenols	Polyphenols (tannin and coloring)	7–10	Free (non-dissociated)

to form ethanol during alcoholic fermentation. Its enzymic, microbial or even chemical oxidation produces acetic acid.

Another acid that develops during fermentation due to the action of yeast is succinic or 1-4-butanedioic acid. Concentrations in wine average 1 g/l. This acid is produced by all living organisms and is involved in the lipid metabolism and the Krebs cycle, in conjunction with fumaric acid. It is a di-acid with a high pK_a (Table 1.3). Succinic acid has an intensely bitter, salty taste that causes salivation and accentuates a wine's flavor and vinous character (Peynaud and Blouin, 1996).

Like succinic acid, citramalic or α-methylmalic acid, confused with citric acid in chromatography for many years, is of yeast origin.

In conclusion, it is apparent from this description that, independently of their origins, most of the main organic acids in must and wine consist of poly-functional molecules, and many are hydroxy acids. These two radicals give these acids polar and hydrophilic characteristics. As a result, they are soluble in water, and even in dilute alcohol solutions, such as wine. Their polyfunctional character is also responsible for the chemical reactivity that enables them to develop over time as wine ages. In this connection, results obtained by monitoring ethyl lactate levels in Champagne for 2 years after malolactic fermentation are highly convincing. Indeed, after 2 years aging on the lees, concentrations reach 2 g/l and then decrease. The degree of acidity, indicated by their pK_a values, controls the extent to which these acids are present in partial salt form in wine (Table 1.3).

A final property of the majority of organic acids in wine is that they have one or more asymmetrical carbons. This is characteristic of biologically significant molecules.

1.3 DIFFERENT TYPES OF ACIDITY

The fact that enologists need to distinguish between total acidity, pH and volatile acidity demonstrates the importance of the concept of acidity in wine. This is due to the different organoleptic effects of these three types of acidity. Indeed, in any professional tasting, the total acidity, pH and volatile acidity of the wine samples are always specified, together with the alcohol and residual sugar contents.

The importance of total acidity is obvious in connection with flavor balance:

$$\begin{array}{c}\text{sweet taste}\\\text{(sugars, alcohols)}\end{array} \rightleftharpoons \begin{array}{c}\text{acid taste}\\\text{(organic and inorganic acids)}\end{array}$$
$$+ \begin{array}{c}\text{bitter taste}\\\text{(phenols)}\end{array}$$

Looking at this balance, it is understandable that dry white wines have a higher total acidity than red wines, where phenols combine with acids to balance the sweet taste of the alcohols. Volatile acidity indicates possible microbial spoilage.

1.3.1 Total Acidity

Total acidity in must or wine, also known as 'titratable acidity', is determined by neutralization, using a sodium hydroxide solution of known normality. The end point of the assay is still often determined by means of a colored reagent, such as bromothymol blue, which changes color at pH 7, or phenolphthalein, which changes color at pH 9. Using one colored reagent to define the end point of the assay rather than the other is a matter of choice. It is also perfectly conventional to use a pH meter and stop the total acidity assay of a wine at pH 7, and, indeed, this is mandatory in official analyses. At this pH, the conversion into salts of the second acid function of the di-acids (malic and succinic) is not completed, while the neutralization of the phenol functions starts at pH 9.

The total acidity of must or wine takes into account all types of acids, i.e. inorganic acids such as phosphoric acid, organic acids including the main types described above, as well as amino acids whose contribution to titratable acidity is not very well known. The contribution of each type of acid to total acidity is determined by its strength, which defines its state of dissociation, as well as the degree to which it has combined to form salts. Among the organic acids, tartaric acid is mainly present in must and wine as monopotassium acid salt, which still contributes towards total acidity. It should, however, be noted that must (an aqueous medium) and wine (a dilute alcohol medium), with the same acid composition and thus the same total acidity, do not have the same titration curve and, consequently, their acid–alkaline buffer capacity is different.

Even using the latest techniques, it is difficult to predict the total acidity of a wine on the basis of the acidity of the must from which it is made, for a number of reasons.

Part of the original fruit acids may be consumed by yeasts and, especially, bacteria (see 'malolactic fermentation'). On the other hand, yeasts and bacteria produce acids, e.g. succinic and lactic acids. Furthermore, acid salts become less soluble as a result of the increase in alcohol content. This is the case, in particular, of the monopotassium form of tartaric acid, which causes a decrease in total acidity on crystallization, as potassium bitartrate still has a carboxylic acid function.

In calculating total acidity, a correction should be made to allow for the acidity contributed by sulfur dioxide and carbon dioxide. Sulfuric acid is much stronger ($pK_{a_1} = 1.77$) than carbonic acid ($pK_{a_1} = 6.6$).

In fact, high concentrations of carbon dioxide tend to lead to overestimation of total acidity, especially in slightly sparkling wines, and even more so in sparkling wines. This is also true of young wines, which always have a high CO_2 content just after fermentation.

Wines must, therefore, be degassed prior to analyses of both total and volatile acidity.

1.3.2 Volatile Acidity

Volatile acidity in wine is considered to be a highly important physicochemical parameter, to be monitored by analysis throughout the winemaking process. Although it is an integral part of total acidity, volatile acidity is clearly considered separately, even if it only represents a small fraction in quantitative terms.

On the other hand, from a qualitative standpoint, this value has always been, quite justifiably, linked to quality. Indeed, when an enologist tastes a wine and decides there is excessive volatile acidity, this derogatory assessment has a negative effect on the wine's value. This organoleptic characteristic is related to an abnormally high concentration of acetic acid, in particular, as well as a few homologous carboxylic acids. These compounds are distilled when wine is evaporated. Those which, on the contrary, remain in the residue constitute fixed acidity.

Volatile acidity in wine consists of free and combined forms of volatile acids. This explains why the official assay method for volatile acidity, by steam distillation, requires combined fractions to be rendered free and volatile by acidifying the wine with tartaric acid (approximately 0.5 g per 20 ml). Tartaric acid is stronger than the volatile acids, so it displaces them from their salts.

In France, both total and volatile acidity are usually expressed in g/l of sulfuric acid. An *appellation d'origine contrôlée* wine is said to be 'of commercial quality' if volatile acidity does not exceed 0.9 g/l of H_2SO_4, 1.35 g/l of tartaric acid or 1.1 g/l of acetic acid. Acetic acid, the principal component of volatile acidity, is mainly formed during fermentation.

Alcoholic fermentation of grapes normally leads to the formation of 0.2–0.3 g/l of H_2SO_4 of volatile acidity in the corresponding wine. The presence of oxygen always promotes the formation of acetic acid. Thus, this acid is formed both at the beginning of alcoholic fermentation and towards the end, when the process slows down. In the same way, an increase in volatile acidity of 0.1–0.2 g/l of H_2SO_4 is observed during malolactic fermentation. Work by Chauvet and Brechot (1982) established that acetic acid was formed during malolactic fermentation due to the breakdown of citric acid by lactic bacteria.

Abnormally high volatile acidity levels, however, are due to the breakdown of residual sugars, tartaric acid and glycerol by anaerobic lactic bacteria. Aerobic acetic bacteria also produce acetic acid by oxidizing ethanol.

Finally, acescence in wine is linked to the presence of ethyl acetate, the ethyl ester of acetic acid, formed by the metabolism of aerobic acetic bacteria (Section 2.5.1).

1.3.3 Fixed Acidity

The fixed acidity content of a wine is obtained by subtracting volatile acidity from total acidity. Total acidity represents all of the free acid functions and volatile acidity includes the free and combined volatile acid functions. Strictly speaking, therefore, fixed acidity represents the free fixed acid functions plus the combined volatile acid functions.

When fixed acidity is analyzed, there is a legal obligation to correct for sulfur dioxide and carbon dioxide. In practice, these two molecules have a similar effect on total acidity and volatile acidity, so the difference between total acidity and volatile acidity is approximately the same, with or without correction (Ribéreau-Gayon *et al.*, 1982).

1.4 THE CONCEPT OF pH AND ITS APPLICATIONS

1.4.1 Definition

The concept of pH often appears to be an abstract, theoretical concept, defined mathematically as log subscript ten of the concentration of hydroxonium ions in an electrically conductive solution, such as must or wine:

$$pH = -\log_{10}[H_3O^+]$$

Furthermore, the expression of pH shows that it is an abstract measure with no units, i.e. with no apparent concrete physical significance.

The concepts of total or volatile acidity seem to be easier to understand, as they are measured in milliliters of sodium hydroxide and expressed in g/l of sulfuric or tartaric acid. This is rather paradoxical, as the total acidity in a wine is, in fact, a complex function with several variables, unlike pH which refers to only one variable, the true concentration of hydroxonium ions in must and wine.

The abstract character generally attributed to pH is even less justified as this physicochemical parameter is based on the dissociation equilibrium of the various acids, AH, in wine, at fixed temperature and pressure, as shown below:

$$AH + H_2O \rightleftharpoons A^- + H_3^+O$$

The emission of H_3^+O ions defines the acidity of the AH molecule. Dissociation depends on the value of the equilibrium constant, K_a, of the acid:

$$K_a = \frac{[A^-][H_3^+O]}{[AH]} \quad (1.1)$$

To the credit of the concept of pH, otherwise known as true acidity, it should be added that its value fairly accurately matches the impressions due to acidity frequently described as 'freshness' or even 'greenness' and 'thinness', especially in white wines.

A wine's pH is measured using a pH meter equipped with a glass electrode after calibration with two buffer solutions. It is vital to check the temperature.

The pH values of wines range from 2.8 to 4.0. It is surprising to find such low, non-physiological values in a biological, fermentation medium such as wine. Indeed, life is only possible thanks to enzymes in living cells, and the optimum activity of the vast majority of enzymes occurs at much higher intra-cellular pH values, close to neutral, rather than those prevailing in extra-cellular media, i.e. must and wine. This provides some insight into the role of cell membranes and their ATPases in regulating proton input and output.

On the other hand, it is a good thing that wines have such low pH values, as this enhances their microbiological and physicochemical stability. Low pH hinders the development of microorganisms, while increasing the antiseptic fraction of sulfur dioxide. The influence of pH on physicochemical stability is due to its effect on the solubility of tartrates, in particular potassium bitartrate but, above all, calcium tartrate and the double salt calcium tartromalate.

Ferric casse is also affected by pH. Indeed, iron has a degree of oxidation of three and produces soluble complexes with molecules such as citric acid. These complexes are destabilized by increasing pH to produce insoluble salts, such as ferric phosphates (see 'white casse') or even ferric hydroxide, $Fe(OH)_3$.

1.4.2 Expression of pH in Wine

Wines are mixtures of weak acids, combined to form salts to a greater or lesser extent according to their pK_a (Table 1.3). The proportion of salts also depends on geographical origin, grape variety, the way the vines are trained, and the types of winepress and winemaking methods used.

Due to their composition, musts and wines are acidobasic 'buffer' solutions, i.e. a modification in their chemical composition produces only a limited variation in pH. This explains the relatively small variations in the pH of must during alcoholic and malolactic fermentation.

The pH of a solution containing a weak monoprotic acid and its strong basic salt proves the Anderson Hasselbach equation:

$$\begin{aligned} pH &= pK_a + \log \frac{[\text{salt formed}]}{[\text{remaining acid}]} \\ &= pK_a + \log \frac{[A^-]}{[AH]} \end{aligned} \quad (1.2)$$

This equation is applicable to must and wine, where the strongest acids are di-acids. It is an approximation, assuming the additivity of the acidity contributed by each acid to the total. The application of Eqn (1.2) also makes the 'simplifying' assumption that the degree to which the acids are combined in salts is independent.

Organic Acids in Wine

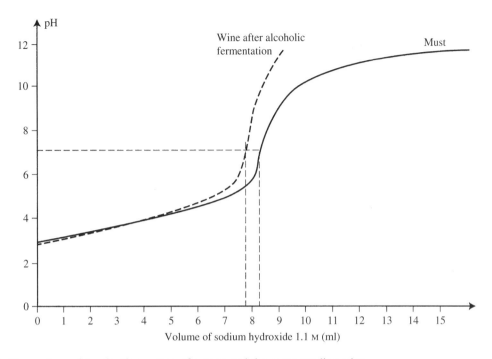

Fig. 1.3. Comparison of the titration curves of a must and the corresponding wine

These assumptions are currently being challenged. Indeed, recent research has shown that organic acids react among themselves, as well as with amino acids (Dartiguenave *et al.*, 2000).

Comparison (Table 1.3) of the pK_a of tartaric (3.01), malic (3.46), lactic (3.81) and succinic (4.18) acids leads to the conclusion that tartaric acid is the 'strongest', so it will take priority in forming salts, displacing, at least partially, the weaker acids. In reality, all of the acids interact. Experimental proof of this is given by the neutralization curve of a must, or the corresponding wine, obtained using sodium or potassium hydroxide (Figure 1.3). These curves have no inflection points corresponding to the pH of the pK of the various acids, as there is at least partial overlapping of the maximum 'buffer' zones (p$K_a \pm 1$). Thus, the neutralization curves are quasi-linear for pH values ranging from 10 to 90% neutralized acidity, so they indicate a constant buffer capacity in this zone. From a more quantitative standpoint, a comparison of the neutralization curves of must and the corresponding wine shows that the total acidity,
assessed by the volume of sodium hydroxide added to obtain pH 7, differs by 0.55 meq. In the example described above, both must and wine samples contained 50 ml and the total acidity of the wine was 11 meq/l (0.54 g/l of H_2SO_4) lower than that of the must. This drop in total acidity in wine may be attributed to a slight consumption of malic acid by the yeast during alcoholic fermentation, as well as a partial precipitation of potassium bitartrate.

The slope of the linear segment of the two neutralization curves differs noticeably. The curve corresponding to the must has a gentler slope, showing that it has a greater buffer capacity than the wine.

The next paragraph gives an in-depth description of this important physicochemical parameter of wine.

1.4.3 The "Buffer" Capacity of Musts and Wines

Wines' acidobasic buffer capacity is largely responsible for their physicochemical and microbiological stability, as well as their flavor balance.

For example, the length of time a wine leaves a fresh impression on the palate is directly related to the salification of acids by alkaline proteins in saliva, i.e. the expression of the buffer phenomenon and its capacity. On the contrary, a wine that tastes "flat" has a low buffer capacity, but this does not necessarily mean that it has a low acidity level. At a given total acidity level, buffer capacity varies according to the composition and type of acids present. This point will be developed later in this chapter.

In a particular year, a must's total acidity and acid composition depend mainly on geography, soil conditions, and climate, including soil humidity and permeability, as well as rainfall patterns, and, above all, temperature. Temperature determines the respiration rate, i.e. the combustion of tartaric and, especially, malic acid in grape flesh cells. The predominance of malic acid in must from cool-climate vineyards is directly related to temperature, while malic acid is eliminated from grapes in hotter regions by combustion.

Independently of climate, grape growers and winemakers have some control over total acidity and even the acid composition of the grape juice during ripening. Leaf-thinning and trimming the vine shoots restrict biosynthesis and, above all, combustion, by reducing the greenhouse effect of the leaf canopy. Another way of controlling total acidity levels is by choosing the harvesting date. Grapes intended for champagne or other sparkling wines must be picked at the correct level of technological ripeness to produce must with a total acidity of 9–10 g/l H_2SO_4. This acidity level is necessary to maintain the wines' freshness and, especially, to minimize color leaching from the red-wine grape varieties, Pinot Noir and Pinot Meunier, used in champagne. At this stage in the ripening process, the grape skins are much less fragile than they are when completely ripe. The last method for controlling the total acidity of must is by taking great care in pressing the grapes and keeping the juice from each pressing separate (Volume 1, Section 14.3.2). In champagne, the *cuvée* corresponds to cell sap from the mid-part of the flesh, furthest from the skin and seeds, where it has the highest sugar and acidity levels.

Once the grapes have been pressed, winemakers have other means of raising or lowering the acidity of a must or wine. It may be necessary to acidify "flat" white wines by adding tartaric acid after malolactic fermentation in years when the grapes have a high malic acid content. This is mainly the case in cool-climate vineyards, where the malic acid is not consumed during ripening. The disadvantage is that it causes an imbalance in the remaining total acidity, which, then, consists exclusively of a di-acid, tartaric acid, and its monopotassium salt.

One method that is little-known, or at least rarely used to avoid this total acidity imbalance, consists of partially or completely eliminating the malic acid by chemical means, using a mixture of calcium tartrate and calcium carbonate. This method precipitates the double calcium salt, tartromalate, (Section 1.4.4, Figure 1.9) and is a very flexible process. When the malic acid is partially eliminated, the wine has a buffer capacity based on those of both tartaric and malic acids, and not just on that of the former. Tartrate buffer capacity is less stable over time, as it decreases due to the precipitation of monopotassium and calcium salts during aging, whereas the malic acid salts are much more soluble.

Another advantage of partial elimination of malic acid followed by the addition of tartrate over malolactic fermentation is that, due to the low acidification rate, it does not produce wines with too low a pH, which can be responsible for difficult or stuck second fermentation in the bottle during the champagne process, leaving residual sugar in the wine.

Standard acidification and deacidification methods are aimed solely at changing total acidity levels, with no concern for the impact on pH and even less for the buffer capacity of the wine, with all the unfortunate consequences this may have on flavor and aging potential.

This is certainly due to the lack of awareness of the importance of the acid-alkali buffer capacity in winemaking. Changes in the acid-alkaline characteristics of a wine require knowledge of not only its total acidity and real acidity (pH), but also of its buffer capacity. These three parameters may be measured using a pH meter. Few articles in the literature deal with the buffer capacity of wine: Genevois and Ribéreau-Gayon, 1935; Vergnes, 1940; Hochli, 1997; and Dartiguenave et al., 2000. This lack of knowledge is probably related to the fact that buffer capacity cannot be measured directly, but requires recordings of 4 or 5 points on a neutralization curve (Figure 1.3), and this is not one of the regular analyses carried out by winemakers.

It is now possible to automate plotting a neutralization curve, with access to the wine's initial pH and total acidity, so measuring buffer capacity at the main stages in winemaking should become a routine.

Mathematically and geometrically, buffer capacity, β, is deduced from the Henderson-Hasselbach equation [equation (1.2), (Section 1.4.2)]. Buffer capacity is defined by equation (1.3).

$$\beta = \frac{\Delta B}{\delta pH} \quad (1.3)$$

where ΔB is the strong base equivalent number that causes an increase in pH equal to ΔpH. Buffer capacity is a way of assessing buffer strength. For an organic acid alone, with its salt in solution, it may be defined as the pH interval in which the buffer effect is optimum [equation (1.4)].

$$pH = pK_a \pm 1 \quad (1.4)$$

Buffer capacity is normally defined in relation to a strong base, but it could clearly be defined in the same way in relation to a strong acid. In this case, the pH = f (strong acid) function decreases and its β differential is negative, i.e.:

$$B = -\frac{\Delta(acid)}{\Delta pH}$$

Strictly speaking, buffer capacity is obtained from the differential of the Henderson-Hasselbach expression, i.e. from the following derived formula:

$$pH = pK_a + \frac{1}{2.303} \cdot \text{Log}_e[A^-]$$
$$- \frac{1}{2.303} \cdot \text{Log}_e[HA]$$

as only the Naperian logarithm is geometrically significant, and provides access to the slope of the titration curve around its pK_a (Figure 1.4).

Both sides of the equation are then differentiated, as follows:

$$dpH = \frac{1}{2.303} \cdot \frac{d[A^-]}{[A^-]} - \frac{1}{2.303} \cdot \frac{d[HA]}{[HA]}$$

Making the assumption that the quantity of strong base added, d[B], generates the same variation in acidity combined as salts, $d[A^-]$, and leads to an equal decrease in free acidity d[HA], per unit, now

$$d[B] = d[A^-] = d[HA]$$

the differential equation for pH is then:

$$dpH = \frac{1}{2.303} \cdot \frac{d[B]}{[A^-]} + \frac{1}{2.303} \cdot \frac{d[B]}{[HA]}$$
$$= \frac{1}{2.303} \cdot d[B] \left\{ \frac{1}{[A^-]} + \frac{1}{[HA]} \right\}$$

or,

$$dpH = \frac{d[B]}{2.303} \cdot \left\{ \frac{[HA] + [A^-]}{[A^-] \cdot [HA]} \right\}$$

Dividing both sides of the equation by d[B] gives the reverse of equation (1.3), defining the buffer capacity. Equations (1.2) and (1.3) have been defined for monoproteic acids, but are also applicable as an initial approximation to di-acids, such as tartaric and malic acids.

Theoretically, variations ΔB and ΔpH must be infinitely small, as the value of the $\Delta B/\Delta pH$ ratio at a fixed pH corresponds geometrically to the tangent on each point on the titration curve (Figure 1.4). More practically, buffer capacity can be defined as the number of strong base equivalents required to cause an increase in pH of 1 unit per liter of must or wine. It is even more practical to calculate smaller pH variations in much smaller samples (e.g. 30 ml). Figure 1.4 clearly shows the difference in buffer capacity of a model solution between pH 3 and 4, as well as between pH 4 and 5.

This raises the issue of the pH and pK_a at which buffer capacity should be assessed. Champagnol (1986) suggested that pH should be taken as the mean of the pK_a of the organic acids in the must

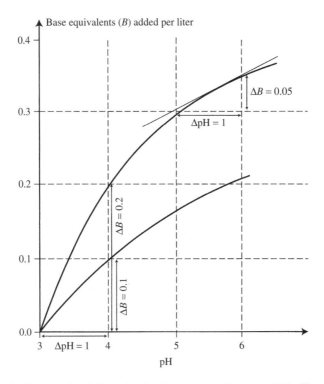

Fig. 1.4. Determining the buffer capacity β from the titration curves of two model buffer solutions

or wine, i.e. the mean pK_a of tartaric and malic acids in must and tartaric and lactic acids in wine that has completed malolactic fermentation.

This convention is justified by its convenience, provided that (Section 1.4.2) there are no sudden inflection points in the neutralization curve of the must or wine at the pK_a of the organic acids present, as their buffer capacities overlap, at least partially. In addition to these somewhat theoretical considerations, there are also some more practical issues. An aqueous solution of sodium hydroxide is used to determine the titration curve of a must or wine, in order to measure total acidity and buffer capacity. Sodium, rather than potassium, hydroxide is used as the sodium salts of tartaric acid are soluble, while potassium bitartrate would be likely to precipitate out during titration. It is, however, questionable to use the same aqueous sodium hydroxide solution, which is a dilute alcohol solution, for both must and wine.

Strictly speaking, a sodium hydroxide solution in dilute alcohol should be used for wine to avoid modifying the alcohol content and, consequently, the dielectric constant, and, thus, the dissociation of the acids in the solution during the assay procedure. It has recently been demonstrated (Dartiguenave *et al.*, 2000) that the buffer capacities of organic acids, singly (Table 1.4 and 1.5) or in binary (Table 1.6) and tertiary (Table 1.7) combinations, are different in water and 11% dilute alcohol solution. However, if the solvent containing the organic acids and the sodium hydroxide is the same, there is a close linear correlation between the buffer capacity and the acid concentrations (Table 1.4).

Table 1.5 shows the values (meq/l) calculated from the regression line of the buffer capacities for acid concentrations varying from 1–6 g/l in water and 11% dilute alcohol solution. The buffer capacity of each acid alone in dilute alcohol solution was lower than in water. Furthermore, the buffer capacity of a 4-carbon organic acid varied more as the number of alcohol functions increased (Table 1.8). Thus, the variation in buffer capacity of malic acid, a di-acid with one alcohol function,

Table 1.4. Equations for calculating buffer capacity (meq/l) depending on the concentration (mM/l) of the organic acid in water or dilute alcohol solution (11% vol.) between 0 and 40 mM/l. (Dartiguenave *et al.*, 2000)

Solvent	Water	Dilute alcohol solution
Tartaric acid	$Y = 0.71\,x + 0.29;\ R^2 = 1$	$Y = 0.60\,x + 1.33;\ R^2 = 1$
Malic acid	$Y = 0.56\,x + 0.43;\ R = 0.998$	$Y = 0.47\,x + 0.33;\ R^2 = 0.987$
Succinic acid	$Y = 0.56\,x - 1.38 \cdot 10^{-2};\ R^2 = 0.993$	$Y = 0.53\,x + 0.52;\ R^2 = 0.995$
Citric acid	$Y = 0.57\,x + 0.73;\ R^2 = 1$	$Y = 0.51\,x + 0.62;\ R^2 = 1$

Table 1.5. Buffer capacity (meq/l) depending on the concentration (g/l) of organic acid in water and dilute alcohol solution. (Dartiguenave *et al.*, 2000)

Acid concentration and type of medium		Tartaric acid	Malic acid	Succinic acid	Citric acid
1 g/l	Water	5.0	4.6	4.7	3.7
	Dilute alcohol	5.3	3.8	4.0	3.5
2 g/l	Water	9.7	8.8	9.5	6.7
	Dilute alcohol	9.3	7.3	9.4	5.9
4 g/l	Water	16.4	17.1	19.0	12.6
	Dilute alcohol	14.9	14.3	17.5	11.3
6 g/l	Water	28.7	25.5	28.4	18.5
	Dilute alcohol	25.3	21.3	26.4	16.6

in a dilute alcohol medium, was 1.4 meq/l higher than that of succinic acid. When the hydroxyacid had two alcohol functions, the increase was as high as 5.3 meq/l (17.7%), e.g. between tartaric and malic acids, even if the buffer capacities of the three acids were lower than in water.

However, the fact that the buffer capacities of binary (Table 1.6) or tertiary (Table 1.7) combinations of acids in a dilute alcohol medium were higher than those measured in water was certainly unexpected. This effect was particularly marked when citric acid was included, and reached spectacular proportions in a T.M.C. blend (Table 1.7), where the buffer capacity in dilute alcohol solution was 2.3 times higher than that in water.

These findings indicate that the acids interact among themselves and with alcohol, compensating for the decrease in buffer capacity of each individual acid when must (an aqueous solution) is converted into wine (a dilute alcohol solution). From a purely practical standpoint, the use of citric acid to acidify dosage liqueur for bottle-fermented sparkling wines has the doubly positive effect of enhancing the wine's aging potential, while maintaining its freshness on the palate.

Table 1.6. Demonstration of interactions between organic acids and the effect of alcohol on the buffer capacity of binary combinations (Dartiguenave *et al.*, 2000)

Medium	Buffer capacity (meq/l)	Composition of equimolar mixes of 2 acids Total acid concentration (40 mM/l)		
		Tartaric acid Malic acid	Tartaric acid Succinic acid	Tartaric acid Citric acid
Water	Experimental value	21	20	23.5
	Calculated value	25.7	25.7	26.3
	Difference (Calc. − Exp.)	4.7	5.7	2.8
EtOH (11% vol.)	Experimental value	18.3	20.1	29
	Calculated value	24	23.3	24
	Difference (Calc. − Exp.)	5.7	3.2	−5
Effect of ethanol	(EtOH − H$_2$O) Exp.	−2.7	0.1	5.5

Table 1.7. Demonstration of interactions between organic acids and the effect of alcohol on the buffer capacity of tertiary combinations (Dartiguenave *et al.*, 2000)

Medium	Buffer capacity (meq/l)	Composition of equimolar mixes of 3 acids (13.3 mM/l) Total acid concentration (40 mM/l)	
		Tartaric acid Malic acid Succinic acid	Tartaric acid Malic acid Citric acid
Water	Experimental value	9.4	11.6
	Calculated value	25.4	25.5
	Difference (Calc. − Exp.)	16.0	13.9
EtOH (11% vol.)	Experimental value	21.7	26.4
	Calculated value	22.8	23.2
	Difference (Calc. − Exp.)	1.1	−3.2
Effect of ethanol	(EtOH − H_2O) Exp.	12.3	14.8

Table 1.8. Effect of hydroxyl groups in the structure of the 4-carbon di-acid on buffer capacity (meq/l) (Dartiguenave *et al.*, 2000)

Medium	1 hydroxyl group			2 hydroxyl groups		
	Malic acid	Succinic acid	Δ (Mal.− Suc.)	Tartaric acid	Malic acid	Δ (Tart.− Mal.)
Water	23.8	23.4	0.4	29	23.8	5.2
11% vol. dilute alcohol solution	22,0	20.6	1.4	25.9	22	3.9

Table 1.9. Changes in the buffer capacity of must from different pressings of Chardonnay grapes at various stages in the winemaking process. (Buffer capacity is expressed in meq/l). (Dartiguenave, 1998)

	Cuvée		Second pressing	
	1995	1996	1995	1996
Initial value of must	77.9	72.6	71.2	65.9
After alcoholic fermentation	60.7	63.6	57.5	ND
After malolactic fermentation	51.1	60.1	48.4	ND
After cold-stabilization	48.1	50.3	ND	42.4

Table 1.9 shows the changes in buffer capacity in successive pressings of a single batch of Chardonnay grapes from the 1995 and 1996 vintages, at the main stages in the winemaking process.

The demonstration of the effect of alcohol and interactions among organic acids (Table 1.6, 1.7, and 1.8) led researchers to investigate the precise contribution of each of the three main acids to a wine's buffer capacity, in order to determine whether other compounds were involved. The method consisted of completely deacidifying a wine by precipitating the double calcium tartromalate salt. After this deacidification, the champagne-base wine had a residual total acidity of only approximately 0.5 g/l H_2SO_4, whereas the buffer capacity was still 30% of the original value. This shows that organic acids are not the only compounds involved in buffer capacity, although they represent 90% of total acidity.

Among the many other compounds in must and wine, amino acids have been singled out for two reasons: (1) in champagne must and wine, the total concentration is always over 1 g/l and may even exceed 2 g/l, and (2) their at least bifunctional character gives them a double-buffer effect. They form salts with carboxylic acids via their ammonium group and can become associated with a non-dissociated acid function of an organic acid via their carboxyl function, largely dissociated from wine pH, thus creating two buffer couples (Figure 1.5).

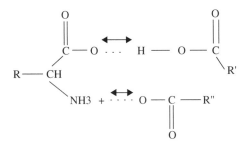

Fig. 1.5. Diagram of interactions between amino acids and organic acids that result in the buffer effect

An in-depth study of the interactions between amino acids and tartaric and malic acids focused on alanine, arginine, and proline, present in the highest concentrations in wine, as well as on amino acids with alcohol functions, i.e. serine and threonine (Dartiguenave et al., 2000).

The findings are presented in Figures 1.6 and 1.7. Hydrophobic amino acids like alanine were found to have only a minor effect, while amino acids with alcohol functions had a significant impact on the buffer capacity of an aqueous tartaric acid solution (40 mM/l). An increase of 0.6 meq/l was obtained by adding 6.7 mM/l alanine, while addition of as little as 1.9 mM/l produced an increase of 0.7 meq/l and addition of 4.1 mM/l resulted in a rise of 2.3 meq/l.

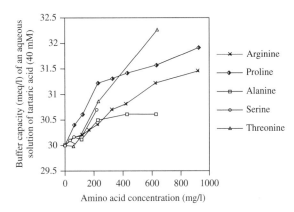

Fig. 1.6. Variations in the buffer capacity of an aqueous solution of tartaric acid (40 mM) in the presence of several amino acids. (Dartiguenave et al., 2000)

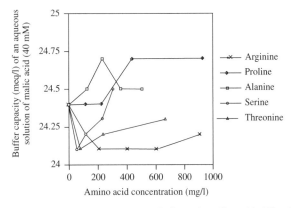

Fig. 1.7. Variations in the buffer capacity of an aqueous solution of malic acid (40 mM in the presence of several amino acids. (Dartiguenave et al., 2000))

The impact of amino acids with alcohol functions was even more spectacular in dilute alcohol solutions (11% by volume). With only 200 mg/l serine, there was a 1.8 meq/l increase in buffer capacity, compared to only 0.8 meq/l in water. It was also observed that adding 400 mg/l of each of the five amino acids led to a 10.4 meq/l (36.8%) increase in the buffer capacity of a dilute alcohol solution containing 40 mM/l tartaric acid.

It is surprising to note that, on the contrary, amino acids had no significant effect on the buffer capacity of a 40 mM/l malic acid solution (Figure 1.7).

All these observations highlight the role of the alcohol function, both in the solvent and the amino acids, in interactions with organic acids, particularly tartaric acid with its two alcohol functions.

The lack of interaction between amino acids and malic acid, both in water and dilute alcohol solution, can be interpreted as being due to the fact that it has one alcohol function, as compared to the two functions of tartaric acid. This factor is important for stabilizing interactions between organic acids and amino acids via hydrogen bonds (Figure 1.8).

1.4.4 Applying Buffer Capacity to the Acidification and Deacidification of Wine

The use of tartaric acid (known as 'tartrating') is permitted under European Community (EC) legislation, up to a maximum of 1.5 g/l in must and 2.5 g/l in wine. In the USA, acidification is permitted, using tartrates combined with gypsum ($CaSO_4$) (Gomez-Benitez, 1993). This practice seems justified if the buffer capacity expression (Eqn 1.3) is considered. The addition of tartaric acid (HA) increases the buffer capacity by increasing the numerator of Eqn (1.3) more than the denominator. However, the addition of $CaSO_4$ leads to the precipitation of calcium tartrate, as this salt is relatively insoluble. This reduces the buffer capacity and, as a result, ensures that acidification will be more effective.

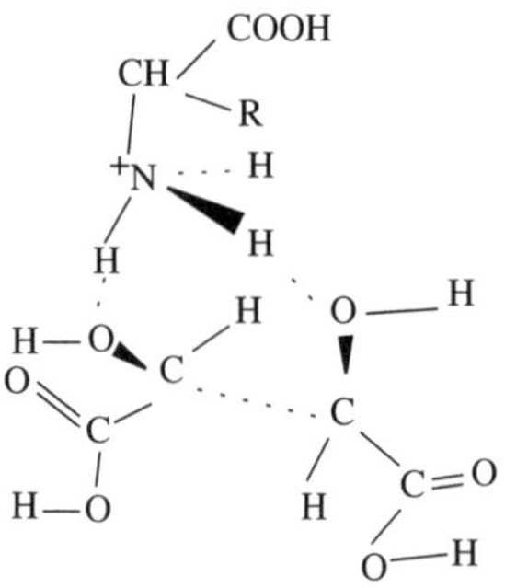

Whenever tartrating is carried out, the effect on the pH of the medium must also be taken into account in calculating the desired increase in total acidity of the must or wine. Unfortunately, however, there is no simple relationship between total acidity and true acidity.

An increase in true acidity, i.e. a decrease in pH, may occur during bitartrate stabilization, in spite of the decrease in total acidity caused by this process. This may also occur when must and, in particular, wine is tartrated, due to the crystallization of potassium bitartrate, which becomes less soluble in the presence of alcohol.

The major difficulty in tartrating is predicting the decrease in pH of the must or wine. Indeed, it is important that this decrease in pH should not be incompatible with the wine's organoleptic qualities, or with a second alcoholic fermentation in the case of sparkling wines. To our knowledge, there is currently no reliable model capable of accurately predicting the drop in pH for a given level of tartrating. The problem is not simple, as it depends on a number of parameters. In order to achieve the required acidification of a wine, it is necessary to know the ratio of the initial concentrations of tartaric acid and potassium, i.e. crystallizable potassium bitartrate.

It is also necessary to know the wine's acido-basic buffer capacity. Thus, in the case of wines from northerly regions, initially containing 6 g/l of malic acid after malolactic fermentation, tartrating may be necessary to correct an impression of 'flatness' on the palate. Great care must be taken in acidifying this type of wine, otherwise it may have

Fig. 1.8. Assumed structure of interactions between tartaric acid and amino acids. (Dartiguenave *et al.*, 2000)

a final pH lower than 2.9, which certainly cures the 'flatness' but produces excessive dryness or even greenness. White wines made from red grape varieties may even take on some red color. The fact that wine has an acidobasic buffer capacity also makes deacidification possible.

Table 1.10 shows the values of the physicochemical parameters of the acidity in champagne-base wines, made from the *cuvée* or second pressing of Chardonnay grapes in the 1995 and 1996 vintages. They were acidified with 1 g/l and 1.5 g/l tartaric acid, respectively, after the must had been clarified.

Examination of the results shows that adding 100 g/hl to a *cuvée* must or wine only resulted in 10–15% acidification, corresponding to an increase in total acidity of approximately 0.5 g/l (H_2SO_4). Evaluating the acidification rate from the buffer capacity gave a similar result. The operation was even less effective when there was a high potassium level, and potassium bitartrate precipitated out when the tartaric acid was added.

Adding the maximum permitted dose of tartaric acid (150 g/hl) to second pressing must or wine was apparently more effective, as total acidity increased by 35% and pH decreased significantly (−0.14), producing a positive impact on wine stability and flavor. The effect on pH of acidifying *cuvée* wines shows the limitations of adding tartaric acid, and there may also be problems with the second fermentation in bottle, sometimes resulting in "hard" wines with a metallic mouth feel.

It would be possible to avoid these negative aspects of acidification by using L(-)lactic acid. This is listed as a food additive (E270) and meets the requirements of both the Food chemical Codex and the European Pharmacopoeia. Lactic acid is commonly used in the food and beverage industry, particularly as a substitute for citric acid in carbonated soft drinks, and is even added to some South African wines.

Its advantages compared to tartaric acid are the pK_a of 3.81 (tartaric acid: 3.01), and the fact that both its potassium and calcium salts are soluble. This enhances the acidification rate while minimizing the decrease in pH. Finally, lactic acid is microbiologically stable, unlike tartaric, malic, and citric acids. Until recently, one disadvantage of industrial lactic acid was a rather nauseating odor, which justified its prohibition in winemaking. The lactic acid now produced by fermenting sugar industry residues with selected bacteria no longer has this odor.

Current production quality, combined with low prices, should make it possible to allow experimentation in the near future, and, perhaps, even a lifting of the current ban on the use of lactic acid in winemaking.

The additives authorized for deacidifying wines are potassium bicarbonate ($KHCO_3$) and calcium carbonate ($CaCO_3$). They both form insoluble salts with tartaric acid and the corresponding acidity is eliminated in the form of carbonic acid (H_2CO_3) which breaks down into CO_2 and H_2O. A comparison of the molecular weights of these two salts and the stoichiometry of the neutralization reactions leads to the conclusion that, in general, one gram of $KHCO_3$(PM = 100) added to one liter of wine produces a drop in acidity of 0.49 g/l, expressed in grams of H_2SO_4(PM = 98). Adding one gram of $CaCO_3$(PM = 100) to a liter of wine produces a decrease in acidity equal to its own weight (exactly 0.98 g/l), expressed in grams of sulfuric acid.

In fact, this is a rather simplistic explanation, as it disregards the side-effects of the precipitation of insoluble potassium bitartrate salts and, especially, calcium tartrate, on total acidity as well as pH. These side-effects of deacidification are only fully expressed in wines with a pH of 3.6 or lower after cold stabilization to remove tartrates. It is obvious from the pH expression (Eqn 1.2) that, paradoxically, after removal of the precipitated tartrates, deacidification using $CaCO_3$ and, more particularly, $KHCO_3$ is found to have reduced the [salt]/[acid] ratio, i.e. increased true acidity. Fortunately, the increase in pH observed during neutralization is not totally reversed.

According to the results described by Usseglio-Tomasset (1989), a comparison of the deacidifying capacities of potassium bicarbonate and calcium carbonate shows that, in wine, the maximum deacidifying capacity of the calcium salt is only 85% of that of the potassium salt. Consequently, to bring a wine to the desired pH, a larger

Table 1.10. Composition of Chardonnay wines after tartaric stabilization, depending on the time of acidification (addition to must or wine after malolactic fermentation). Cuvées were acidified with 1 g/l tartaric acid and second pressings with 1.5 g/l. (Dartiguenave, 1998)

	Cuvée						Second pressing		
	1995			1996			1996		
	Control	Acidified must	Acidified wine	Control	Acidified must	Acidified wine	Control	Acidified must	Acidified wine
pH	3.06	2.97	2.97	3.06	2.99	2.97	3.18	3.04	3.00
Total acidity (g/l, H_2SO_4)	5.2	6.0	5.6	5.4	5.9	5.8	4.1	4.9	5.0
Tartaric acid (g/l)	3.6	4.0	4.3	4.4	5.2	5.0	3.4	4.6	4.8
Malic acid (g/l)	0.1	0.1	0.1	0.1	0.1	0.1	0.1	0.1	0.1
Lactic acid (g/l)	4	4.3	4.4	4.2	4.1	4.1	3	3	2.7
Total nitrogen (mg/l)	274.7	221.9	271	251.6	280.3	289.8	245.9	250.4	254.4
Amino acids (mg/l)	1051.4	703.7	1322.6	1254.2	1422.7	1471.7	1177.5	1350.4	1145
Potassium (mg/l)	390	345	320	345	290	285	380	305	300
Calcium (mg/l)	71.5	90	79	60	64	61	50	55	48
Buffer capacity (NAOH, H_2O)	48.1	56.6	56.2	50.3	55.5	56.9	42.4	49.1	47.7
Buffer capacity (NAOH.EtOH 11% vol)	55.6	59.2	55.9	47.1	51.9	50.2	37.9	44.3	42

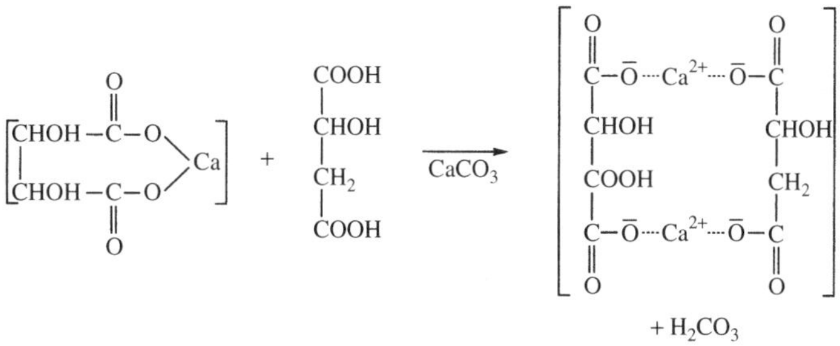

Organic Acids in Wine

Fig. 1.9. Formation of insoluble calcium tartromalate when calcium tartrate reacts with malic acid in the presence of calcium carbonate

quantity of $CaCO_3$ than $KHCO_3$ must be used, as compared to the theoretical value. On the other hand, $CaCO_3$ has a more immediate effect on pH, as the crystallization of CaT is more complete than that of KTH, a more soluble salt.

Another side-effect of deacidification using calcium carbonate, and especially potassium bitartrate, is a decrease in the alkalinity of the ash.

Finally, deacidification with these two carbonic acid salts only affects tartaric acid. This accentuates the tartromalic imbalance in the total acidity in wines that have not completed malolactic fermentation, as the potassium and calcium salts of malic acid are soluble.

There is a way of deacidifying these wines while maintaining the ratio of tartaric acid to malic acid. The idea is to take advantage of the insolubility of calcium tartromalate, discovered by Ordonneau (1891). Wurdig and Muller (1980) used malic acid's property of displacing tartaric acid from its calcium salt, but at pHs above 4.5 (higher than the pK_{a_2} of tartaric acid), in a reaction (Figure 1.9) producing calcium tartromalate.

The technology used to implement this deacidification known as the DICALCIC process (Vialatte and Thomas, 1982) consists of adding volume V, calculated from the following equation, of wine to be treated, to obtain the desired deacidification of the total volume (V_T):

$$V = V_T \frac{A_i - A_f}{A_i - 1} \quad (1.5)$$

In Eqn (1.5), A_i and A_f represent initial and final acidity, respectively, expressed in g/l of H_2SO_4, of the total volume V_T. The volume V of wine to be deacidified by crystallization and elimination of the calcium tartromalate must be poured over an alkaline mixture consisting, for example, of calcium carbonate (1 part) and calcium tartrate (2 parts). Its residual acidity will then be very close to 1 g/l of H_2SO_4.

It is important that the wine should really neutralize the $CaCO_3$/CaT mixture and not the reverse, as the formation of the stable, crystallizable, double tartromalate salt is only possible above pH 4.5. Below this pH, precipitation of the endogenous calcium tartrate occurs, promoted by homogeneous induced nucleation with the added calcium tartrate, as well as precipitation of the potassium bitartrate by heterogeneous induced nucleation (Robillard *et al.*, 1994).

The addition of calcium tartrate is necessary to ensure that the tartaric acid content in the wine does not restrict the desired elimination of malic acid by crystallization of the double tartromalic salt, but also to maintain a balance between the remaining malic and tartaric acid.

1.5 TARTRATE PRECIPITATION MECHANISM AND PREDICTING ITS EFFECTS

1.5.1 Principle

At the pH of wine, and in view of the inevitable presence of K^+ and Ca^{2+} cations, tartaric acid is mainly salified in the following five forms, according to its two dissociation balances:

potassium bitartrate (KTH)
potassium tartrate (K_2T)

calcium tartrate (CaT) with the formula $CaC_4H_4O_6 \cdot 4H_2O$

potassium calcium tartrate

calcium tartromalate

In wine, simple salts are dissociated into TH^- and T^{2-} ions. The last two tartrates (Figure 1.10) share the property of forming and remaining stable at a pH of over 4.5. On the other hand, in terms of solubility, they differ in that potassium calcium tartrate is highly soluble, whereas the tartromalate is relatively insoluble and crystallizes in needles. The properties of this mixed salt may be used to eliminate malic acid, either partially or totally. Table 1.11 shows the solubility, in water at 20°C, of tartaric acid and the salts that cause the most problems in terms of crystalline deposits in wine.

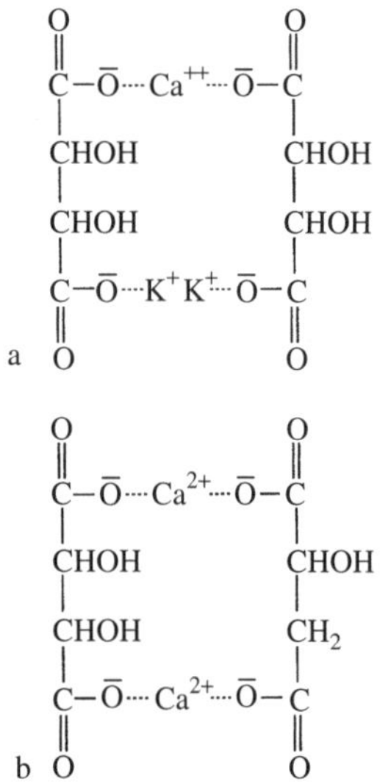

Fig. 1.10. Structure of (a) double potassium calcium tartrate and (b) calcium tartromalate

Table 1.11. Solubility in water at 20°C in g/l of L-tartaric acid and the main salts present in wine

Tartaric acid	Potassium bitartrate	Neutral calcium tartrate
L(+)-$C_4H_6O_6$	$KHC_4H_4O_6$	$CaC_4H_4O_6 \cdot 4H_2O$
4.9 g/l	5.7 g/l	0.53 g/l

While potassium bitartrate is perfectly soluble in water, it is relatively insoluble in alcohol. Thus, in a dilute alcohol solution at 10% v/v and 20°C, its solubility (S) is only 2.9 g/l.

The potassium concentration in wine is frequently as high as 780 mg/l or 20 meq/l, i.e. 3.76 g/l of potassium bitartrate. Therefore, the concentration (C) of the salt is greater than its solubility (S). It follows that the product CP of the real concentrations (r)

$$CP = [TH^-]_r[K^+]_r \quad (1.6)$$

is greater than the solubility product SP defined by

$$SP = [TH^-]_e[K^+]_e \quad (1.7)$$

according to the solubility balance:

$$\underset{\text{solid}}{KTH} \overset{K_e}{\rightleftharpoons} \underset{\text{in solution}}{TH_e^- + K_e^+} \quad (1.8)$$

In this equation, the concentrations (e) of TH^- anions and K^+ cations are theoretically obtained at the thermodynamic equilibrium of the solid KTH/dissolved KTH system, under the temperature and pressure conditions in wine.

The diagram (Figure 1.11) presenting the states of potassium bitartrate in a system correlating the temperature/concentration axes with conductivity shows three fields of states, 1, 2 and 3, with borders defined by the solubility (A) and hypersolubility (B) exponential curves. The exponential solubility curve (A) is obtained by adding 4 g/l of crystallized KTH to a wine. The increase in the wine's electrical conductivity according to temperature is then recorded. This corresponds to the dissolving and ionization of tartrates. As explained in Section 1.6.4, conductivity values correspond to saturation temperatures (T_{Sat}), since wine is capable of dissolving increasing amounts of KTH as the temperature rises. The exponential solubility curve represents the boundary between two possible states of KTH in a wine according to temperature. Thus, at a constant concentration (or conductivity), when the temperature of the wine rises, KTH changes from state 2, where it is supersaturated and surfused, to state 1, i.e. dissolved, where

Organic Acids in Wine

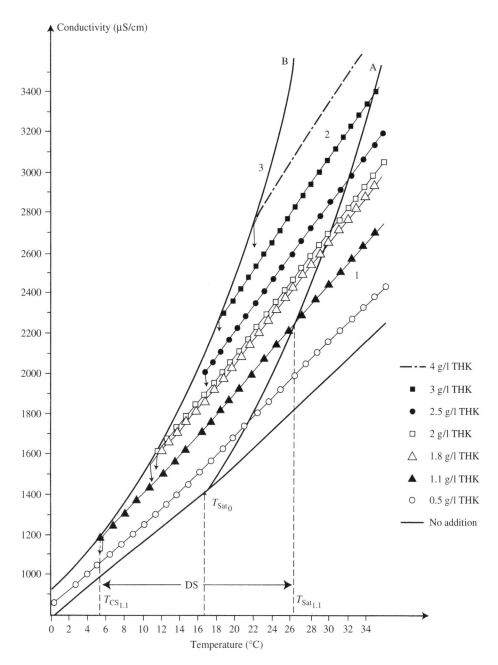

Fig. 1.11. Determining the solubility (A) and hypersolubility (B) exponential curves of potassium bitartrate in a wine. Defining the hyper-saturation and instability fields according to the KTH content (Maujean *et al.*, 1985). DS = saturation field; 1, dissolved KTH; 2, supersaturated, surfused KTH; 3, crystallized KTH; $T_{CS_{1.1}}$, spontaneous crystallization temperature when 1.1 g/l KTH is added; $T_{Sat_{1.1}}$, saturation temperature of a wine in which 1.1 g/l KTH have been dissolved

its concentration product CP is lower than its solubility product SP.

The exponential hypersolubility curve (B) is obtained experimentally and geometrically from the envelope linking the spontaneous crystallization temperature (TCS_i) points of a wine brought to various states of supersaturation by completely dissolving added KTH and then reducing the temperature of the wine until crystallization is observed. The exponential hypersolubility curve represents the boundary between state 2, where potassium bitartrate is in a state of supersaturation ($C - S$) and surfusion, and state 3, where it is crystallized.

Once the solubility (A) and hypersolubility (B) exponential curves have been defined, it is possible to determine the state of a wine at a known temperature with considerable accuracy. Indeed, any wine with a KTH concentration, or conductivity, above that defined by the intersection of the vertical line drawn upwards from the temperature of the wine and the exponential solubility curve (A) is in a supersaturated state so, theoretically, there is a probability of spontaneous crystallization. The crystallization phenomenon will, in fact, be observed at the intersection of the same vertical line and the exponential hypersolubility curve (B). It appears, therefore, that supersaturation is necessary, but not sufficient, for primary nucleation phenomena and spontaneous crystallization to occur in a wine.

The delay in crystallization of a salt in relation to its solubilization, which is partially responsible for the supersaturated state in superfused form, is due to lack of energy.

The formation of a small crystal, known as a nucleus, in a liquid phase corresponds to the creation of an interface between two phases. This requires a great deal of energy, known as interfacial surface energy. In a wine, the width DS of the supersaturation field (Figure 1.7), expressed in degrees Celsius, is increased by the presence of macromolecules that inhibit the growth of nuclei and crystallization of the KTH. These macromolecules, known as 'protective colloids', include proteins and condensed tannins, and also glucide polymers, such as pectins and gums, i.e. neutral polysaccharides. Besides these chemical macromolecules, there are also more complex polymers, such as glycoproteins, e.g. mannoproteins of yeast origin (Lubbers et al., 1993).

The impact of the protective colloid effect on the bitartrate stabilization of a wine varies according to the winemaking methods used. Red wines have a higher phenol content than white wines, and their condensed tannins have a strong inhibiting effect.

In its natural state, wine is always supersaturated and therefore unstable. This situation may be more or less durable, depending on the reorganization of the colloids that occurs during aging. Storage temperatures may be decisive in triggering bitartrate crystallization.

It is certainly true that spontaneous crystallization, under natural conditions, is an unreliable, unpredictable phenomenon. This is why the production process for many red and white wines includes artificial cold stabilization before bottling. This type of treatment is justified, especially as consumers will not tolerate the presence of crystals, even if they do not affect quality.

Furthermore, artificial cold stabilization is indispensable for sparkling wines. Indeed, microcavities in the surface of the glass or in solid particles in suspension, especially microcrystals of potassium bitartrate, may lead to the formation of too many bubbles when the bottle is opened, causing excessive effervescence known as 'spraying'. This is sometimes responsible for the loss of large quantities of wine during disgorging, or when bottles are opened by consumers (Volume 1, Section 14.3.4). The origin of this effervescence and spraying is given by the repetitive bubble formation model (Casey, 1988) (Figure 1.12). This bubble degassing model is based on the phenomenon of heterogeneous induced nucleation.

However, nucleation may be induced and the microcavities are efficient only if they have a radius R_1 greater than a critical radius R_c defined by Laplace's law. Indeed, below this value, the excess pressure in the bubble is such that carbon dioxide passes from the gas phase to the liquid phase and so the bubble disappears.

On the other hand, if R_1 is greater than R_c, carbon dioxide diffusion occurs in the opposite

Organic Acids in Wine

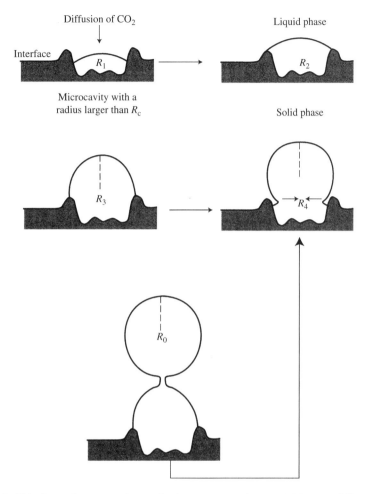

Fig. 1.12. Repetitive bubble formation on a microcavity in a tartrate microcrystal in a sparkling wine. Heterogeneous induced nucleation, according to the Casey model (1988)

direction and the bubble increases in size, reaching the values R_2, R_3 and R_4. At this last stage, the bubble is subjected to the laws of gravity and starts to rise when its radius reaches the value R_0, leaving behind a new bubble that has started to form. This is how the phenomenon of durable effervescence is achieved.

The fact that the phenomenon of effervescence may be exacerbated due to a large number of microcavities in tartrate microcrystals is an additional reason for ensuring the thorough tartrate stabilization of still wine intended for sparkling wine production. Treatment parameters at this stage must take into account the destabilizing effect of the increase in alcohol content following the second alcoholic fermentation in vat or in bottle.

There are two main types of must and wine treatment technologies for preventing bitartrate instability based on the phenomenon of low-temperature crystallization. The first uses traditional slow stabilization technology (Section 1.7.2), as opposed to the more recent Müller-Späth rapid contact stabilization process (1979), where the wine is seeded with cream of tartar crystals. There are two variants of the short process, one static and the other dynamic, known as 'continuous treatment'.

Besides these two systems, a new separation technique, electrodialysis, is also applied to the bitartrate stabilization of wine (Section 12.5). The use of ion-exchange resins is also permitted in certain countries, including the USA (Section 12.4.3). Finally, it is possible to prevent the precipitation of these salts by adding crystallization inhibitors, such as metatartaric acid or yeast mannoprotein extracts (Section 1.7.7), or carboxymethylcellulose (Section 1.7.8).

1.5.2 Tartrate Crystallization and Precipitation

The two artificial cold stabilization technologies described elsewhere (Sections 1.7.1. and 1.7.2) do not use the same crystallization mechanism. The traditional stabilization process involves spontaneous, primary nucleation, a long process that produces large crystals because the nuclei grow slowly. In rapid stabilization processes, the awkward stage of primary nucleation is replaced by a fast, homogeneous secondary nucleation. This is induced by adding massive quantities of small exogeneous tartrate crystals, which also considerably boost supersaturation $(C - S)$.

Furthermore, in this technique, the temperature of the wine is reduced abruptly, promoting the formation of small endogeneous tartrate nuclei, i.e. significantly increasing the surface area (A) of the liquid/solid interface by maximizing the diffusion of bitartrate aggregates with pre-crystalline structures, thus ensuring faster growth of the nuclei (Figure 1.13).

It has been experimentally verified (Maujean *et al.*, 1986) that the crystallization rate, monitored by measuring the electrical conductivity of wine, is directly proportional to the surface area of the liquid/solid interface represented by the nuclei. This result is consistent with the following equation, proposed by Dunsford and Boulton (1981), defining the mass velocity at which the precrystalline aggregates of potassium bitartrate diffuse towards the surface (A) of the adsorption interface:

$$\frac{dm}{dt} = k_d(A)(C - C_i) \qquad (1.9)$$

where C is the concentration of the solution and C_i is the concentration of the interface.

One practical application of these theoretical results is that producers and distributors have been

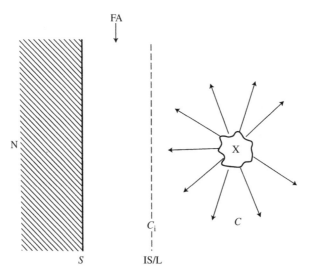

Fig. 1.13. Diagram illustrating the importance of the diffusion speed of THK aggregates towards the solid/liquid adsorption interface for the growth of nuclei: FA, adsorption film; X, molecular aggregate of THK diffusing towards the interface; IS/L, solid/liquid interface; N, nuclei; C, THK concentration in the liquid phase; C_i, THK concentration at the solid/liquid interface; S, theoretical solubility of THK; $C - S$, supersaturation of the wine; $C > C_i > S$

obliged to ensure that their cream of tartar particles have a radius of less than 40 µm. This parameter is also important as nuclei with a radius greater than 200 µm grow much more slowly than smaller nuclei.

This confirms the findings of Devraine (1969), who also concluded that large nuclei stop growing as they release 'fines', i.e. 'daughter' nuclei. This observation explains the continued effectiveness in stabilizing white wines of cream of tartar that has been recycled five times, provided that the particles were initially very small. On the other hand, it is not possible to recycle cream of tartar so many times in red wines due to the affinity between tartaric acid and phenols, known to be powerful crystallization inhibitors.

Another advantage of the contact process is that seeding with small cream of tartar particles enhances the state of supersaturation ($C - C_i$). This is important as the crystallization rate is not only proportional to the interface value (A), but also to the state of supersaturation ($C - C_i$) (Eqn 1.9).

The added cream of tartar must be maintained in suspension homogeneously, throughout the vat by appropriate agitation, so that the nuclei provide a maximum contact interface with the aggregates of endogeneous tartrate. As soon as the cream of tartar is added, the crystallization rate depends solely on the interface factor (A), as ($C - C_i$) is so large that it may, at least in the first hour of contact, be considered constant. It may therefore be stated that, during the first hour, the crystallization rate depends solely on the rate of diffusion of the aggregates (Eqn 1.9).

After this initial contact time, the nuclei have grown but, more importantly, ($C - C_i$) has decreased, as the very high crystallization rate has consumed large quantities of exogeneous tartrate. In other words, A, i.e. the diffusion rate, is no longer the limiting factor, but rather the state of supersaturation ($C - C_i$). As C tends towards C_i, the situation in the wine approaches the theoretical solubility (S) of tartrate under these treatment conditions. Therefore, by the end of the treatment process, the crystallization rate is controlled more by thermodynamics than kinetics.

These theoretical considerations, applied to a short treatment involving seeding with tartrate crystals, show that great care and strict supervision is required to ensure the effectiveness of artificial cold stabilization. The following factors need to be closely monitored: the wine's initial state of supersaturation, the particle size of the added tartrates, the seeding rate, the effectiveness of agitation at maintaining the crystals in suspension, treatment temperature and, finally, contact time.

1.5.3 Using Electrical Conductivity to Monitor Tartrate Precipitation

Wurdig and Muller (1980) were the first to make use of the capacity of must and wine to act as electrolytes, i.e. solutions conducting electricity, to monitor tartrate precipitation. Indeed, during precipitation, potassium bitartrate passes from the dissolved, ionized state, when it is an electrical conductor, to a crystalline state, when it precipitates and is no longer involved in electrical conductivity:

$$HT^- + K^+ \longrightarrow KTH$$
$$\downarrow$$

The principle of measuring conductivity consists of making the wine into an 'electrical conductor', defined geometrically by the distance l separating two platinum electrodes with S-shaped cross-sections. The resistance R (in ohms) of the conductor is defined by the relation:

$$R = \rho \frac{l}{S}$$

In this equation, ρ is the resistivity. Its inverse (γ) is the conductivity expressed in siemens per meter (S/m) or microsiemens per centimeter (µS/cm = 10^{-4} S/m).

The expression of resistivity $\rho = RS/l$ involves the term S/l, known as the cell 'k' constant. This constant is particular to each cell, according to its geometry, and may also vary with use, due to gradual deterioration of the electrodes or the effect of small impacts.

It is therefore necessary to check this constant regularly and to determine it at a conductivity close

Table 1.12. Resistivity and conductivity of a KCl (0.02 M) solution according to temperature (in °C)

Temperature (°C)	15	16	17	18	19	20	21	22	23	24	25
Resistivity (Ω/cm)	446	436	426	417	408	400	392	384	376	369	362
Conductivity (μS/cm)	2242	2293	2347	2398	2451	2500	2551	2604	2659	2710	2769

to that of wine. In practice, a 0.02 M KCl solution is used. The temperature of the KCl (0.02 M) solution must be taken into account in checking the cell constant. The resistivity and conductivity values of this solution according to temperature are specified in Table 1.12.

The conductivity meter cell is subjected to an alternating current. The frequency is set at 1 kHz for the standardized solution (KCl = 0.02 M) and wine, to avoid polarizing the electrodes. A conductivity meter is used for continuous monitoring of tartrate precipitation in wine (see Section 1.6.4, Figure 1.16).

1.6 TESTS FOR PREDICTING WINE STABILITY IN RELATION TO CRYSTAL PRECIPITATION AND MONITORING THE EFFECTIVENESS OF ARTIFICIAL COLD STABILIZATION TREATMENT

1.6.1 The Refrigerator Test

This traditional test is somewhat empirical. A sample (approximately 100 ml) of wine, taken before or after artificial cold stabilization, is stored in a refrigerator for 4–6 days at 0°C and then inspected for crystals. In the case of wines intended for a second fermentation, alcohol may be added to increase the alcohol content by 1.3–1.5% v/v. This simulates the effects of the second fermentation and makes it possible to assess the bitartrate stability of the finished sparkling wine.

The advantages of this test are that it is simple and practical, and requires no special equipment. On the other hand, it is mainly qualitative, and does not provide an accurate indication of the wine's degree of instability. Its major disadvantage is that it takes a long time and is incompatible with short contact stabilization technologies, where rapid results are essential to assess the treatment's effectiveness in real time.

Finally, this test is neither reliable, nor easily repeatable, as it is based on the phenomenon of spontaneous, non-induced crystallization—a slow, undependable process.

1.6.2 The 'Mini-contact' Test

A sample of wine with 4 g/l added potassium bitartrate is maintained at a temperature of 0°C for 2 hours, and constantly agitated. The wine sample is cold-filtered and the weight increase of the tartrate collected (exogenous tartrate + wine tartrate) is assessed. It is also possible to dissolve the precipitate in a known volume of hot water and measure the increase in acidity as compared to that of the 4 g/l exogenous potassium bitartrate added to the wine.

The mini-contact test is based on homogeneous induced nucleation, which is faster than primary nucleation. However, this test does not take into account the particle size of the seed tartrate, although the importance of its effect on the crystallization rate is well known. The operative factor in this test is the surface area of the liquid/solid contact interface. Furthermore, this test defines the stability of the wine at 0°C and in its colloidal state at the time of testing. In other words, it makes no allowance for colloidal reorganization in wine, especially red wine, during aging.

It is normal to find potassium bitartrate crystals, associated with precipitated condensed coloring matter, in wine with several years' aging potential. When phenols condense, they become bulky, precipitate and are no longer able to express their 'protective colloid' effect.

It should be noted that mini-contact test results tend to overestimate a wine's stability and therefore the effectiveness of prior treatment. This statement is based on work by Boulton (1982). After 2 hours' contact, only 60–70% of the endogeneous tartrate has crystallized and therefore the increase in weight of the crystal precipitate is minimized. These results are interpreted to mean that the treatment was more effective, or the wine more stable, than was actually the case. In order to make the mini-contact test faster, more reliable and compatible with the dynamic contact process, the Martin Vialatte Company proposed the following variant in 1984: seeding a wine sample with 10 g/l of cream of tartar and measuring the drop in conductivity at 0°C.

The rules governing stability under the extreme supersaturation conditions prevailing in wine are as follows:

1. If, in the 5–10 min after seeding, the drop in conductivity is no more than 5% of the wine's initial conductivity (measured before adding potassium bitartrate), the wine may be considered to be properly treated and stabilized.
2. If the drop in conductivity is over 5%, the wine is considered unstable.

As this test is based on measuring the wine's electrical conductivity, it has the tremendous advantage that there is no need to collect the precipitate by filtration and determine the increase in weight. This new mini-contact test, measuring conductivity, is much faster (5–10 min instead of 2 h). Furthermore, by comparison with the first variant of the mini-contact test, as the contact surface (A) and, consequently, the state of supersaturation of the wine are multiplied by 2.5 (adding 10 g/l of KTH instead of 4 g/l), it gives a more accurate assessment of a wine's stability.

In spite of these improvements, this test remains open to criticism and its reliability is limited. Indeed, as is the case with the preceding test, it does not always take into consideration the effect of particle size, and is based on excessively small variations in conductivity and too short a contact time. The results in Tables 1.13, 1.14 and 1.15 corroborate this point of view. In Table 1.13, results indicate that a variation of over 5 units in the concentration product PC_K (see samples A and D) only caused a decrease of 1% from the wine's initial conductivity. In this instance, a white wine with a PC_K close to 13 was considered unstable, but this assessment was not confirmed by the percentage conductivity.

Table 1.13. Values of the concentration products of wines and the corresponding percentage drop in conductivity produced by the mini-contact test

Samples	$PC_K \times 10^5$	Drop in conductivity at 0°C (%)
A	7.28	0.5
B	11.62	1.0
C	11.84	0.0
D	12.96	1.5

The unreliability of this result is confirmed by the experiment described in Table 1.14, involving a wine with an initial PC_K of 9.17×10^5, maintained at 30°C, in which increasing concentrations of commercial cream of tartar were dissolved. It was observed that, when the PC_K of a wine was doubled (e.g. wine +0.2 g/l of dissolved KTH and wine +1 g/l of dissolved KTH) the percentage drop in conductivity was the same, although there was obviously a difference in stability.

Table 1.8 shows that the effects of variations in cream of tartar particle size and contact time in the same wine were capable of causing a difference of 5% in the drop in initial conductivity, which is the benchmark for deciding whether a wine is stable or not.

In practice, a rapid-response test is required for monitoring the effectiveness of artificial cold stabilization. The preceding results show quite clearly that the tests based on induced crystallization are relatively unreliable for predicting the stability of a wine at 0°C.

1.6.3 The Wurdig Test and the Concept of Saturation Temperature in Wine

Wurdig *et al.* (1982) started with the idea that the more KTH a wine is capable of dissolving at low

Table 1.14. Demonstrating the limitations of the reliability of the mini-contact test in assessing the stability of a wine by adding increasing quantities of potassium bitartrate and measuring the percentage drop in conductivity

Samples	pH	K^+ (mg/l)	$PC_K \times 10^5$	Drop in initial conductivity (%)
Control	3	390	9.17	1.5
Wine + 0.2 g/l KTH	3	420	10.85	11.5
Wine + 0.5 g/l KTH	3.03	469	13.33	7.5
Wine + 0.7 g/l KTH	3.05	513	15.26	12.5
Wine + 1 g/l KTH	3.06	637	21.16	11.5

Table 1.15. Influence of tartrate particle size and mini-contact test time on the percentage drop in conductivity of the wine

Drop in conductivity (%)	Commercial KTH	KTH: particle size greater than 100 μm	KTH: particle size smaller than 63 μm
After 10 min	12	9	14
After 20 min	13	11	16

temperatures, the less supersaturated it is with this salt and, therefore, the more stable it should be in terms of bitartrate precipitation. The authors defined the concept of saturation temperature (T_{Sat}) in a wine on the basis of this approach.

The saturation temperature of a wine is the lowest temperature at which it is capable of dissolving potassium bitartrate. In this test, temperature is used as a means of estimating the bitartrate stability of a wine, on the basis of the solubilization of a salt.

In comparison with the previously described tests, based on crystallization, this feature seems very convincing. Indeed, the solubilization of a salt is a spontaneous, fast, repeatable phenomenon, much less dependent on the particle size of the added tartrate crystals. The solubilization of KTH is also much less affected by the colloidal state of the wine at the time of testing. It has been observed that 'protective colloids' act as crystallization inhibitors, but do not affect the solubilization of salts. Consequently, estimating the bitartrate stability of a wine by testing the solubilization of KTH, i.e. saturation temperature, is a more reliable measurement in the long term as it is independent of any colloidal reorganization during storage and aging.

The saturation temperature of a wine was determined by measuring electrical conductivity (Figure 1.14) in a two-stage experiment.

In the first experiment, the wine was brought to a temperature of approximately 0°C in a thermostat-controlled bath equipped with sources of heat and cold. The temperature was then raised to 20°C in 0.5°C increments and the wine's conductivity measured after each temperature change. In this way, it was observed that the variation in conductivity according to the temperature of a wine containing no KTH crystals was represented by a roughly straight line.

In the second experiment, a volume (100 ml) of the same wine was brought to a temperature close to 0°C, 4 g/l of KTH crystals were added and the temperature was once again raised to 20°C in 0.5°C increments. The wine was agitated constantly and its conductivity measured after each temperature change. Two patterns were observed:

1. Subsequent to the addition of 4 g/l of KTH, the wine (Figure 1.14a) showed a linear variation in conductivity at low temperatures that could almost be superimposed on that of the wine without crystals until a temperature T_{Sat}, where the conductivity left the straight line and followed the exponential solubility curve.

Organic Acids in Wine

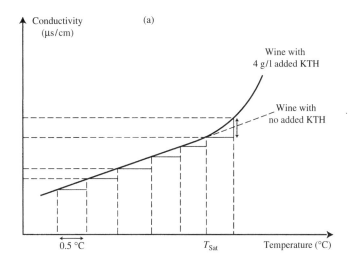

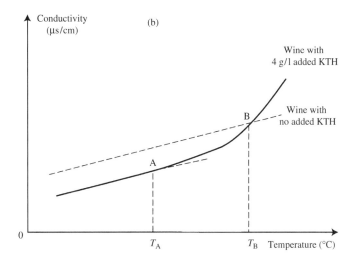

Fig. 1.14. Experimental determination of the saturation temperature of a wine by the temperature gradient method (Wurdig *et al.*, 1982). (a) Example of a wine that is not highly supersaturated, in which no induced crystallization occurs after the addition of tartrate crystals at low temperature. (b) Example of a highly supersaturated wine, in which induced crystallization occurs immediately after the addition of calcium potassium tartrate crystals

2. Following the addition of 4 g/l of KTH, the wine's conductivity (Figure 1.14b) at temperatures around 0°C was below that of the wine alone. This meant that low-temperature induced crystallization had occurred, revealing a state of supersaturation with high endogeneous KTH levels in the wine. Its conductivity then increased in a linear manner until temperature T_A; then the KTH started to dissolve and the conductivity followed the exponential solubility curve. At temperature T_B, the exponential solubility curve crossed the straight line showing the conductivity of the wine alone. This intersection corresponds to the wine's true saturation temperature. The temperature T_A corresponds to that of the same wine after a 'contact', leading

to desaturation caused by induced crystallization. It is therefore normal that, following desaturation, the wine should solubilize more KTH, at a temperature lower than its true saturation temperature, T_B.

On a production scale, where rapid stabilization technologies are used, experimental determination of the saturation temperature by the temperature gradient method is incompatible with the rapid response required to monitor the effectiveness of ongoing treatment.

On the basis of statistical studies of several hundred wines, Wurdig et al. (1982) established a linear correlation defined by:

$$T_{Sat} = 20 - \frac{(\Delta L)_{20°C}}{29.3} \quad (1.10)$$

This straight-line correlation (Figure 1.15) between the variation in conductivity of a wine at 20°C before and after the addition of 4 g/l of potassium bitartrate (ΔL) and the saturation temperature has only been verified for wines where the solubilization temperature of KTH is between 7 and 20°C. The practical advantage of using this equation is that the saturation temperature of a wine may be determined in just a few minutes, using only two measurements.

In some wines, crystallization may be induced by adding cream of tartar at 20°C. This means that they have a lower conductivity after the addition of tartrate, i.e. a saturation temperature above 20°C. This is most common in rosé and red wines. In order to determine their precise saturation temperature, the samples are heated to 30°C. Cream of tartar is added and the increase in conductivity at this temperature is measured. The saturation temperature is deduced from (Maujean et al., 1985):

$$T_{Sat} = 29.91 - \frac{(\Delta L)_{30°C}}{58.30} \quad (1.11)$$

Calculating the saturation temperature of a wine prior to cold stabilization provides information on the optimum seeding rate for that wine. Indeed, it is not necessary to seed at 400 g/hl, as often recommended, if 40 g/hl are sufficient.

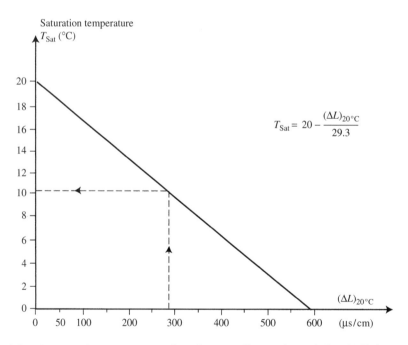

Fig. 1.15. Determining the saturation temperature of a wine according to the variation (ΔL) in conductivity at 20°C before and after the addition of potassium bitartrate (KHT) (Wurdig et al., 1982)

1.6.4 Relationship Between Saturation Temperature and Stabilization Temperature

The temperature at which a wine becomes capable of dissolving bitartrate is a useful indication of its state of supersaturation. However, in practice, enologists prefer to know the temperature below which there is a risk of tartrate instability. Maujean et al. (1985, 1986) tried to determine the relationship between saturation temperature and stability temperature.

The equations for the solubility (A) and hypersolubility (B) curves (Section 1.5.1, Figure 1.11) were established for this purpose by measuring electrical conductivity. They follow an exponential law of the following type: $C = a\, e^{bt}$, where C is the conductivity, t is the temperature and a and b are constants.

The experiment to obtain the exponential hypersolubility curve (B) consisted of completely dissolving added cream of tartar in a wine at 35°C and then recording the conductivity as the temperature dropped. This produced an array of straight-line segments (Figure 1.11) whose intersections with the exponential solubility curve (A) corresponded to the saturation temperatures (T_{Sat_i}) of a wine in which an added quantity i of KTH had been dissolved. The left-hand ends of these straight-line segments corresponded to the spontaneous crystallization temperatures (T_{CS_i}). For example, if 3 g/l of KTH is dissolved in wine, the straight line representing its linear decrease in conductivity stops at a temperature of 18°C, i.e. the temperature where spontaneous crystallization occurs (T_{CS_3}).

Of course, if only 1.1 g/l of KTH is dissolved in the same wine, crystallization occurs at a lower temperature, as the wine is less supersaturated ($T_{CS_{1.1}} = 4.5°C$). It is therefore possible to obtain a set of spontaneous crystallization temperatures based on the addition of various quantities i of KTH (Figure 1.11).

The envelope covering this set of spontaneous crystallization temperatures (T_{CS_i}) defines the exponential hypersolubility curve (B). The exponential solubility and hypersolubility curves, representing the boundaries of the supersaturation field, are parallel. This property, first observed in champagne-base wines, is used to deduce the spontaneous crystallization temperature of the initial wine.

Indeed, projecting from the intersections between the straight lines indicating conductivity and the two exponentials (A) and (B) to the temperature axis, produces temperatures T_{Sat_i} and T_{CS_i}, respectively. The difference, $T_{Sat_i} - T_{CS_i}$, defines the width of the supersaturation field of the wine in which i added KTH has been dissolved, expressed in degrees Celsius. The width of the supersaturation field is independent of the addition value i, as exponents (A) and (B) are roughly parallel. Thus, in the example described (Figure 1.11), the width of the supersaturation field is close to 21°C, whether 1.1 g/l ($T_{Sat_{1.1}} - T_{CS_{1.1}} = 25.2 - 4.5 = 20.7°C$) or 1.8 g/l ($T_{Sat_{1.8}} - T_{CS_{1.8}} = 30.2 - 10.4 = 20.8°C$) of KTH is added. If 21°C is subtracted from the true saturation temperature of the wine (T_{Sat_0}), i.e. no added KTH ($i = 0$), it may be deduced that spontaneous crystallization is likely to occur in this wine at temperature $T_{CS_0} = T_{Sat_0} - 21 = -5°C$.

The experimental method for finding the width of the supersaturation field has just been described, and the relationship between the saturation temperature and the temperature below which there is a risk of crystallization has been deduced. The width of the supersaturation field, corresponding to the delay in crystallization, must be linked, at least partially, to the phenomenon of surfusion (the effect of alcohol), as well as the presence of macromolecules in the wine which inhibit the growth of the nuclei. These macromolecules include carbohydrate, protein and phenol colloids. It seems interesting, from a theoretical standpoint, to define the contribution of these protective colloids to the width of the supersaturation field. It also has a practical significance, and should be taken into account in preparing wines for tartrate stabilization. For this purpose, aliquots of the same white wine at 11% v/v alcohol were subjected to various treatments and fining (Table 1.16). At the same time, a model dilute alcohol solution was prepared: 11% v/v buffered at pH 3, containing 4 g/l of

Table 1.16. Influence of pre-treatment on the physicochemical parameters of a cold-stabilized white wine. Wines treated with slow cold-stabilization (10 days at −4°C). Assessment of protective effects (Maujean et al., 1985)

Samples		Total acidity (g/l H_2SO_4)	pH	Potassium (mg/l)	Tartaric acid (g/l H_2SO_4)	PCK × 10^5	T_{Sat} measured (°C)	T_{Sat} calculated (Wurdig) (°C)	T_{CS} calculated (°C)	$T_{Sat} - T_{CS}$ measured (°C)[a]
Control	Before cold	7.03	3.13	970	1.46	19.67	18.19	17.85	−2.60	20.8
	After cold	7	3.05	730	0.98	9.21	9.55	11.06	−12.7	22.25
Bentonite (30 g/hl)	Before cold	7.29	3.09	985	1.59	20.97	17.05	17.14	−1.15	18.2
	After cold	6.97	3.04	740	0.77	7.26	9.6	9.77	−9.4	19
Charcoal decolorant (30 g/hl)	Before cold	7.21	3.1	940	1.59	20.97	17.05	17.2	−2.7	19.75
	After cold	6.89	3.1	750	1.01	10.24	9.1	10.33	−11.3	20.4
Gum arabic (3 g/hl)	Before cold	7.31	3.08	940	1.45	18.07	16.8	16.98	−3.8	20.6
	After cold	7.04	3.03	730	0.91	8.37	11	11.32	−10.95	21.95
Tanin (6 g/hl) and Gelatin (3 g/hl)	Before cold	7.25	3.08	970	1.42	18.26	18	17.97	−4.9	22.9
	After cold	7.2	3.08	970	1.32	17.46	16	16.16	−5.5	21.05
Metatartaric acid (5 g/100 bottles)	Before cold	7.19	3.01	975	1.23	20.35	19.25	18.91	<−3.75	>23
	After cold	7.26	3.09	975	0.23	16.06	18.65	18.61	−6.09	24.7
Filtered membrane 10^3 Da	Before cold	6.51	3.08	955	1.25	15.83	16.9	16.54	2.85	14.05
	After cold	5.67	3.01	535	0.3	2.24	1.8	0.63	−12.8	14.6
Filtered membrane 0.22 μm	Before cold	7.22	3.08	970	1.54	19.8	17	17.06	−3.65	20.65
	After cold	7	3.03	970	0.94	9.08	11.6	11.21	−8.5	20.1

[a] The differences, $T_{Sat} - T_{CS}$, were determined by dissolving 1 and 2 g/l of THK in the wine. Conductivity was then recorded at decreasing temperatures until crystallization occurred; the T_{CS} values were deduced.

Organic Acids in Wine

KTH, with a saturation temperature of 22.35°C. The spontaneous crystallization temperature of the same solution was also determined after 1.4 g/l of KTH had been dissolved in it, $T_{CS_{1.4}} = 7.4°C$. It was thus possible to find the width of the supersaturation field, i.e. 15°C.

The spontaneous crystallization temperature of each sample of treated wine (Table 1.16) was also determined using the same procedure. Examination of the results shows that a wine filtered on a 10^3 Da Millipore membrane, i.e. a wine from which all the colloids have been removed, has the lowest value for the supersaturation field ($T_{Sat} - T_{CS_0}$), closest to that of the model dilute alcohol solution. Therefore, the difference between the results for this sample and the higher values of the supersaturation fields of 'fined' samples define the effect of the protective colloids. It is interesting to note that the sample treated with metatartaric acid had the widest supersaturation field, and cold stabilization was completely ineffective in this case. This clearly demonstrates the inhibiting effect this polymer has on crystallization and, therefore, its stabilizing effect on wine (Section 1.7.6). Stabilization by this method, however, is not permanent.

On the basis of these results evaluating the protective effects of colloids and saturation temperatures before and after cold stabilization, it is possible to determine the most efficient way to prepare a white wine for bitartrate stabilization. It would appear that tannin–gelatin fining should not be used on white wines, while bentonite treatment is the most advisable. The effect of tannin–gelatin fining bears out the findings of Lubbers *et al.* (1993), highlighting the inhibiting effect of yeast-wall mannoproteins on tartrate precipitation.

There are quite tangible differences in the performance of slow stabilization when wines have no protective colloids (cf. wine filtered on a membrane retaining any molecule with a molecular weight above 1000 Da). These effects ought to be even more spectacular in the case of rapid stabilization technologies. Indeed, the results presented in Figure 1.16 show the impact of prior preparation on the effectiveness of the contact process.

It was observed that the crystallization rate during the first hour of contact, measured by the slope of the lines representing the drop in conductivity of the wine in μS/cm per unit time, was highest for the wine sample filtered on a 10^3 Da membrane, i.e. a wine containing no protective colloid macromolecules. On the contrary, the addition of metatartaric acid (7 g/hl) completely inhibited the crystallization of potassium bitartrate, even after four hours. In production, bentonite and charcoal decolorant are the best additives for preparing wine for tartrate stabilization using the contact process.

1.6.5 Applying the Relationship between Saturation Temperature (T_{Sat}) and Stabilization Temperature (T_{CS}) to Wine in Full-scale Production

In practice, the saturation temperature is obtained simply by two electrical conductivity measurements, at 20°C for white wines and 30°C for red wines. The first is measured on the wine alone, the other after the addition of 4 g/l of KHT crystals. Equations (1.10) and (1.11) are used to calculate T_{Sat} for white wines and for red wines, respectively. The relationship between saturation temperature T_{Sat} and true stability temperature in various types of wine is yet to be established.

In order to define a rule that would be reliable over time, i.e. independent of the colloidal reorganizations in white wine during aging, Maujean *et al.* (1985, 1986) proposed the following equation:

$$T_{CS} = T_{Sat} - 15°C$$

Note that this equation totally ignores protective colloids, and is valid for a wine with an alcohol content of 11% v/v. For white wines with an alcohol content of 12.5% v/v, or those destined for a second fermentation that will increase alcohol content by 1.5% v/v, the equation becomes:

$$T_{CS} = T_{Sat} - 12°C$$

Thus, if stability is required at −4°C, the saturation temperature should not exceed 8°C. The stability normally required in Champagne corresponds to the temperature of −4°C used in

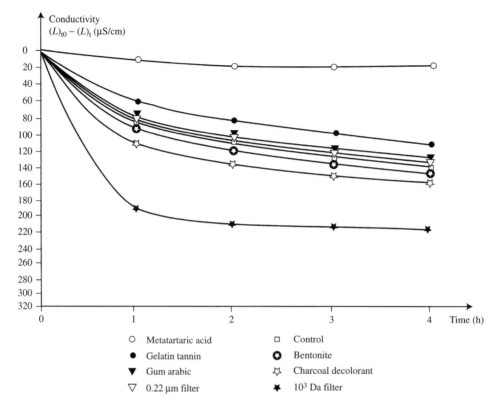

Fig. 1.16. Crystallization kinetics of potassium bitartrate analyzed by measuring the drop in conductivity of a wine according to the type of treatment or fining. Samples were stored at 2°C, seeded with 5 g/l of KTH and subjected to the static contact process for four hours (Maujean *et al.*, 1986)

the slow artificial cold stabilization process. It is questionable whether such a low temperature is necessary to minimize the probability of tartrate crystallization.

In the case of a rosé champagne-base wine, the equation is as follows:

$$T_{CS} = T_{Sat} - 15°C$$

This equation shows that, if stability is required at −4°C, the saturation temperature must be 11°C or lower.

In the case of red wines, it is possible to be less demanding, due to the presence of phenols. To simplify matters, Gaillard and Ratsimba (1990) relate the tartrate stability of wines uniquely to saturation temperature. They estimate that stability is achieved if:

1. In white wines, $T_{Sat} < 12.5°C$.
2. In red wines, $T_{Sat} < (10.81 + 0.297 \text{ IPT})°C$, where IPT represents the total polyphenol number.

These methods, based on the solubilization of KHT, independent of the medium's composition, are applicable to monitoring cold stabilization treatments.

1.6.6 Using Mextar® Calculation Software

This is a completely different approach to forecasting tartrate instability, still one of the main problems in winemaking.

By transposing methods used for crystallization in solution, Devatine *et al.* (2002) developed

Mextar®, a software program that offers a reliable measure of the stability or degree of instability of a wine, by means of calculations using analysis data on the constituents of the wine's acidity. It is, thus, theoretically possible to obtain an accurate assessment of the need to subject a wine to stabilization treatment. The calculation also predicts changes in chemical composition during spontaneous or induced transformations. Finally, Mextar® can be used to model changes in a wine's acidity, by simulating acidification and deacidification operations, as well as malolactic fermentation, and predicting the pH and total acidity values following these processes.

It will be interesting to monitor the development of this system and its application to different types of wine.

1.7 PREVENTING TARTRATE PRECIPITATION

1.7.1 Introduction

This section will describe the main bitartrate stabilization technologies used for wine (see also Section 12.3.2).

Whatever the technology used, and regardless of any treatment used preparatory to bitartrate stabilization, wine treated with artificial cold must be clean, i.e. not excessively contaminated with yeast or bacteria, as is often the case with wines stored in large vats. These wines should, therefore, be filtered on a simple continuous earth filter. Another advantage of filtration is the elimination of part of the protective colloids. Fine filtration is not useful at this stage, and is certainly not recommended, as there is a risk of eliminating microcrystals likely to act as crystallization nuclei.

1.7.2 Slow Cold Stabilization, Without Tartrate Crystal Seeding

This is the traditional technology for the bitartrate stabilization of wine. Before wineries were equipped with refrigeration and air-conditioning systems, wines were simply exposed to natural cold by opening the vat room doors during the coldest winter weather.

The temperature may decrease at varying rates. It is gradual if the wine is chilled by means of a submerged refrigerating rod in the vat. It may be much faster in a normal installation (Section 12.3.4, Figure 12.1) including a plate heat exchanger to recover energy from the treated wine and reduce the temperature of wine to −4°C more rapidly prior to treatment. It is known (Section 12.3.4) that faster cooling promotes more complete precipitation of the tartrate in the form of small crystals.

Heat-insulated vat rooms, equipped with heating/cooling systems, are also used. The wines are stored in uninsulated vats with a high heat-transfer coefficient, such as stainless steel. The entire room is maintained at the desired temperature, keeping the wine at a negative temperature for 8–10 days (white wines) or up to several weeks, in the case of red wines (Blouin, 1982).

The treatment temperature is generally defined by the following rule:

$$\text{Treatment temperature} = -\frac{\text{Alcohol content}}{2} - 1 \quad (1.12)$$

This rule is deduced from the equation defining the freezing temperature of wine according to its alcohol content:

$$\text{Freezing temperature} = -\frac{\text{Alcohol content} - 1}{2} \quad (1.13)$$

Slow stabilization is tending to evolve towards pseudo-contact technology by seeding with 30–40 g/hl of cream of tartar, agitating for 36 hours and ensuring that the wine does not oxidize. Paddle agitators with variable-speed motors are the most efficient, also ensuring that only a minimal amount of oxygen is dissolved in the wine. There is a significant risk of excessive oxidation as gases dissolve more readily at low temperatures. It is recommended that the agitation rate is monitored by measuring the optical density at 420 nm. In a white wine that has not suffered oxidation, this value decreases by 10% during cold stabilization. Seeding with 20–40 g/hl of KTH should be envisaged if, for example, natural chilling of the wine has produced some crystallization, so that it is in a less-saturated state.

Slow stabilization often causes loss of color (OD at 520 nm) in both red and white wines. It is therefore recommended that the length of treatment is reduced by adding small particles of cream of tartar, which are easier to maintain in a homogeneous suspension. Another advantage of seeding is that the wine may be maintained at a less cold temperature ($-2°C$ instead of $-4°C$).

It has been demonstrated on a production scale (360 hl vats) that the stabilization time for a white wine treated with 30 g/hl of bentonite, maintained at $-2°C$ and seeded with 30 g/hl of cream of tartar, may be reduced to 62 hours (including 24 hours without agitation before filtration), instead of 6 days for the standard treatment. Under these conditions, the wine was found to be perfectly stabilized ($T_{Sat} = 7°C$).

1.7.3 Rapid Cold Stabilization: Static Contact Process

This technique has the major advantage of reducing the artificial cold treatment of wine to 4 hours, and sometimes less for white wines. Furthermore, the wine no longer has to be maintained at negative temperatures, but only at 0°C, which minimizes not only energy consumption but also frost accumulation on the equipment. A heat-insulated, conical-bottomed vat known as a crystallizer is used. It is equipped with a drain to remove excess crystals at the end of the cycle.

Such high-performance levels can only be achieved with this type of rapid stabilization treatment by seeding with large quantities of cream of tartar (400 g/hl). This large mass of crystals, with a small initial particle size, must absolutely be maintained in suspension by an agitator, taking care to avoid any unwanted aeration (Section 1.5.2). It is also advisable to blanket the wine with inert gas, or at least use an airtight crystallizer.

Treatment effectiveness is monitored by the rapid response analysis technique described in Section 1.6.4. If the results are satisfactory, agitation is stopped to allow most of the tartrate to settle

Table 1.17. Changes in the physicochemical parameters of cold-stabilized wine when the contact tartrate was recycled (Maujean et al., 1986)

Number of times used	K$^+$ (mg/l)	Total acidity (g/l H$_2$SO$_4$)	Tartaric acid (g/l H$_2$SO$_4$)	pH	pC $\times 10^5$
1	315	4.93	1.59	3.11	6.83
2	325	4.92	1.54	3.12	6.88
3	320	4.90	1.59	3.11	6.84
4	300	4.98	1.83	3.09	7.35
5	320	4.94	1.55	3.08	6.57

in the conical bottom of the crystallizer. Complete clarification is not easy to obtain. Great care must be taken in using centrifugation as the crystals are highly abrasive. Good results are obtained with horizontal plate filters, using the crystals themselves as the filter layer. Of course, all these operations must be carried out at 0°C.

The static contact process is a very flexible system. It is possible to run 2–3 cycles per day with volumes of 50–100 hl in each batch. This technology is advisable for small and medium-sized wineries. The weak point of this system is the price of cream of tartar, but costs may be reduced by recycling tartrate.

In the case of white champagne-base wines, it has proved possible to recycle the tartrate four times, with almost constant treatment effectiveness (Table 1.17). The continued effectiveness of the treatment, even when the tartrate has been recycled four times, has been explained (Maujean et al., 1986). They showed that the smallest particle size after treatment (<50 µm) was larger than the initial size in the commercial product.

Of course, recycling is not possible when red wines are treated, as the crystals become coated with phenols and coloring matter and rapidly lose their effectiveness.

1.7.4 Rapid Cold Stabilization: Dynamic Continuous Contact Process

Unlike the preceding 'batch' technology, the process described in Figure 1.17 is a continuous

Organic Acids in Wine

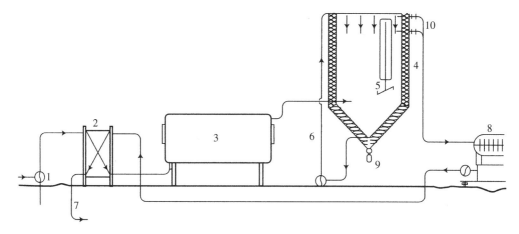

Fig. 1.17. Schematic diagram of a continuous cold stabilization system: 1, intake of wine to be treated; 2, heat exchanger; 3, refrigeration system (with compressor, condenser, etc.); 4, insulation; 5, mechanical agitator; 6, recycling circuit (optional); 7, outlet of treated wine; 8, filter (earth); 9, drain; 10, overflow

bitartrate stabilization process, where the length of time the crystals are in contact with the wine, i.e. the treatment time, is defined by the throughput in relation to the volume of the crystallizer. Thus, for example, if the throughput is 60 hl/h and the volume of the crystallizer is 90 hl, the average time the wine spends in the system is 1 h 30 min.

This emphasizes the need for a method of monitoring effectiveness with a very short response time. There is, of course, a system for recycling wine through the crystallizer if the treatment is insufficiently effective, but the results must be determined very rapidly, as the energy required to treat these quantities of wine is expensive, and unnecessary extra treatment will by no means improve quality.

Continuous treatment is understandably more demanding than the other processes, because it requires close monitoring, but it is also more efficient. For example, the particle size of the contact tartrate and the level in the crystallizer must be monitored by sampling after a few hours, using the drain system.

Agitation is partly provided by a tangential input of wine into the crystallizer. This creates turbulence in the mass of the liquid and maintains at least the smallest crystals in suspension. The wine may also be mechanically agitated.

The throughput, i.e. the average time in the crystallizer, is defined according to the wine's initial state of supersaturation, as well as the type of preparatory treatment (fining, bentonite, etc.) the wine received prior to artificial cold stabilization. The importance of preparation has already been mentioned (Section 1.6.4).

The effectiveness of the three processes described above is generally satisfactory, although results depend on the type of wine (white or red), its alcohol content and any previous treatment or fining.

It is true that, in contact treatments involving large-scale seeding, the wine's background is less important. Indeed, enologists do not always have this information if the wine has been purchased from another winery. In any event, wine must be well prepared and, above all, properly clarified, to ensure the effectiveness of rapid artificial cold stabilization treatments.

1.7.5 Preventing Calcium Tartrate Problems

Calcium tartrate is a relatively insoluble salt, ten times less soluble than potassium bitartrate (see 1.5.1, Table 1.11). Independently of any accidental contamination, calcium added in the form of calcium bentonite for treating must or wine, calcium carbonate for deacidification purposes, or even as a contaminant in saccharose used for chaptalization, may cause an increase in the calcium tartrate content of wine. Combined with an increase in pH, this may put the wine into a state of supersaturation for this salt, leading to crystal deposits. Robillard *et al.* (1994) reported that crystallization of TCa was even observed in champagne-base wines with a particularly low pH. There is considered (Ribéreau-Gayon *et al.*, 1977) to be a real risk of tartrate deposits in the bottle when the calcium content is over 60 mg/l in red wine and 80 mg/l in white wine.

Stabilizing wines to prevent precipitation of calcium tartrate is not easy, as the crystallization of potassium bitartrate does not induce that of calcium tartrate, despite the fact that these two salts should logically syncrystallize as they have the same crystal systems. On the contrary, crystallization of TCa may induce that of KTH. The prevention of calcium tartrate precipitation is further complicated by the fact that the solubility of TCa (Postel, 1983) is not very temperature-sensitive. Thus, TCa is hardly three times more soluble at 20°C than at −4°C.

Furthermore, according to Abgueguen and Boulton (1993), although the crystallization kinetics of TCa should be higher than those of KTH, the time required for spontaneous nucleation of TCa is much longer. It is therefore easier to understand why calcium tartrate precipitation generally occurs in wine after several years' aging.

On the basis of research into potassium bitartrate (Figure 1.7), Vallée (1995) used measurements of electrical conductivity to define the width of the supersaturation field expressed in degrees Celsius, as well as the calcium tartrate saturation temperature of various types of wines. The low solubility of calcium tartrate indicates that saturation temperatures are likely to be much higher than those of potassium bitartrate.

In order to avoid the risk of calcium tartrate precipitation, the saturation temperature of white, rosé and *vins doux naturels* must be lower than 26°C to ensure that calcium tartrate deposits will not be formed if the wine is kept at 2°C for one month. The calcium tartrate saturation temperature for red wines must be below 35°C.

According to Postel (1983), the addition of 100 mg/l of metatartaric acid is capable of stabilizing a wine stored at 4°C for several months, so that it does not suffer from crystalline deposits of TCa. Furthermore, the use of racemic acid (D-L-tartaric acid) or left-calcium tartrate has been suggested for eliminating excess calcium (Ribéreau-Gayon *et al.*, 1977). In both cases, the precipitation of calcium racemate, a highly insoluble salt, totally eliminates the cation. The treatment's effectiveness depends on the colloid content of the wine, as it hinders precipitation of the salt. These treatments are used to varying degrees in different wine regions according to the types of wines produced.

Finally, ion exchange (Section 12.4.3) and electrodialysis (Section 12.5) are also processes for preventing calcium tartrate deposits.

1.7.6 The Use of Metatartaric Acid

In the processes described above, tartrate precipitations are prevented by eliminating the corresponding salts. It is also possible to envisage the addition of crystallization inhibitors.

The first positive results were obtained with hexametaphosphate, which certainly proved to be effective (Ribéreau-Gayon *et al.*, 1977). However, very high doses were necessary in certain wines and, above all, the increase in phosphate content led to the formation of a ferric complex that caused instability on contact with air (phosphatoferric casse).

Metatartaric acid is currently the product most widely used for this purpose. Carboxymethylcellulose (Section 1.7.8) and mannoproteins extracted from yeast (Section 1.7.7) have also been suggested as stabilizers.

The use of carboxymethylcelluloses has also been suggested. These are a group of complex, poorly-defined products with various properties. Their effectiveness seems to vary according to the type of wine, but especially in relation to the presence of protective colloids. Carboxymethylcelluloses modify a wine's viscosity. They have not as yet been developed on an industrial scale.

Organic Acids in Wine

The possibility of using mannoproteins extracted from yeast seems worth considering, since this product is both effective and stable (Section 1.7.7).

Metatartaric acid is a polyester resulting from the inter-molecular esterification of tartaric acid at a legally imposed minimum rate of 40%. It may be used at doses up to a maximum of 10 g/hl to prevent tartrate precipitation (potassium bitartrate and calcium tartrate) (Ribéreau-Gayon et al., 1977).

When tartaric acid is heated, possibly at low pressure, a loss of acidity occurs and water is released. A polymerized substance is formed by an esterification reaction between an acid function of one molecule and a secondary alcohol function of another molecule. Tartaric acid may be formed again if the metatartaric acid is subjected to hydrolysis. In reality, however, not all of the acid functions react (Figure 1.18).

Metatartaric acid is not a single compound, but rather a dispersed polymer, i.e. a mixture of polymers with different molecular weights. There are many metatartaric acid preparations with different anti-crystallizing properties, depending on the average esterification rate of their acid functions. It is possible to obtain an esterification rate higher than the theoretical equilibrium rate (33% for a secondary alcohol) by heating tartaric acid to 160°C in a partial vacuum. Under these conditions, the thermodynamic esterification equilibrium is shifted by eliminating water.

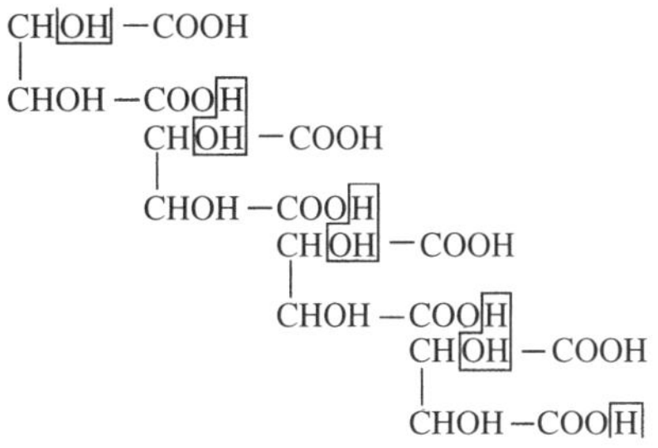

Fig. 1.18. Metatartaric acid polyesterification reaction

The esterification number of different metatartaric acid preparations may be determined by acidimetric assay, before and after saponification. Table 1.18 shows the importance of the preparation conditions in determining this value.

Metatartaric acid is by no means a pure product: solutions are slightly colored and oxidizable. They may contain oxaloacetic acid, but the main impurity is pyruvic acid, representing 1–6% by weight of the metatartaric acid, according to the preparation conditions (Table 1.18). It is, therefore, important to correct the esterification number to compensate for this impurity. The formation of

Table 1.18. Detailed analysis of various metatartaric acid preparations (Peynaud and Guimberteau, 1961)

Preparation method	For 1 g of chemical			Esterification number (%)	Pyruvic acid (%)	Corrected esterification number (%)
	Acidity (meq)	Esters (meq)	Acidity$^+$ esters (meq)			
Reduced pressure, 160°C						
15 min	10.67	3.13	13.80	22.6	0.9	22.8
40 min	8.77	5.14	13.91	36.9	4.2	37.5
45 min	8.63	5.57	14.20	39.2	4.4	40.6
50 min	8.48	5.70	14.18	40.2	4.1	41.5
55 min	8.32	5.74	14.06	40.8	5.6	42.7
Normal pressure, 175°C						
20 min	9.91	3.65	13.46	27.1	5.2	28.3
90 min	9.56	3.76	13.32	28.2	2.3	28.7
105 min	9.11	4.58	13.69	33.4	5.4	35.0

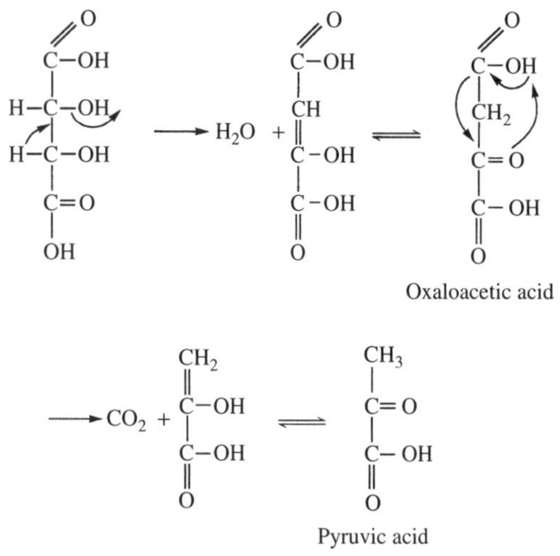

Fig. 1.19. Impurities in metatartaric acid

these two acids results from the intra-molecular dehydration of a tartaric acid molecule, followed by decarboxylation (Figure 1.19).

There are many laboratory tests for assessing the effectiveness of a metatartaric acid preparation. Table 1.19 presents an example of a procedure where a saturated potassium bitartrate solution is placed in 10 ml test tubes and increasing quantities of metatartaric acid preparations with different esterification numbers are added. This inhibits the precipitation of potassium bitartrate induced by adding 1 ml EtOH at 96% vol and leaving the preparation overnight at 0°C. Only 1.6 mg of a preparation with an esterification number of 10 is required to inhibit crystallization, while 4.0 mg are necessary if the preparation has an esterification number of 26.6.

Metatartaric acid acts by opposing the growth of the submicroscopic nuclei around which crystals are formed. The large uncrystallizable molecules of metatartaric acid are in the way during the tartrate crystal building process, blocking the 'feeding' phenomenon, i.e. crystal growth. If the dose is too low, inhibition is only partial, and anomalies and unevenness are observed in the shape of the crystals.

The fact that metatartaric acid solutions are unstable has a major impact on their use in winemaking. They deteriorate fairly rapidly and are also sensitive to temperature. Hydrolysis of the ester functions occurs, accompanied by an increase in acidity. After 20 days at 18–20°C, there is a considerable decrease in the esterification number (Figure 1.20). Under experimental conditions, total hydrolysis of a 2% metatartaric acid solution took three months at 23°C and 10 months at 5°C. Consequently, it is necessary to ensure that metatartaric acid solutions for treating wine are prepared just prior to use.

Furthermore, the same phenomenon occurs in wine and is detrimental to the treatment's effectiveness. Ribéreau-Gayon *et al.* (1977) demonstrated that stability in terms of tartrate precipitations may be considered effective for the following lengths of time, depending on temperature:

Several years at 0°C
Over two years at 10–12°C

Table 1.19. Inhibition of potassium bitartrate precipitation by various metatartaric acids (Peynaud and Guimberteau, 1961)

Number	Esterification number	Metatartaric acid added in each tube (in mg)					
		0.4	0.8	1.6	2.4	3.2	4.0
1	40.8	12.0	15.8	17.2	17.2	17.2	17.2
2	38.2	12.0	15.6	17.2	17.2	17.2	17.2
3	37.3	12.0	15.3	17.2	17.2	17.2	17.2
4	33.4	9.6	12.0	16.3	17.0	17.2	17.2
5	31.5	8.6	11.0	15.3	15.9	16.5	17.2
6	26.6	7.9	10.5	12.7	15.0	16.0	17.2
7	22.9	6.4	7.6	11.2	13.6	15.6	16.8

Potassium remaining in solution (in mg) in each tube containing 10 ml of a saturated potassium bitartrate solution. The original amount was 17.2 mg. Only 5 mg of potassium was left in the tube without metatartaric acid.

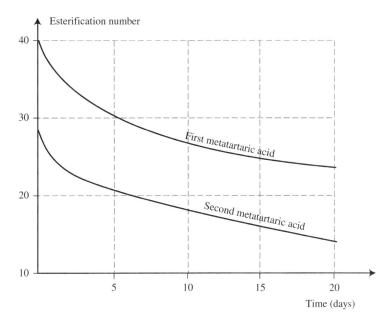

Fig. 1.20. Hydrolysis rate of two qualities of metatartaric acid in 2% solution ($t = 18$–$20°C$), followed by a decrease in the esterification number (Ribéreau-Gayon *et al.*, 1977)

One year to eighteen months at temperatures varying between 10°C in winter and 18°C in summer

Three months at 20°C

One month at 25°C

One week at 30°C

A few hours between 35 and 40°C

Metatartaric acid instability accounts for initially surprising observations concerning wines treated in this way. One sample, stored at 0°C in a refrigerator, had no precipitation, while calcium tartrate precipitation occurred in another sample stored at 20–25°C when it was no longer protected due to hydrolysis of the metatartaric acid.

The conditions for using metatartaric acid depend on its properties. A concentrated solution, at 200 g/l, should be prepared in cold water at the time of use. As metatartaric acid is strongly hygroscopic, it must be stored in a dry place.

Metatartaric acid is added after fining, as there is a risk of partial elimination due to flocculation. It is particularly affected by bentonite and potassium ferrocyanide treatments. Although there was some cause for concern that high-temperature bottling would reduce the effectiveness of metatartaric acid, in fact, under the actual conditions where it is used, this technique has little or no negative impact (Section 12.2.4). Incidentally, a slight opalescence may be observed after a wine has been treated, especially when the most efficient products, with high esterification numbers, have been used. It is therefore recommended that metatartaric acid be added before the final clarification.

1.7.7 Using Yeast Mannoproteins

It is well known that wine, especially red wine, naturally contains macromolecules that act as protective colloids (Section 9.4.2). At concentrations present in wine, these substances tend to hinder tartrate crystallization, but do not completely inhibit it (Section 3.6.5). Little research has been done into isolating these crystallization inhibitors in wine and making use of their stabilizing properties. On the contrary, for many years, major efforts were made to eliminate these colloids, by drastic fining and filtration, as they reduce the effectiveness of physical stabilization treatments, especially cold stabilization.

It is known, however, that the traditional practice of barrel-aging white wines on yeast lees for several months often gives them a high level of tartrate stability, so that cold stabilization is unnecessary (Section 12.3.2). Although, in practice, this phenomenon is very widespread, very little mention of it has been made until now in enology theory. Thus, in Bordeaux, most dry white wines aged on the lees are not stable in March after their first winter, but become stable by June or July without any further treatment. When the same wines are not aged on the lees, they must be systematically cold-stabilized to protect them from tartrate crystallization. As it was known that white wines are enriched with mannoproteins released by the yeast during aging on the lees, it was reasonable to suppose that these macromolecules contributed to the tartrate stabilization of wine.

Yeast mannoproteins were first found to have a certain inhibiting effect on tartrate crystallization in a model medium by Lubbers *et al.* (1993). However, these experiments used mannoproteins extracted by heat in alkaline buffers, under very different conditions from those accompanying the spontaneous enzymic release of mannoproteins during aging on the lees. Furthermore, the effectiveness of mannoproteins extracted by physical processes in preventing tartrate precipitation has not been established in most wines, despite demonstrations in a model medium.

The discovery of the crystallization-inhibiting effect of mannoproteins extracted by the enzymic treatment of yeast walls (Dubourdieu and Moine-Ledoux, 1994) adds a new dimension to this subject. The mannoprotein preparations are obtained by digesting yeast walls with an industrial preparation of β-(1–3)- and β-(1–6)-glucanases (Glucanex™), permitted in winemaking as a clarifying enzyme for improving the filtrability of wines made from botrytized grapes (Sections 3.7.2 and 11.5.2). These preparations inhibit tartrate crystallization in white, red and rosé wines, whereas the same dose (25 g/hl) of heat-extracted mannoproteins does not have this stabilizing effect (Moine-Ledoux and Dubourdieu, 1995).

The inhibiting effect of mannoproteins extracted from yeast on tartrate crystallization is not due to compound MP32, the invertase fragment responsible for protein stabilization in wine (Section 5.6.4) (Dubourdieu and Moine-Ledoux, 1996). The mannoproteins in question are more highly glycosylated, with an average molecular weight of approximately 40 kDa. They have been purified (Moine-Ledoux *et al.*, 1997) from the same mannoprotein preparations, obtained by the enzymic treatment of yeast walls.

Furthermore, it has been demonstrated that these mannoproteins share covalent bonds with glucane (Moine-Ledoux and Dubourdieu, 1999). They remain in the cell walls treated simultaneously with sodium dodecyl sulfate (SDS) (which cuts the hydrogen bonds) and β-mercaptoethanol (Figure 1.21), which do not affect osidic bonds.

The presence of peak 2, corresponding to elution of the mannoprotein responsible for tartrate stabilization, confirms that the bond is covalent. Some of the mannoproteins that share covalent bonds with glucane also have a special type of glycosylation, leading to a glycosyl-phosphatidyl-inositol (GPI). The use of a mutant strain (FBYII), deficient in GPI-anchored mannoproteins when cultured at 37°C (FBYII-37), showed that the mannoproteins responsible for tartrate stabilization had this type of glycosylation. Two types of mannoprotein extracts were obtained by enzyme hydrolysis of yeast cell walls (FBYII), cultured at 24°C or 37°C.

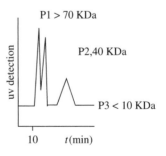

Fig. 1.21. HPLC analysis of molecular-screened mannoprotein extract obtained by enzyme digestion of cell walls treated simultaneously with SDS and β-mercaptoethanol

Organic Acids in Wine

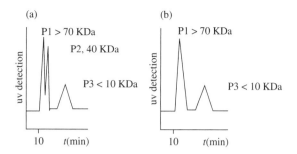

Fig. 1.22. HPLC analysis of molecular-screened mannoprotein extract obtained by enzyme digestion of (a) FBYII-24 and (b) FBYII-37 yeast cell walls, cultured at 24°C and 37°C, respectively

HPLC analysis of these two extracts (Figure 1.22) showed that peak 2 was absent when the cell walls came from yeast cultured at 37°C, i.e. deficient in GPI-anchored mannoproteins. These results: (1) show that the mannoproteins responsible for tartrate stabilization are GPI-anchored and (2) explain why they are only extractible by enzyme digestion.

An industrial preparation (Mannostab™) has been purified from yeast-wall mannoprotein. It is a perfectly soluble, odorless, flavorless, white powder. This product has been quite effective (Table 1.20) in preventing tartrate precipitation in white wine samples taken before the normal cold stabilization prior to bottling. Initial results show that Mannostab™ inhibits potassium bitartrate crystallization at doses between 15 and 25 g/hl. However, in certain wines in Table 1.13 (1996 white Bordeaux and 1996 white Graves), larger quantities apparently reduced the stabilizing effect. A similar phenomenon has been reported with a protective colloid used to prevent protein precipitation (Pellerin *et al.*, 1994). The dose of Mannostab™ necessary to stabilize a wine must be determined by preliminary testing. It is very clear that the use of excess amounts of this additive is inefficient.

The addition of this product could replace current stabilization methods (Moine-Ledoux *et al.*, 1997). With this in mind, its effectiveness has been compared to that of two other tartrate stabilization methods: continuous contact cold stabilization and the addition of metatartaric acid (Table 1.21). This comparison was carried out by measuring spontaneous crystallization after the addition of KHT (Section 1.6.4). The values obtained indicate the effectiveness of protective colloids, even if they do not necessarily correspond to the instability temperatures. The addition of 15 g/hl of Mannostab™ to wine 2 and 25 g/hl

Table 1.20. Tartrate stabilization of various wines by adding Mannostab™. Visual observation of potassium crystallization after 6 days at −4°C (Moine-Ledoux *et al.*, 1997)

Wines		Mannostab™ (g/hl)				
		0	15	20	25	30
1996 Blanc de Blanc	Visual test	a	0	0	0	0
	Δ(K⁺) (mg/l)	52	72	17	0	0
White *vin de table*	Visual test	a	0	0	0	0
	Δ(K⁺) (mg/l)	104	53	33	0	0
1996 white Bordeaux	Visual test	a	0	0	0	0
	Δ(K⁺) (mg/l)	62	21	0	0	21
1996 white Graves	Visual test	a	a	0	0	0
	Δ(K⁺) (mg/l)	155	52	0	0	62
1996 white Bordeaux	Visual test	a	0	0	0	0
	Δ(K⁺) (mg/l)	51	0	0	0	0
1996 Entre Deux Mers	Visual test	0	0	0	0	0
	Δ(K⁺) (mg/l)	52	0	0	0	11

[a] precipitation; 0, no precipitation.

Table 1.21. Effect of different treatments on the spontaneous crystallization temperature of various wines (Moine-Ledoux et al., 1997)

Stabilization treatments	Wine 1	Wine 2
Control	−10°C	−11°C
Mannostab™ (15 g/hl)	−21°C	−18°C
Mannostab™ (25 g/hl)	−31°C	−13°C
Continuous contact cold	−28°C	−17°C
Metatartaric acid (10 g/hl)	−40°C	−40°C

Wine 1, 1996 Entre Deux Mers; Wine 2, 1996 white Bordeaux.

Table 1.22. Influence of keeping a white wine supplemented with metatartaric acid or Mannostab™ at 30°C for 10 weeks on the tartrate stability, estimated by the decrease in potassium concentration after 6 days at −4°C (Moine-Ledoux et al., 1997)

	$\Delta(K^+)$ mg/l, after 6 days at −4°C
Control	200
Metatartaric acid (10 g/hl)	260
Mannostab™ (25 g/hl)	0

to wine 1 produced the same spontaneous crystallization temperature, i.e. a stability comparable to that obtained by continuous cold stabilization (Table 1.21). The addition of metatartaric acid, however, considerably reduced the crystallization temperature.

However, metatartaric acid is hydrolyzed in wine, and loses its effectiveness, while adding tartaric acid may even facilitate potassium bitartrate crystallization. Under the same conditions, mannoproteins are stable and have a durable protective effect on tartrate crystallization. To demonstrate this difference, white wines treated with metatartaric acid or Mannostab™ and kept at 30°C for 10 weeks were then subjected to a cold test. Crystallization occurred in the sample treated with metatartaric acid, while the Mannostab™ sample remained stable (Table 1.22).

This new treatment process to protect wines from tartrate precipitation has been used experimentally in France since 1997 (Moine-Ledoux and Dubourdieu, 2002). Mannoprotein preparation treatment of white wine is registered in the OIV International Code of Oenological Practice. Their findings are likely to lead to the authorization of this type of treatment in the near future.

1.7.8 The Use of Carboxymethylcellulose

Carboxymethylcellulose (CMC) is a polysaccharide. Like metatartaric acid and mannoproteins, its polymer structure gives it "protective colloid" characteristics. It is obtained by priority etherification of the primary alcohol functions of the glucopyranose units (Figure 1.23) linked by β-type stereochemical 1–4 etheroxide bonds. A CMC is, therefore, characterized partly by the degree of etherification of its alcohol functions, known as the degree of substitution (DS), and partly by its degree of polymerization (DP), i.e. the average number of glucopyranose units per polymer molecule. This mean number indicates that a given CMC, such as metatartaric acid, is a polymer with a dispersed molecular weight.

A DS of 0.65 means that, out of 100 glucopyranose units, 65 have been etherified by sodium

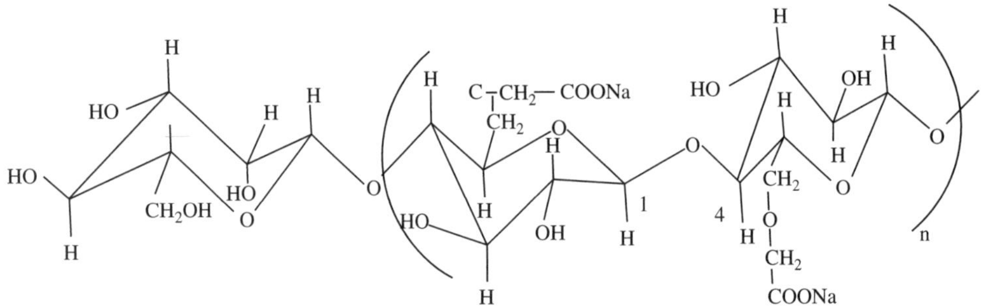

Fig. 1.23. Structure of a carboxymethylcellulose (CMC) chain

$$\text{R} - \text{cellulose (OH)}_3 + 2\text{Cl} - \text{CH}_2 - \text{COONa} \xrightarrow{2\text{NaOH}} \text{R} - (\text{OH}) - (\text{OCH}_2 - \text{COONa})_2 + 2\,\text{NaCl} + 2\text{H}_2\text{O}$$

Fig. 1.24. Formula for the etherification of celluloses (R-[OH]$_3$) by sodium chloroacetate

chloroacetate in an alkaline medium, as shown in the reaction diagram (Figure 1.24).

The DP determines the viscosity of a CMC and increases with molecular weight. The viscosity of a CMC also varies according to the cation—divalent cations (calcium, magnesium, iron, etc.) reduce viscosity. The DP determines the molecular weight, which may vary from 17,000 to 1,500,000 Daltons.

For a CMC with a given DP, the higher its DS, the more cation anchor sites it has, and the more effective it is as a protective colloid (Lubbers *et al.*, 1993).

In the past, CMCs were poorly-defined compounds, with relatively heterogeneous DPs. Their viscosity was unreliable, to the extent that they could modify the viscosity of a wine. The CMCs currently on the market have much more clearly-defined characteristics, and quality control is more effective, resulting in purer products. Minimum purity is 99.5%, with a sodium content between 7 and 8.9%. Viscosity varies from 25,000–50,000 mPa at 25°C, depending on the type of CMC selected, and cannot, therefore, alter the viscosity of the finished beverage.

The production and use of CMCs as a gelatin substitute dates back to the 1940s to 1950s. They are now used in the food and beverage industry (code: E466), at levels up to 10 g/l or 10 g/kg, as well as in cosmetics and pharmaceuticals. The CMC content of alcoholic and non-alcoholic beverages may be as high as 500 mg/l.

Water solubility of CMCs is variable, depending on their degree of substitution and polymerization. They owe their hydrophilic qualities to their highly hydric carbohydrate character. CMCs used in very sweet beverages are less viscous, probably due to the formation of hydrogen bonds between the sugar and the gum. CMC-saccharose interactions depend on the order in which the products are added: if the sugar is dissolved in the water first, its hydrophilic character reduces the solubility of CMC (Federson and Thorp, 1993). This should be taken into account in preparing the concentrated CMC solutions (20–40 g/l) used to treat beverages, such as wine, that require a restricted addition of water (0.05–0.1 l/l).

CMCs are also reputed to promote solubilization of proteins and stabilize solutions containing them (Federson and Thorp, 1993). This property is useful in winemaking for the purpose of preventing protein casse. These CMC–protein interactions may be compared to the carbohydrate–protein association in glycoproteins and yeast mannoproteins.

CMCs are available in the form of powder or white granules. As these absorb humidity from the air, they must be stored in a dry place. They are not yet authorized for use in winemaking in the EU but an application is pending. Recent results, indicating that low-viscosity CMCs are effective in preventing tartrate crystallization at doses 12–250 times lower than those currently used in the food industry (Crachereau *et al.*, 2001), should lead to an authorization in the near future. A dose of 2 g/hl is often ineffective, but good results have been obtained in wines supersaturated with potassium bitartrate without exceeding 4 g/hl. Details of the results are given in Table 1.23 and Figure 1.25.

These results demonstrate comparable effectiveness for metatartaric acid (10 g/hl) and CMC (4 g/hl). Furthermore, a comparison of the stability and effectiveness of these two additives, following prolonged heat treatment at 55–60°C for 5–30 days and one month at −4°C, showed that CMC was perfectly stable. It was still perfectly effective, whereas wine treated with metatartaric acid became totally unstable after only 5 days at 55–60°C (Peynaud and Guimberteau, 1961; Ribéreau-Gayon *et al.*, 1977).

The effectiveness of CMC is due to its property of significantly reducing the growth rate of crystals: a dose of 2 mg/l reduces crystal growth by a

Table 1.23. Treating various wines with CMC (Results after 1 month at −4°C; see Figure 1.25)

Wine treated	Dose of CMC used	Comments
Red A.O.C. Bordeaux	2 g/hl	Unfiltered
Red A.O.C. Buzet	4 g/hl	Filtered prior to treatment
White A.O.C. Bordeaux	4 g/hl	Fined, treated with CMC, then filtered
White *vin de pays* (Gers)	4 g/hl	Fined, treated with CMC, then filtered
White *vin de pays* (Loire)	4 g/hl	Fined, treated with CMC, then filtered
Sparkling wine (Gers)	4 g/hl	Treated prior to second fermentation

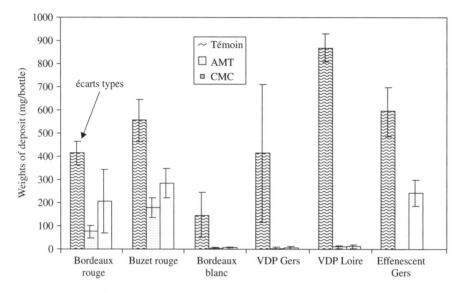

Fig. 1.25. Comparison of the effectiveness of metatartaric acid and carboxymethylcellulose on turbidity due to tartrate crystals (Crachereau *et al.*, 2001) (See Table 1.23 for treatment conditions)

ratio of 7 (Gerbaud, 1996). CMC also modifies the shape of potassium bitartrate crystals.

In the case of wines destined for a second fermentation, three different CMCs produced more stable, persistent bead. Only the CMC with the highest molecular weight caused a slight increase in bubble size. A similar inhibition of crystallization has also been observed in champagne-base wines (Maujean, 1997).

All these positive results, combined with the fact that they are easy to use, relatively inexpensive, and do not require special investments, should lead to their authorization for use in winemaking in the very near future, as is already the case in the food and beverage industry. Further research is required to assess the effectiveness in different types of wine, especially tannic red wines, which have a particularly complex colloidal structure.

(See Table 1.23 for treatment conditions)

REFERENCES

Abgueguen O. and Boulton R. (1993) *Am. J. Enol. Viticult.*, 44, 65.
Blouin J. (1982) *Conn. Vigne et Vin*, 16, 63.
Boulton R. (1982) *Rev. Fr. Œnol.*, 87, 97.
Casey J.A. (1988) *The Australian Grapegrower and Winemaker*, 19.
Champagnol F. (1986) *Rev. Fr. Oenol.*, 104, 26.
Chauvet J. and Brechot P. (1982) *Sciences des Aliments*, 2, 495.
Crachereau J.C., Gabas N., Blouin J., Hebrard S. and Maujean A. (2001) *Bull. OIV*, 841–842, 151.
Dartiguenave C. (1998) Contribution à la maîtrise de l'acidité par l'étude de l'évolution de la capacité

tampon acido-basique des vins de champagne au cours de leur élaboration. Thesis at l'Université de Reims-Champagne-Ardenne.

Dartiguenave C., Jeandet P. and Maujean A. (2000) *Am. J. Enol. Viticult.*, 51–54, 347; Dartiguenave C., Jeandet P. and Maujean A. (2000) *Am. J. Enol. Viticult.*, 51–54, 352.

Devatine A., Gerbaud V., Gabas N. and Blouin J. (2002) *J. Int. Sci. Vigne Vin*, 36–32, 77.

Devraine J. (1969) Thèse Doctorat ès Sciences, Université de Lille.

Dubourdieu D. and Moine-Ledoux V. (1994) Brevet d'Invention Français 2 726 284.

Dubourdieu D. and Moine-Ledoux V. (1996) Brevet d'Invention Français 9608187.

Dunsford P. and Boulton R. (1981) *Am. J. Enol. Viticult.*, 2, 100.

Federson R.L. and Thorp S.N. (1993) *Industrial Gums*, 3graveeme edn. Academic Press, Londres.

Gaillard M. and Ratsimba B. (1990) *Rev. Fr. Œnol.*, 123, 11.

Genevois L. and Ribéreau-Gayon J. (1935) *Bull. Soc. Chim. Fr.*, 21286, 87.

Gerbaud V. (1996) Détermination de l'état de sursaturation et effet des polysaccharides sur la cristallisation du bitartrate de potassiu dans les vins. Thèse de l'Université de Toulouse (I.N.P.)

Gomez-Benitez J. (1993) *Am. J. Enol. Viticult.*, 4, 400.

Hochli U. (1997) *J. Int. Sci. Vigne Vin*, 31, 139.

Jackson R.S. (1994) *Wine Science. Principles and Applications.* Academic Press, San Diego.

Kalathenos P., Sutherland J.P. and Roberts T.A. (1995) *J. Appl. Bacteriol.* 78, 245.

Lubbers S., Léger B., Charpentier C. and Feuillat M. (1993) *J. Int. Sci. Vigne Vin*, 271, 13.

Maujean A., Sausy L. and Vallée D. (1985) *Rev. Fr. Œnol.*, 100, 39.

Maujean A., Vallée D. and Sausy L. (1986) *Rev. Fr. Œnol.*, 104, 34.

Maujean A. (1997) Personal Communication.

Moine-Ledoux V. and Dubourdieu D. (1995) *Ve Symposium International D'Œnologie*. Tec. et Doc. Lavoisier, Paris.

Moine-Ledoux V. and Dubourdieu D. (1999) *60 Symposium International D'oenologie*, A. Lonvaud-Funel editor. Tec et Doc, Lavoisier, Paris, p 527.

Moine-Ledoux V. and Dubourdieu D. (2002) *Bull. OIV*, 75, 857.

Moine-Ledoux V., Perrin A., Paladin I. and Dubourdieu D. (1997) *J. Int. Sci. Vigne Vin*, 31 (1), 23.

Müller-Späth H. (1979) *Rev. Fr. Œnol.*, 41, 47.

Ordonneau C. (1891) *Bull. Soc. Chim. France*, 261.

Pellerin P., Waters E., Brillouet J.-M. and Moutounet M. (1994) *J. Int. Sci. Vigne Vin*, 23 (3), 213.

Peynaud E. and Guimberteau G. (1961) *Ann. Falsif. Fraudes*, 53, 567.

Peynaud E. and Blouin J. (1996) *Le Goût du vin*. Dunod, Paris.

Postel W. (1983) *Bull. OIV*, 629–630, 554.

Prelog V. (1953) *Helv. Chim. Acta*, 36, 308.

Ribéreau-Gayon J., Peynaud E., Sudraud P. and Ribéreau-Gayon P. (1977) *Sciences et Techniques du Vin*, Vol. IV: *Clarification et Stabilization*. Dunod, Paris.

Ribéreau-Gayon J., Peynaud E., Sudraud P. and Ribéreau-Gayon P. (1982) *Sciences et Techniques du Vin*, Vol. I: *Analyse et Contrôle du Vin*, 2nd edn. Dunod, Paris.

Robillard B., Baboual S. and Duteurtre B. (1994) *Rev. Fr. Œnol.*, 145, 19.

Usseglio-Tomasset L. (1989) *Chimie Œnologique*. Tec. et Doc., Lavoisier, Paris.

Vallée D. (1995) *J. Int. Sci. Vigne et Vin*, 29, 143.

Vergnes P. (1940) Coefficient tampon des vins. PhD. Thesis at l'Université de Montpellier.

Vialatte C. and Thomas J.-C. (1982) *Rev. Fr. Œnol.*, 87, 37.

Wurdig G. and Muller T. (1980) *Die Weinwirtschaft*, 116, 720.

Wurdig G., Muller T. and Fiedrich G. (1982) *Bull. OIV*, 613, 220.

2

Alcohols and Other Volatile Compounds

2.1 Ethyl alcohol	51
2.2 Other simple alcohols	53
2.3 Polyols	55
2.4 Fatty acids in the aliphatic series	58
2.5 Esters	59
2.6 Miscellaneous compounds	61

2.1 ETHYL ALCOHOL

Besides water, ethanol (ethyl alcohol) is the most plentiful compound in wine. A wine's strength is expressed in terms of alcohol content, or the percentage of alcohol by volume. As ethanol has a density of 0.79, a wine with an alcohol content of 10% vol contains 79 g/l of ethanol by weight. The alcoholic strength of wine is generally 100 g/l (12.6% vol), although it may exceptionally be as high as 136 g/l (e.g. an alcohol content of 16% vol).

Due to the low density of ethanol, dry wines, containing negligible amounts of sugar, have densities below that of water (1.00), ranging from 0.91 to 0.94. This value decreases as the alcohol content increases.

Ethanol in wine is mainly produced by the alcoholic fermentation of sugar in must. However, grape cells are also capable of forming small quantities, mainly under anaerobic conditions (carbonic maceration; see Volume 1, Section 12.9.3). The appearance of traces of ethanol in grapes results from alcohol dehydrogenase activity, which acts as a marker for ripeness.

As approximately 18 g/l of sugar is required to produce 1% vol of ethanol during alcoholic fermentation, grape must has to contain 180, 226 and 288 g/l of sugar to produce wines with 10, 12.6 and 16% ethanol by volume. The latter is considered to be the maximum ethanol content yeast can survive, although, under certain laboratory conditions, some strains have been found capable of resisting up to 18% vol. Some types of wine may, of course, have even higher alcoholic strengths, but this results from the addition of ethanol.

Many consumers place excessive importance on alcohol content as an essential quality factor. Most countries legally require the alcohol content to be displayed on barrels in retail shops and wine labels. In some cases, wines are priced on the basis of alcohol content. It is nevertheless true that there is a relationship between vinous character and alcohol content, on the one hand, and soft, full-bodied flavor, on the other hand.

In temperate climates, the natural alcohol content depends directly on grape ripeness. Wines only have a high alcohol level in years when the weather is particularly good, if vineyard conditions and exposure to the sun are favorable. Great vintages are often years when wines reach high alcoholic strengths. It would, however, be ridiculous to suppose that alcohol is the only quality factor. Some excellent Médoc wines have an alcohol content of around 10% vol while other, high-alcohol wines are heavy, undistinguished and unattractive.

In chemical terms, ethanol is a primary alcohol (Figure 2.1), i.e. its carbon 1 is tetrahedrally hybridized sp^3, and carries two hydrogen atoms twinned with the hydroxyl radical (Figure 2.1). Alcohol functions should not be confused with enol functions

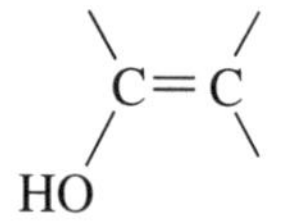

$$\text{HO} \diagdown \text{C} = \text{C} \diagup$$

or phenol functions (OH bonded to a carbon in a benzene cycle). In both of these cases carbon, the hydroxyl radical carrier, is hybridized sp^2 and the acid character of the hydrogen atom, carried by oxygen, is much more pronounced. Consequently, the function is more easily salifiable.

In ethyl alcohol produced by fermentation, some hydrogen atoms on carbons 1 and 2 in certain

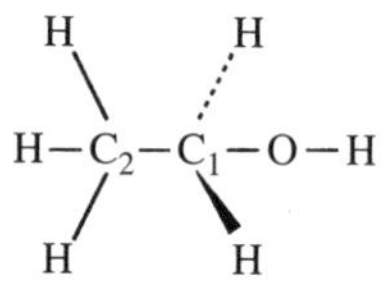

Fig. 2.1. Structure of ethanol and definition of the alcohol function

molecules are replaced by deuterium, an isotope of hydrogen. These molecules are present in very small amounts, and the exact proportion depends on the origin of the sugar that was fermented (grapes, beets and/or sugarcane). A method for controlling the addition of sugar to must (chaptalization) and detecting fraud has been developed on the basis of this property (Martin and Brun, 1987). This method has been officially recognized by the OIV (Office International de la Vigne et du Vin).

Ethanol's affinity for water and its solubility, by forming hydrogen bonds, makes it a powerful dehydrant. This property is useful in flocculating hydrophilic colloids, proteins and polysaccharides. It also gives ethanol disinfectant properties that are particularly valuable in aging wine. The combination of ethanol and acidity makes it possible to keep wine for a long time without any noticeable spoilage. The addition of ethanol to stabilize certain wines is a long-standing winemaking tradition (Port, *Vins doux naturels*). However, ethanol is toxic for humans, affecting the nerve cells and liver. The lethal dose (LD_{50}) by oral consumption is 1400 mg/kg body weight.

Ethanol's solvent properties are also useful for dissolving phenols from pomace during fermentation. This capacity is involved in solubilizing certain odoriferous molecules and certainly contributes to the expression of aromas in wine.

Ethanol has all the chemical properties of an alcohol function. In particular, it esterifies with tartaric, malic and lactic acids (Section 2.5.3). Ethyl acetate gives wine an unpleasant odor and is a sign of bacterial spoilage (Section 2.5.1). Ethanol may also react with aldehydes, especially ethanal, if the latter is present. This is never the case in sulfured wine as SO_2 reacts very strongly with ethanal, producing an acetal: diethoxyethane (Figure 2.2) (Section 2.6.2).

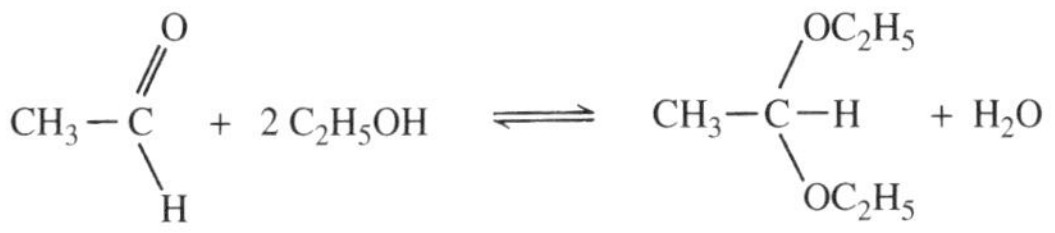

Alcohols and Other Volatile Compounds

$$CH_3-\overset{\overset{O}{\|}}{\underset{H}{C}} + 2\,C_2H_5OH \rightleftharpoons CH_3-\overset{OC_2H_5}{\underset{OC_2H_5}{C-H}} + H_2O$$

Fig. 2.2. Acetalization of ethanal and formation of diethoxyethane

$$CH_3-CH_2OH + H_2S \rightleftharpoons CH_3-CH_2-SH + H_2O$$

Fig. 2.3. Reaction between hydrogen sulfide and ethanol

$$2\,C_2H_5SH \rightleftharpoons C_2H_5-S-S-C_2H_5 + 2\,H^+ + 2\,e^-$$

Fig. 2.4. Oxidation–reduction balance of the thiol/disulfide system

Ethanol may also react with hydrogen sulfide, produced by fermenting yeast or resulting from the residues of some vineyard treatment products (Section 8.6.3). This reaction generates ethanethiol (Figure 2.3), which has a very unpleasant smell. As this compound is much less volatile than H_2S, it is more difficult to eliminate. It is, therefore, advisable to rack wines as soon as alcoholic fermentation is completed and again immediately after malolactic fermentation, since hydrogen sulfide may also be produced by lactic bacteria. Furthermore, the oxidation–reduction balance may also cause ethanethiol to form diethyl disulfide (Figure 2.4). This compound is even less volatile and has a very unpleasant smell which spoils a wine's flavor.

2.2 OTHER SIMPLE ALCOHOLS

2.2.1 Methyl Alcohol

Methanol is always present in wine in very small quantities, between 30 and 35 mg/l. It has no organoleptic impact. Methanol is not formed by alcoholic fermentation, but results exclusively from enzymic hydrolysis of the methoxyl groups of the pectins during fermentation:

$$-OCH_3 + H_2O \longrightarrow -OH + CH_3OH$$

As grapes have a relatively low pectin content, wine is the fermented beverage with the lowest methanol concentration.

The methanol content depends on the extent to which the grape solids, especially skins that have a high pectin content, are macerated. Red wines have a higher concentration (152 mg/l) than rosés (91 mg/l), while white wines have even less (63 mg/l) (Ribéreau-Gayon *et al.*, 1982). Wines made from hybrid grape varieties have a higher methanol content than those made from *Vitis vinifera*. This is due to the higher pectin content of hybrid grape skins. The use of pectolytic enzymes to facilitate extraction or clarification of the must may cause an increase in methanol as a result of the pectin esterase activity.

Methanol's toxicity is well-known. Following ingestion, it oxidizes to produce formic aldehyde and formic acid, both toxic to the central nervous system. Formic aldehyde deteriorates the optical nerve, causing blindness. Wines made in the normal way never have methanol concentrations anywhere near dangerous levels (LD_{50} = 350 mg/kg).

2.2.2 Higher Fermentation Alcohols

Alcohols with more than two carbon atoms are known as higher alcohols (Table 2.1). Several of these are produced during fermentation and reach concentrations on the order of 150–550 mg/l in wine (Ribéreau-Gayon *et al.*, 1982; Jackson, 1994). These alcohols and their esters have intense odors that play a role in wine aromas. The main higher fermentation alcohols, components of Fusel oils, are isobutyl (methyl-2-propanol-1) and amyl alcohols (a mixture of methyl-2-butanol-1 and methyl-3-butanol-1). At low concentrations (less than 300 mg/l), they contribute to a wine's aromatic complexity. At higher levels, their penetrating odors mask the wine's aromatic finesse. Acetic esters of these alcohols, especially isoamyl acetate, have a banana odor that may play a positive role in the aroma of some young red wines (*primeur* or *nouveau*).

Higher alcohols are formed by yeast, either directly from sugars or from grape amino acids by the Ehrlich reaction (Figure 2.5). This reaction

Table 2.1. Simple alcohols originating from plants and yeast (Ribéreau-Gayon et al., 1982)

Formula	Name	Boiling point (°C)	Concentration (g/l)	Comments
H–CH$_2$OH	Methanol	65	0.1	Produced by hydrolysis of pectins, not fermentation
CH$_3$–CH$_2$–OH	Ethanol	78	100	
CH$_3$–CH$_2$–CH$_2$OH	Propanol-1	97	0.03	
CH$_3$–CHOH–CH$_3$	Propanol-2	82	Traces	Isopropyl alcohol
CH$_3$–CH$_2$–CH$_2$–CH$_2$OH	Butanol-1	117	Traces	
CH$_3$–CH(CH$_3$)–CH$_2$OH	Methyl-2-propanol-1	107	0.1	Isobutyl alcohol
CH$_3$–COH(CH$_3$)–CH$_3$	Methyl-2-propanol-2	82	7	
CH$_3$–CH$_2$–CHOH–CH$_3$	Butanol-2	99	Traces	
CH$_3$–CHOH–CHOH–CH$_3$	Butanediol-2,3	183	1	
CH$_3$–CH$_2$–CH$_2$–CH$_2$–CH$_2$OH	Pentanol-1	137	Traces	
CH$_3$–CH$_2$–CH$_2$–CHOH–CH$_3$	Pentanol-2	119	Traces	
CH$_3$–CH$_2$–CHOH–CH$_2$–CH$_3$	Pentanol-3	115	?	
CH$_3$–CH(CH$_3$)–CH$_2$–CH$_2$OH	Methyl-3-butanol-1	131	0.2	Isoamyl alcohol
CH$_3$–CH$_2$–CH(CH$_3$)–CH$_2$OH	Methyl-2-butanol-1	129	0.05	Active amyl alcohol
CH$_3$–CH(CH$_3$)–CHOH–CH$_3$	Methyl-3-butanol-2	112	?	
CH$_3$–CH$_2$–C(H)=C(H)–CH$_2$–CH$_2$OH	cis-Hexene-3-ol-1	156		Present only in grapes with herbaceous odors
CH$_3$–(CH$_2$)$_4$–CH$_2$OH	Hexanol-1	158	0.01	
CH$_3$–(CH$_2$)$_3$–CHOH–CH$_3$	Hexanol-2	138	?	
CH$_3$–(CH$_2$)$_5$–CH$_2$OH	Heptanol-1	177	Traces	
CH$_3$–(CH$_2$)$_4$–CHOH–CH$_3$	Heptanol-2	160	?	
CH$_3$–(CH$_2$)$_6$–CH$_2$OH	Octanol-1	194	?	
CH$_3$–(CH$_2$)$_5$–CHOH–CH$_3$	Octanol-2	180	?	
CH$_3$–(CH$_2$)$_7$–CH$_2$OH	Nonanol-1	212	?	
CH$_3$–(CH$_2$)$_6$–CHOH–CH$_3$	Nonanol-2		?	
CH$_3$–(CH$_2$)$_8$–CH$_2$OH	Decanol-1	229	?	
Φ–CH$_2$–CH$_2$OH	Phenyl-2-ethanol	219	0.05	Fermentation alcohol (rose odor)
HO–Φ–CH$_2$–CH$_2$OH	Tyrosol			
CH$_3$–(CH$_2$)$_4$–CHOH–CH=CH$_2$	Octene-1-ol-3			Mushroom odor

Φ = benzene cycle

Alcohols and Other Volatile Compounds

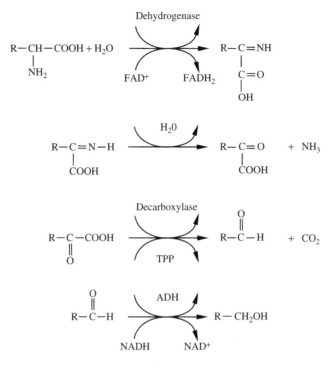

Fig. 2.5. Biosynthesis of higher alcohols, according to Ehrlich

is caused by the activity of a FAD^+ dehydrogenase, which oxidizes amino acids into imino acids. These are hydrolyzed into α-ketoacid, then subjected to the action of a decarboxylase with thiamin pyrophosphate coenzymes (TPP). Via this channel, leucine produces isoamyl alcohol and isoleucine results in optically active amyl alcohol with an asymmetrical carbon. The higher fermentation alcohol content of wine varies according to fermentation conditions, especially the species of yeast. In general, factors that increase the fermentation rate (yeast biomass, oxygenation, high temperature and the presence of matter in suspension) also increase the formation of higher alcohols.

The higher alcohol content of a wine may increase due to microbial spoilage involving yeast or bacteria. In these cases, the amylic odor may become excessive.

Higher alcohols are retained in brandies after distilling and contribute to their individual characters. Distillation techniques have a major impact on their overall concentration.

2.2.3 Miscellaneous Alcohols

These molecules originate from grapes. One group consists of C_6 alcohols, hexanols and hexenols from plant tissues that give the herbaceous smells so characteristic of wines made from unripe grapes (Volume 1, Sections 11.6.2 and 13.3.4).

Another of these compounds is octen-1-ol-3, with an odor reminiscent of mushrooms. Its presence in wine is due to the action of *Botrytis cinerea* on grapes.

Finally, terpenols, the main components in the distinctive Muscat aroma, are described in full elsewhere (Section 7.2.1).

2.3 POLYOLS

Polyols are characterized by the presence of several 'hydroxyl' radicals in the same linear or cyclic molecule. In general, an accumulation of hydroxyl radicals in a compound raises the boiling point considerably (Table 2.2) due to the large number

Table 2.2. Impact of the number of hydroxyl groups on the boiling point of alcohols

	Alcohols and polyols	Boiling point (°C)
Ethanol	CH_3-CH_2OH	78
Ethyleneglycol	CH_2OH-CH_2OH	198
Glycerol	$CH_2OH-CHOH-CH_2OH$	290

of hydrogen bonds, as well as increasing its viscosity. Parallel increases are observed in solubility and sweetness. Sugars are good examples of polyols.

2.3.1 C_3 Polyol: Glycerol

Besides water and ethanol, glycerol (Table 2.3) is probably the chemical compound with the highest concentration in wine. It is the most important by-product of alcoholic fermentation. The minimum glycerol concentration in wine is 5 g/l but it may reach values as high as 15–20 g/l, depending on the fermentation conditions (especially the must sulfuring levels). Grapes affected by noble rot already contain a few grams of glycerol, which is added to the quantity produced by fermentation.

Glycerol is formed by yeast at the beginning of the fermentation process. It is generally considered to be produced 'with the first 50 grams of sugars fermented'. This corresponds to the start of the glyceropyruvic fermentation. The only way for yeast to ensure the reoxidation of the NADH + H^+ coenzyme is by reducing dihydroxyacetone to glycerol. At this stage, the ethanal level is too low for this reoxidation to occur as well as producing ethanol. When must is treated with high doses of SO_2, this molecule combines with ethanal, thus increasing the glyceropyruvic fermentation rate and the amount of glycerol formed.

Glycerol in wine may act as a carbohydrate nutrient for the growth of various microorganisms, e.g. the yeast *flor* in Sherry production (Volume 1, Section 14.5.2). Also, certain detrimental bacteria are capable of breaking down glycerol, with a double dehydration reaction that produces acrolein (Figure 2.6). Acrolein interacts with tannins to

Table 2.3. Highest concentrations of polyols found in wines (Ribéreau-Gayon *et al.*, 1982)

	Formula	Highest concentrations (mg/l)
Glycerol	$CH_2OH-CHOH-CH_2OH$	5000–20 000
Butanediol-2,3	$CH_3-CHOH-CHOH-CH_3$	330–1350
Erythritol	$CH_2OH-(CHOH)_2-CH_2OH$	30–200
Arabitol	$CH_2OH-(CHOH)_3-CH_2OH$	25–350
Mannitol	$CH_2OH-(CHOH)_4-CH_2OH$	90–750
Sorbitol	$CH_2OH-(CHOH)_4-CH_2OH$	30–300
meso-Inositol	$(CHOH)_6$	220–730

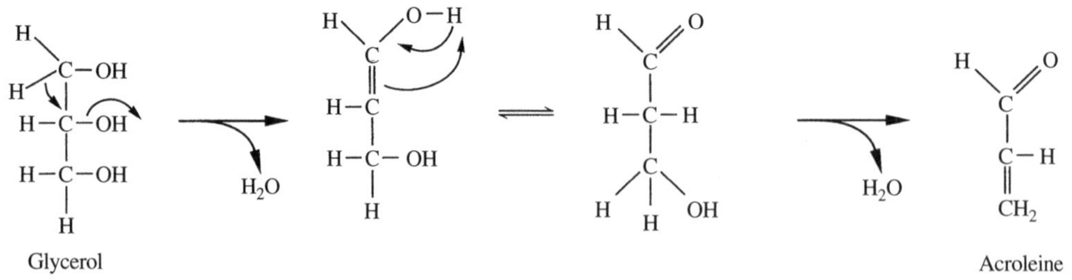

Fig. 2.6. Mechanism for the formation of acrolein by double dehydration of glycerol

Alcohols and Other Volatile Compounds 57

reinforce bitterness, giving its name to this type of microbial spoilage (Section 8.3.2).

In view of its high concentration, it was thought that glycerol affected wine flavor, giving an impression of fullness and softness. In fact, doses much higher than those occurring naturally in wine are required to affect flavor to any significant extent. Glycerol has a sweet taste that reinforces the sweetness of ethyl alcohol in dry wines, but it is probably not responsible for any of the sweetness in sweet wines.

2.3.2 C_4 Polyols: 2,3-butanediol and Erythritol

Although 2,3-butanediol is a C_4 molecule (Table 2.3) it is really a diol. It is a by-product of alcoholic fermentation and is probably also formed by malolactic fermentation. This compound has little odor and its flavor is slightly sweet and bitter at the same time, but it does not have much organoleptic impact in wine. It is stable, and, above all, unaffected by bacteria.

The most significant role of 2,3-butanediol is in maintaining an oxidation–reduction balance with acetoin (or acetylmethyl carbinol) and diacetyl (Figure 2.7). This compound (2,3-butanediol) is formed following the reduction of acetoin, produced by the condensation of two ethanal molecules.

Acetoin has a slight milky odor and is present at concentrations on the order of 10 mg/l. Diacetyl has a pleasant odor of butter and hazelnuts which may be perceptible at low concentrations (2 mg/l). The diacetyl concentration in wine is generally on the order of 0.3 mg/l.

These two volatile compounds are distilled into brandy. The concentration in brandy depends on that in the wine, and also the distillation technique, making it possible to distinguish between Cognac, made by double distillation, and Armagnac, which is distilled only once.

Erythritol (Table 2.3) is also a C_4 molecule, but it has four alcohol functions. Small quantities, 30–200 mg/l, are formed by yeast. It has no known properties.

2.3.3 C_5 Polyol: Arabitol

Small quantities (25–350 mg/l) of arabitol are also known to be formed by yeast (Table 2.3). This compound has five alcohol functions and is directly derived from arabinose. Small quantities may also be produced by lactic bacteria and larger quantities by *Botrytis cinerea*.

2.3.4 C_6 Polyols: Mannitol, Sorbitol and Meso-inositol

These three compounds (Table 2.3) have six alcohol functions. The first two are linear while the third is cyclic.

Mannitol is derived from reduction of the C_1 on mannose. In wine, it is produced by the reduction of the C_2 on fructose by lactic bacteria. Mannitol is usually present in very small quantities. Higher concentrations are due to lactic bacteria or possibly *Botrytis cinerea*. Abnormally high concentrations indicate severe lactic spoilage.

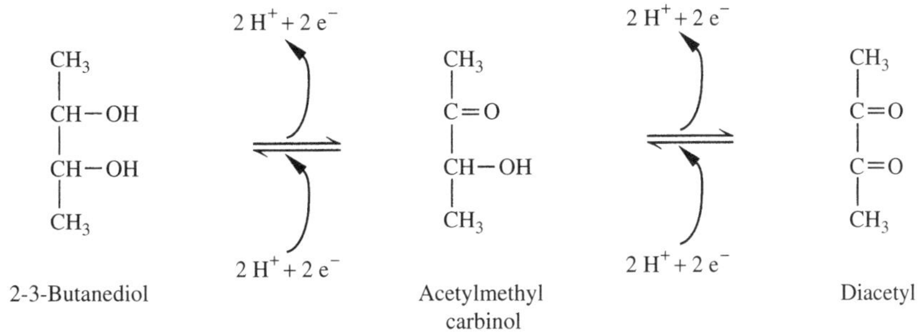

$$\begin{array}{c} CH_3 \\ | \\ CH-OH \\ | \\ CH-OH \\ | \\ CH_3 \end{array} \underset{2H^++2e^-}{\overset{2H^++2e^-}{\rightleftarrows}} \begin{array}{c} CH_3 \\ | \\ C=O \\ | \\ CH-OH \\ | \\ CH_3 \end{array} \underset{2H^++2e^-}{\overset{2H^++2e^-}{\rightleftarrows}} \begin{array}{c} CH_3 \\ | \\ C=O \\ | \\ C=O \\ | \\ CH_3 \end{array}$$

2-3-Butanediol   Acetylmethyl carbinol   Diacetyl

Fig. 2.7. Oxidation–reduction balances of 2-3-butanediol

Sorbitol results from the reduction of the C_1 on glucose. This diastereoisomer of mannitol is totally absent from healthy grapes. Varying quantities are formed when *Botrytis cinerea* develops. Alcoholic fermentation produces approximately 30 mg/l. Lactic bacteria do not form this compound. Large quantities of sorbitol indicate that wine has been mixed with wines made from other fruits. Besides rowan berries (*Sorbus aucuparia*), from which it takes its name, apples, pears and cherries also have a high sorbitol content.

meso-Inositol is a normal component of grapes and wine. It is a cyclic polyol with six carbon atoms, each carrying a hydroxyl radical. Among the nine inositol stereoisomers, all diastereoisomers of each other, *meso*-inositol has a plane of symmetry passing through carbons 1 and 4. Its *meso* property makes it optically inactive. This polyol is widespread in the animal and plant kingdoms. It is a vital growth factor for many microorganisms, especially certain yeasts.

It is difficult to attribute any organoleptic role to C_6 polyols.

2.4 FATTY ACIDS IN THE ALIPHATIC SERIES

This series is shown in Table 2.4. The most important of these compounds is acetic acid, the essential component of volatile acidity. Its concentration, limited by legislation, indicates the extent of bacterial (lactic or acetic) activity and the resulting spoilage of the wine. As yeast forms a little acetic acid, there is some volatile acidity in all wines. Other C_3 (propionic acid) and C_4 acids (butyric acids) are also associated with bacterial spoilage.

The C_6, C_8 and C_{10} fatty acids are formed by yeast. As they are fermentation inhibitors at concentrations of only a few mg/l, they may be responsible for stuck fermentations (Volume 1, Section 3.6.2).

Unsaturated long-chain fatty acids (C_{18}, C_{20}) are related to the sterol family. These compounds are fermentation activators, mainly under anaerobic conditions. The most important of these are oleic (C_{18} with one double bond) and linoleic acids (C_{18} with two double bonds). They are active in

Table 2.4. Fatty acids in the aliphatic series among the volatile components in wine (Ribéreau-Gayon *et al.*, 1982)

Formula	Name	Boiling point (°C)	Concentration (g/l)	Comments
H–COOH	Formic	101	0.05	
CH_3–COOH	Acetic	118	0.5	
CH_3–CH_2–COOH	Propionic	141	Traces	
CH_3–CH_2–CH_2–COOH	Butyric	163	Traces	
CH_3–CH(CH$_3$)–COOH	Isobutyric	154	Traces	Methyl-2-propionic acid
CH_3–CH_2–CH_2–CH_2–COOH	Valerianic	186	Traces	
CH_3–CH(CH$_3$)–CH_2–COOH	Isovalerianic	177	?	Methyl-3-butyric acid
CH_3–CH_2–CH(CH$_3$)–COOH	Methyl-2-butyric		?	
CH_3–$(CH_2)_4$–COOH	Caproic	205	Traces	Hexanoic acid
CH_3–$(CH_2)_5$–COOH	Oenanthic	223	Traces	Heptanoic acid
CH_3–$(CH_2)_6$–COOH	Caprylic		Traces	Octanoic acid
CH_3–$(CH_2)_7$–COOH	Pelargonic	253	?	Nonanoic acid
CH_3–$(CH_2)_8$–COOH	Capric	270	Traces	Decanoic acid

trace amounts and come from the waxy cuticle of grape skins.

2.5 ESTERS

Esters are formed when an alcohol function reacts with an acid function and a water molecule is eliminated (Figure 2.8). It is a reversible reaction, limited by the inverted reaction of hydrolysis of the ester. When the system is in balance, there is a constant correlation between the concentrations of the substances present, governed by the mass action law.

There are a large number of different alcohols and acids in wine, so the number of possible esters is also very large. Ethyl acetates are the most common for kinetic reasons, i.e. the large quantities of ethanol present and the fact that primary alcohols are the most reactive.

Very few esters are present in grapes. Odoriferous molecules such as methyl anthranilate are responsible for the foxy odor in *Vitis labrusca* grapes and wines made from them. There are also methoxyl groups in pectins that release methanol by hydrolysis (Section 2.2.1).

Esters in wine have two distinct origins: enzymic esterification during the fermentation process and chemical esterification during long-term aging. The same esters may be synthesized in either way.

2.5.1 Ethyl Acetate

The most prevalent ester in wine is certainly ethyl acetate. A small quantity is formed by yeast during fermentation, but larger amounts result from the activity of aerobic acetic bacteria, especially during aging in oak barrels. Apparently, lactic bacteria are not capable of synthesizing this ester. Ethyl acetate is responsible for the olfactory characteristics in wines affected by 'acescence'—a suffocating, vinegary odor. These wines also have high volatile acidity, but acetic acid is not responsible for acescence. In a simple solution, ethyl acetate is perceptible at concentrations approximately 200 times lower than the perception threshold of acetic acid.

The olfactory perception threshold of ethyl acetate is approximately 160 mg/l. Even below this value, while it may not be identifiable, it may spoil the bouquet with an unpleasant, pungent tang. It is, however, possible that at very low doses (50–80 mg/l) ethyl acetate contributes to a wine's olfactory complexity and thus has a positive impact on quality.

Furthermore, ethyl acetate affects wine flavor. At relatively high concentrations (above 120 mg/l) that are still below the olfactory perception threshold, it gives red wines a hot flavor which reinforces the impression of bitterness on the aftertaste. Ethyl acetate contributes to harshness and hardness in red wines. An acetic acid concentration of at least 0.90 g/l (a volatile acidity of 0.95 g/l expressed in H_2SO_4) is required to produce a noticeable bitter, sour aftertaste. Even at these high levels, however, it does not have a strong odor, whereas ethyl acetate is perceptible at much lower concentrations.

2.5.2 Ethyl Acetates of Fatty Acids and Acetic Esters of Higher Alcohols

Ethyl acetates of fatty acids, mainly ethyl caproate and caprylate, are produced by yeast during alcoholic fermentation. They are synthesized from forms of the acids activated by the coenzyme A (HS-CoA), acyl-S-CoA. Acetyl-S-CoA, from pyruvic acid, may be involved in a Claisen reaction with malonyl-S-CoA, producing a new acyl-S-CoA with two additional carbon atoms (Figure 2.9). Acetyl-S-CoA thus produces butyryl-S-CoA, then hexanyl-S-CoA, etc. Specific enzymes then catalyze the alcoholysis of acyl-S-CoA into ethyl acetates of fatty acids. At the same time, the coenzyme A is regenerated.

$$R-\overset{O}{\underset{\|}{C}}-OH + CH_3-CH_2-OH \rightleftharpoons R-\overset{O}{\underset{\|}{C}}-O-CH_2-CH_3 + H_2O$$

Fig. 2.8. Esterification balance of an alcohol

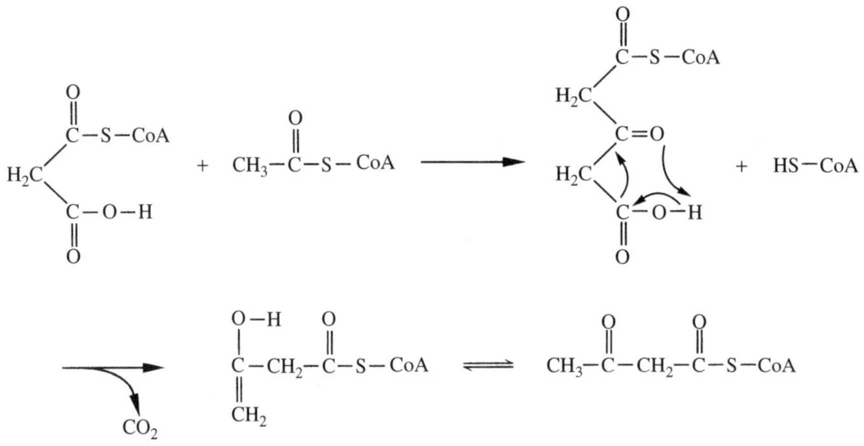

Fig. 2.9. Biosynthesis mechanism of fatty acids

Table 2.5. Changes in fatty acid ester concentrations (in μmol/l) depending on the aging time at 25°C and at two different pH values (Garofolo and Piracci, 1994)

Compounds	pH = 3.00				pH = 3.50			
	0 months	2 months	5 months	29 months	0 months	2 months	5 months	29 months
Isobutyl acetate	0.70	0.40	0.00	0.00	0.70	0.60	0.00	0.00
Hexyl acetate	1.90	1.20	0.00	0.00	1.70	1.50	0.40	0.00
Isoamyl acetate	36.60	13.30	3.10	0.40	36.50	20.60	14.00	2.50
Phenyl-2-acetate	11.00	2.40	0.50	0.50	4.80	3.40	2.60	0.88
Ethyl hexanoate	12.20	8.70	6.40	4.30	11.00	8.80	8.40	4.60
Ethyl octanoate	9.30	9.00	7.40	6.40	5.70	5.50	5.50	3.69
Ethyl decanoate	2.70	3.40	3.10	2.00	1.20	1.20	1.40	0.79

In general, the concentrations of esters increase during aging (Section 2.5.3). Ethyl acetates of fatty acids are formed by yeast, under anaerobic conditions, in quantities greater than those predicted by the mass action law. Consequently, they are hydrolyzed during aging and concentrations tend to decrease (Table 2.5). Garofolo and Piracci (1994) determined the kinetics equations for the hydrolysis of esters of fatty acids and isoamyl acetate in model media and in wines, at various pH values, over a period of 29 months.

Ethyl acetates of fatty acids have very pleasant odors of wax and honey which contribute to the aromatic finesse of white wines. They are present at total concentrations of a few mg/l.

Acetic esters of higher alcohols (isoamyl acetate and phenylethyl acetate) should also be included among the fermentation esters. These compounds are present in moderate quantities, but have intense, rather unusual odors (banana, acid drops and apple). They contribute to the aromatic complexity of naturally neutral wines, but may mask some varietal aromas. The formation of all these esters is promoted when fermentation is slow (Bertrand, 1983; Dubois, 1993) and difficult, due to absence of oxygen, low temperatures and clarified must.

2.5.3 Esters of Chemical Origin

The formation of esters continues throughout the aging process thanks to the presence of many different acids in wine, together with large quantities of ethanol. Research into esterification mechanisms in wine (Ribéreau-Gayon et al., 1982) showed that, under normal cellar conditions, none of the acids ever reach the balance predicted in theory. The ester content represents approximately 30% of the theoretical limit after one year, 50% after 2 or 3 years and 80% after 50 years. The total ester concentration (formed by chemical or enzymic reactions) is governed by the wine's composition and age. It varies from 2 or 3 meq/l in young wines up to 9 or 10 meq/l in old wines, in which approximately 10% of the acids are esterified.

Mono-acids react with ethanol to form only neutral esters, whereas di-acids may produce one neutral and one acid ester (e.g. ethyl tartrate and ethyl tartaric acid). On average, wine contains approximately the same quantity of neutral and acid esters. The latter contribute to wine acidity.

Ethyl acetates of the main organic acids seem to play only a limited role in the organoleptic qualities of healthy wines. In any case, they cannot be considered to contribute to improving wines during aging, as they develop in the same way in all wines.

Ethyl lactate is a special case. Its formation is linked to malolactic fermentation and the involvement of an esterase of bacterial origin cannot be excluded. Concentrations of ethyl lactate increase throughout aging via chemical reactions. In Champagne that has completed malolactic fermentation, the ethyl lactate concentration has been observed to increase to a maximum of 2 g/l after two years and then decrease during further aging on the lees. According to Arctander (1969), ethyl lactate has an odor reminiscent of butter, or even sour milk. Other authors think that the odor of ethyl lactate has been confused with that of other odoriferous compounds.

2.6 MISCELLANEOUS COMPOUNDS

Among the other volatile products likely to contribute to wine aroma are volatile phenols and sulfur derivatives. The latter are responsible for olfactory defects whose causes and consequences are now well-known and are described elsewhere in this book (Sections 8.4 and 8.6). There are also several compounds that contribute to the varietal aromas of different grape varieties, e.g. terpenes in Muscats. These compounds are also discussed elsewhere (Section 7.2.1).

This section is thus exclusively devoted to carbonylated compounds, lactones and acetols.

2.6.1 Carbonylated Compounds (Aldehydes and Ketones)

Ethanal is the most important of these compounds (Table 2.6). The many ways it can be produced and its high reactivity (the CHO radical has extensive chemical affinities), as well as its rapid combination with sulfur dioxide at low temperatures and its organoleptic properties, make ethanal a very important component of wine. The presence of ethanal, produced by the oxidation of ethanol, is closely linked to oxidation–reduction phenomena. It is involved in the alcoholic fermentation mechanism. Furthermore, ethanal plays a role in the color changes occurring in red wines during aging by facilitating the copolymerization of phenols (anthocyanins and catechins) (Section 6.3.10).

In wine preserved with regular, light sulfuring, the sulfite combination of ethanal (CH_3–CHOH–SO_3H), stable in an acid medium, is the most prevalent form (Volume 1, Section 8.4.1). When grapes have been heavily sulfured, the ethanal concentration increases and may exceed 100 mg/l, also combined with sulfite. This sulfite combination of ethanal protects yeast from the antiseptic effects of SO_2.

Wines containing excess ethanal as compared to the quantity of SO_2, i.e. free (non-combined) ethanal, are described as 'flat' (Section 8.2.3). A slight trace of free ethanal is sufficient to produce a characteristic odor, reminiscent of freshly cut apple. This problem disappears rapidly if a little SO_2 is added, as it combines with the free ethanal. This is one of the reasons for sulfuring barrels during racking (Section 10.3.3).

A few other aldehydes are present in wine in trace amounts (Table 2.6). Higher aldehydes contribute to the bouquets of some wines. The neutralizing effect of sulfur dioxide on the fruitiness of

Table 2.6. Aldehydes and ketones in wine

Formula	Name	Boiling point (°C)	Concentration (g/l)	Comments
H–CHO	Methanal	21	?	Formic aldehyde
CH_3–CHO	Ethanal	21	0.1	In combined state with SO_2. Only oxidized wines (Rancio, Sherry, etc.) contain free ethanal
CH_3–CH_2–CHO	Propanal	49	Traces	
CH_3–CH_2–CH_2–CHO	Butanal	76	?	Valerianic aldehyde
CH_3–CH(CH_3)–CHO	Methyl-2-propanal	92	Traces	Isovalerianic aldehyde
CH_3–CH_2–CH_2–CH_2–CHO	Pentanal	102	?	Valerianic aldehyde
CH_3–CH(CH_3)–CH_2–CHO	Methyl-3-butanal	92	Traces	Isovalerianic aldehyde
CH_3–CH_2–CH_2–CH_2–CH_2–CHO	Hexanal	128	Traces	Caproic aldehyde
CH_3–CH_2–CH_2–CH=CH–CHO	Hexene-2-al		?	Only present in grapes
CH_3–$(CH_2)_5$–CHO	Heptanal	155	Traces	Oenanthic aldehyde
CH_3–$(CH_2)_6$–CHO	Octanal	167	?	Caprylic aldehyde
CH_3–$(CH_2)_7$–CHO	Nonanal	185	?	Pelargonic aldehyde
CH_3–$(CH_2)_8$–CHO	Decanal	208	?	Capric aldehyde
CH_3–$(CH_2)_{10}$–CHO	Dodecanal		?	Lauric aldehyde
CH_3–CO–CH_3	Propanone	56	Traces	Acetone
CH_3–CH_2–CO–CH_3	Butanone	80	?	Methylethyl ketone
CH_3–CH_2–CH_2–CO–CH_3	Pentanone-2	102	?	
CH_3–CHOH–CO–CH_3	Acetylmethyl carbinol	143	0.01	Acetoin
CH_3–CO–CO–CH_3	Diacetyl	87	Traces	
CH_3–C(SH)–CH_2–C(=O)–CH_3	Mercaptopentanone			Sauvignon Blanc aroma
C₆H₅–CHO	Benzoic aldehyde	178	?	
(3-CH_3O, 4-HO)C₆H₃–CHO	Vanillin	285	?	
C₆H₅–CH=CH–CHO	Cinnamic aldehyde	253	?	
HO–H_2C–(furan)–CHO	Hydroxymethyl furfural			Grape juice or wine subjected to heat treatment

certain white wines is due to the fact that it combines with the aldehyde fraction in the bouquet.

Aldehydes in the aromatic series are also present in wine. The most significant of these is vanillin, associated with barrel aging, which has a distinctive vanilla aroma.

Grapes apparently contain few aldehydes. Hexenal and hexenol have, however, been identified as contributing to the herbaceous odors of C_6 compounds (Section 2.2.3).

Several molecules with ketone functions have been identified, including propanone, butanone and pentanone. As previously mentioned, the most important of these are acetylmethyl carbinol and diacetyl (Section 2.3.2).

Finally, a mercaptopentanone has been identified among the specific components of Sauvignon Blanc aroma.

The following molecules with several aldehyde or ketone functions have also been identified: glyoxal, methyl-glyoxal, and hydroxypropanedial (Volume 1, Section 8.4.4).

2.6.2 Acetals

Acetal is formed every time an aldehyde comes into contact with an alcohol. The reaction involves two alcohol molecules and one aldehyde molecule, as shown in Figure 2.10.

About twenty compounds of this type have been reported in wine. The most important of these, diethoxyethane, results from a reaction between ethanal and ethyl alcohol (Figure 2.10). Acetalization is a slow, reversible reaction, catalyzed by H^+ ions. The reaction is completed in a few hours at pH 2–3, while it takes several days at pH 4. In 10% vol alcohol solution, 3% of the ethanal may react, while 6.5% reacts if the alcohol content is 20% vol.

In view of the very small quantities of free ethanal present in still wine, their acetal content is practically zero. Only wines with a high ethanal content have a significant concentration of acetal. Sherry, with an ethanol concentration on the order of 280 mg/l, contains 45–60 mg/l, while the concentration in *Vin Jaune* from the Jura may be as high as 150 mg/l.

Acetals have a herbaceous odor that may add to the aromatic complexity of Sherry. Diethoxyethane is described by Arctander (1969) as having a pleasant, fruity odor.

2.6.3 Lactones

Lactones are formed by an internal esterification reaction between an acid function and an alcohol function in the same molecule. This reaction produces an oxygen heterocycle.

Volatile lactones, produced during fermentation, are likely to contribute to wine aroma. The best known is γ-butyrolactone, present in wine at concentrations on the order of a mg/l. This compound results from the lactonization of γ-hydroxybutyric acid, an unstable molecule produced by deamination and decarboxylation of glutamic acid, according to the Ehrlich reaction (Figures 2.5 and 2.11). It does not seem to play a major role in the organoleptic characteristics of wine. Lactones may

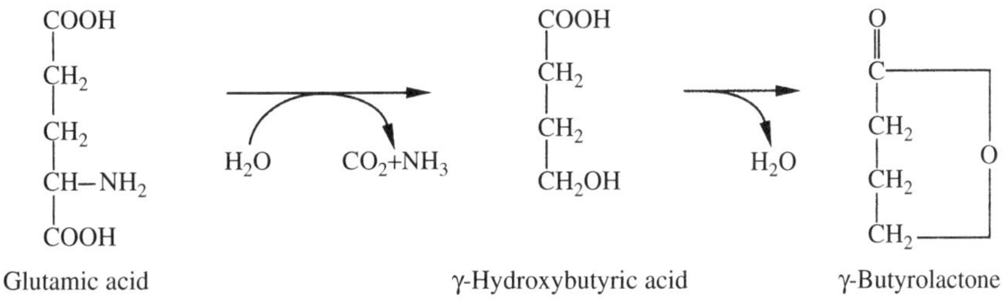

$$R-CHO + 2R'-OH \rightleftharpoons R-HC{\overset{O-R'}{\underset{O-R'}{\diagup}}} + H_2O$$

Fig. 2.10. Formation of an acetal

Glutamic acid → γ-Hydroxybutyric acid → γ-Butyrolactone

Fig. 2.11. Formation of γ-butyrolactone

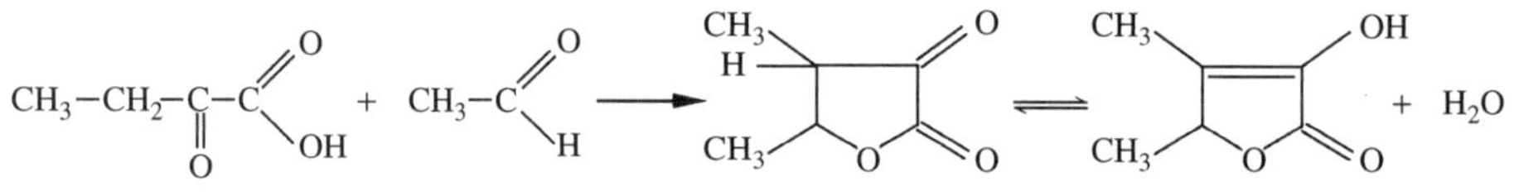

Fig. 2.12. Formation of sotolon

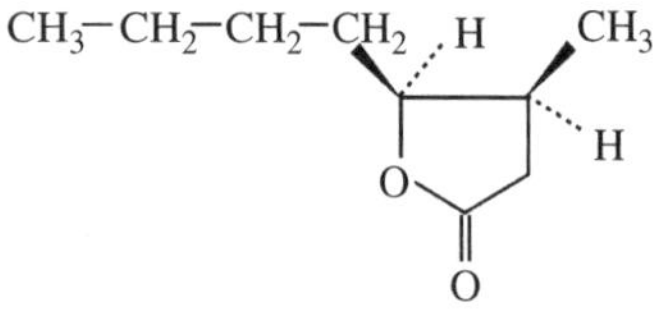

Fig. 2.13. β-methyl-γ-octalactone

also come from grapes, as is the case in Riesling, where they contribute to the varietal aroma.

Infection of grapes by *Botrytis cinerea* probably produces sotolon (4.5-dimethyl-3-hydroxy-2-furanone) (Figure 2.12), involved in the toasty aroma characteristic of wines made from grapes with noble rot (Masuda *et al.*, 1984). Concentrations present, on the order of 5 µg/l, are above the perception threshold.

Sotolon (Volume 1, Section 10.6.4) also results from a condensation reaction, not catalyzed by enzymes, between α-keto butyric acid and ethanal. It is also present in *Vin Jaune* from the Jura, where it is responsible for the 'walnut' aroma so typical of this wine (Martin *et al.*, 1992).

Finally, oak releases lactones into wine during barrel aging. The *cis* and *trans* isomers of 3-methyl-γ-octalactone (Figure 2.13) are known as 'oak lactones' or 'whisky lactones'. The pure compounds have a coconut odor, and when diluted they are reminiscent of oaky wines. Concentrations in wine are on the order of a few tens of mg/l, considerably higher than the perception threshold (a few tens of µg/l).

REFERENCES

Arctander S. (1969) *Perfume and Flavor Chemicals* (ed. N.J. Montclair.).
Bertrand A. (1983) Volatiles from grape must fermentation, in *Flavor of Distilled Beverages, Origin and Development* (ed. J.R. Piggot). Ellis Horwood Ltd.
Dubois P. (1993) *Rev. Fr. Œnol.*, 144, 63.
Garofolo A. and Piracci A. (1994) *Bull. OIV*, 67 (757–758), 225.
Jackson R.S. (1994) *Wine Sciences. Principles and Applications*. Academic Press, San Diego.
Martin B., Etievant P., Le Quere J.-L. and Schlich P. (1992) *J. Agric. Food. Chem.*, 40, 475.
Martin G. and Brun S. (1987) *Bull. OIV*, 671–672, 131.
Masuda M., Okawa E., Nishimura K. and Yunone H. (1984) *Agr. Biol. Chem.*, 48 (11), 2707.
Ribéreau-Gayon J., Peynaud E., Ribéreau-Gayon P. and Sudraud P. (1982) *Sciences et Techniques du Vin*, Vol. I: *Analyse et Contrôle des Vins*, 2nd edn. Dunod, Paris.
Sauvage F.-X., Romieux C.-G., Sarris J., Pradal M., Robin J.-P. and Flanzy C. (1991) *Rev. Fr. Œnol.*, 132, 14.

3

Carbohydrates

3.1 Introduction	65
3.2 Glucose and fructose	66
3.3 Other sugars	68
3.4 Chemical properties of sugars	72
3.5 Sugar derivatives	75
3.6 Pectic substances in grapes	77
3.7 Exocellular polysaccharides from microorganisms	83

3.1 INTRODUCTION

Sugars are normally known as carbohydrates (Ribéreau-Gayon *et al.*, 1982; Jackson, 1994). It is clear from the chemical formula below that, for example, in a hexose such as glucose or fructose, each carbon atom corresponds to a water molecule:

$$C_6H_{12}O_6 = C_6(H_2O)_6$$

The name 'carbohydrate' is also justified by the way these sugars are produced by photosynthesis in vine leaves, which shows the importance of water in ripening grapes. The equation is as follows:

$$6CO_2 + 6H_2O = C_6H_{12}O_6 + 6O_2$$

Calling sugars 'carbohydrates' also indicates their affinity for water. Their hydrophilic character predicts their great solubility in water. In fact, at 25°C, 1.1 kg of glucose will dissolve in 1 liter of water. The great solubility of simple sugars in water explains their cryoprotective capacity. This is how, thanks to the hydrolase activity of α and β amylases, starch in plants is broken down into simple sugars, which mobilize water. They therefore protect cells in vine shoots and buds from winter and spring frosts.

Another characteristic of carbohydrates is that they consist of polyfunctional molecules, capable of participating in a large number of chemical, biochemical and metabolic reactions (Volume 1,

Section 10.3). Carbohydrates are the precursors of organic acids. Glucose is the precursor of citric, malic and succinic acids via the aerobic glycolyse channel. Glucose is also the precursor of tartaric and shikimic acids via the pentose channel. Sugars are precursors of phenols, and even of aromatic amino acids such as tyrosine, phenylalanine and tryptophan.

During alcoholic fermentation, glucose and fructose generate ethanol and various by-products. The production of 1° (% vol) of ethanol requires 16.5–18.0 g/l of sugar. These same hexoses may be attacked by lactic bacteria, producing lactic acid. Mannitol may also be produced from fructose. The main by-product of this lactic spoilage, however, is acetic acid. For this reason, it is only advisable to encourage these bacteria to start malolactic fermentation when all of the sugars have been broken down by yeast.

The expressions 'reducing sugar' and 'fermentable sugar' are frequently used in winemaking. Reducing sugars have an aldehyde or ketone function (more precisely, α-hydroxyketone) that reduces the alkaline cupric solutions used to assay them. Reducing sugars consist of hexoses and pentoses. The two elementary molecules in the structure of saccharose and other disaccharides are bonded by their aldehyde or ketone functions. They are not reducing sugars and must be hydrolyzed before they can be assayed.

Yeast uses fermentable sugars as nutrients. These sugars are the direct precursors of ethanol. Glucose and fructose are readily fermentable, while saccharose is fermentable after chemical or enzymic hydrolysis into glucose and fructose. Pentoses are not fermentable.

Although for many years they were known in winemaking by the generic term 'carbohydrate colloids', polysaccharides in must and wine are a complex, heterogeneous group of compounds. They consist of polymers of neutral oses and/or uronic acids, linked together by O-glycoside bonds, with α or β anomeric configurations. Most of these compounds may be isolated by precipitation. The procedure consists of adding 5 volumes of alcohol at 95% to one volume of must or wine acidified with 5% HCl. Polysaccharides may be present from a few hundred milligrams to several grams per liter.

The two main sources of polysaccharides in wine are grapes and yeast. When grapes are affected by *Botrytis cinerea*, polysaccharides secreted by this fungus into the grapes are also present in the wine. Finally, certain lactic bacteria in the genus *Pediococcus*, responsible for 'graisse' or 'ropy wine', are also capable of producing polysaccharides. Instead of using the 'old' classification, which distinguishes between 'pectins' and 'neutral polysaccharides', we prefer to make a distinction between grape polysaccharides (Section 3.6) and those produced by microorganisms (yeasts, *Botrytis cinerea* and bacteria) (Section 3.7).

3.2 GLUCOSE AND FRUCTOSE

3.2.1 Presence in Grapes and Wine

The two main hexoses in the vacuolar juice of grape pulp cells are:

1. D-Glucose, also known as dextrose because it deflects polarized light to the right $(\alpha) = +52.5°$.

2. D-Fructose, also known as levulose because it deflects polarized light to the left $(\alpha) = -93.0°$.

During ripening, the glucose/fructose ratio changes due to the action of an epimerase, with a significant increase in the proportion of fructose. This ratio is monitored as a marker for grape ripening, in the same way as alcohol dehydrogenase activity. The glucose/fructose ratio is on the order of 1.5 at color change and drops below 1 at full maturity.

The overall glucose and fructose concentration in ripe grape juice is between 150 and 250 g/l. It may be higher in overripe or dried grapes, or those affected by noble rot. Must made from botrytized grapes is characterized by the presence of oxo-5-fructose, a molecule that combines with sulfur dioxide (Barbe *et al.*, 2001).

Dry wines, where the sugars have been completely fermented, contain small quantities of

Carbohydrates

hexose (on the order of 1 g/l). This is mainly fructose, because glucose is preferentially fermented by the great majority of yeasts. For this reason, the G/F ratio, which is around 1 in grape must, decreases regularly during fermentation. Indeed, in sweet wines containing several tens of grams of sugar per liter, there may be 2–4 times as much fructose as glucose.

These two sugars are also differentiated by their sweetness and their effect on a wine's flavor. If saccharose has a rating of 1 on a scale of sweetness, fructose rates 1.73 and glucose 0.74. Consequently, if the residual sugar content is the same, the apparent sweetness of a wine depends on the G/F ratio.

3.2.2 Chemical Structure

Examining the chemical formulae in the Fischer projection (Figure 3.1) shows that, in open, aliphatic form, glucose has an aldehyde function on carbon 1, whereas fructose has a ketone function on carbon 2. These two sugars are interchangeable by chemical or enzymic epimerization via enediol, and are thus function isomers. The absolute configuration of the asymmetrical carbon in position 5 is the same, and corresponds conventionally to a D structure. In general, whether an ose belongs to the D or L series depends on the absolute configuration of the asymmetrical carbon closest to the primary alcohol function. This is determined in relation to D(+)-glyceraldehyde, whose D configuration has been verified by X-ray diffraction (Table 3.1). When this rule is applied to hexoses, in glucose and fructose, the distinction is made in relation to carbon 5 (Figure 3.1).

These two hexoses are in the D series. The absolute configuration of this asymmetrical carbon does not, however, predict its optical activity assessed by polarimetry. Thus, D-fructose is strongly anticlockwise in water while D-glucose is clockwise. The optical activity of a chiral molecule is determined by experimental measurement. The value depends on both the molecule's structure and the experimental conditions (temperature, solvent, pH, etc.).

The open form of glucose (Figure 3.1) has four asymmetrical carbons and, therefore, $2^4 = 16$ stereoisomers. The eight in the D series are listed in Table 3.1.

Hemiacetalization consists of a reaction between the aldehyde function and the alcohol function in position 5 (Figure 3.2). It produces a cyclic pyranic form, with the addition of a new chiral center on carbon 1. Its absolute configuration defines the α or β character of the D-glucopyranose stereoisomers, both diastereoisomers of each other.

For each cyclic stereoisomer, there is an equilibrium between two 'chair' conformations. The absolute configuration of each carbon is maintained during the transformation from one conformer to the other (Figure 3.3).

Under epimerization conditions, otherwise known as balance conditions, dissolving specific concentrations of, for example, α-D-glucopyranose alone will create an equilibrium between the two stereoisomers. Each one is a diastereoisomer of the other (α and β). These two epimers of D-glucopyranose (Figure 3.4) have different physical and chemical properties, including their optical activity, assessed by measuring the optical rotation with a polarimeter.

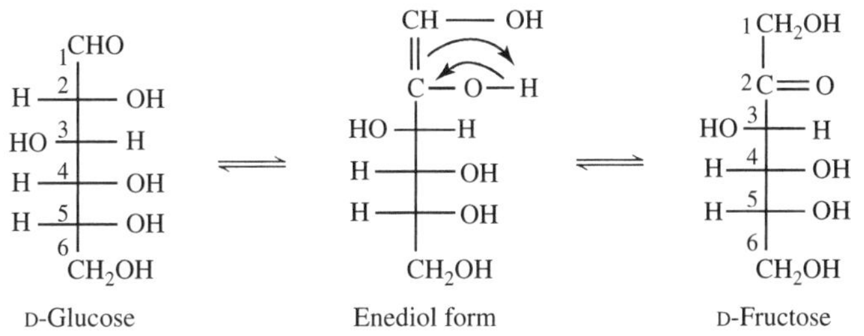

Fig. 3.1. Epimerization of glucose into fructose by enolization

Table 3.1. Fischer projection of the D series of homologous aldoses

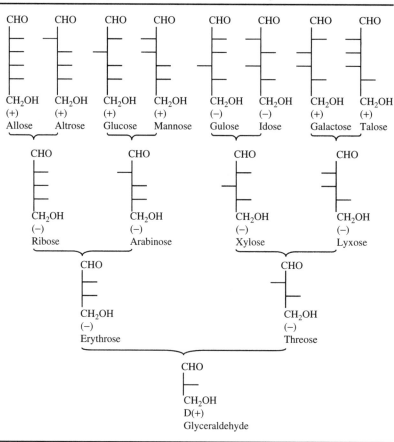

It is remarkable that α- and β-D-glucopyranose, which only differ by the absolute configuration of carbon 1, should have optical rotations with the same sign, but very different values (113.4° and 19.7°, respectively). When D-glucopyranose is dissolved in solution, the α and β forms are not immediately in balance, and the optical rotation only stabilizes at +52.5° after some time (mutarotation phenomenon).

The intra-molecular hemiacetalization reaction, corresponding to cyclization, may also take place on the alcohol function on carbon 4, rather than carbon 5, to produce a five-link cycle such as furan. This reaction produces β-D-glucofuranose instead of β-D-glucopyranose (Figure 3.5).

3.3 OTHER SUGARS

3.3.1 Simple Oses

The first line of Table 3.1 shows seven aldohexose isomers of D-glucose. D-Galactose is the only other isomer identified in wine, but in very small quantities (0.1 g/l). Each of these D isomers has an L stereoisomer that is of no major biological importance. Similarly, ketohexoses, with three asymmetrical carbons, have eight isomers. D-Fructose is the only one present in grapes and wine.

Wine always contains small quantities of pentoses (0.3–2 g/l in rare cases). These are reducing substances and may be assayed using cupric solutions. Pentoses are not fermentable by yeast and are

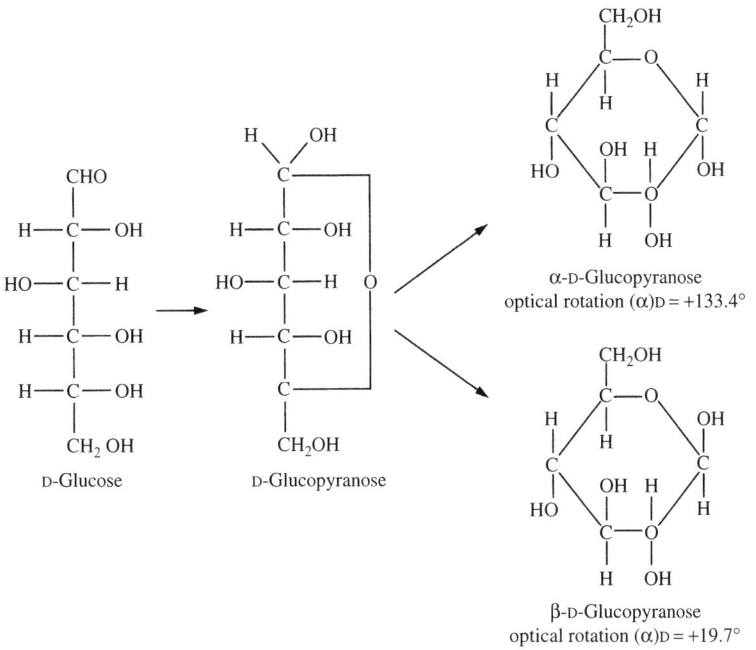

Carbohydrates 69

Fig. 3.2. Intra-molecular hemiacetalization reaction with formation of two stereoisomers, α- and β-D-glucopyranose

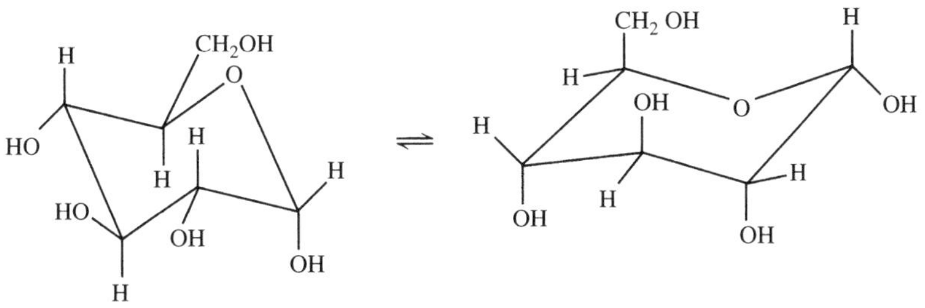

Fig. 3.3. Conformation equilibrium of α-D-glucopyranose

more common in red than white wines. They contribute less to sweet flavors than hexoses. If saccharose has a sweetness rating of 1 (Section 3.2.1), pentoses only rate 0.40.

Aldopentoses with three asymmetrical carbons have eight isomers. The four isomers with a D configuration are shown in the second line of Table 3.1. The main pentoses identified in grapes and wine are L-arabinose and D-xylose (Figure 3.6), representing a few hundreds of mg/l.

D-ribose and L-rhamnose, a methylated pentose, are present at concentrations below 100 mg/l.

D-xylose has a pyranose structure, while D-ribose has a furanose structure (Figure 3.7). The first is very widespread in wood, where it is associated with cellulose in a polysaccharide (xylane) form. It is also present in glycoside form. D-ribose is an essential component of nucleotides and nucleic acids. Arabinose is widespread in the plant kingdom, and the polysaccharide (gum)

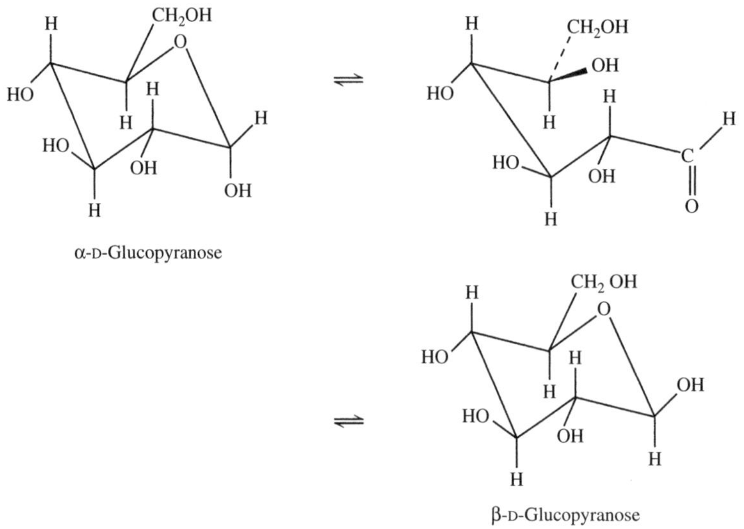

Fig. 3.4. Epimerization equilibrium of α-D-glucopyranose and β-D-glucopyranose

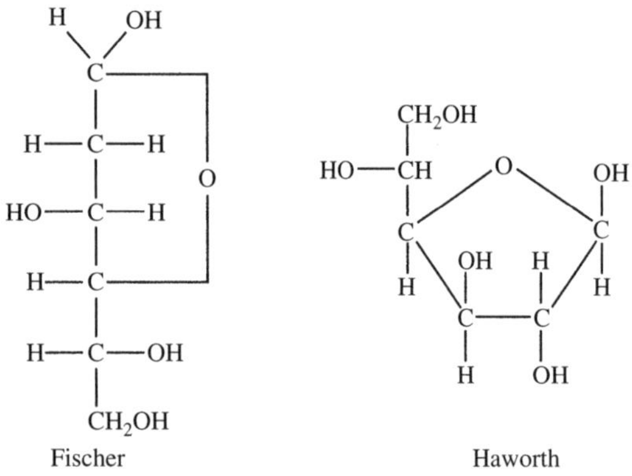

Fig. 3.5. Fischer and Haworth representations of β-D-glucofuranose

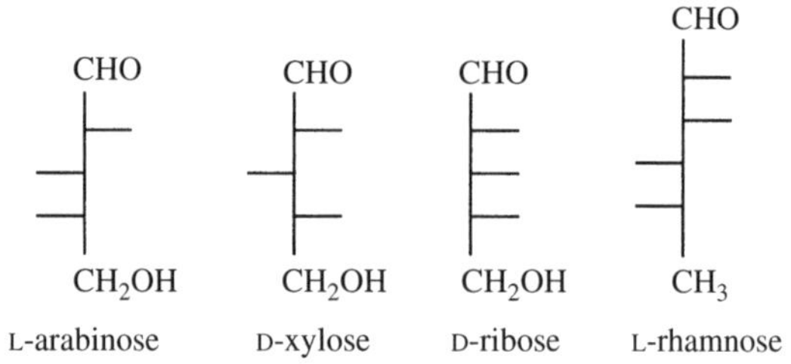

Fig. 3.6. Fischer projections of the main aldopentoses

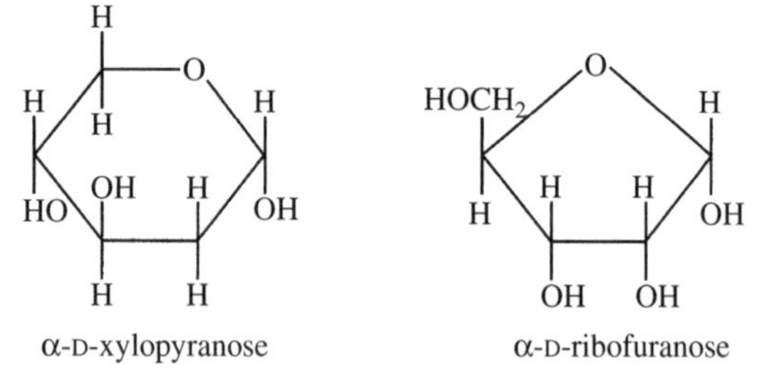

Fig. 3.7. Cyclic structures of α-D-xylopyranose and α-D-ribofuranose

form found in grapes is often associated with pectins. L-rhamnose is a methyl pentose and there is also L-deoxymannose, which results from the deoxygenation of the carbon 6 of L-mannose. L-deoxymannose is present in heteroside form in a wide range of plants. The terpene glycoside aroma precursors in Muscat grapes contain L-rhamnose (Sections 3.5.1 and 7.2.2).

The C_4 sugars in Table 3.1 have not been identified in grapes or wine. Glyceraldehyde is merely shown as a standard configuration for the D series. Furthermore, together with dihydroxyacetone, glyceraldehyde is one of the first fermentation breakdown products of hexoses. It also participates in their formation by photosynthesis (Figure 3.8).

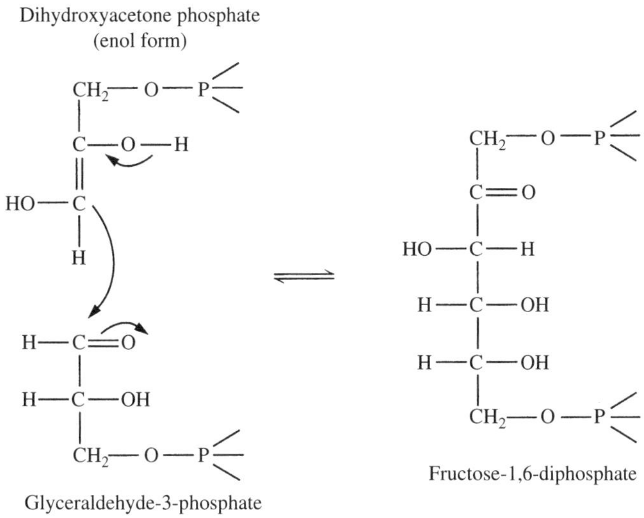

Carbohydrates

Fig. 3.8. Formation of glyceraldehyde-3-phosphate and dihydroxyacetone-1-phosphate from fructose-1,6-diphosphate during fermentation

3.3.2 Disaccharides

Various disaccharides have been identified in grapes or wine, generally in small quantities:

melibiose: galactose + glucose (reducing)
maltose: glucose + glucose (reducing)
lactose: glucose + galactose (reducing)
raffinose: fructose + melibiose (non-reducing)
trehalose: glucose + glucose (non-reducing)
saccharose: glucose + fructose (non-reducing)

The reducing property is due to the presence of an aldehyde function or free α-hydroxyketone. When disaccharides are non-reducing, all of the reducing functions of simple sugars are engaged in the bonds between them.

Trehalose is absent from grape must, but present in wine at concentrations on the order of 150 mg/l. It is produced by yeast autolysis at the end of fermentation.

Saccharose is the most important disaccharide. It is produced when a bond is made between carbon 1 of α-D-glucopyranose and carbon 2 of β-D-fructofuranose, according to the reaction shown in Figure 3.9. Its presence in grape juice was not confirmed until relatively recently. Concentrations are usually low, frequently between 2 and 5 g/l, although they may occasionally be slightly higher. Saccharose accumulates in vine leaves due to photosynthesis, but is hydrolyzed during transfer to the grapes, forming the essential sugars, glucose and fructose. Saccharose in grape juice may also come from hydrolyzable carbohydrate reserves in vine

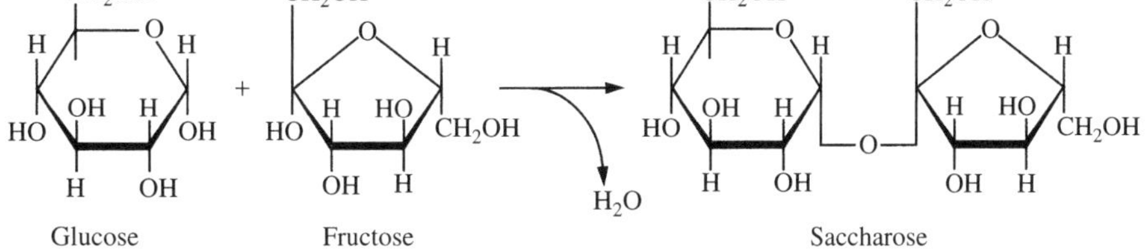

Fig. 3.9. Formation of saccharose from glucose and fructose

branches that contain 40–60 g/kg fresh wood, in the form of cellulose and starch.

Saccharose is an essential food sugar produced from beets and sugarcane. It is easily crystallized, which facilitates purification. Saccharose is perfectly soluble in water. Its optical rotation is $(\alpha) = +66.5°$. Saccharose may be hydrolyzed, producing a mixture of glucose and fructose with a negative optical rotation. This is due to the strongly anti-clockwise character of fructose as compared to the weak clockwise character of glucose. For this reason, the equimolecular mixture of glucose and fructose resulting from the chemical or enzymic hydrolysis of saccharose is known as 'invert sugar'.

Saccharose is fermented by yeast, after hydrolysis into glucose and fructose, under the influence of yeast invertase. Saccharose cannot, therefore, be present in wine, unless it has been added illegally after fermentation. Saccharose is the main sugar used to add potential alcohol to grapes (chaptalization), due to its purity and low cost (Volume 1, Section 11.5.2).

3.4 CHEMICAL PROPERTIES OF SUGARS

Sugar molecules are predictably reactive due to their polyfunctional character, especially the presence of a carbonyl, aldehyde or ketone radical. A certain number of these reactions play a role in winemaking, especially in assaying sugar levels in wine. Addition reactions with sulfur dioxide are described elsewhere (Volume 1, Sections 8.3.2 and 8.4.4).

3.4.1 Specific Properties

Oxidizability is a chemical property characteristic of aldehyde functions. The aldehyde function of glucose makes it capable of reducing copper salts, and this property is widely used to assay sugars. Furthermore, many aldehydes oxidize to form carboxylic acid with oxidizing agents as weak as silver oxide:

$$R-\underset{\underset{O}{\|}}{C}-H \xrightarrow{Ag_2O} R-\underset{\underset{O}{\|}}{C}-O-H$$

The oxidation of aldoses produces onic acids. Thus, the oxidation of glucose and mannose produces gluconic and mannonic acids, respectively. These should not be confused with the uronic acids (glucuronic and mannuronic) that result from the oxidation of the primary alcohol function.

Fructose has a ketone function, which is less reactive with nucleophilic agents than aldehydic carbonyl. Strictly speaking, this ketone function should not make fructose a reducing agent. However, fructose is also an α-hydroxyketone (like acetoin), which gives it a definite reducing property:

$$-\underset{\underset{O}{\|}}{C}-\underset{\underset{OH}{|}}{C}H- + 2Cu(OH)_2$$
$$\downarrow$$
$$-\underset{\underset{O}{\|}}{C}-\underset{\underset{O}{\|}}{C}- + Cu_2O + 3H_2O$$

We have seen that, due to their carbonyl radical, these sugars may be balanced and converted by enolization. The labile, acid character of hydrogen atoms in the α of the carbonyl makes aldolization and ketolization reactions predictable. These involve the condensation of two sugar molecules or, on the contrary, the breakdown of one molecule (Figure 3.8). These reactions play a vital role in the synthesis mechanisms of hexoses in photosynthesis and their breakdown during fermentation. These aldolization and hydroxyketonization reactions have been observed *in vivo*, but they also take place *in vitro*.

Glucose and fructose may submit to nucleophilic additions. Indeed, the addition of nucleophilic units, such as amino acids, to the carbonyl of the aldehydes and ketones (Figure 3.10) produces aldimines ($R_1 = H$) (or Schiff base) and ketimines ($R_1 \# H$), via carbinolamine.

This very widespread reaction is used in the assimilable nitrogen assay for assessing the potential fermentability of must and wine according to the Sörensen method, or formaldehyde titration ($R_1 = R_2 = H$). In a cold, aqueous phase, the condensation reaction stops at the carbinolamine stage, i.e. methylol, when the aldehyde is formaldehyde. However, if the carbonylated compound is glucose or fructose and the temperature of the medium

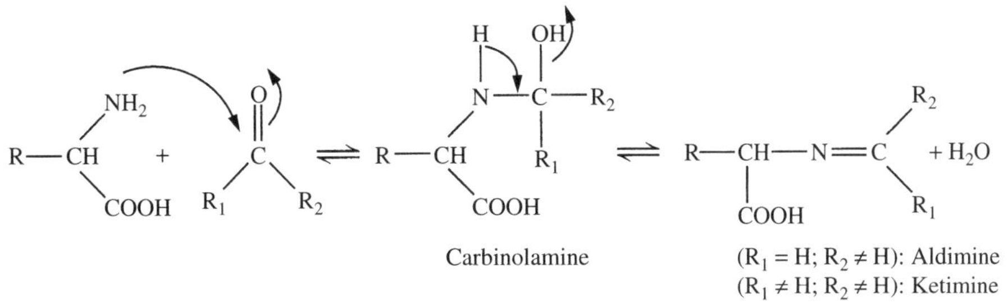

Carbohydrates

Fig. 3.10. Aldimine and ketimine formation mechanism by the addition of an amino acid on the aldehyde or ketone function of a sugar

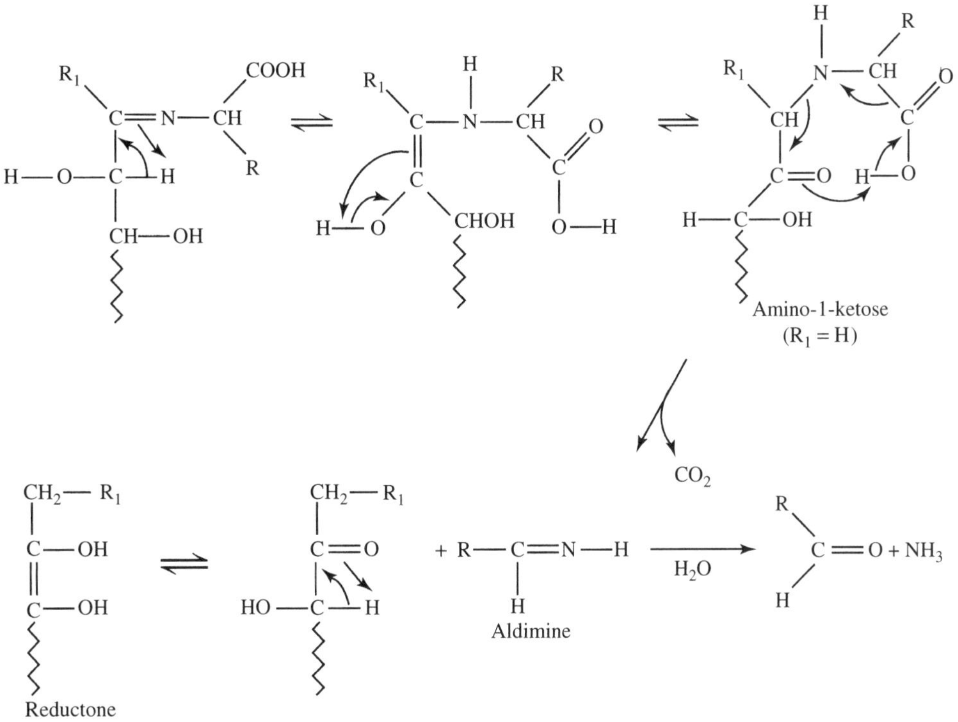

Fig. 3.11. Breakdown, according to the Strecker reaction, of an aldimine resulting from the reaction of an amino acid with a sugar

is high, as is the case when the grapes are heated (thermovinification), the carbinolamine may be rearranged. Under these conditions, the first products of the Maillard reaction (amino-1-ketose and amino-1-aldose) are formed (Figure 3.11). These unstable intermediaries are rearranged again, producing the oxidizing–reducing systems (reductones) responsible for non-enzymic browning.

Furthermore, as the amine function is that of an amino acid, amino-1-ketose or amino-1-aldose may decarboxylate by the Strecker breakdown mechanism (Figure 3.11), developing into an aldimine derivative of amino acid and, finally, after hydrolysis, into a carbonylated compound (aldehyde or ketone). Thus, alanine (R = CH_3), the amino acid present in the highest concentrations in must and wine, and also among the most reactive compounds, produces aldimine. The aldimine is then hydrolyzed to ethanal, responsible for a flat, oxidized odor. If the initial amino acid is methionine, the reaction produces thermally and photochemically unstable methional. This in turn evolves into methanethiol, which smells of wet dog, cabbage and reduction odors.

3.4.2 Reduction of Alkaline Cupric Solutions (Fehling's Solution)

The chemical assay of sugars is usually based on their capacity to reduce a cupro-alkaline solution. Hot cupric hydroxide (CuO) produces a brick red Cu_2O precipitate. It is possible either to assay the excess copper ions that have not reacted using iodometry (Luff method) or to measure the quantity of sweet solution necessary to remove the color from the cupric solution (Fehling's solution). A third method consists of collecting the red Cu_2O precipitate in a ferric sulfate solution and assaying the ferrous sulfate formed, using potassium permanganate (Bertrand method).

The products formed from sugars vary according to reaction conditions, so great care must be taken to obtain reproducible results.

3.4.3 Chemical Identification by Adding Phenylhydrazine

Due to the presence of carbonyl, aldehyde or ketone radicals, sugars are capable of addition reactions with nucleophilic reagents such as phenylhydrazine (C_6H_5–NH–NH_2). The addition of three phenylhydrazine molecules to an aldose (Figure 3.12) leads to the formation of osazone, a crystallized product with specific physicochemical characteristics, especially its melting point. This makes it possible to identify the corresponding sugar.

3.4.4 Methylation and Acetylation Reactions Producing Volatile Derivatives Identifiable by Gas-Phase Chromatography

In the presence of an acid catalyst (BF3), methanol has a selective addition reaction on carbon 1, producing methyl ether (Figure 3.13). In the case of glucose, for example, this hemiacetalic ether–oxide then evolves, by an intramolecular cyclizing acetalization reaction, into α- and β-methyl-D-glucopyranoside (Figure 3.13). After balancing under epimerizing conditions,

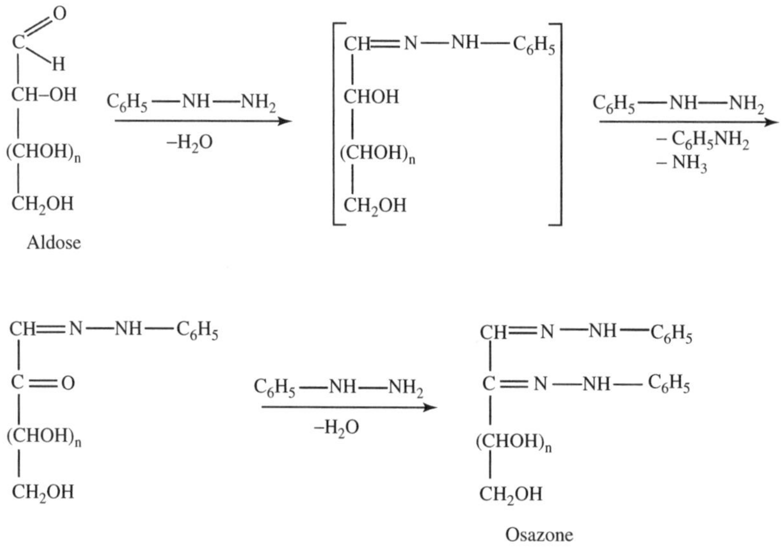

Fig. 3.12. Osazone formation mechanism by the addition of three phenylhydrazine molecules to one aldose molecule

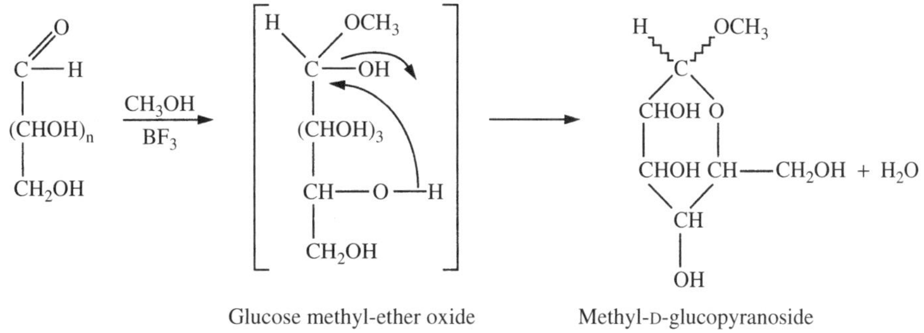

Carbohydrates

Fig. 3.13. Methyl D-glucopyranoside formation mechanism

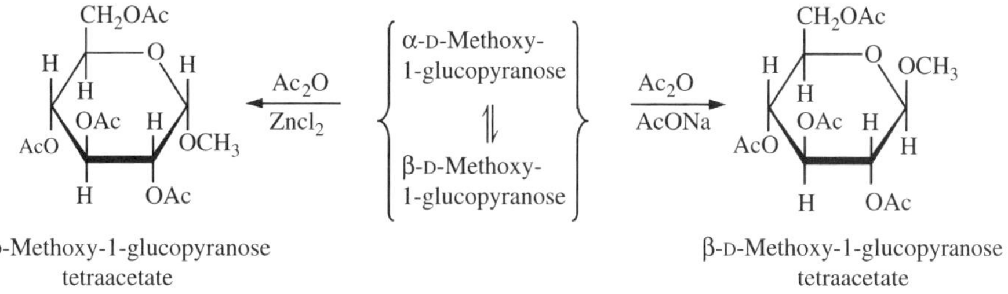

Fig. 3.14. Derivatization of α- and β-methoxy-1-glucopyranoses in tetraacetate form

the cyclizing reaction produces a mixture of the two α and β diastereoisomers of methyl-D-glucopyranose.

The methoxylation reaction of carbon 1 on the glucose is used as the first stage in volatilization, preparatory to gas-phase chromatography analysis of the simple sugars in the macromolecules of colloidal osides in must and wine, including neutral polysaccharides and gums (Dubourdieu and Ribéreau-Gayon, 1980a). This etherification reaction is followed by a second derivation reaction of the sugars in polyester state, such as tetraacetate.

The esterification reaction (Figure 3.14), using acetic anhydride or trifluoroacetic anhydride, may be catalyzed either by a Lewis acid ($ZnCl_2$) or a base (AcONa). If the catalyst is acid, only the acetylated derivative of α-D-methylglucopyranose is formed. However, when a basic catalyst is used, only the most thermodynamically stable β stereoisomer is formed.

In this form, which is both an ether and an ester, each simple sugar molecule in a polysaccharide is identifiable and quantifiable by gas-phase chromatography. It is also possible to carry out an elementary analysis of oside macromolecules.

3.5 SUGAR DERIVATIVES

3.5.1 Glycosides

Glycosides are produced when a non-carbohydrate compound reacts with the semi-acetalic function of an ose. The non-oside part of the glycoside is known as an aglycone.

There are two groups: O-glycosides and N-glycosides. In O-glycosides, the carbohydrate part is linked to the aglycone part by an oxygen atom. The ether–oxide bond is easily broken by enzymic or chemical hydrolysis. In N-glucosides, the bond is on a nitrogen atom. The nucleotide fractions of nucleic acids are the best-known molecules in this category.

O-Glycosides are very widespread in plants. A large number of these compounds have pharmacodynamic characteristics. Glycosides in grapes have a variety of interesting enological properties.

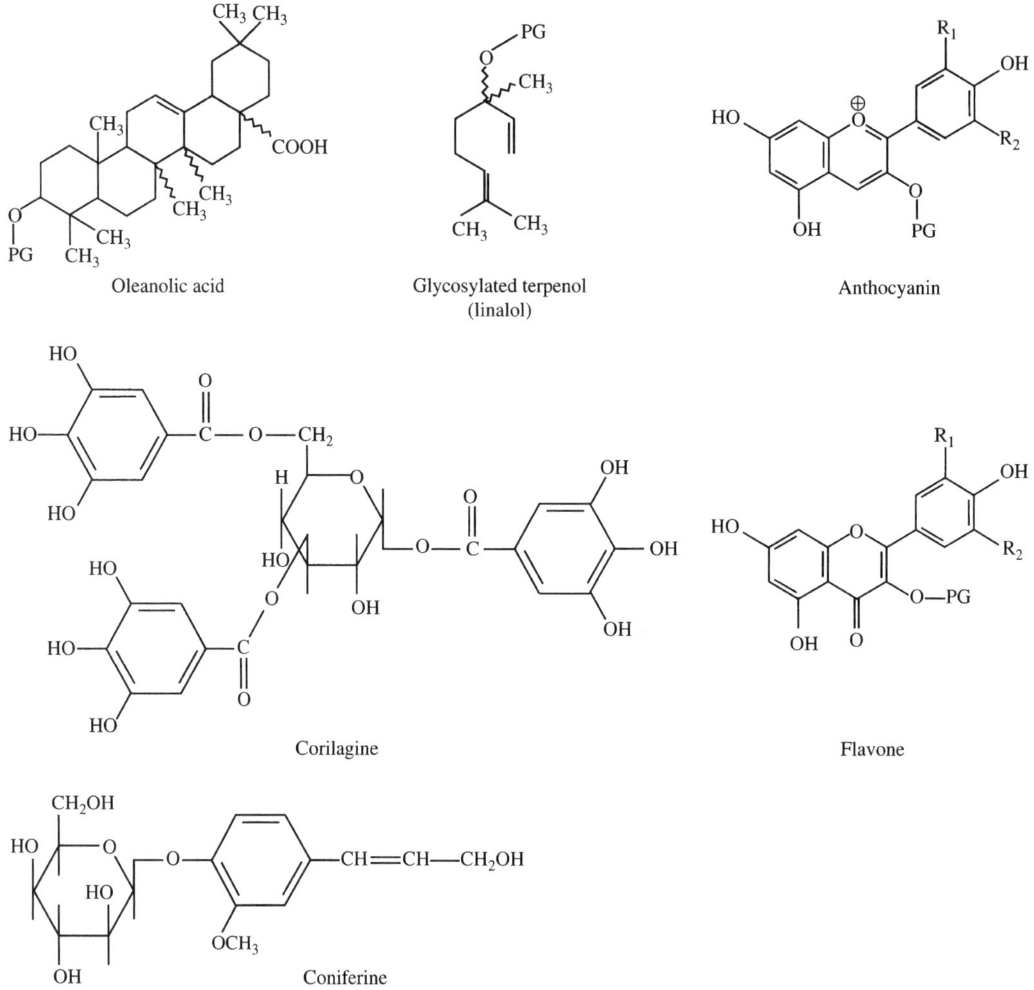

Fig. 3.15. Examples of O-glycosides significant in enology

Thus, a glycoside found in the cuticle has a triterpenoid in homosteroid form as its aglycone: oleanolic acid (Figure 3.15). Oleanolic acid also forms the aglycone part of saponins, known for their foaming properties. For enologists, the most useful attribute of oleanolic acid is certainly its role as a survival factor for yeast. It may even be an anaerobic growth factor (Lafon-Lafourcade et al., 1979), facilitating completion of fermentation (Volume 1, Section 3.5.2).

Some glycosides in the skin are also varietal aroma precursors, especially those with terpenol as their aglycone (Section 7.2.1). Prolonged skin contact, promoting glycosidase activity, is a technique for releasing these terpenic alcohols, such as linalol, geraniol, nerol, citronellol, etc. (Cordonnier and Bayanove, 1980; Crouzet, 1986).

Another very important category of glycosides is found in the grape hypoderm. These flavones, especially anthocyanins, are responsible for color in red grapes and wines (Section 6.2.3). Their heteroside nature is known to contribute to their stability. Like other phenols, (proanthocyanidins, tannins, etc.) (Section 6.2.4), they are considered to have antioxidant and antiseptic properties, as well as specific flavors.

Carbohydrates

Other phenol glycosides are derived from benzoic acids (Section 6.2.1), like corilagine, or cinnamic acids (Section 6.2.1), like coniferine (Figure 3.15).

3.5.2 Oxidation Products

The sugar oxidation products most significant in winemaking are mainly acids resulting from oxidation of one of the terminal functions of the carbon chain (Section 1.2.2). The most important of these is gluconic acid, produced by oxidation of the aldehyde function of glucose. Its presence in grapes and wine is directly linked to the effects of *Botrytis cinerea*, in the form of noble rot, and rot in general (Volume 1, Section 10.6.4).

Spoilage of grapes by various parasites and bacteria may result in sugar oxidation products with one or more ketone functions. One of their properties is to combine with sulfur dioxide, making the wines difficult to store (Volume 1, Section 8.4.3.).

3.6 PECTIC SUBSTANCES IN GRAPES

3.6.1 Terminology and Monomer Composition of Pectic Substances

Grape polysaccharides result from the breakdown and solubilization of some of the pectic substances in the skin and flesh cell walls. The terminology describing these compounds may be confusing, as the definition of certain terms (pectins, gums, neutral pectic substances, acid pectic substances, etc.) varies from one author to another.

Peynaud (1952), Buchi and Deuel (1954), Usseglio-Tomasset (1976, 1978), Ribéreau-Gayon *et al.* (1982) and Dubourdieu (1982) make a distinction between pectins and gums according to the following criteria:

1. Pectins are chains formed almost exclusively of galacturonic acid (α-D-galacturopyranoside acid) units partially esterified by methanol. The degree of esterification of grape pectins is high (70–80%). The terms 'polygalacturonic acid' or 'homogalacturonane' are synonyms describing these chains. Homogalacturonanes (Figure 3.16) have α-(1,4)-type oside bonds. These compounds are easily separated from the other total soluble polysaccharides in grape must by saponification, converting the pectin into pectic acid. This is precipitated in the form of pectate by adding calcium chloride. Hydrolysis of the precipitate (several hundreds of mg/l) produces only galacturonic acid. These compounds are absent from must made from grapes affected by rot because they are hydrolyzed by the *endo*-polygalacturonases in *Botrytis cinerea* (Dubourdieu, 1978). In healthy grapes, pectins also disappear due to the action of endogenous pectolytic enzymes (Volume 1, Section 11.6.1) or those added by winemakers (Volume 1, Section 11.7). For this reason, by the end of alcoholic fermentation, wines contain practically no homogalacturonanes.

2. Gums are soluble polysaccharides. Once the pectins (homogalacturonanes) have been eliminated, gums are obtained either by saponification and precipitation or by breakdown using

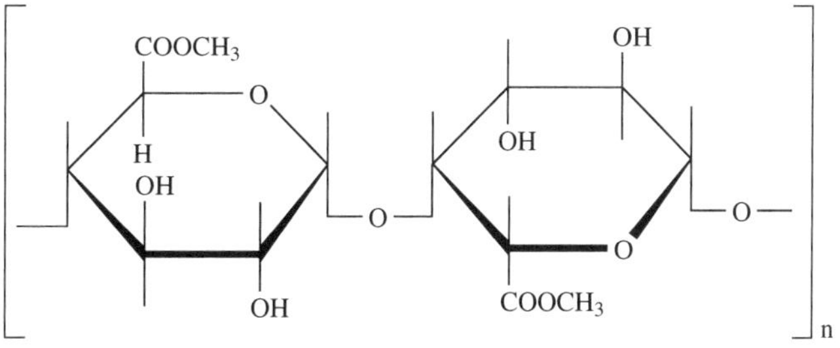

Fig. 3.16. Basic structure of α-homogalacturane. Chain of partially methylated galacturonic acid units, linked by α-(1,4)-type oside bonds

Table 3.2. Fractionation of the gums in grape must on DEAE Sephadex A25 (percent composition of the oses in the resulting fractions) (Dubourdieu and Ribéreau-Gayon, 1980a)

	Fraction 1	Fraction 2	Fraction 3	Fraction 4
Total gums (%)	43	17	13	25
Arabinose	22	27	30	19
Rhamnose	1	2	4	12
Xylose	Traces	Traces	Traces	0
Mannose	14	7	2	5
Galactose	54	61	58	23
Glucose	8	2	2	2
Galacturonic acid	0	0	4	39

exogeneous pectinases (Dubourdieu *et al.*, 1981a). Besides galacturonic acid, gums consist of neutral oses, mainly arabinose, rhamnose, galactose and small quantities of xylose, mannose and glucose. Gums are sometimes described as non-pectic polysaccharides. This description is unsatisfactory as, in their natural state, all of the soluble polysaccharides in healthy grapes are components of the middle lamella and primary pectocellulose walls of plant cells. They are all, therefore, pectic substances. Basically, gums are residues from the transformation of pectic substances in must by endogeneous or exogeneous pectinases.

From the findings of Usseglio-Tomasset (1978), it is known that gums are a complex mixture of heteropolysaccharides, with a wide range of molecular weights (from less than 10 000 to over 200 000), and highly variable monomer compositions, according to the molecular weight. Gums have been separated into fractions using DEAE Sephadex ion exchangers (Dubourdieu and Ribéreau-Gayon, 1980a). These fractions consist of neutral gums or osanes (mainly arabinose and galactose) and acid gums, made up of galacturonic acid, rhamnose, arabinose, and galactose. The more strongly acid gums have a high rhamnose content (Table 3.2).

A more recent, clearly preferable classification (Brillouet, 1987; Brillouet *et al.*, 1989) simply divides soluble polysaccharides in must into neutral pectic substances and acid pectic substances, depending on whether or not the molecules contain galacturonic acid. In this context, the term 'pectin' describes all acid pectic substances and not just homogalacturonanes.

3.6.2 Variations in Total Polysaccharides in Must During Ripening

Ripe grapes have relatively low concentrations of pectic substances compared to other fruits. These compounds are the main component of the fine cell walls in grape flesh. In addition to pectic substances, the thicker skin cell walls also contain hemicelluloses and larger amounts of cellulose.

The variations in total soluble polysaccharide concentrations in must from healthy grapes during ripening (Table 3.3) show the importance of hydrolysis phenomena in the cell walls during this period (Dubourdieu *et al.*, 1981a; Dubourdieu, 1982). Initially, at color change, the protopectins (insoluble pectins) in the middle lamella are solubilized. This increases the concentration of soluble pectic substances in must. Galacturonic acid is the main component of these pectic substances. Later in the ripening process, the breakdown of the pectocellulose wall solubilizes pectic substances, while the chains of galacturonic acid units in the water-soluble acid pectic substances are hydrolyzed. During this second period, a decrease is generally observed in the total soluble polysaccharide concentration in must, as well as a declining proportion of uronic acid. Mourgues (1981) also observed that total soluble polysaccharides in Carignan grapes increased

Table 3.3. Variations in the soluble polysaccharide content of must as grapes ripened in 1980 (Dubourdieu *et al.*, 1981a)

	Grape variety					
	Sauvignon Blanc			Semillon		
Sample date	09/15	09/24	10/16	09/15	09/24	10/16
Total polysaccharides (mg/l)	728	1024	426	860	1163	355
Neutral sugars (%)	29	42	54	27	47	70
Uronic acids (%)	81	58	46	73	52	30

$$\alpha\text{-D-GalA}p\text{-}(1 \to 2)\text{-}\alpha\text{-L-Rha}p\text{-}(1 \to 4)\text{-}\alpha\text{-D-GalA}p\text{-}(1 \to 2)\text{-}\alpha\text{-L-Rha}p$$

Fig. 3.17. Structure of a rhamnogalacturonan I (RG-I). α-D-GalAp: α-D-galactopyranoside acid (galacturonic acid); α-L-Rha$p$: α-L-rhamnopyranose (rhamnose)

sharply at color change and that the pectin content, estimated by assaying galacturonic acid, decreased at the end of ripening. In some cases, however, due to a lack of endogeneous pectinase activity in grapes, the concentration of soluble acid pectic substances in must increases throughout ripening. This often occurs when vines are subjected to severe drought conditions. It is then necessary to use exogeneous pectinases to achieve satisfactory clarification of white wine musts during settling (Volume 1, Section 13.5).

3.6.3 Molecular Structures of Pectic Substances in Must

The first structural studies of soluble pectic substances in grapes are relatively recent. Pioneering work was carried out by Villetaz *et al.* (1981), Brillouet (1987), Saulnier and Thibault (1987) and Saulnier *et al.* (1988). Continuing studies have identified detailed molecular structures and linked them to known pectic substances in plants.

Acid pectic substances in grape must

These substances, like those in other higher plants, have long chains of galacturonic acid (homogalacturonan) units, interrupted by rhamnogalacturonan structures, where rhamnose units (Rha) alternate with galacturonic acid units (GalA) (Figure 3.17). This entire structure is currently known as rhamnogalacturonan (RG-I).

The α-(1,4)-type oside bonds in the homogalacturonan chain form a secondary open-helix structure, where each coil consists of three galacturonic acid units. When the α-GalA-(1,2)-α-Rha-(1,4)-α-GalA sequence is inserted, the axis of the helix pivots through 90°, producing what is known as the 'pectic bend'.

The rhamnogalacturonan zones of acid pectic substances generally have lateral chains of neutral oses, connected to carbon 4 of the rhamnose. For this reason, they are known as 'bristled zones', as opposed to the 'smooth zones' of homogalacturonan that do not branch off (Figure 3.18). There are two types of lateral chains in the 'bristled zones': arabinogalactans in the majority (AG-II) and arabinans (A).

The arabinogalactans in pectic substances from plants are divided into two categories (I and II) (Figure 3.19):

1. Arabinogalactan I (AG-I) (Figure 3.19a), the most widespread, consists of a main chain of galactose units, bonded on β-(1,4), with ramifications (on carbon 3) made up of individual units or arabinose oligomers bonded on α-(1,5).

2. Arabinogalactan II (AG-II) (Figure 3.19b) has a more complex structure. It has a main chain of galactose units linked by β-(1,3) oside bonds, with short lateral chains of galactose, bonded on β-(1,6). In C_3, these are replaced by individual

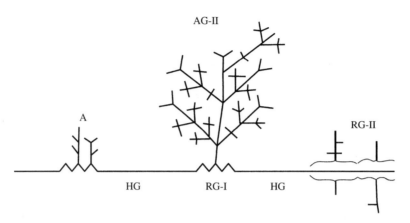

Fig. 3.18. Suggested structural model for acid pectic substances in grapes (Doco *et al.*, 1995). A, arabinan; HG, homogalacturonase; AG-II, arabinogalactan II; RG-I, rhamnogalacturonan I; RG-II, rhamnogalacturonan II

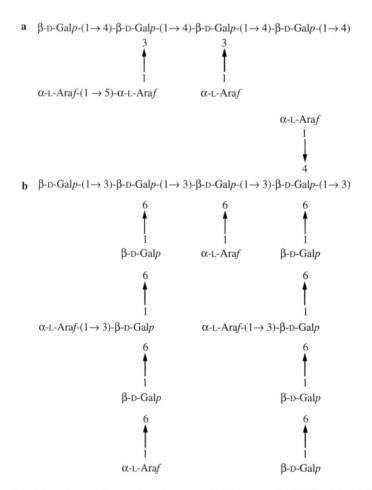

Fig. 3.19. Structure of arabinogalactans in pectic substances (Brillouet *et al.*, 1989). (a) arabinogalactan I (AG-I), (b) arabinogalactan II (AG-II). α-L-Ara$f$: α-L-arabinofuranose (arabinose); α-D-Gal$p$: α-D-galactopyranose (galactose)

Carbohydrates

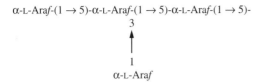

Fig. 3.20. Structure of arabinans in pectic substances (Villetaz *et al.*, 1981). α-L-Ara*f*: α-L-arabinofuranose (arabinose)

arabinose units. Type II arabinogalactan also has a few individual arabinose units on carbon 6 or 4 of the galactose in the main and secondary chains (Figure 3.20). The arabinogalactans on the 'bristled zones' of acid pectic substances in grape must are type II.

Arabinans, minor lateral branches of acid pectic substances, consist of short arabinose chains, bonded in α-(1,5), with ramifications of individual arabinose units on C_3 (Figure 3.20).

One particular rhamnogalacturonan, rhamnogalacturonan II (RG-II), has recently been identified in plant tissues (O'Neill *et al.*, 1990; Albersheim *et al.*, 1994). Its structure is remarkably well preserved in the cell walls of all higher plants. RG-II is a very complex polysaccharide with a low molecular weight (approximately 5400) (Figure 3.21), and consists of a rather short main chain of galacturonic acid units with four lateral oligosaccharide chains. These contain not only arabinose, rhamnose, fucose and galactose, as well as galacturonic and glucuronic

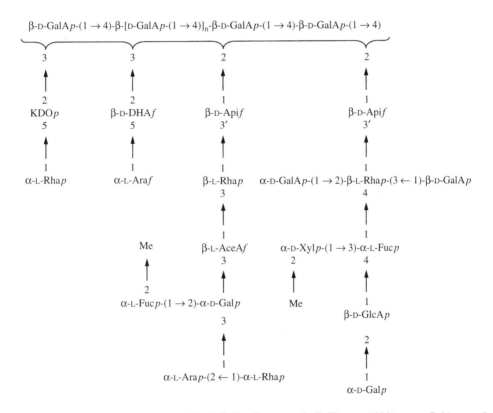

Fig. 3.21. Structure of rhamnogalacturonan II (RG-II) (Doco and Brillouet, 1993). α-D-GalA*p*: galacturonic acid; α-D-Gal*p*: galactopyranoside; β-L-Rha*p*: β-L-rhamnopyranose (rhamnose); α-L-Ara*f*: α-L-arabinofuranose (arabinose); α-L-Fuc*p* = α-L-fucopyranose (fucose); 2-*O*-Me-α-D-xylose: 2-*O*-methyl-α-D-xylose; β-D-GlcA*p*: β-D-glucuropyranosic acid (glucuronic acid); β-D-Api*f*: β-D-apiofuranose (apiose); KDO*p*: 3-dioxy-D-manno-2-octululopyranosic acid; β-D-DHA*f*: β-deoxy-D-lyxo-2-heptulafuronosic acid; β-L-AceA*f*: β-3-C-carboxy-5-deoxy-L-xylofuranose (aceric acid); Me: methyl

acids, but also various rare sugars, such as 2-*O*-methyl-fucose, apiose, 2-*O*-methyl-xylose, KDO (2-keto-3-deoxyD-manno-octulosonic acid) DHA (3-deoxy-D-lyxo-heptulosaric acid) and aceric acid (3-*C*-carboxy-5-deoxy-*L*-xylose). RG-II has been found in grape must (Doco and Brillouet, 1993; Pellerin *et al.*, 1996; Doco *et al.*, 1997) at concentrations of several tens of mg/l. Its large number of rare oside bonds make it resistant to pectolytic enzymes. RG-II is found in fruit juice obtained by total liquefaction with pectinases, as well as in wine. Recent research has shown that, thanks to its boric acid diester cross-bonds, RG-II may form dimers (dRG-II-B) in plant cell walls (Ishii and Matsunaga, 1996; Kobayashi *et al.*, 1996; O'Neill *et al.*, 1996).

Neutral pectic substances in grape must

These substances have molecular structures similar to those of the lateral branches of acid pectic substances, consisting of arabinan and type II arabinogalactan.

Arabinans are small polymers (6000) not precipitated by ethanol. They were first isolated from Pinot Noir must (Villetaz *et al.*, 1981).

The type II arabinogalactans isolated from the flesh of Carignan grapes (Saulnier and Brillouet, 1989) contained 88% neutral sugar, 3% uronic acid and 8% protein. Arabinose and galactose were present in a molar ratio of 0.66. The structure of type II arabinogalactans in neutral pectic substances in must is slightly less ramified than that of the 'bristled zones' of acid pectic substances (Brillouet *et al.*, 1990). The substitution rate of the galactose chains on 1–6 is lower, as galactose is only replaced by arabinose on C_3. These compounds have an average molecular weight of 165 000. They generally have a small peptide section, consisting mainly of the amino acids hydroxyproline, serine, glycine and alanine. The glycane–peptide binding point involves an *O*-glycoside bond with the threonine on the polypeptide chain. Type II arabinogalactans are in fact arabinogalactan proteins (AGP). This type of proteoglycanes is widespread in higher plants (Aspinall, 1980; Fincher *et al.*, 1983).

3.6.4 Molecular Structures of Pectic Substances in Wine

Pectic substances are considerably modified when must is turned into wine, under the influence of natural grape pectinases or commercial exogeneous enzymes made from *Aspergillus niger*, especially those that break down acid pectic substances (Volume 1, Section 11.6.1): *endo-* and *exo-*polygalacturonase, *endo*-pectinlyase, *endo-* and *exo*-pectatelyase, and pectinmethylesterase. Thus, although homogalacturonanes may be isolated from must, they are never present in wine. In the same way, the homogalacturonan zones in acid pectic substances in must are absent from the pectic substances in wine. These, in fact, consist exclusively of neutral pectic substances, the 'bristled zones' of acid pectic substances from the must and type II rhamnogalacturonans.

Arabinogalactan proteins (AGP) represented a major proportion (40%) of the total soluble polysaccharides (300 mg/l) in a red Carignan wine (Pellerin *et al.*, 1995; Vidal *et al.*, 2001). These macromolecules were first fractionated on a DEAE Sephacel anion exchanger, then separated from the yeast mannoproteins by affinity chromatography on Concanavaline A and finally purified, by exclusion chromatography on a Sephacryl S400 column. Table 3.4 shows the molecular characteristics of the five species obtained.

Type II rhamnogalacturonan is present in wine as a dimer (dRG-II-B). It may be isolated by adsorption chromatography on a polystyrene and divinylbenzene copolymer resin column (Pellerin *et al.*, 1997). The average RG-II concentrations in white wines are between 20 and 60 mg/l, while those of red wines range from 100 to 160 mg/l.

3.6.5 Impact of Pectic Substances on Wine Character

Pectic substances have often been attributed a role in the softness and full-bodied character of wines (Muntz and Lainé, 1906). It is true that wines that have these qualities almost always contain large quantities of pectic substances, but these are apparently not directly responsible for these sensory impressions. Indeed, when they are isolated from

Carbohydrates

Table 3.4. Composition and characteristics of the various fractions isolated from AGP in wine (Pellerin *et al.*, 1995)

	AGP0	AGP1	AGP2	AGP3	AGP4
Total soluble Polysaccharides in wine (%)	30	1.3	2.1	5.6	2.1
Average molecular weight	184 000	262 000	261 000	236 000	237 000
Protein[a]	3.6	2.4	3.0	2.4	0.8
Uronic acids[a]	2.7	6.5	7.4	12.4	20.4
Neutral sugars[a]	79.5	75	76.2	77.0	61.3
Rhamnose[b]	1.1	2.2	3.1	7.1	10.8
Arabinose[b]	40.5	43.8	39.2	43.2	28.5
Xylose[b]					1.3
Mannose[b]	0.5				
Galactose[b]	53.8	48.4	1.3	1.0	1.9
Glucose[b]	1.0	0.7	1.1	0.9	1.4
Glucuronic acid[b]	3.1	4.9	5.4	6.1	13.3
Galacturonic acid[b]				1.9	2.3

[a] Percentage dry weight.
[b] Molar %.

wine and added to model solutions or white wines (even in quantities as high as g/l), they do not affect their fullness or softness. These compounds also have a low intrinsic viscosity, in keeping with their rather compact structure. However, it is accepted that pectic substances (polysaccharides) may alter the tasting impression of wine when they combine with phenolic compounds (tannins). In particular, astringency is attenuated (Section 6.7.2).

The role of carbohydrate colloids in wine clarification and stability has been studied much more extensively. Pectic substances in wine foul filter layers (Castino and Delfini, 1984) during filtration. This phenomenon is particularly marked in tangential microfiltration (Section 11.5.2) (Feuillat and Bernard 1987; Serrano *et al.*, 1988) as the carbohydrate colloid retention rate may be over 50%. A much higher proportion of AGPs than yeast mannoproteins is retained in this process (Brillouet *et al.*, 1989; Belleville *et al.*, 1990).

AGPs also have a protective effect against protein casse in white wines (Pellerin *et al.*, 1994; Waters *et al.*, 1994). They are, however, less effective than yeast mannoproteins (Sections 3.7.1 and 5.6.3).

Both rhamnogalacturonans (RG-I and RG-II) act as tartrate crystallization inhibitors in wine (Gerbaud *et al.*, 1997), while AGPs have no effect (Section 1.7.7). The natural inhibition of tartrate crystallization at low temperatures is more marked in red than white wines. This difference is due to the effect of polyphenols, which are also crystallization inhibitors, as well as the presence of higher quantities of RG-I and RG-II in red wines.

Finally, rhamnogalacturonan dimers (dRG-II-B) may form coordination complexes with specific di- and trivalent cations, especially Pb^{2+} ions (Pellerin *et al.*, 1997). For this reason, 85–95% of the lead assayed in wine is in the form of a stable complex with dRG-II-B. The effect of these complexes on the toxicity of lead in the body is not known.

3.7 EXOCELLULAR POLYSACCHARIDES FROM MICROORGANISMS

3.7.1 Exocellular Polysaccharides from Yeast

Yeasts are the second major source of polysaccharides in wine. A great deal of research has been devoted to the structure of yeast cell walls (Volume 1, Section 1.2), but much less to the type of carbohydrate colloids released into wine.

The polysaccharide content of a sweet, colloid-free model medium inoculated with

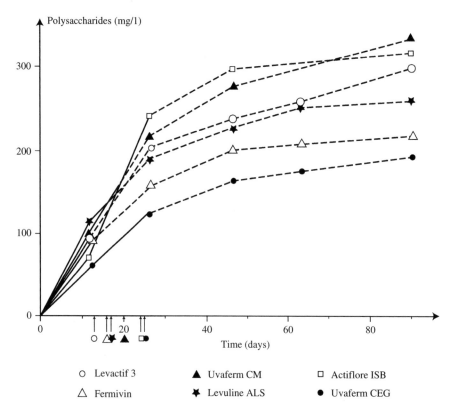

Fig. 3.22. Production of exocellular polysaccharides by commercial yeast strains in a model medium at 20°C. Solid lines (———) indicate the length of alcoholic fermentation for each strain. Uvaferm CEG, Actiflore ISB, Levactif 3, Uvaferm CM, Fermivin, Lévuline ALS, end of alcoholic fermentation (Llaubéres *et al.*, 1987)

Saccharomyces cerevisiae increases during fermentation (Usseglio-Tomasset, 1961, 1976). The level continues to rise after alcoholic fermentation and throughout aging of the fermented medium on yeast biomass (Llaubéres *et al.*, 1987) (Figure 3.22). The total amount may represent several hundreds of mg/l. The quantity of polysaccharides released by the yeast depends on the strain, as well as the fermentation and aging conditions. The yeast releases more polysaccharides at high temperatures (Figure 3.23), in an agitated medium (Table 3.5), following prolonged aging on the biomass. Yeast polysaccharides are mainly released into dry white wines during aging on the lees, especially if they are regularly stirred into suspension (Volume 1, Section 13.8.1). This phenomenon is slow, as the temperature is low (12–16°C). Red wines mainly acquire yeast colloids during high-temperature (30–35°C) maceration after fermentation. This only continues for a very limited period because most of the yeast lees are separated from the wine when it is run off.

The exocellular polysaccharides released by yeast during fermentation and aging on the lees may be isolated by precipitation with ethanol, or membrane ultrafiltration at a cutoff of 10 000. They may be fractionated by two processes: precipitation with hexadecyltrimethylammonium bromide (Cetavlon) and affinity chromatography on a Concanavaline A-Sepharose gel column. The composition of the fractions obtained by these two methods is very similar, consisting of the following:

1. A group of mannoproteins form the majority (80%) of all exocellular polysaccharides,

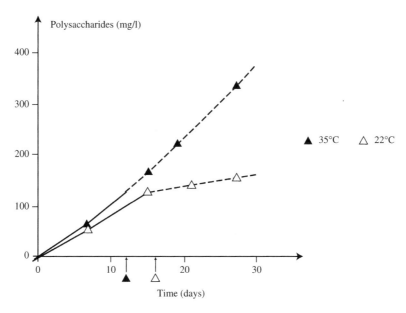

Fig. 3.23. Influence of temperature on the production of exocellular polysaccharides. Solid lines (———) indicate the length of alcoholic fermentation. (Llaubéres *et al.*, 1987)

containing approximately 90% mannose and 10% protein. Shortly after fermentation, this fraction contains small quantities of glucose that disappear after a few months of aging on the lees. Yeast exocellular mannoproteins have a wide range of molecular weights (from 100 000 to over 2 million). The average molecular weight is estimated at 250 000.

2. A much smaller quantity of a glucomannoprotein complexes (20% of total exocellular polysaccharides) containing 25% glucose, 25% mannose and 50% protein. The molecular weights of glucomannoproteins are lower (20 000–90 000) than those of mannoproteins.

Table 3.5. Impact of agitation on exocellular polysaccharide production (Llaubëres, 1988)

	TSP content[a] (mg/l)	
	Agitated medium	Non-agitated medium
First month	200	117
Third month	250	123

[a]TSP: total soluble polysaccharides

The general molecular structure of yeast exocellular mannoproteins is similar to that of mannoproteins in the cell wall (Volume 1, Section 1.2) (Villetaz *et al.*, 1980; Llaubéres *et al.*, 1987; Llaubéres, 1988). It consists of a peptide chain connected to short side chains made up of four mannose units and a high molecular weight, branched α-D-mannane (Figure 3.24).

These oligosaccharides have α-(1,2) and α-(1,3) oside bonds. They are linked to the serine and threonine residues of the peptide by O-glycoside bonds. These may be broken by weak alkaline hydrolysis (β-elimination), releasing mannose, mannobiose, mannotriose and mannotetrose.

Most of the mannoproteins released during fermentation are excreted by the yeast as unused cell-wall material. Yeast polysaccharides are also released due to yeast parietal enzyme activity: endo-β-(1,3)- and endo-β-(1,6)-glucanases (Volume 1, Section 1.2). These enzymes are very active during alcoholic fermentation and, at a lower level, for several months after the death of the yeast cells (Llaubéres, 1988). The resulting parietal autolysis (Charpentier *et al.*, 1986) leads to the release of mannoproteins fixed on the parietal glucane

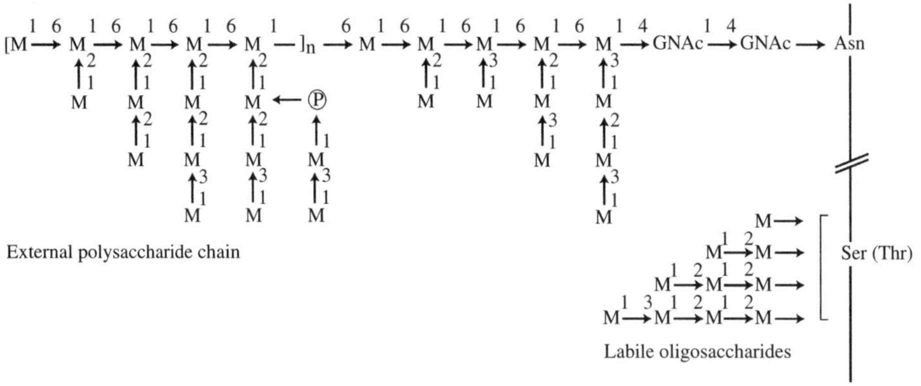

Fig. 3.24. Model of the molecular structure of exocellular mannoproteins produced by yeast. M, mannose; Asn, asparagine; Ser, serine; Thr, threonine; GNAc, *N*-acetyl-glucosamine; Ⓟ, phosphate (Llaubéres, 1988).

and partial hydrolysis of the glucomannoprotein fraction. These glucanase activities also explain the decrease in the proportion of glucose in yeast polysaccharides isolated from fermented media stored on the biomass.

Similarly to pectic substances, the direct organoleptic role of mannoproteins on the impression of body and softness in wine is certainly negligible. Mannoproteins may, however, have an indirect effect on astringency when they combine with phenolic compounds from grapes or oak (Section 13.7.3 and Volume 1, Section 13.8.2).

Yeast polysaccharides also cause filtration problems (Wucherpfennig *et al.*, 1984). The fouling effect of these colloids during membrane filtration is most severe in young wines separated prematurely from their lees. This is due to the presence of β-glucane fractions associated with mannoproteins. Filtrability of white wines aged on their lees improves rapidly during aging due to hydrolysis of the yeast glucanes by glucanases in the lees. A commercial β-glucanase preparation (Glucanex) may be used to achieve the same result (Llaubéres, 1988) (Section 11.5.2).

The most important (and most recently demonstrated) role of yeast mannoproteins is their stabilizing effect on protein precipitation in white wine (Section 5.6.4) and tartrate crystallization (in both red and white wines) (Section 1.7.7). The most 'protective' mannoproteins are extracted from yeast cell walls during aging on the lees.

They may also be prepared commercially by a patented process (Dubourdieu and Moine-Ledoux, 1994) for digesting yeast cells with a β-glucanase (Glucanex) preparation.

3.7.2 Polysaccharides from *Botrytis Cinerea*

It has been known for many years that wines made from botrytized grapes are difficult to clarify. Laborde called the colloid responsible for these problems 'dextrane' (1907), and it was long confused in winemaking with bacterial dextrane from *Leuconostoc dextranicum*, an α-(1,6)-glucane. In fact, *Botrytis cinerea* produces a specific β-glucane. This glucane's molecular structure, development in grapes and technological properties, as well as its breakdown by exogenous enzymes, have been studied in detail (Dubourdieu, 1978, 1982; Dubourdieu *et al.*, 1981b, 1981c, 1985).

When *Botrytis cinerea* is cultivated in a liquid model medium, it forms two categories of polysaccharides from sugars. These may be separated by fractionated precipitation with ethanol into a high molecular weight glucane (10^5–10^6) and a heteropolysaccharide complex with a molecular weight between 10^5 and 5×10^5 containing mainly mannose (60%), but also rhamnose (5%), galactose (30%) and glucose (5%). A small protein fraction (a few %) is also associated with this heteropolysaccharide complex.

The glucane precipitates in a characteristic filament form in the presence of ethanol (0.5 volume per volume of must or wine). Concentrations in must or wine may reach several hundreds of mg/l, but a few mg/l are enough to cause serious clarification problems during fining or filtration (Section 11.5.2). The fouling effect of glucane is aggravated by the presence of ethanol, as it promotes the formation of hydrogen bonds between the chains, creating a three-dimensional lattice structure. Ultrasonic treatment or vigorous mechanical agitation (ultra dispersion) breaks these bonds and reduces the fouling capacity of glucane.

The heteropolysaccharides produced by *Botrytis cinerea* precipitate when larger amounts of alcohol are added (4 volumes). They foul filter layers much less than glucane. Unfortunately, they have an inhibiting effect on the metabolism of *S. cerevisiae* (Volume 1, Section 2.3.4). On the one hand, they slow down alcoholic fermentation, while, on the other hand, they increase the formation of acetic acid and glycerol by the yeast. These heteropolysaccharides are partially or totally identified with the substance known in the 1950s as 'botryticine' (Ribéreau-Gayon *et al.*, 1952). These polysaccharides also play a biological role in plant cells. They cause symptoms of phytotoxycity and are elicitors of phytoalexines, natural fungicides produced by plants in response to fungal attacks (Kamoen *et al.*, 1980).

A comparison of the total polysaccharides in juice from healthy grapes and those affected by rot shows that the latter contain glucane but no homogalacturonan, due to the intense pectinase activity of the fungus (Table 3.6).

The percent composition of pectic substances in healthy grape must (after elimination of the homogalacturonans) is different from that of polysaccharides in grape must affected by rot (excepting glucane). The latter has a lower galactose content and a much higher mannose content. This is due to the fact that *Botrytis cinerea* in grapes produces heteropolysaccharides with a high mannose content (Table 3.7).

Botrytis cinerea produces a β-(1,3:1,6)-type glucane, sometimes known as cinereane. This molecule consists of a main chain of glucose units, bonded on β-(1,3), with branches made up of individual glucose units bonded on β-(1,6). Two units out of five in the main chain are thus substituted (Figure 3.25). This compound is similar to sclerotane, an exocellular glucane produced by the fungus *Sclerotium rolfsii*. Its structure is different, however, from that of mycolaminaranes, glucanes characteristic of *Oomycetes* fungi.

Glucane provides an exocellular reserve for *Botrytis cinerea*. Its mobilization requires a β-1,3-glucanase (Dubourdieu and Ribéreau-Gayon, 1980b), and the synthesis of this substance by the fungus is subjected to catabolic repression by

Table 3.6. Polysaccharide content of juice from healthy grapes and grapes affected by *Botrytis cinerea* (mg/l) (Dubourdieu, 1978)

	Glucane	Homogalacturonan	Other polysaccharides
Juice from healthy grapes	0	670	340
Juice from grapes affected by rot	387	0	627

Table 3.7. Polysaccharide monomer composition (percentage) of juice from healthy grapes and grapes affected by rot, after elimination of the *Botrytis cinerea* glucane and pectin (cf. Table 3.6, other polysaccharides) (Dubourdieu, 1982)

	Ara	Rha	Gal	Man	Glc	GalA
Polysaccharides in juice from healthy grapes	24.1	4.0	52.1	2.3	3.8	13.4
Polysaccharides in juice from grapes affected by rot	17.4	9.4	36.0	18.0	5.7	13.2

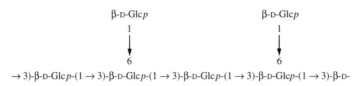

Fig. 3.25. Molecular structure of the exocellular β-glucane produced by *Botrytis cinerea* (Dubourdieu *et al.*, 1981b). β-D-Glc*p*: β-D-glucopyranose

glucose. This enzyme is, therefore, absent from grape musts affected by rot.

A β-(1,3)-glucanase that affects the *Botrytis cinerea* glucane has been isolated from a commercial enzyme preparation (Glucanex) made from *Trichoderma* sp. This fungus is a natural antagonist to *Botrytis* in soil. Glucanex is authorized for use in winemaking (Section 11.5.2). The active fraction of the preparation is an *exo-β*-1,3-glucanase that hydrolyzes the glucane from its non-reducing end, producing glucose and gentiobiose (a disaccharide where both glucose units are bonded on β-1,6), as well as a β-glucosidase that hydrolyzes the gentiobiose into glucose. Glucanex contains other glucanases (*endo-β*-1,3- and *exo-β*-1,6-) as well as an acid protease (Dulau, 1990) which have no effect on the *Botrytis cinerea* glucane. These enzymes are involved in digesting yeast cell walls. This property is used in the industrial process for producing mannoproteins to inhibit protein and tartrate precipitation. They are also involved in hydrolyzing yeast glucanes, which foul filter membranes (Section 13.7.1).

Glucane is located in the sub-epidermic cambium of grape skins affected by rot. Mechanical handling (crushing, pumping, etc.) which seriously damages the skins of grapes affected by rot promotes the dispersion of glucane in the must. The resulting wines are difficult to clarify (Section 11.5.2 and Volume 1, Section 14.2.4). Careful pressing can minimize the glucane concentrations in must made from grapes with noble rot and, consequently, the difficulty in clarifying these sweet wines. When wines still contain glucane in spite of these precautions, a β-glucanase (Glucanex) preparation may be used to ensure that the wine can be filtered under normal conditions. The enzyme preparation should, preferably, be added at the end of alcoholic fermentation, at a temperature above 10°C. It may be eliminated by standard bentonite treatment as soon as the glucane is hydrolyzed.

3.7.3 Polysaccharides in 'Graisse'

The problem of 'graisse' or 'ropy wine', which gives affected wines a viscous, oily consistency, was described by Pasteur. Laborde (1907) postulated that this spoilage was caused by anaerobic micrococci, producing a mucilage (dextrane) that he assimilated to the substance produced by *Botrytis cinerea*. This problem may apparently have diverse bacterial causes, and two types of polysaccharides are involved in producing the oily consistency: heteropolysaccharides and β-glucane.

According to Lúthy (1957) and Buchi and Deuel (1954), this phenomenon may be caused by certain streptococci, capable of breaking down malic acid into lactic acid. At the same time, they produce extremely viscous exocellular heteropolysaccharides, containing galactose, mannose, arabinose and galacturonic acid.

Certain lactic bacteria in the genus *Pediococcus* may also metabolize traces of glucose to produce a viscous polysaccharide responsible for 'graisse' (Volume 1, Section 5.4.4).

This polysaccharide is a glucose polymer, with a structure similar to that of the *Botrytis cinerea* glucane, but different from the bacterial dextrane produced by *Leuconostoc dextranicum*. It is a β-(1,3:1,2)-glucane consisting of a main chain of glucose units, bonded on β-1.3, with lateral branches of individual glucose units bonded on β-1,2 (Llaubéres *et al.*, 1990) (Figure 3.26). The type of branches on the glucane responsible for 'graisse' make it resistant to the β-glucanases which are effective in treating the *Botrytis cinerea*

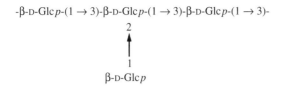

Fig. 3.26. Molecular structure of the exocellular β-(1,3: 1,2)-glucane produced by *Pediococcus* sp responsible for the 'ropy' texture of wines affected by 'graisse' (Llaubéres *et al.*, 1990). β-D-glc$p$: β-D-glucopyranose

glucane. Only a few mg/l of *Pediococcus* glucane are required to produce this oiliness in wine.

The remedy for this spoilage, if it occurs during barrel aging, has been known for many years. It consists of racking the affected wine, agitating it vigorously with a whisk or a powerful agitator and then carrying out earth filtration with average porosity to remove the colloid. The final stage consists of sterilizing plate filtration to eliminate the bacteria.

REFERENCES

Albersheim P., An J., Freshourg G., Fuller M.S., Guillern R., Ham K.S., Hahn M.G., Huang J., O'Neill M.A., Whitcombe A., Williams M.V., York W.S. and Darvill A.G. (1994) *Biochem. Soc. Trans.*, 22, 374.

Aspinall G.O. (1980) In *The Biochemistry of Cell-Wall Polysaccharide* (ed. J. Preiss). Academic Press, London, p. 473.

Barbe J.C., de Revel G., Perello M.C., Lonvaud-Funel A. and Bertrand A. (2001) *Rev. Fr. Oenol.*, 190, 16.

Belleville M.-P., Brillouet J.-M., Tarodo de la Fuente B. and Moutounet M. (1990) *J. Food Sci.*, 55, 1598.

Brillouet J.-M. (1987) *Biochimie*, 69, 713.

Brillouet J.-M., Saulnier L. and Moutounet M. (1989) *Bull. OIV*, 62, 339.

Brillouet J.-M., Saulnier L. and Moutounet M. (1990) *Rev. Fr. OEnol.*, 122, 43.

Buchi W. and Deuel H. (1954) *Helv. Chemica Acta*, 37, 1392.

Castino M. and Delfini C. (1984) *Vini d'Italia*, 26, 45.

Charpentier C., Nguyen Van Long T., Bonaly R. and Feuillat M. (1986) *Appl. Microbiol. Biotechnol.*, 24, 405.

Cordonnier R. and Bayanove C. (1980) *Conn. Vigne et Vin*, 15 (4), 269.

Crouzet J. (1986) *Rev. Fr. OEnologie*, 102, 42.

Doco T. and Brillouet J.-M. (1993) *Carbohydr. Res.*, 243, 333.

Doco T., Williams P., Vidal S. and Pellerin P. (1997) *Carbohydr. Res.*, 297, 89.

Dubourdieu D. (1978) etude des polysaccharides sécrétés par Botrytis cinerea dans la baie de raisin. Incidence sur les difficultés de clarification des vins de vendanges pourries. Thése Docteur-Ingénieur, Inversité de Bordeaux II.

Dubourdieu D. (1982) Recherches sur les polysaccharides sécrétés par Botrytis cinerea dans la baie de raisin. Thése Doctorat d'Etat ès Sciences, Université de Bordeaux II.

Dubourdieu D., Desplanques C., Villetaz J.C. and Ribéreau-Gayon P. (1985) *Carbohydr. Res.*, 144, 277–287.

Dubourdieu D., Hadjinicolau D. and Ribéreau-Gayon P. (1981a). *Conn. Vigne Vin*, 15 (1), 29.

Dubourdieu D. and Moine-Ledoux V. (1994) Brevet d'invention 2 726 284.

Dubourdieu D. and Ribéreau-Gayon P. (1980a) *Conn. Vigne Vin*, 14 (1), 29.

Dubourdieu D. and Ribéreau-Gayon P. (1980b) *C.R. Acad. Sci. (Paris)*, 290D, 25.

Dubourdieu D., Ribéreau-Gayon P. and Fournet B. (1981b) *Carbohydr. Res.*, 93, 295.

Dubourdieu D., Villetaz J.-C., Desplanques C. and Ribéreau-Gayon P. (1981c) *Conn. Vigne Vin*, 3, 161.

Dulau L. (1990) Recherche sur les protéines responsables de la casse protéique des vins blancs. Thése de Doctorat, Université de Bordeaux II.

Feuillat M. and Bernard P. (1987) *Bull. OIV*, 60, 227.

Fincher G.B., Stone B.A. and Clarke A.E. (1983) *Ann. Rev. Plant Physiol.*, 34, 47.

Gerbaud V., Gabas N., Blouin J., Pellerin P. and Moutounet M. (1997) *J. Int. Sci. Vigne et Vin*, 31 (2), 65.

Ishii T. and Matsunaga T. (1996) *Carbohyd. Res.*, 284, 1–9.

Jackson R.S. (1994) *Wine Science—Principles and Applications*. Academic Press, San Diego.

Kamoen O., Jamart G., Declercq H. and Dubourdieu D. (1980) *Ann. Phytopathol.*, 12, 4.

Kobayashi M., Matoh T. and Azuma J.L. (1996) *Plant. Physiol.*, 110, 1017.

Laborde J. (1907) *Cours d'OEnologie*. Malo, Paris.

Lafon-Lafourcade S., Larue F. and Ribéreau-Gayon P. (1979) *Appl. Environ. Microbiol.*, 38 (6), 1069.

Llaubéres R.M. (1988) Les polysaccharides sécrétés dans les vins par S. cerevisiae et Pediococcus sp. Thése Doctorat, Université de Bordeaux II.

Llaubéres R.M., Dubourdieu D. and Villetaz J.-C. (1987) *J. Sci. Food. Agric.*, 41, 277.

Llaubéres R.M., Lonvaud-Funel A., Dubourdieu D. and Fournet B. (1990) *Carbohyd. Res.*, 203, 103–107.

Lüthy H. (1957) *Am. J. Enol. Viticult.*, 8 (4), 176.

Mourgues J. (1981) *Sciences des Aliments*, 1 (3), 377.

Muntz A. and Lainé E. (1906) *Ann. Brass. Diot.*, 9, 157.

O'Neill M.A., Albersheim P. and Darvill G. (1990) *The Pectic Polysaccharides of Primary Cell Walls*.

Methods in Plant Biochemistry, Vol. 2 (ed. P.M. Dey). Academic Press, London.

O'Neill M.A., Warenfeltz D., Kates K., Pellerin P., Doco T., Darvill A. and Albersheim P. (1996) *J. Biol. Chem.*, 271, 2923.

Pellerin P., Waters E., Brillouet J.-M. and Moutounet M. (1994) *J. Int. Sci. Vigne Vin*, 28 (3), 213.

Pellerin P., Vidal S. Williams P. and Brillouet J.-M. (1995) *Carbohyd. Res.*, 135–143.

Pellerin P., Doco T., Vidal S., Williams P., Brillouet J.-M. and O'Neill M.A. (1996) *Carbohydr. Res.*, 290, 183.

Pellerin P., O'Neill, Pierre C., Cabanis M.T., Darvill A.G., Albersheim P. and Moutounet M. (1997) *J. Int. Sci. Vigne Vin*, 31 (1), 33–41.

Peynaud E. (1952) *Ann. Fals. Fraudes*, 45, 11.

Ribéreau-Gayon J., Peynaud E. and Lafourcade S. (1952) *C. R. Acad. Sci. (Paris)*, 234, 423.

Ribéreau-Gayon J., Peynaud E., Sudraud P. and Ribéreau-Gayon P. (1982) *Sciences et Techniques du vin*, Vol. I: *Analyse et Contrôle des vins*, 2nd edn, Dunod, Paris.

Saulnier L. and Brillouet J.-M. (1989) *Carbohydr. Res.*, 188, 137.

Saulnier L. and Thibault J.F. (1987) *Carbohydr. Polym.*, 7, 329.

Saulnier L, Brillouet J.-M. and Joseleau J.-P. (1988) *Carbohydr. Res.*, 182 (1), 63–78.

Serrano M., Vannier A.-C. and Ribéreau-Gayon P. (1988) *Conn. Vigne Vin*, 22, 49.

Usseglio-Tomasset L. (1961) *Riv. Viticult. Enol.*, 14, 63.

Usseglio-Tomasset L. (1976) *Conn. Vigne Vin*, 10 (2), 193.

Usseglio-Tomasset L. (1978) *Ann. Technol. Agric.*, 27, 261.

Vidal S., Williams P., O'Neill M.A. and Pellerin P. (2001) *Carbohydr. Polym.* 45, 315.

Villetaz J.-C., Amado R., Neukom H. Horisberger M. and Horman I. (1980) *Carbohydr. Res.*, 81, 341.

Villetaz J.-C., Amado R. and Neukom H. (1981) *Carbohydr. Polym.*, 1, 101.

Waters E., Pellerin P. and Brillouet J.-M. (1994) *Biosci. Biotech. Biochem.*, 58, 43.

Wucherpfennig K., Dietrich H. and Schmitt H. (1984) *Z. Lebensm. Unters. Forsch.*, 179, 119.

4

Dry Extract and Minerals

4.1 Introduction	91
4.2 Dry extract	92
4.3 Ash	93
4.4 Inorganic anions	94
4.5 Inorganic cations	95
4.6 Iron and the ferric casse mechanism	96
4.7 Copper and copper casse	102
4.8 Heavy metals	104

4.1 INTRODUCTION

Total dry extract, or dry matter, consists of all substances that are non-volatile under physical conditions which deliberately avoid altering their nature. Dry extract in wine includes non-volatile organic substances and mineral compounds. During combustion of the extract, the organic compounds are converted into CO_2 and H_2O. The inorganic substances produce carbonates and inorganic anionic salts that form ash.

Organic acids in grapes are partly neutralized by potassium and calcium ions, forming salts. This salification of tartaric and malic acids gives must and wine an acidobasic buffer capacity. Besides these two most prevalent cations, minerals such as iron, copper, magnesium and manganese are also essential for cell metabolism, as they have an aprotic cofactor function in the activity of certain enzymes, e.g. oxidoreductases and kinases.

All of the inorganic cations are naturally present in must and wine at non-toxic concentrations. However, certain metals, such as lead, zinc, tin and mercury, may occur in higher concentrations as a result of the social and economic environment and/or vineyard cultivation methods. Fertilizers and pesticides are responsible for increases in anion levels, e.g. phosphates and nitrates, as well as cations, such as copper, zinc, manganese, etc. Furthermore, due to their acidity, must and wine are capable of dissolving certain metals, such as copper, nickel, zinc and even chrome, from winemaking equipment containing alloys, e.g. bronze in pumps, taps, hose connections, etc.

Lastly, certain techniques, such as prolonged skin contact before fermentation in white wines and prolonged vatting in reds, promote the dissolving of mineral and organic salts, mainly present in the solid parts of grapes (stalks, skins, seeds, and cell walls). Some techniques for stabilizing wine and preventing crystal turbidity have the opposite effect, causing major reductions in potassium, or even calcium content, and consequently in the amount of dry extract and ash.

4.2 DRY EXTRACT

A simple medium containing no organic matter, e.g. mineral water, has approximately 4770 mg/l of dry extract at 180°C. This consists only of anions (bicarbonates, chlorides, sulfates, fluorides, etc.) and cations (sodium, potassium, calcium, magnesium, etc.). The dry extract from water therefore corresponds to ash.

In an organic medium such as wine, the total dry extract does not consist only of inorganic substances. Even a dry wine, with no fermentable sugars, has considerably more dry extract than mineral water. Furthermore, it is only to be expected that the weight of the total dry extract should vary according to winemaking techniques. Crushing, destemming, skin contact and long vatting promote the extraction of mineral and organic substances from grapes, so they are bound to have an impact on the dry matter content.

The weight of total dry matter also depends on conditions during concentration of the wine and evaporation of the liquid phase. It is important, of course, not to eliminate certain volatile compounds. It is also necessary to ensure that other compounds in wine should not be subject to chemical transformations during the elimination of water and alcohol, e.g. oxidation, decarboxylation and, even more importantly, the breakdown of volatile compounds. The ratio of alcohol content to the weight of extract is used to detect certain fraudulent practices such as adding alcohol or sugar.

Three types of extract may be distinguished in wine:

1. Total dry extract (or total dry matter), measured under carefully controlled conditions by spreading the wine on a spiral of blotting paper until it is completely absorbed and evaporating it in an oven at 70°C, under a partial pressure of 20–25 mm of mercury. This corresponds to a dry air current of 40 l/h. Total dry extract is expressed in g/l, and results must be determined to within 0.5 g/l. Red wines generally contain approximately 25–30 g/l dry extract. Dry white wines contain less than 25 g/l. In sweet wines, this value depends on the sugar content.

2. The non-reducing extract is obtained by subtracting total sugar content from total dry extract.

3. The reduced extract consists of the total dry extract minus total sugars in excess of 1 g/l, potassium sulfate in excess of 1 g/l, as well as any mannitol and any chemicals added to the wine.

Besides these three types of extract, there is also the 'remaining extract'. This consists of the non-reducing extract minus fixed acidity, expressed in grams of tartaric acid per liter.

The weighing method takes a long time, but is more accurate than the reference method. In the reference method, the total dry extract is calculated from the relative density d_r, measured by areometry. Relative density is determined by measuring the density (l_v) of a wine at 20°C and that of a dilute alcohol mixture with the same alcohol content (l_a). The equation for relative density is as follows: $d_r = 1.0018(l_v - l_a) + 1.000$.

Table 4.1 is used to calculate the total dry extract from the first three decimal places of the relative density. Another table is used to take the fourth decimal place into account.

Dry Extract and Minerals

Table 4.1. Calculating total dry extract from relative density (in order to determine the total dry extract content in g/l)

Density to 2 decimal places	Third decimal place									
	0	1	2	3	4	5	6	7	8	9
	Grams of dry extract per liter									
1.00	0	2.6	5.1	7.7	10.3	12.9	15.4	18.0	20.6	23.2
1.01	25.8	28.4	31.0	33.6	36.2	38.8	41.3	43.9	46.5	49.1
1.02	51.7	54.3	56.9	59.5	62.1	64.7	67.3	69.9	72.5	75.1
1.03	77.7	80.3	82.9	85.5	88.1	90.7	93.3	95.9	98.5	101.1
1.04	103.7	106.3	109.0	111.6	114.2	116.8	119.4	122.0	124.6	127.2
1.05	129.8	132.4	135.0	137.6	140.3	142.9	145.5	148.1	150.7	153.3
1.06	155.9	158.5	161.2	163.8	166.4	169.0	171.6	174.3	176.9	179.5
1.07	182.1	184.8	187.4	190.0	192.6	195.2	197.8	200.5	203.1	205.8
1.08	208.4	211.0	213.6	216.2	218.9	221.5	224.1	226.8	229.4	232.0
1.09	234.7	237.3	239.9	242.5	245.2	247.8	250.4	253.1	255.7	258.4
1.10	261.0	263.6	266.3	268.9	271.5	274.2	276.8	279.5	282.1	284.8
1.11	287.4	290.0	292.7	295.3	298.0	300.6	303.3	305.9	308.6	311.2
1.12	313.9	316.5	319.2	321.8	324.5	327.1	329.8	332.4	335.1	337.8
1.13	340.4	343.0	345.7	348.3	351.0	353.7	356.3	359.0	361.6	364.3
1.14	366.9	369.6	372.3	375.0	377.6	380.3	382.9	385.6	388.3	390.9
1.15	393.6	396.2	398.9	401.6	404.3	406.9	409.6	412.3	415.0	417.6
1.16	420.3	423.0	425.7	428.3	431.0	433.7	436.4	439.0	441.7	444.4
1.17	447.1	449.8	452.4	455.2	457.8	460.5	463.2	465.9	468.6	471.3
1.18	473.9	476.6	479.3	482.0	484.7	487.4	490.1	492.8	495.5	498.2
1.19	500.9	503.5	506.2	508.9	511.6	514.3	517.0	519.7	572.4	525.1
1.20	527.8	"	"	"	"	"	"	"	"	"
1.21	555.0	557.7	560.4	563.1	565.8	568.5	571.2	573.9	576.6	579.3
1.22	582.0	584.8	587.5	590.2	593.0	595.7	598.4	601.1	603.9	606.6
1.23	609.3	612.1	614.8	617.5	620.3	623.0	625.7	628.4	631.2	633.9
1.24	636.6	639.9	642.1	644.9	647.6	650.3	653.1	655.8	658.6	661.3
1.25	664.0	666.8	669.5	672.3	675.0	677.7	680.5	683.2	686.0	688.7
1.26	691.4	694.2	697.0	699.8	702.5	705.3	708.8	710.8	713.6	716.4
1.27	719.1	721.9	724.7	727.4	730.2	732.9	735.7	738.5	741.2	744.0
1.28	746.7	749.5	752.3	755.1	757.8	760.6	763.4	766.1	768.9	771.7
1.29	774.4	777.2	780.0	782.8	785.6	788.3	791.1	793.9	796.7	799.5
1.30	802.3	805.0	807.8	810.6	813.4	816.2	819.0	821.8	824.6	827.4
1.31	830.2	833.1	835.9	838.7	841.5	844.3	847.1	849.9	852.7	855.5
1.32	858.3	861.2	864.0	866.8	869.6	872.4	875.3	878.1	880.9	883.7
1.33	886.5	889.4	892.2	895.0	897.9	900.7	903.5	906.4	909.2	912.0
1.34	914.8	917.7	920.5	923.3	926.2	929.0	931.8	934.7	937.5	940.3
1.35	943.1	—	—	—	—	—	—	—	—	—

4.3 ASH

4.3.1 Preparing Ash

There is more organic than inorganic matter in dry extract from wine. The latter is represented by ash, which contains all the products resulting from burning the evaporation residue of wine in such a way as to obtain all the cations in the form of carbonates and other anhydrous mineral salts. The ammonium cation is capable of sublimating in certain salts. This is why it is excluded from the cations.

The weight of ash obtained from normal wines varies from 1.5 to 3 g/l. There is a fairly constant weight ratio of about one to ten between ash and reduced extract. Inorganic substances are mainly located in grape solids: skins, seeds and cellulose–pectin in the flesh cell walls. In red wine, some of these compounds are dissolved, increasing the concentration. Others are at least partially eliminated during fermentation due to the formation of insoluble salts. Treatment and handling operations involving contact with certain materials may increase the concentrations of specific substances.

4.3.2 Assaying the Alkalinity of Ash

Ash includes all the cations, other than ammonium, that combine with organic acids in wine. Normally, ash is white or grayish. Green ash that turns red in an acid medium indicates a high manganese concentration. Yellow is a sign that there is a high iron content.

Ash is alkaline, as it is obtained by carbonizing dry extract at 525°C ± 25°C with continuous aeration. Under these conditions, all the salified organic acids are converted into salts, mainly potassium carbonate, but also calcium and magnesium carbonate, etc. The strong inorganic acids present in a salified state in wine (HCl, H_2SO_4, HNO_3 and H_3PO_4) are unchanged in ash.

The alkalinity of ash is measured by titration, after adding a known excess of sulfuric acid and heating. The sulfuric acid that has not reacted is titrated with 0.1 M sodium hydroxide, using methyl orange as a neutralization indicator. Alkalinity is therefore expressed in milliequivalents per liter, i.e. millimoles of OH^- ions per liter and determined to within 0.5 meq/l.

One reason for measuring the alkalinity of ash (C) and total acidity (T) is to obtain an initial approximation (Section 1.4.4) of the quantity of milliequivalents of acid or alkali necessary (ΔB) if acidification or deacidification is required. The following formula is used to calculate the desired variation in $\Delta |pH|$:

$$\frac{\Delta |B|}{\Delta |pH|} = 2.303 \frac{TC}{T+C}$$

The alkalinity of ash is also used to establish the acidimetric balance. When the NH_4^+ concentration is added, this gives the total quantity of salified organic acids. Consequently, the sum of alkalinity of ash plus NH_4^+ plus total acidity (expressed in meq/l) gives the total organic anions (also expressed in meq/l), according to the following equation:

[Alkalinity of ash] + [NH_4^+] + [total acidity]

= Σ[organic anions]

It is necessary to correct for sulfur dioxide and phosphoric acid, as one of its functions is titrated in the total acidity.

4.4 INORGANIC ANIONS

The main inorganic anions in must and wine correspond to the presence of more-or-less soluble salts. A good assessment of the total inorganic anions may be obtained from the total cations (Section 4.5), which also represent the total anions. When the alkalinity of the ash, representing the organic anions, is subtracted, the quantity of inorganic anions is obtained.

All nitrates are soluble. However, they are only present in trace amounts in wine. It is also possible to predict the presence of chlorides, as only lead, silver and mercury chlorides are insoluble. In most wines, the chloride concentration is below 50 mg/l, expressed in sodium chloride. It may exceed 1 g/l in wine made from grapes grown by the sea. Sodium chloride is sometimes added during fining, especially when egg whites are used.

Phosphoric and sulfuric anions are also present. According to Ribéreau-Gayon et al. (1982), natural wine, i.e. made from grapes, contains only low concentrations of sulfates, between 100 and 400 mg/l expressed in K_2SO_4. Concentrations gradually increase during aging due to repeated sulfuring and oxidation of the SO_2. In heavily sulfured sweet wines, concentrations may exceed 2 g/l after a few years of barrel aging.

Phosphorus is naturally present in wine in both organic and inorganic forms. Ferric casse in white wine, known as 'white casse', is caused by ferric phosphate. White wines contain 70–500 mg/l of phosphate, whereas concentrations in red wines range from 150 mg/l to 1 g/l. These wide variations are related to the addition of diammonium phosphate to must to facilitate alcoholic fermentation. In view of the negative role of phosphoric ions in ferric casse, it is preferable to use ammonium sulfate.

Other inorganic substances present in trace amounts in wine include bromides, iodides and fluorides, as well as silicic and boric acids.

4.5 INORGANIC CATIONS

Cations play a major role in winemaking. However, they must be monitored in view of the risk of turbidity, especially bitartrate with potassium, tartrate with calcium, ferric with trivalent iron, cuprous with the copper cation with a degree of oxidation of one. In sparkling wines, alkaline earth cations, especially magnesium, may have an effect on effervescence (Maujean *et al.*, 1988).

Potassium is the dominant cation in wine, as it is in all plants. Concentrations are between 0.5 and 2 g/l, with an average of 1 g/l. Wines made from grapes concentrated by noble rot have the highest potassium content. Red wines contain more than dry whites due to the capacity of phenols to inhibit the precipitation of potassium bitartrate.

The calcium cation produces many relatively insoluble salts. The most insoluble is calcium oxalate. Oxalic acid is used to demonstrate the presence of calcium in a liquid as it causes turbidity and precipitation. Calcium tartrate is also relatively insoluble, especially in the presence of ethanol (Section 1.6.5). In the same way, calcium gluconate and mucate, present in wine made from botrytized grapes, are reputed to be responsible for crystalline turbidity (Section 1.2.2). Calcium concentrations in white wines are between 80 and 140 mg/l, while they are slightly lower in red wines. The calcium content may increase following deacidification with calcium carbonate. As calcium is divalent, it is more energetically involved than potassium in colloid flocculation and precipitation, e.g. ferric phosphate, tannin-gelatin complexes, etc.

Although sodium is the most widely represented cation in the universe, only small quantities are present in wine. Concentrations range from 10 to 40 mg/l, although higher values may be found in wines treated with sodium bisulfite or insufficiently purified bentonites. As in the case of chloride, wines produced near the sea have a higher sodium content.

Wine contains more magnesium (60–150 mg/l) than calcium, and concentrations do not decrease during fermentation and aging, as all magnesium salts are soluble. Small quantities of manganese are present in all wines (1–3 mg/l). The concentration depends on the manganese content of the vineyard soil. Winemaking techniques for red wines increase this concentration, as seeds contain three times as much manganese as skins and thirty times as much as grape flesh.

Iron and copper are present in small quantities, but they are significant causes of instability (ferric casse and copper casse), so they are described in separate sections (Sections 4.6 and 4.7). Heavy metals, mainly lead, even in trace amounts, also affect toxicity and deserve a separate description (Section 4.8).

Bonastre (1959) developed a technique for assaying the total cations in wine, using a strong cation sulfonic resin exchanger. In this assay, the cations in wine shifted the salification balance of an initially acid sulfonic resin completely to the right, i.e.:

$$R-SO_3H + cation^+ \rightleftharpoons R-SO_3-cation + H^+$$

As there was an excess of resin compared to the amount of cations in the wine, stoichiometric H^+ ions were released in exchange and assayed using sodium hydroxide. Wine contains the cations naturally present in grapes, with the addition of manganese and zinc, originating from vine sprays containing fungicides such as dithiocarbamates.

Finally, in view of its acidity, wine is likely to corrode metal winemaking equipment, thus dissolving some toxic cations and others responsible for metallic casse. For example, copper, nickel and even lead may be extracted from bronze equipment (pumps, vat taps, hose connections, etc.). This property may also be exploited when wine is racked, by using copper equipment to eliminate certain thiols and especially hydrogen sulfide, in the form of copper salts, which are particularly insoluble.

4.6 IRON AND THE FERRIC CASSE MECHANISM

4.6.1 Presence and State of Iron in Wine

Iron is very widespread in the earth's crust, representing a little over 5% of total mass. It is soluble in the form of ferrous and ferric chloride. Both forms occur in wine, maintaining an oxidation–reduction balance according to the electroactive redox system below:

$$Fe^{2+} \rightleftharpoons Fe^{3+} + e^-$$

The normal oxidation–reduction potential E_0 of this redox couple in relation to the hydrogen electrode is 771 mV. The oxidation–reduction potential of still wines, even when young, is often much lower, around 500 mV. This value explains why iron is present in both ferrous and ferric forms. If all the iron in wine were in ion form, the potential would be higher. It is obvious that much of the iron is involved in complexes, and is thus more difficult to identify.

The $(Fe^{3+})/(Fe^{2+})$ ratio in wine depends on storage conditions, especially the free sulfur dioxide concentration. For this reason, wine is more susceptible to ferric casse after aeration, as this increases the proportion of the Fe^{3+} form responsible for this phenomenon.

As regards the redox potential, it is quite predictable that iron in wine is not totally in ion form. Part of the iron is involved in soluble complexes with organic acids, especially citric acid. Ferric iron is much more likely to form complexes than ferrous iron. Ferric and ferrous iron, expressed as Fe^{III} and Fe^{II}, constitute total iron, in both ions and complexes, i.e. non-reactive forms. A total iron assay therefore requires the complete destruction of these complexes by acidification. The use of potassium thiocyanate, a specific reagent for ferric iron, in this assay also presupposes that all the iron must previously have been oxidized using hydrogen peroxide.

Wine always contains a few mg/l of iron. A small percentage comes from grapes (2 to 5 mg/l). The rest comes from soil on the grapes, metal winemaking, handling and transportation equipment, as well as improperly coated concrete vats. The general use of stainless steel has considerably reduced the risk of excess iron and, consequently, of ferric casse.

4.6.2 Ferric Casse Mechanism

Ferric casse occurs in white wines due to the formation of an unstable colloid resulting from a reaction between Fe^{3+} ions and phosphoric acid (white casse). This colloid then flocculates and precipitates, in a reaction involving proteins.

Ribéreau-Gayon *et al*. (1976) emphasized that, in phosphatoferric casse, the wine does not become turbid due to clusters of ferric phosphate molecules, which are rather small and remain in the colloidal state in a clear solution. Turbidity, i.e. white casse, occurs when proteins, positively charged at the pH of wine, 'neutralize' the negative charge of these phosphatoferric clusters, making them hydrophobic and therefore insoluble. Flocculation can only occur under these conditions.

Ferric iron may react with phenols in red wines, producing a soluble complex that leads to an increase in color intensity. This phenomenon may be significant in some young wines. The color develops a darker, more purplish hue. This complex later flocculates and precipitates. Iron generally reacts preferentially with phenols, present at high concentrations in red wine, due to the mass action law. White casse is therefore impossible, unless the wine has both high acidity and a high phosphoric acid content. A blackish tinge is frequently observed in new red wines exposed to air. This may be accompanied by precipitation containing varying quantities of iron.

The diagram in Figure 4.1 shows the various reactions involving iron in aerated wines, including the reaction with potassium ferrocyanide (Section 4.6.5). This diagram shows that the total

Dry Extract and Minerals

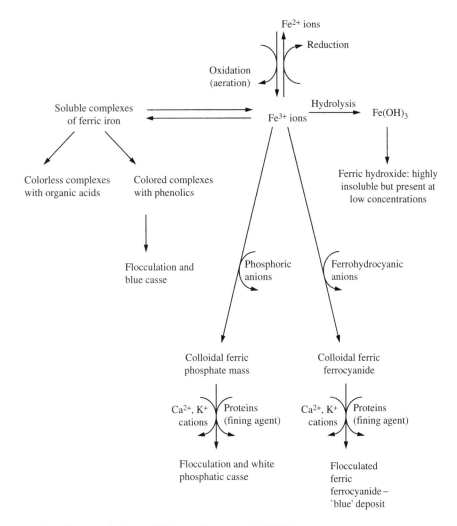

Fig. 4.1. Iron reactions in aerated wines (Ribéreau-Gayon *et al.*, 1976)

iron concentration alone is not sufficient to predict the risk of casse. Some wines become turbid with only 6–8 mg/l of iron, whereas others remain clear with concentrations of 25 mg/l.

Besides the iron concentration, a wine's oxidation–reduction state and the possible presence of oxidation catalysts are also involved. The quantity of soluble complexes with organic acids, i.e. iron that is not involved in the casse mechanism, is also significant. This factor is, however, impossible to measure.

Acidity has a complex effect on ferric casse, not only due to the quantity of acids but also the type. Figure 4.2 shows the proportion of the various forms of Fe^{III} in an initially reduced wine containing none of this form of iron. Samples were saturated with oxygen and adjusted to different pHs. Total Fe^{III} and the proportion of Fe^{III} in the form of soluble complexes both increased with pH. The Fe^{III} in the form of phosphatoferric colloids increased up to pH 3.3, which corresponds to the maximum risk of casse. The risk then decreased due to the insolubility of these complexes at higher pHs. It is therefore understandable that, as ferric casse does not occur above pH 3.5, it is unknown in certain winemaking regions.

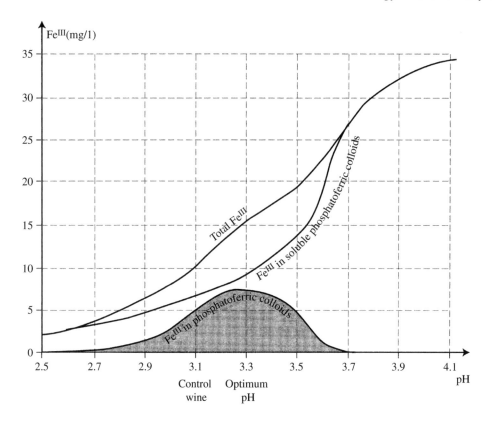

Fig. 4.2. The various forms of ferric iron after saturation with oxygen, in several initially reduced samples of the same wine, containing no Fe^{III} and adjusted to different pHs. (Ribéreau-Gayon, 1976)

These ferric colloids are less soluble at low temperatures, which tends to facilitate casse, especially in winter. For example, a wine may be aerated at 20°C without developing turbidity, while slight turbidity occurs at 15°C and serious ferric casse at 10°C.

It was easy to devise a test to assess the risk of ferric casse and the effectiveness of various treatments. A clear glass bottle, half filled with wine, is injected with oxygen. It is corked, agitated and placed, cork down, in a refrigerator in the dark. A wine highly susceptible to casse will become turbid within 48 hours. If the wine stays clear for a week, it will not suffer from ferric casse.

In view of the ferric casse mechanism (Figure 4.1), there are various treatment processes based on different principles:

1. Increasing the Fe^{III} in soluble complexes by adding citric acid. Other products (polyphosphates, sodium salts and ethylenediaminetetraacetic acid) are also effective, but not authorized.

2. Decreasing the ferric iron, by reducing it with ascorbic acid.

3. Stopping precipitation of the ferric colloid by adding gum arabic, which acts as a protective colloid.

4. Precipitating the iron through deliberate ferric casse caused by oxygenation. This process is too brutal and affects wine quality. It is no longer used.

5. Eliminating the excess iron using potassium ferrocyanide for white wines and calcium

phytate for red wines. Cation exchangers may fix the iron in exchange for Na^+ or Mg^{2+} ions (Section 12.4.3). Their effectiveness is debatable and they are not permitted in many countries.

4.6.3 Citric Acid and Gum Arabic Treatment

Citric acid solubilizes iron, forming soluble iron citrate. Citric acid is an authorized additive at doses up to 0.5 g/l. The total concentration must never exceed 1 g/l. This treatment may only be envisaged for wines that have been sufficiently sulfured to protect them from bacterial activity that would otherwise break down the citric acid, producing volatile acidity. In practice, this treatment is used exclusively for white wines that are not very susceptible to ferric casse (with no more than 15 mg/l of iron) and that will not be damaged by this acidification. Doses of 20–30 g/hl are usually sufficient.

Citric acid treatment to prevent ferric casse may be reinforced by adding gum arabic, which acts as a protective colloid (Section 9.4.3). This is especially effective in preventing the flocculation of colloidal ferric phosphate. The doses of gum arabic generally used are on the order of 5–20 g/hl. This additive is available in aqueous solutions, at concentrations ranging from 15–30%. Gum arabic must be used in perfectly clear wines that are ready for bottling. It not only stabilizes clarity, but also turbidity, and has a very high capacity for fouling filter surfaces.

4.6.4 Ascorbic Acid Treatment

Ascorbic acid has a strong reducing effect and acts instantaneously (Volume 1, Section 9.5.4). It is therefore capable of preventing iron in wine from oxidizing due to aeration (during racking, bottling, etc.) when it would normally be particularly susceptible to ferric casse. Ascorbic acid is unstable, however, and only provides temporary protection. It is most effective (10 g/hl) just before bottling, when a wine has shown a slight tendency towards ferric casse and there is not enough time to carry out any other treatment.

4.6.5 Potassium Ferrocyanide Treatment

Treating wines with potassium ferrocyanide or 'fining blue' (Figure 4.1) was recommended in Germany as long ago as 1923. In France, this treatment was authorized for white and rosé wines, including sparkling wines, in 1962. It is also permitted for *vins doux naturels*.

In France, winemakers who decide to use potassium ferrocyanide must declare their intention to the authorities at least eight days beforehand. Each treatment must be supervised by a qualified enologist. The enologist must carry out an analysis of each vat or barrel to be treated, including preliminary tests to determine the doses required. The winemaker is then issued with a purchasing slip for the appropriate amount of reagent. Establishments selling potassium ferrocyanide are obliged to keep a record of all purchases and sales and must also keep the purchasing forms issued by enologists. Winemakers must also keep two sets of records for inspection by the authorities, one indicating the quantities of ferrocyanide received and the other describing the conditions under which this substance was used.

Potassium ferrocyanide reacts with both ferrous (Fe^{2+}) and ferric (Fe^{3+}) ions, producing several insoluble salts of different colors. The ferrous iron salt is white, while ferric iron produces a blue precipitate (Prussian blue). Other metals are also precipitated, mainly copper and zinc as well as, to a lesser extent, lead and tin.

If all the iron is ferric, 5.65 mg of ferrocyanide is required to eliminate 1 mg of iron and the reaction is as follows:

$$3Fe(CN)_6^{4-} + 4Fe^{3+} \longrightarrow [Fe(CN)_6]_3Fe_4$$

The reaction involving ferrous iron is more complex and produces insoluble salts, $Fe(CN)_6FeK_2$ and $Fe(CN)_6Fe_2$. Between 3.78 and 7.56 mg of ferrocyanide are theoretically needed to eliminate 1 mg of iron. In practice, however, it is generally considered that between 6 and 9 mg of potassium ferrocyanide are required to eliminate 1 mg of iron from wine.

The hydroferrocyanic derivatives produced are in colloidal form. Their flocculation in wine is accelerated by adding a protein fining agent. Furthermore, precipitation of ferric ferrocyanide at least partially eliminates proteins (Vogt, 1931). This may be advantageous in white wines susceptible to protein turbidity. The precipitation of proteins is not due to the ferrocyanide itself, but rather to an insoluble ferric complex. Indeed, when potassium ferrocyanide is added to a wine containing no iron, no protein turbidity is observed.

It should also be taken into account that only Fe^{3+} ions react with ferrocyanide, and that most of the ferric iron Fe^{III} is combined in soluble complexes with organic acids. As the ferrocyanide reacts with the Fe^{3+} ions, the soluble complexes break down to reestablish an equilibrium, generating new Fe^{3+} ions that react in turn. This series of reactions may continue for several hours, or even days. Reaction time is longest at high pH, as larger quantities of soluble complexes are present (Figure 4.2).

This leads to two major consequences. On the one hand, there is a risk that the potassium ferrocyanide may break down in the wine, forming hydrocyanic acid. On the other hand, if the ferric ferrocyanide is eliminated by fining before the reaction has finished, any potassium ferrocyanide that has not yet reacted remains in solution. This may produce turbidity at a later stage and there is a risk of bluish deposits occurring in bottled wine.

The reaction with potassium ferrocyanide is not nearly as slow when the iron is in ferrous form, as less ferrous iron is combined in complexes than ferric iron. It is, therefore, clear that wine should be in a reduced state when it is treated with ferrocyanide. Prior treatment (24 hours before) with ascorbic acid (5–6 g/hl) considerably improves effectiveness.

The above considerations indicate not only how difficult it is to predict the quantity of ferrocyanide required to achieve stabilization, but also how hard it is to avoid the two associated risks: breakdown of ferrocyanide and potassium ferrocyanide that has not reacted remaining in solution in the wine.

The use of standard doses, for example 10 g/hl to treat ferric casse and 3 g/hl for copper casse, is certainly not recommended. It is also true that an assay of the wine's iron content is not sufficient to predict the quantity of iron that needs to be eliminated.

Preliminary trials must be carried out (Ribéreau-Gayon et al., 1977). These trials consist of determining, under laboratory conditions, the quantity of ferrocyanide that is immediately precipitated by the iron in the wine. Increasing quantities of potassium ferrocyanide, e.g. doses between 5 and 25 g/hl, are added to test samples. Once the solution has been homogenized, 1 ml of a fining agent solution, corresponding to 2 g/hl^{-1} of casein, is added. After a few minutes, the sample is filtered, or preferably centrifuged. Iron alumina is used to test the clear wine for excess ferrocyanide that has not reacted. The solution turns blue if potassium ferrocyanide is present. It is thus possible to measure the highest dose of ferrocyanide that is completely precipitated under these conditions. A second, more accurate, trial may then be carried out using the same procedure, based on the results of the first test, e.g. adding between 17 and 23 g/hl of potassium ferrocyanide. As a precautionary measure, 3 g/hl less potassium ferrocyanide is used in treatment than the maximum dose identified in the preliminary trial.

Once the reaction has started, the ferric ferrocyanide should be eliminated rapidly by fining and filtration (or centrifugation). In a preliminary trial (Ribéreau-Gayon et al., 1976), it was observed that 30 g/hl of ferrocyanide reacted in two hours, but when contact time was reduced to ten minutes, only 14 g/hl reacted. If, in practice, the same wine were treated with 30 g/hl and the ferric ferrocyanide eliminated immediately, some potassium ferrocyanide would remain in solution, with all of the risks that this entails.

Once the appropriate dose has been established, the entire batch of wine can be treated. It is essential that it should be in the same oxidation–reduction state as the sample used for the preliminary trial. Treatment should therefore take place shortly after the test and the wine must not be handled in the meantime. If the wine to be treated is kept in several containers, separate trials must be carried out for each one.

Dry Extract and Minerals

The potassium ferrocyanide is dissolved in cold water (50–100 g/l), then added to the wine and mixed well. The fining agent (casein or blood albumin) solution is added after a few minutes, under the same conditions. The blue deposit settles out quickly. It may be removed by racking and filtration after four days.

Although this treatment may be highly effective, it is by no means a universal solution to all types of stability problems. Furthermore, it has been criticized for modifying the development of fine wines in bottle. Wines containing no metals have had problems with their bouquet due to very low oxidation–reduction levels.

The use of stainless steel in wineries is definitely a positive factor, as it has reduced iron concentrations in wine to such an extent that this treatment is now rarely necessary.

4.6.6 Calcium Phytate Treatment

Potassium ferrocyanide treatment is restricted to white and rosé wines, at least in France. Excess iron is eliminated from red wines using calcium phytate.

Phytic acid (Figure 4.3) is the hexaphosphoric ester of *meso*-inositol. The affinity of ferric iron for phosphoric anions, already described in connection with the ferric casse mechanism, is responsible for calcium phytate's effectiveness in eliminating iron from red wines. Under these conditions, phytic acid produces a mixed calcium–iron salt, known as Calciphos, with the following composition: Ca, 20%, P, 14% and Fe^{3+}, 2%. This mixed salt is not very soluble in water and easily precipitates, thus eliminating the excess ferric iron. Phytic acid is very widespread in plants. It acts as a phosphorus reserve, located in the seed coat, i.e. in wheat, rice and corn bran. Wheat bran may be used directly to eliminate iron from wine.

Standard doses of calcium phytate, for example 20 g/hl, are sometimes recommended. At this concentration, the phytate never precipitates completely and, in spite of its low solubility, some of it remains in the wine. This is an unsatisfactory situation, although calcium phytate is quite harmless to health.

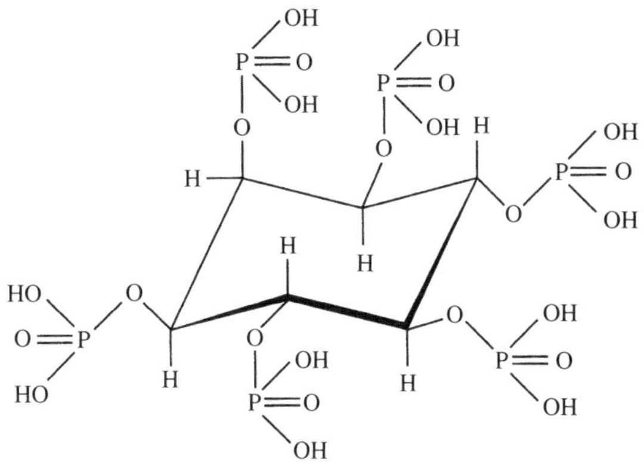

Fig. 4.3. Phytic acid

As only Fe^{III} reacts with calcium phytate, the first stage in treatment is to aerate the wine by racking or injecting oxygen. It is then left to rest for four days, so that the ferric iron concentration reaches a maximum level. The wine must be protected by sulfuring at 3–5 g/hl to avoid extensive spoilage due to oxidation. At this concentration, SO_2 does not prevent iron from oxidizing.

The ferric iron is then titrated to determine the dose of phytate required, given that 5 mg of phytate eliminates 1 mg of Fe^{III}. The dose used is 1 g/hl less than the calculated amount, to provide a safety margin.

White calcium phytate powder is dissolved in hot, concentrated, citric acid solution. The resulting solution is thoroughly mixed into the wine. The ferric phytate starts to flocculate a few hours later. Three to four days are necessary to complete the treatment. The wine should then be fined (with gelatin, casein or blood albumin) to ensure that all the colloidal ferric phytate flocculates and, finally, filtered.

Calcium phytate is an efficient treatment for ferric casse in white, but above all red wines. Its effectiveness may be enhanced by adding citric acid or gum arabic (Section 4.6.3). If the above procedure is properly implemented, no residue is left in the wine, and so there can be no objections on health grounds.

Calcium phytate treatment has been criticized for increasing the calcium content (20–30 mg/l). Its main disadvantage, however, lies in the fact that the wine must be thoroughly oxygenated, which requires extensive handling operations, always entailing the risk of a negative impact on organoleptic quality.

4.7 COPPER AND COPPER CASSE

4.7.1 Presence and State of Copper in Wine

Grape must always contain relatively large amounts of copper (5 mg/l). A few tens of mg/l come from vines and grapes, but most of it originates from sprays, based on the disinfectant properties of copper sulfate, used to treat vines for mildew. It is well known that this excess copper is eliminated by reduction during fermentation, forming sulfides that are among the most insoluble salts known to man. These compounds are eliminated with the yeasts and lees that also have the property of adsorbing copper. Ultimately, new wines contain only 0.3–0.4 mg/l of copper, which is insufficient to cause turbidity in wine. Concentrations may increase during aging due to contact with equipment made of copper, tin or bronze. The concentration may even exceed 1 mg/l in some cases, leading to a risk of copper casse. Certain countries authorize the use of copper sulfate (2 g/hl) to eliminate noxious sulfur derivatives. This treatment, however, increases the risk of excess copper and, thus, of copper casse. The maximum permitted copper concentration in wine in the EU is 1 mg/l.

Copper is in an oxidized state, divalent Cu^{2+}, in aerated wines. However, when white wines are kept in the absence of air and the oxidation–reduction potential reaches a sufficiently low level, the copper is reduced to Cu^+ in the presence of sulfur dioxide. This is likely to cause turbidity at concentrations of around 1 mg/l. Unlike ferric casse, copper casse develops after a long period of aging in the absence of air, at high temperatures and in bright light. It may disappear in contact with air.

Furthermore, even at low doses, copper acts as an oxidation catalyst. It is involved in the oxidative transformations that take place in red wines during aging. Copper also promotes oxidation of iron and white casse, which would be much less common if no copper were present.

Copper is an indispensable trace element for normal functions in plant tissues. It is an active component of certain enzymes such as oxydases (laccase). At high doses, however, it is toxic, which justifies the legal limit of 1 mg/l.

The prolonged aging of wine on its yeast lees causes a significant decrease in oxidation–reduction potential, which favors the reduction of copper and, consequently, the appearance of copper casse. At the same time, the presence of yeast lees promotes the fixing of copper, which tends to prevent copper casse. For example, a copper concentration of 0.1–0.3 mg/l in a champagne-base wine dropped to zero after second fermentation and aging in the bottle in a horizontal position, as this increased the lees/wine interface and promoted exchanges.

4.7.2 Copper Casse Mechanisms

Turbidity may appear in bottled white wines containing free SO_2 and no air. The precipitate gradually settles out to form a brownish-red deposit. This is a two-stage reaction, with the initial formation of an unstable copper colloid, followed by the flocculation and precipitation of this colloid on contact with proteins in the wine.

These deposits contain colloidal copper sulfide and copper, as well as proteins. This led to the hypothesis that two mechanisms were involved:

1. The first mechanism has four stages:
   (a) Reduction of the copper ions:
   $$Cu^{2+} + RH \longrightarrow Cu^+ + H^+ + R$$
   (b) Reduction of the sulfur dioxide:
   $$6Cu^+ + 6H^+ + SO_2 \longrightarrow 6Cu^{2+} + SH_2 + 2H_2O$$
   (c) Formation of copper sulfide and flocculation of the SCu on contact with proteins:
   $$Cu^{2+} + SH_2 \longrightarrow SCu + 2H^+$$

Dry Extract and Minerals

2. The second mechanism assumes that Cu^{2+} is reduced until it becomes Cu metal, part of which precipitates as a colloidal element, while the rest reduces SO_2 to H_2S, leading to the formation of SCu. The colloid then flocculates. Proteins are also thought to be involved in copper casse, not only by causing flocculation of the copper colloid, but also forming a new colloid by creating bonds between copper ions and SH groups in the cysteine that they contain (Figure 4.4). It is well known that turbidity and copper deposits are not readily formed in the absence of colloidal protein.

Once the mechanism of copper casse had been elucidated, a test was developed for predicting this type of instability. White wine in full clear glass bottles is exposed to sunlight or ultraviolet radiation (Section 4.7.3) for seven days. If it does not become turbid under these conditions, it will remain clear during aging and storage. Copper casse also develops after three to four weeks in an oven at 30°C.

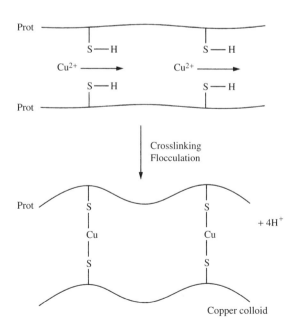

Fig. 4.4. Protein cross-linking by copper and copper casse

4.7.3 Preventing Copper Casse

Copper casse is a serious problem because it may occur when the wine has been in the bottle for a long time. Affected bottles must be uncorked, the wine treated and then re-bottled.

Copper casse is specific to white wines. They are not as well protected from oxidation and reduction phenomena as red wines, where phenols have a redox buffer capacity. Furthermore, the colloidal cupric derivative contains proteins, while red wines have a low protein content due to combination reactions with phenols.

Bentonite treatment (Section 10.9.3) is a simple method for protecting wines from copper casse by eliminating proteins. Gum arabic also has a protective effect, by preventing flocculation of the colloid (Section 9.4.3). This method is effective if the copper concentration is below 1 mg/l, otherwise the excess copper must be eliminated.

The most efficient process for eliminating excess copper is potassium ferrocyanide treatment (Section 4.6.5), which produces an insoluble complex with the following formula: $Fe(CN_6)Cu_2$. This treatment generally brings the copper content down to 0.1 or 0.2 mg/l. Copper is present in much less complex forms than iron, so it precipitates faster. When small quantities of ferrocyanide are added to wines with high copper and low iron concentrations, the precipitate is sometimes reddish instead of blue.

However, if wines with a low iron content are treated with normal, i.e. moderate, doses of ferrocyanide, the copper will not be completely eliminated (Table 4.2). A sufficient quantity of iron, 5–10 mg/l, must be present to eliminate the copper properly. Ferric ferrocyanide treatment acts by cation exchange and is much more effective (Table 4.2), but it is not authorized. Before the use of potassium ferrocyanide was authorized, copper used to be eliminated by heating or adding sodium sulfide (Ribéreau-Gayon, 1947).

Wine is heated in the absence of air to eliminate copper. It must then be cooled, fined and closely filtered to eliminate the copper, which is combined in a colloid that flocculates on heating. It may also be assumed that heating acts by forming protective colloids, as the probability of turbidity is

Table 4.2. Comparison of the elimination of copper from a *vin de liqueur* using potassium ferrocyanide and ferric ferrocyanide, according to the initial iron content (Ribéreau-Gayon *et al.*, 1977)

Wine to be treated		Wine treated with	
		$Fe(CN)_6K_4$	$(Fe(CN)_6)_3Fe_4$
Iron	Copper	Copper	Copper
20	5	0.2	0
10	5	0.5	0
5	5	1.0	0
2.5	5	1.5	0
1	5	2.0	0
Traces	5	3.0	0

much lower, or even non-existent, in heated wines (Ribéreau-Gayon, 1947).

The use of sodium sulfide to eliminate not only excess copper but also lead and even arsenic from white wines has been systematically investigated (Ribéreau-Gayon, 1935). Sodium sulfide is used in hemisulfide form. This salt is highly soluble in water and crystallizes with nine water molecules ($Na_2S \cdot 9H_2O$). As copper concentrations in wine are no more than a few mg/l, the quantity of sodium hemisulfide required to eliminate it is generally no more than 25 mg/l, which corresponds to 3.5 mg/l of hydrogen sulfide (H_2S). This reaction is theoretically capable of producing 9.55 mg/l of copper sulfide (SCu) and precipitating 9.3 mg/l, in view of that fact that this salt's solubility is approximately 0.2 mg/l.

This treatment may initially seem surprising, as it corresponds, especially in wines with a low pH, to displacing noxious-smelling hydrogen sulfide from its salt, especially due to the action of tartaric acid. In fact, yeast always produces hydrogen sulfide, and its reaction with alcohols, particularly ethanol (Maujean *et al.*, 1993), is as follows:

$$H_2S + C_2H_5OH \longrightarrow C_2H_5SH + H_2O$$

This reaction is much slower than the precipitation of copper sulfide. Consequently, mercaptans, which have much lower perception thresholds than hydrogen sulfide, are very unlikely to form in the presence of copper. Sodium sulfide treatment is rendered even more harmless by the fact that wine contains sulfur dioxide, which destroys hydrogen sulfide according to the following reaction:

$$2SO_2 + H_2S + H_2O \longrightarrow S + H_2O_5S_2 + H_2$$

This results in the formation of colloidal sulfur and pentathionic acid, which is unstable and produces a salt, sodium *meta*-bisulfite.

It is therefore clear that adding sodium sulfide, even at doses slightly higher than those strictly necessary to eliminate the copper, does not represent any significant organoleptic risk. According to Ribéreau-Gayon *et al.* (1977), sodium sulfide treatment is neither as easy to implement nor as effective as ferrocyanide treatment, so it is no longer used.

Copper may also be eliminated by exposing wines to sunlight (see the test for copper casse, Section 4.7.2). Haye and Maujean (1977) and Maujean and Seguin (1983) showed that wavelengths of 370 and 440 nm rapidly decrease the redox potential of white wine, which may drop as low as 100 mV. This reaction involves the reduction of riboflavin (vitamin B_2), which is photosensitive and absorbs light at the above wavelengths, which are emitted by sunlight and most fluorescent lamps. Although copper precipitated in a model medium under these conditions, this process may not be used for wine, as light triggers the photo-oxidative breakdown of cysteine and especially methionine, producing volatile thiols. This fault, known as 'sunlight flavor', makes the wine completely undrinkable (Section 8.6.5).

4.8 HEAVY METALS

4.8.1 Definition

The term 'heavy metals' covers a wide range of elements, such as copper, lead, mercury, cadmium, manganese, zinc, etc. These are naturally present in the environment at low concentrations. Some heavy metals are indispensable for plant and animal growth in very small quantities (trace elements), but become toxic at higher doses. Besides the fact that all heavy metal sulfides are insoluble, their chemical properties are dissimilar. Copper is treated separately in winemaking in view

Dry Extract and Minerals

Table 4.3. Eliminating heavy metals from wine using potassium ferrocyanide (50 mg/l was the dose indicated by prior testing—Section 4.6.5) (Ribéreau-Gayon et al., 1977)

	Untreated	Treated: 50 mg/l	Treated: 90 mg/l
Iron	14	7	1
Copper	4	0.4	0.2
Zinc	2.5	1.0	0.2
Lead	2.5	2.0	0.8
Manganese	1.5	1.5	0.5
Aluminum	10	10	10

of its role in causing instability. Besides iron, potassium ferrocyanide may partially eliminate some heavy metals, especially copper, zinc and, possibly, lead (Table 4.3).

4.8.2 Arsenic

Arsenic has properties between those of metals and non-metals, as well as many similarities to phosphorus. Arsenic is present in almost all natural metal sulfides, especially those of copper, tin and nickel. It is a highly toxic element. The toxic dose of arsenic trioxide (As_2O_3) is on the order of 2 mg/kg body weight. Concentrations of 0.01–0.02 mg/l are found in wine; however, when vines have been treated with arsenic salts, amounts may be higher. If the arsenic content is above 1 mg/l, the wine is unfit for consumption. The OIV has set a limit of 0.2 mg/l of arsenic in wine.

4.8.3 Cadmium

Cadmium occurs naturally as an insoluble sulfide (greenockite). It is found in this form combined with zinc in blende. Pure CdS is a yellow salt used in paint and fireworks. In industry, it is a raw material for nickel–cadmium batteries.

Cadmium is toxic for humans at low doses. The main contamination agents are air, water and, above all, food. Indeed, this heavy metal accumulates considerably in the food chain.

The total daily absorption tolerated by subjects in industrialized countries not exposed to cadmium for professional reasons is on the order of 60 µg. Fortunately, only 4 µg remain in the body. The WHO has estimated that the weekly dose for adults should not exceed 0.4 mg and recommends 5 µg/l as the maximum limit in drinking water. The OIV has set the same limit for wine (1981).

Cadmium's toxicity is mainly due to its role as an enzyme inhibitor. This results in disturbances in kidney functions and the phosphocalcium metabolism (decalcification).

4.8.4 Mercury

Mercury is the only metal that remains liquid at normal temperatures. This makes it a rather volatile element and mercury vapor is highly toxic. Air saturated with mercury vapor at 20°C has a mercury content one hundred times higher than the toxicity limit. Mercury ore, or cinnabar (HgS), is a lovely, red-colored sulfide. Mercury stored in sediments is returned to the food chain via microbial activity that converts metallic mercury into organic mercury, in the form of highly toxic, volatile, methyl mercury. The toxicity of mercury gives it useful medical properties as an antiseptic and anti-parasitic. Mercury concentrations in wines are below 5 µg/l. (Brun and Cayrol, 1976). It is not easy to assay mercury by atomic absorption, as the high pressure of mercury vapor means that the temperature cannot exceed 60°C when the wine is being mineralized.

4.8.5 Lead

Lead is the heaviest common metal. The most widespread natural form of lead ore is galena (PbS), a sulfide with a density of 7.60. Pure galena contains 86.6% lead. Lead is found naturally in soil (16 mg/g) and represents 0.002% of the earth's crust. There are four natural isotopes: ^{204}Pb, ^{206}Pb, ^{207}Pb and ^{208}Pb. Three of these (206, 207 and 208) are stable end products, resulting from the breakdown of uranium 238 and 235 and thorium 230, respectively.

Lead is present as a chemical element in all biological systems. It accumulates in living organisms (bioaccumulation), mainly in the bones and teeth. It may be absorbed orally or via the respiratory system.

An EU survey shows that the human environment in industrialized countries, especially in Europe, has an influence on lead concentrations. European wines contain twice as much lead (average: 63 μg/l) as Australian (28 μg/l) or even American wines (24 μg/l). This results from the use of lead in various industries (printing, paint, glass, crystal, engineering, fuel, etc.). Fuel is responsible for 90–95% of the lead in atmospheric pollution, due to the use of octane boosters [$(Pb(CH_3)_4$ and $Pb(C_2H_5)_4$], volatile organometallic compounds that are soluble in oil. In the USA, a reduction in the consumption of gasoline containing lead caused a reduction in the lead concentration in plants.

The use of lead arsenate as an agricultural insecticide has fortunately been abandoned. The majority of treatment products currently on the market are lead-free.

The pathological effects of lead mainly affect the following organs:

The blood system: inhibits hemoglobin synthesis and causes anemia.

The nervous system: chronic encephalopathy, neurological and psychomotor problems.

The renal system: nephropathy and gradual deterioration of kidney functions.

The cardiovascular system.

Most human exposure to lead (approximately 80%) is via food intake. Lead concentrations vary from a few tens of μg/kg in many foods to a few hundreds in certain types of seafood and, particularly, kidneys. Wine does not have an excessively high lead content (60 μg/l). The precautions taken to avoid contamination have led to lower concentrations.

In 1960, a study by Jaulmes et al. based on the analysis of 500 wine samples, showed that French wines had an average lead content of 180 μg/l. In 1983, a survey by the French Ministry of the Environment reported that the average lead concentration in commercially available wines was 118 μg/l. More recently, on the basis of 2733 samples analyzed in 1990, L'Office National Interprofessionnel des Vins (ONIVINS) found an average value of 68 μg/l. Pellerin et al. (1997) showed that lead in wine is partially combined in a stable complex with a pectic polysaccharide, a dimer of rhamnogalacturonan II. This discovery sheds a new light on the toxicological problems related to lead in wine (Section 3.6.4).

A series of articles (Jaulmes et al., 1960; Medina, 1978; Teisseidre, 1993) identified the major sources of lead contamination in must and wine. Rain water is the main source of lead in vine leaves and bunches of grapes; indeed, 90% of the lead in the atmosphere is precipitated by rain (Teisseidre et al., 1993). Table 4.4 shows the indisputable role of fossil fuels in atmospheric pollution. Higher

Table 4.4. Evolution of the lead content from vines to grapes and wines (Teisseidre et al., 1993)

Site number	Soil (μg/dry weight)	Leaves (μg/dry weight)	Grapes (μg/dry weight)		Must (μg/dry weight)	Wine during alcoholic fermentation (μg/l wine)	Pomace (μg/kg fresh weight)	Wine during malolactic fermentation (μg/l wine)	Lees (μg/kg fresh weight)	Decrease in lead content after alcoholic fermentation (%)
			Grapes sampled next to a major road	Grapes sampled from the entire plot						
1	24.94	4.76	141.5	70	80	40	259.5	35	288	50
3	23.03	3.31	136	70	80	25	306	20	348.1	68.8
5	20.07	9.19	202.9	84	95.6	25	263	25	87.2	73.9
13	15.02	1.72	55	53.2	62	30	366.9	47	296	51.6
14	9.17	8.28	104	80	170	80	375.2	80	138.8	52.9
16	12.42	2.5	22	18	24	10	118.7	10	75.1	58.33
18	20.88	1.39	37	34	40	20	163.8	20	118.7	50
22	20.59	6.28	152	115	100	45	660.9	46	297	55
30	23.71	2.67	34	29	170	85	101.3	80	54.9	50
34	8.64	2.44	71.3	45	50	25	121.3	25	162.1	47.6

lead concentrations have been observed in grapes growing beside major roads.

Lead is closely controlled by the hygiene services in all industrialized countries due to the health risks it represents and its bioaccumulable character. The Office International de la Vigne et du Vin (OIV) have regularly lowered the authorized limit in wine. It was reduced in several stages, from 600 µg/l in 1953 to 250 µg/l in 1993. In March 1996, the OIV declared that the limit should be further reduced to a maximum of 200 µg/l. This trend is the consequence of many studies aimed at identifying possible sources of contamination at all stages in the winemaking process: vineyard and winery equipment, enological products, etc.

One major finding of this research (Table 4.4) is that the lead content decreases significantly following the alcoholic and malolactic fermentations. This reduction of approximately 45% is due to precipitation of lead sulfide. It has, however, been observed that these decreases in lead concentration are highly variable. This may be partly explained by the conditions during transport and processing of the grapes. Thus, contact with a crusher–stemmer coated in epoxy resin, or even with painted machinery, may produce an increase in the lead content of grapes and must (Jaulmes *et al.*, 1960; Teisseidre, 1993).

Lead may be leached into must or wines from bronze or brass hose connections, taps and pumps, as these alloys contain 7 and 2% lead, respectively. Wine's acid pH facilitates the dissolving of metallic lead. Every time a wine is pumped through a bronze pump, the lead content of must or wine increases by an average of 10 µg/l (Tusseau *et al.*, 1996).

Materials used for winemaking equipment and storage vats are another non-negligible source of contamination in wine, due to the large surfaces and long contact times involved. Lead concentrations in the vicinity of 600 µg/l. have been found in wines stored in vats lined with ceramic tiles. According to Medina (1978), ceramic-glazed tiles may have a lead content between 0.1 and 0.5%. These results are just an indication of the possible impact of storage containers.

Lead–tin capsules are also a source of lead in wine. Jaulmes *et al.* attracted attention to this source of contamination as early as 1960. Capsules for still wines used to be made of lead covered with a thin layer of tin. Although the permeability of corks is minimal, there may be a slight leakage of wine that oxidizes to form acetic acid. This can erode the capsule, producing lead acetate. Contamination occurs mainly when the wine is poured. One study showed that the first glass poured from a bottle could contain up to 20 mg/l of lead. Lead capsules have since been banned.

4.8.6 Zinc

Zinc occurs naturally in the form of sulfide ores, such as blende, i.e. a mixture of zinc and lead sulfides. Pure ZnS is phosphorescent under bright light.

Zinc is a trace element that plays a major role in the auxin metabolism and therefore in plant growth. Zinc deficiency causes a decrease in plant size, as well as a change in the arrangement and color of the leaves. It also leads to malformations in the root system. Zinc salts have antiseptic properties, so they are toxic.

Traces of zinc are naturally present in must and wine. Higher concentrations of this heavy metal may come from the vineyard, due to galvanized iron wire damaged by mechanical harvesting, or dithiocarbamate-based fungicides. Another source is winemaking equipment made of alloys, such as bronze pumps, hose connections, taps, etc.

Zinc concentrations in wines range from 0.14 to 4 mg/l. Prolonged maceration of grape solids leads to an increase in zinc concentrations. The use of potassium ferrocyanide to treat ferric casse reduces a wine's zinc content (Table 4.3).

REFERENCES

Bonastre J. (1959) Thèse Doctorat ès Sciences Physiques, Université de Bordeaux.
Brun S. and Cayrol M. (1976) *Ann. Falsif. Exp. Chim.*, 737, 361.
Haye B. and Maujean A. (1977) *Conn. Vigne et Vin*, 11 (3), 243.

Jaulmes P., Hamelle G. and Roques J. (1960) *Ann. Tech. Agri.* 9 (3) 189.

Maujean A., Poinsaut P., Dantan H., Brissonet F. and Gossiez E. (1990) *Bull. OIV*, 63 (711–712), 405.

Maujean A., Gomerieux T. and Garnier J.M.C. (1988) *Bull. OIV*, 61 (683–684), 25.

Maujean A., Nedjma Musapha, Cantagrel R. and Vidal J.-P. (1993) Élaboration et Connaissances des Spiritueux. 1st Symposium International de Cognac, Tec. et Doc., Lavoisier, paris.

Maujean A. and Seguin N. (1983) *Sciences des Aliments*, 3 (4), 589–601.

Medina B. (1978) Thèse de Doctorat, Université de Bordeaux II.

Pellerin P., O'Neill M.A., Pierre C., Cabanis M.T., Darvill A.G., Albersheim P. and Moutounet M. (1997). *J. Int. Sci. Vigne Vin*, 31 (1), 33.

Ribèreau-Gayon J. (1935) *Rev. Viticult.*, 82, 367.

Ribèreau-Gayon J. (1947) *Traité d'Œnologie*. Beranger, Paris.

Ribèreau-Gayon J., Peynaud E., Ribereau-Gayon P. and Sudraud P. (1976, 1977, 1982) *Sciences et Techniques du Vin*, Vols. 1, 3 and 4. Dunod, Paris.

Teisseidre P.L. (1993) Thèse de Doctorat, Université de Montpellier I.

Teisseidre P.L., Cabanis M.-T., Daumas F. and Cabanis J.-C. (1993) *Revue Française d'Œnologie*, 140, 6.

Tusseau D., Valade M. and Moncomble D. (1996) *Le Vigneron Champenois* 5, 6.

Vogt E. (1931) *Weinbeau u. Kellerwirt*, 10, 5.

5

Nitrogen Compounds

5.1 Introduction	109
5.2 The various forms of nitrogen	109
5.3 Amino acids	113
5.4 Other forms of nitrogen	119
5.5 Proteins and protein casse	124
5.6 Preventing protein casse	132

5.1 INTRODUCTION

Nitrogen is one of the most plentiful elements in the universe. Indeed, the earth's atmosphere contains nearly 80% nitrogen in molecular form. The presence of this uncombined element indicates that nitrogen's chemical reactivity is extremely low. One illustration of this in winemaking is the use of a blanket of nitrogen to protect wines in partially empty vats. Animal and plant cells cannot assimilate molecular nitrogen, so it must be obtained in mineral or organic forms.

5.2 THE VARIOUS FORMS OF NITROGEN

5.2.1 Total Nitrogen

Total nitrogen in must or wine includes one inorganic form and various organic forms. Total nitrogen is measured by mineralizing it as ammonium sulfate. Sulfuric acid is added to the substance to be analyzed in the presence of a catalyst, and the mixture is heated (Kjeldahl method). All forms of nitrogen are thus converted to ammonium sulfate. The ammonia is then separated from its salt by sodium hydroxide and assayed by titrating with an acid solution. Total nitrogen assays are standard techniques in winemaking. Nitrogen assays are still very widely used due to their reliability, although there are more sophisticated, specific assay techniques for the various forms of nitrogen.

A total nitrogen assay (expressed in g/l of nitrogen) of grape must showed that this value was variable from year to year, probably due to variations in grape ripeness. Total nitrogen may vary by a factor of 4 from one year to the next. The grape variety and region of production also affect the nitrogen concentration. In general, Champagne contains two to three times as much total nitrogen as white Bordeaux wines. This characteristic certainly has an impact on fermentation management and foaming properties.

Red wines have average nitrogen concentrations almost twice as high as those of white wines. This is due to winemaking techniques, including high-temperature maceration, which causes nitrogenated substances to dissolve more readily from the skins and seeds, as well as autolysis of dead yeast cells.

The total nitrogen concentration of Bordeaux wines varies from 70 to 700 mg/l. Values in white wines range from 77 to 377 mg/l, with an average of 185 mg/l. Red wines have nitrogen contents between 143 and 666 mg/l, with an average of 330 mg/l (Ribéreau-Gayon et al., 1982).

The weight of the nitrogenated substances, calculated approximately by using an empirical coefficient of 6.25, has been observed to range from 0.5 to 4.0 g/l. Nitrogenated substances may therefore represent up to 20% of the dry extract in non-sweet (dry) wines.

5.2.2 Mineral Nitrogen

Mineral nitrogen, in the form of ammonia salts, is most prevalent in grape flesh cells during the vegetative growth phase. Ammonia nitrogen represents 80% of the total nitrogen in grapes originating, of course, from nitrates extracted from the soil by the roots. At color change, ammonia nitrogen is still largely predominant. Concentrations decrease rapidly, however, due to transamination, producing more elaborate forms of organic nitrogen. A transamination reaction with α-keto acids, via the Krebs cycle and the respiration of the sugars, initially converts ammonia nitrogen mainly into free amino acids and then into bonded forms of peptides, polypeptides and proteins (Volume 1, Section 10.3.5).

When grapes are fully ripe, mineral nitrogen represents less than 10% of total nitrogen, or a few tens of mg/l expressed in ammonia. Ammonia, or more exactly the NH_4^+ ammonium cation, is the form most directly assimilable by yeasts. Its concentration affects the rapidity with which must starts to ferment as well as its potential fermentability. This form of nitrogen is frequently observed to have totally disappeared from must at the end of alcoholic fermentation. It is, therefore, important to assay ammonia in must, especially in years when the grapes are completely ripe. The Boussingault method is used to assay ammonia nitrogen selectively, once it has been separated out as NH_3 by distillation in the presence of a large excess of weak base (magnesium oxide). A specific enzyme method also gives accurate results.

When the concentration in must is lower than 50 mg/l, it may be advisable to add 10 g/hl of diammonium phosphate or, preferably, diammonium sulfate [$(NH_4)_2HPO_4$ or $(NH_4)_2SO_4$] to ensure that alcoholic fermentation gets off to a rapid start. The systematic addition of this ammonia salt, without analyzing the must to determine whether it is really necessary, is not recommended. This may, in fact, lead to wines with low concentrations of odoriferous compounds, especially higher alcohols, esters and particularly ethyl acetates of fatty acids.

There may be a few tens of mg/l of inorganic nitrogen in wine after aging on the lees, or even after malolactic fermentation. Indeed, lactic bacteria do not assimilate ammonia nitrogen and may even excrete it. It is prudent to add diammonium phosphate in conjunction with thiamin pyrophosphate to wines intended for a second fermentation in sealed vats or in the bottle.

5.2.3 The Various Forms of Organic Nitrogen

The many forms of organic nitrogen are summarized below. The main compounds are described in detail in the following sections:

1. Amino acids with the general formula:

$$R-\underset{\underset{NH_2}{|}}{CH}-COOH$$

Amino acids have molecular weights below 200, and 32 of them have been identified in must and wine. The most important are listed in Table 5.1. Amino acids contribute to the acidobasic buffer capacity

Nitrogen Compounds

Table 5.1. Amino acids in grapes and wine

Amino acid	Structure	Abbreviation
Alanine	$CH_3-CH(NH_2)-COOH$	Ala
Arginine	$NH_2-C(=NH)-NH-(CH_2)_3-CH(NH_2)-COOH$	Arg
Aspartic acid	$HOOC-CH_2-CH(NH_2)-COOH$	Asp
Asparagine	$NH_2-CO-CH_2-CH(NH_2)-COOH$	Asn
Citrulline	$NH_2-CO-NH-(CH_2)_3-CH(NH_2)-COOH$	Cit
Cysteine	$HS-CH_2-CH(NH_2)-COOH$	Cys
Glutamic acid	$HOOC-CH_2-CH_2-CH(NH_2)-COOH$	Glu
Glutamine	$NH_2-CO-CH_2-CH_2-CH(NH_2)-COOH$	Gln
Glycine	$CH_2(NH_2)-COOH$	Gly
Histidine	(imidazole)$-CH_2-CH(NH_2)-COOH$	His
Isoleucine	$CH_3-CH_2-CH(CH_3)-CH(NH_2)-COOH$	Ile
Leucine	$(CH_3)_2CH-CH_2-CH(NH_2)-COOH$	Leu

(*continued overleaf*)

Table 5.1. (*continued*)

Amino acid	Structure	Abbreviation
Lysine	$NH_2-(CH_2)_3-CH(NH_2)(COOH)$	Lys
Methionine	$CH_3-S-(CH_2)_2-CH(NH_2)(COOH)$	Met
Ornithine	$NH_2-(CH_2)_3-CH(NH_2)(COOH)$	Orn
Phenylalanine	$C_6H_5-CH_2-CH(NH_2)(COOH)$	Phe
Proline	pyrrolidine-2-COOH	Pro
Hydroxy 3 proline	3-hydroxy-pyrrolidine-2-COOH	Hypro
Serine	$HO-CH_2-CH(NH_2)(COOH)$	Ser
Threonine	$CH_3-CHOH-CH(NH_2)(COOH)$	Thr
Tryptophan	indolyl-$CH_2-CH(NH_2)(COOH)$	Trp
Tyrosine	$HO-C_6H_4-CH_2-CH(NH_2)(COOH)$	Tyr
Valine	$CH_3-CH(CH_3)-CH(NH_2)(COOH)$	Val

of must and wine (Dartiguenave et al., 2000) (Section 1.4.3). These substances are very useful, due to their antioxidant, antimicrobial, surfactant, and emulsifying properties.

2. Oligopeptides and polypeptides, formed by linking a limited number of amino acids with peptide bonds:

$$R_1-CH(NH_2)-COOH + R_2-CH(NH_2)-COOH \rightarrow R_1-CH(NH_2)-CO-NH-CH(COOH)-R_2 + H_2O$$

Oligopeptides contain a maximum of four amino acids. Polypeptides have molecular weights under 10 000 Dalton. They can be separated out by membrane ultrafiltration and represent a major proportion of the nitrogen in wine. They can be separated out by microfiltration (Desportes et al., 2000) and nanofiltration, followed by chromatography on Sephadex LH20 gel, and represent a major proportion of the nitrogen in wine.

Some small peptides also have interesting sweet or bitter flavors that are useful in winemaking.

3. Proteins are macromolecules produced by linking a large number of amino acids. They have molecular weights above 10 000. Their structure includes other bonds besides peptide bonds that give the chain a three-dimensional configuration: spherical, helix, etc.

Grapes and wine contain many proteins with a wide range of molecular weights (30 000–150 000). Some unstable proteins are responsible for protein casse in white wines. Other proteins are associated with a carbohydrate fraction, e.g. yeast mannoproteins. Other proteins are associated with a carbohydrate fraction, e.g. yeast mannoproteins and isolectin, recently identified in Chardonnay must (Berthier et al., 1999).

4. Amide nitrogen, with the following general formula:

$$R-\underset{\underset{O}{\|}}{C}-NH_2$$

This category is represented by small quantities of asparagine and glutamine (Table 5.1). Urea is also included in this group:

$$NH_2-\underset{\underset{O}{\|}}{C}-NH_2$$

as well as ethyl carbamate:

$$CH_3-CH_2-O-\underset{\underset{O}{\|}}{C}-NH_2$$

which is very strictly controlled for health reasons.

5. Bioamines, with the formula $R-NH_2$, are also particularly closely controlled in view of their harmful effects (histamine).

6. Nucleic nitrogen is present in purine and pyrimidine bases, nucleosides and nucleotides, as well as nucleic acids. This form of nitrogen has not been extensively studied.

7. Amino sugar nitrogen consists of hexoses in which an –OH is replaced by $-NH_2$. Small quantities of glucosamine and galactosamine have been found in protein nitrogen in wine.

8. Pyrazines (Figure 5.1) are heterocycles with six links containing two nitrogen atoms and four carbon atoms bearing radicals, which largely determine their olfactory impact. They contribute to the aroma in Cabernet Sauvignon wines (Section 7.4).

5.3 AMINO ACIDS

5.3.1 Structure

From a structural point of view, the twenty most common amino acids in must and wine, excepting γ-amino-butyric acid and β-alanine, are α-amino

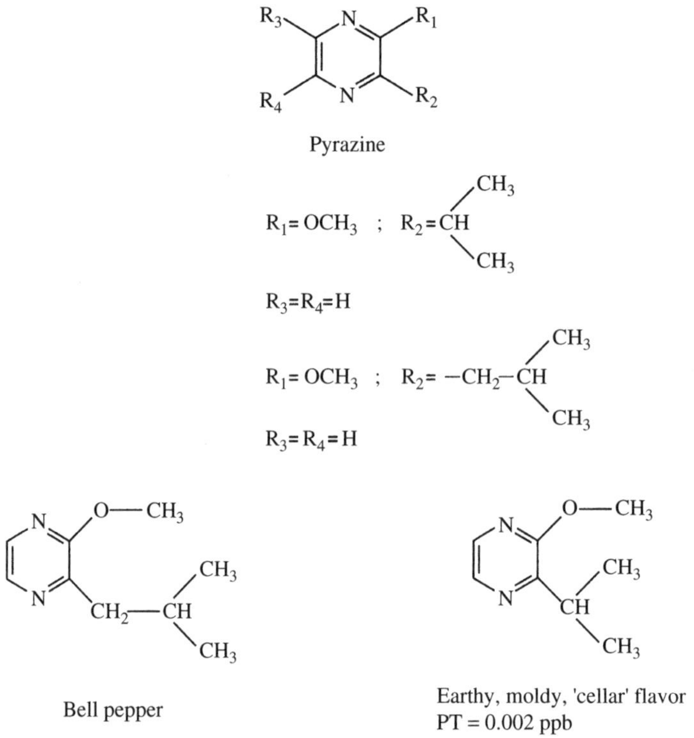

Fig. 5.1. Structure of pyrazines

acids (Table 5.1); i.e. the carboxylic acid radical and the amino basic radical are linked to the same carbon atom. Amino acids are therefore amphoteric molecules. The R radical (Figure 5.2) may be a hydrogen atom, as is the case in one amino acid, glycine, or glycocoll, an amino acid prototype that is a component of many proteins. In other amino acids, the carbon linked to the functional radicals and the R carbohydrate radical is chiral. This feature gives them the asymmetrical property common to biologically significant molecules. Natural amino acids all have a true L configuration. Their optical activity depends on the type of R radical, but also on the solvent, and even the pH of the solution.

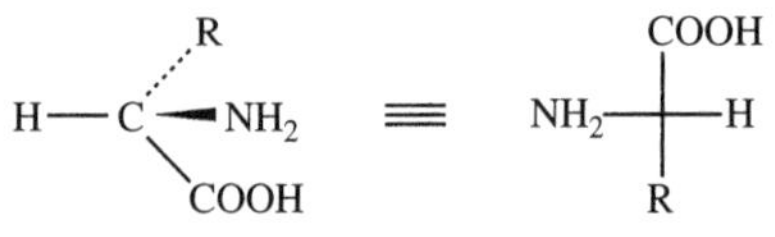

Fig. 5.2. The L configuration α-amino acid

In trifunctional amino acids, the R radical is connected to an acid or basic radical. Aspartic acid and glutamic acid are acid amino acids, while lysine, histidine, ornithine, citrulline and arginine are basic amino acids. Other trifunctional amino acids have no marked acid or basic character. They have hydroxylated (serine, threonine and tyrosine), thiol or sulfide radicals (cysteine and methionine).

Only these trifunctional amino acids are involved in the catalyst property of enzymes. It may seem surprising that only small numbers of this type of amino acid are present at the active enzyme sites, as compared to the very large number of enzymes, at least 2500 per cell in the latent state.

In solution, amino acids must be expressed in a double balance state (Figure 5.3), involving the ionization of the acid or basic function. The population of each form is defined by the pH of the medium. At a pH where the positive and negative charges are in balance, i.e. the isoelectric point, amino acids have minimum solubility and

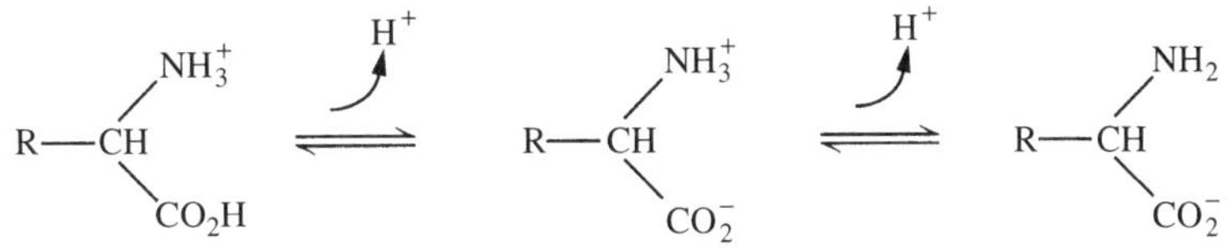

Fig. 5.3. Forms of α-amino acid in an aqueous solution

no longer migrate when subjected to an electric field, as is the case in electrophoresis. The variably charged character of amino acids at a given pH is also useful in fixing, and especially separating and assaying them on strong cation exchange resins (sulfonic resins).

5.3.2 Presence in Must and Wine

Amino acids are the most prevalent form of total nitrogen by weight in grape juice and wine. Indeed, the total free amino acid concentration varies from 1 to 4 g/l, depending on the year. In ripe grapes, amino acids generally represent 30–40% of total nitrogen.

In view of the role of this form of nitrogen as a nutrient for yeast and its importance in the fermentation process, many analysis methods have been suggested, in addition to total nitrogen and ammonium cation assays. Free amino nitrogen (free α-amino nitrogen, FAN) may be assayed using a reaction colored with ninhydrin and assimilable nitrogen by titration with formaldehyde (Sörensen method). The latter method is used to analyze total free amino acids and ammonia forms in must. It is an interesting, useful assay for evaluating the fermentability of must or wine. This assay does not take proline, a preponderant amino acid in must and wine, into consideration. However, yeasts cannot assimilate proline in the absence of oxygen. When a must has completed its alcoholic fermentation, the relative proportion of proline is therefore higher in the wine. The Sörensen method does allow for the neutralization of amino functions by methanal. The acid functions are no longer affected by the presence of basic functions and may be titrated with sodium hydroxide (Ribéreau-Gayon et al., 1982).

The individual assay of each amino acid in must or wine by chromatography on strong cation exchange resin, using the Moore and Stein method (1951) with ninhydrin as a colored reagent, has become a routine technique (Figure 5.4). This assay is carried out automatically and provides very useful data. Table 5.2 gives an example of changes in the concentrations of about twenty amino acids in a free state in the must from three Champagne grape varieties during the ripening period. These data are particularly significant, as must from grapes picked in 1986 had unusually high nitrogen contents.

Indeed, Table 5.2 clearly shows that certain amino acids are predominant, especially α-alanine, serine, arginine, and proline, as well as glutamic acid and its amide form, glutamine, known to be an ammonia transporter. Arginine and proline are characteristic of certain grape varieties. Thus, proline is dominant in Chardonnay, Cabernet Sauvignon, and Merlot, whereas arginine predominates in Pinot Noir and Aligoté. Unlike proline, arginine is used by yeasts and lactic bacteria.

A third phenomenon highlighted by the results in Table 5.2 is the considerable increase in total free amino acids as the grapes ripen. Concentrations more than double between the beginning of color change and the time the grapes are ready to pick. This phenomenon occurs every year, but should not lead to the hasty conclusion that the potential fermentability of must increases as the grapes ripen. Indeed, a more detailed analysis of concentrations of each amino acid, especially arginine and proline (Table 5.2), shows that the proline content increases markedly about two weeks before the grapes are picked.

A three-year study monitoring the two major Champagne grape varieties (Millery et al., 1986) established a close correlation between the proline concentration and the [sugar/acid] ratio, i.e. the ripeness of the grapes (IM). The correlation

116 Handbook of Enology: The Chemistry of Wine

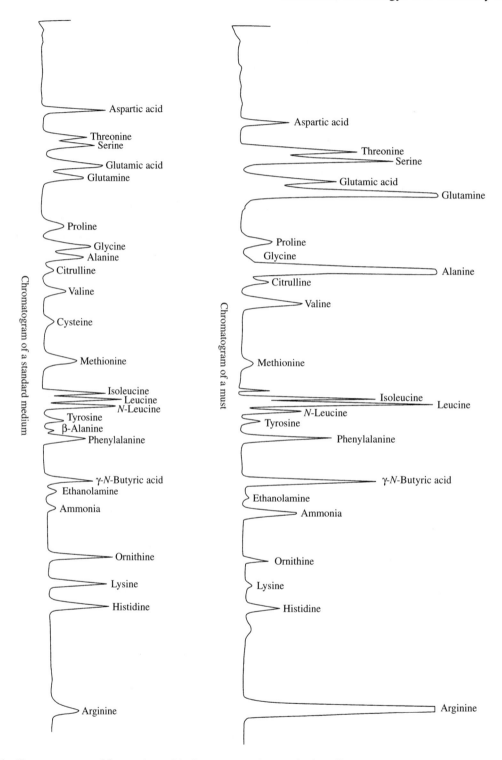

Fig. 5.4. Chromatograms of free amino acids from must and a standard medium

Nitrogen Compounds

Table 5.2. Survey of ripening in Champagne grape varieties in 1986 (amino acid content expressed in mg/l) (Millery, 1988)

	Chardonnay					Pinot Noir				Pinot Meunier		
	September			October		September		October		September		October
Sample dates	8	16	22	7	8	10	22	2	8	16	22	5
Aspartic acid	44	38	16	41	47	33	18	77	33	31	21	31
Threonine	74	136	134	174	91	127	137	219	111	121	146	172
Serine	158	143	119	283	152	165	143	192	192	206	212	176
Glutamic acid	177	173	128	74	108	179	147	116	174	178	68	103
Glutamine	476	361	154	772	286	429	305	638	810	530	730	660
Proline	111	208	187	1123	64	135	147	396	232	365	582	294
Alanine	251	282	248	487	284	338	333	476	325	306	347	539
Citrulline	17	45	32	55	39	47	37	70	16	22	17	68
Valine	36	50	50	106	26	44	46	97	78	70	91	67
Cysteine	0	0	0	0	0	0	0	0	0	0	0	0
Methionine	0	11	0	23	0	4	7	14	9	12	18	15
Isoleucine	17	39	38	97	6	29	35	84	58	45	72	61
Leucine	20	48	46	98	24	38	43	91	61	55	73	67
Tyrosine	7	14	8	28	12	16	12	21	14	12	19	16
Alanine (β)	0	0	0	38	0	0	0	0	0	0	0	2
Phenylalanine	29	39	35	119	-	25	35	85	90	64	108	55
γ-N-Butyric acid	18	18	42	218	12	20	41	118	14	43	100	191
Ethanolamine	5	11	5	20	5	9	9	1	5	5	8	1
Ornithine	1	18	3	3	1	14	9	14	1	1	1	23
Lysine	1	7	3	5	1	5	5	8	1	1	1	10
Histidine	17	27	22	38	34	24	24	30	34	30	52	31
Arginine	299	813	682	790	392	796	816	1379	393	419	569	1415
Total amino acids	1760	2482	1953	4590	1889	2478	2350	4124	2652	2518	3235	3997

Table 5.3. Correlation between the logarithm of the proline concentration and the ripeness index in two Champagne grape varieties in 1983 (Millery, 1988)

Variety	Chardonnay				Pinot Noir			
Ripeness index (IM)	4.5	15	21	22	3.5	9.5	13.5	17
Proline (mg/l)	30	120	290	510	41	80	135	224

Chardonnay: Log [Pro] = 0.151 IM + 2.4; $r = 0.987$. Pinot noir: Log [Pro] = 0.126 IM + 3.22; $r = 0.998$. IM = [sugar]/[acid].

between these parameters corresponds to the following formula:

$$\text{Log [proline]} = a[\text{IM}] + b$$

Table 5.3 shows figures illustrating this correlation, from observations made in 1983.

Proline therefore appears to be a marker for ripeness. It is thus possible to explain the decrease in potential fermentability of must during ripening. In years when the grapes are very ripe, the must needs to be monitored, and diammonium phosphate (or sulfate) may need to be added. This is especially important for grape varieties with a high proline content, such as Chardonnay (Figure 5.5), where there is a spectacular increase in proline during the two weeks before the grapes are picked, while arginine levels remain approximately stationary.

5.3.3 Oligopeptides

Oligopeptide nitrogen is not clearly distinguishable from polypeptide nitrogen in must and wine. It

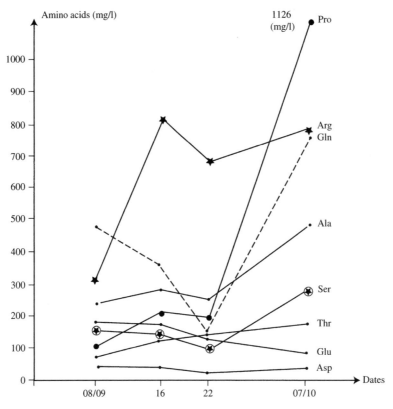

Fig. 5.5. Profiles showing changes in concentrations of individual amino acids during ripening (grape variety: Chardonnay) (Millery, 1988)

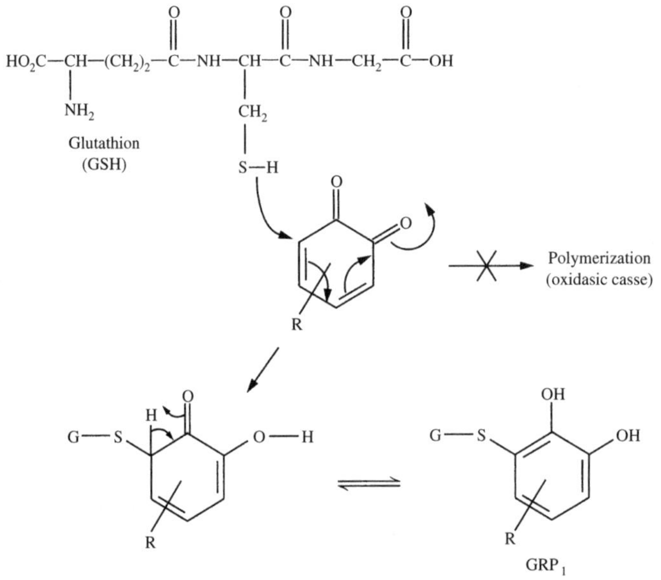

Fig. 5.6. Structure of glutathion and its reaction with quinones produced by the oxidation of phenols

Nitrogen Compounds

is not possible to fractionate these two forms of nitrogen by molecular screening on a dextrane gel column, e.g. Sephadex G25. Oligopeptide nitrogen consists of nitrogen compounds made up of a maximum of four amino acids.

Glutathion is an important tripeptide in must (Figure 5.6). Indeed, its cysteine residue reacts partially with the quinones resulting from oxidation of the phenols. The new derivative (grape reaction product, GRP) is oxidizable in the presence of *Botrytis cinerea* laccase, but not grape tyrosinase (Moutounet, 1990) (Volume 1, Sections 11.6.2 and 13.4.2).

Lactic bacteria (De Roissart and Luquet, 1994) are known to have membrane amino peptidases which enable them to assimilate small peptides, especially tripeptides. These substances have organoleptic properties that are likely to affect wine flavor, but no specific studies have been carried out on this subject.

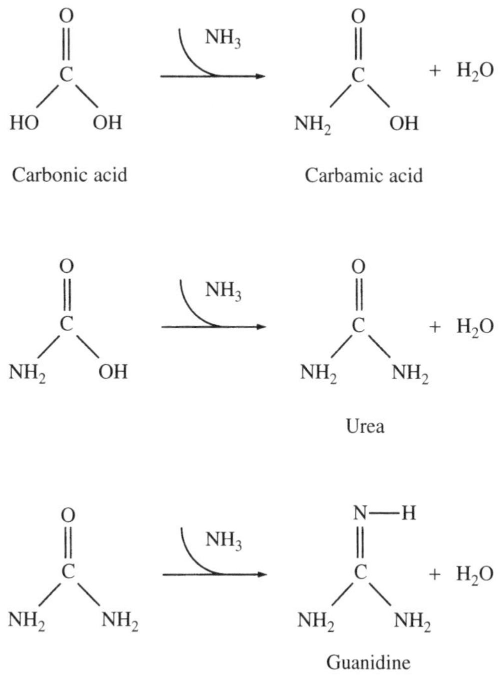

5.4 OTHER FORMS OF NITROGEN

5.4.1 Urea

Urea is a di-amino derivative of carbonic acid, known as carbonic diamide. Urea may also be considered an amide of carbamic acid, according to the formal sequences suggested in Figure 5.7. This figure also indicates that when an ammonia molecule reacts with urea, it produces guanidine, an organic imide molecule.

It has long been known that it is possible to isomerize ammonium cyanate, an inorganic molecule, to form urea:

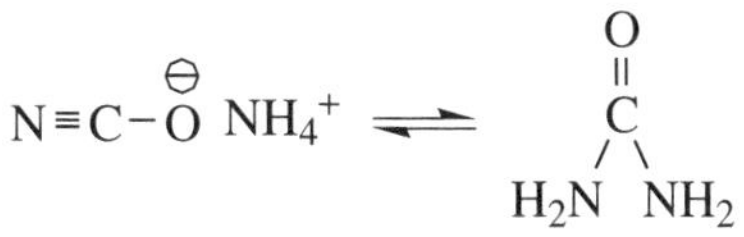

$$N \equiv C - O^{\ominus} \ NH_4^+ \rightleftharpoons \underset{H_2N \ \ NH_2}{\overset{O}{\underset{\|}{C}}}$$

This reaction, together with those in Figure 5.7, creates a bridge between inorganic and organic chemistry.

Urea is a colorless solid that melts at 132°C and is highly soluble in water (1000 g/l) and ethyl alcohol (100 g/l). Its chemical properties are those

Fig. 5.7. Formal reaction diagram showing the formation of amino derivatives of carbonic acid

of an amide. It can be hydrolyzed to form carbonic acid and ammonia:

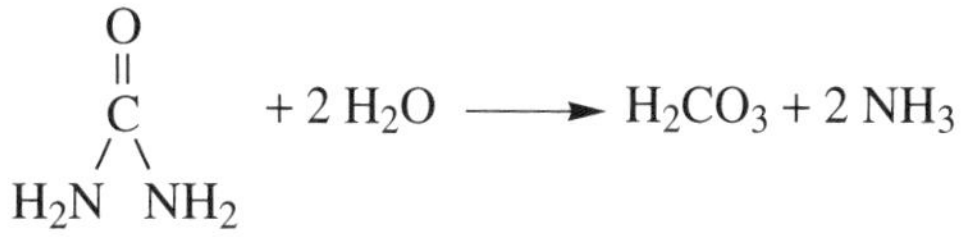

$$\underset{H_2N \ \ NH_2}{\overset{O}{\underset{\|}{C}}} + 2 H_2O \longrightarrow H_2CO_3 + 2 NH_3$$

This chemical property is useful in agriculture, and urea is one of the raw materials in many fertilizers. It also has a wide range of industrial applications.

Urea is assayed using urease, and ammonium carbonate is formed. Wine contains less than 1 mg/l, and it is certainly of microbial origin. Urea is significant in winemaking as it may be a precursor of ethyl carbamate. In spite of certain reservations, L'Office International de la Vigne et du Vin (OIV) authorize the treatment of wines by active urease in an acid medium. This enzyme is extracted from *Lactobacillus fermentum*. The objective is to reduce excessive urea concentrations in wine to avoid the formation of ethyl carbamate as the wine ages.

Urea is a metabolic end product in mammals. It is thought to be a denaturing agent for proteins.

5.4.2 Ethyl Carbamate

Ethyl carbamate is the ethyl ester of carbamic acid:

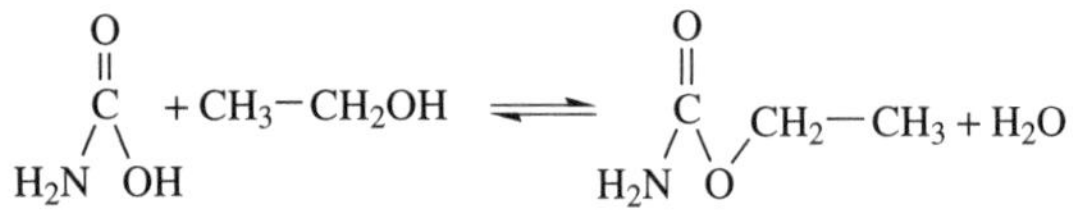

$$\underset{H_2N}{\overset{O}{\underset{|}{\overset{||}{C}}}}\!\!-\!OH + CH_3\!-\!CH_2OH \rightleftharpoons \underset{H_2N}{\overset{O}{\underset{|}{\overset{||}{C}}}}\!\!-\!O\!-\!CH_2\!-\!CH_3 + H_2O$$

Whereas carbamic acids are very unstable and break down rapidly into carbon dioxide and ammonia, esters of this acid are stable. They are known as urethanes.

In winemaking, and more generally in the fermented beverage industry, ethyl carbamate has achieved an unenviable notoriety as a carcinogen. This has, however, probably been exaggerated by the media. Ethyl carbamate is also considered to have tranquilizing, sleep-inducing properties at high doses. It was previously used as a preservative in the food industry, especially in beverages, but has now been banned.

In one survey (Bertrand, 1993), 1600 French wines from different vineyard regions were analyzed using the method described by Bertrand and Barros (1988). The average value for all types of wine was 7.7 µg/l with a standard deviation of 5.5 µg/l. Very few wines, therefore, had values over the 15 µg/l limit specified in an agreement between the American wine industry and the Food and Drug Administration (FDA) in 1988.

A great deal of research examining the origins of ethyl carbamate in wine and brandies is described in the literature. Bertrand et al. (1991) concluded that ethyl carbamate concentrations in wine are linked to grape variety, as well as to excessive nitrogen fertilization in the vineyards, but these factors are not highly significant.

These authors also observed that certain winemaking techniques, such as high temperatures especially during the final maceration, and not removing grape stems, cause increases in the ethyl carbamate content of the wine. Yeasts synthesize ethyl carbamate and also contain a precursor of this compound. The role of lactic bacteria has been demonstrated by the fact that wines that have undergone malolactic fermentation always have slightly higher concentrations of carbamic esters (a few µg/l) than other wines. All strains of lactic bacteria do not seem to have the same effect. Heterofermentative bacteria from the genus *Oenococcus œni* apparently produce lower quantities than homofermentative strains from the genus *Lactobacillus plantarum*. There also seems to be a correlation between the bioamine concentration and that of ethyl carbamate. These bacterial problems are amplified in wines with excessively high pH.

According to Bertrand (1993), the impact of aging on the ethyl carbamate content of wine varies from one vineyard region to another. Concentrations in Champagne, aged on yeast lees, increase more than those in Bordeaux wines.

Storage conditions, especially temperature, during distribution and sale may be a decisive factor affecting increases in ethyl carbamate in wine.

Its development throughout aging seems linked to the presence of urea, and also to the amino acids involved in the urea cycle (Figure 5.8). According to Bertrand (1993), urea is responsible for two-thirds of the ethyl carbamate that developed in a wine during five years' aging. This author adds that a five year old wine containing over 2 mg/l of urea has a theoretical ethyl carbamate concentration of around 20 µg/l.

In brandy, ethyl carbamate may also be formed from hydrocyanic acid. Bertsch (1992) followed the development over a two-hour period of hydrocyanic acid and ethyl carbamate in a distillate made from Baco Noir wine in an experimental still. He observed a negative correlation between these two compounds, which is explained by the following reaction sequence, suggested by Bauman and Zimmerli (1988):

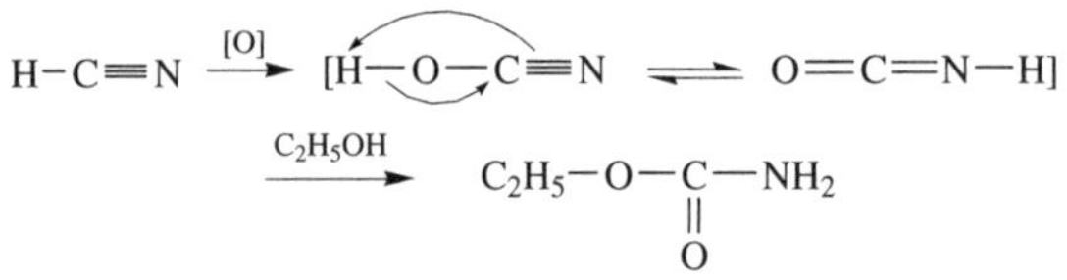

$$H-C\equiv N \xrightarrow{[O]} [H-O-C\equiv N \rightleftharpoons O=C=N-H]$$
$$\xrightarrow{C_2H_5OH} C_2H_5-O-\underset{\underset{O}{\overset{||}{}}}{C}-NH_2$$

This research demonstrated once again that the presence of copper, and especially light, during the distillation of this Baco (Noir) wine promoted a

Nitrogen Compounds

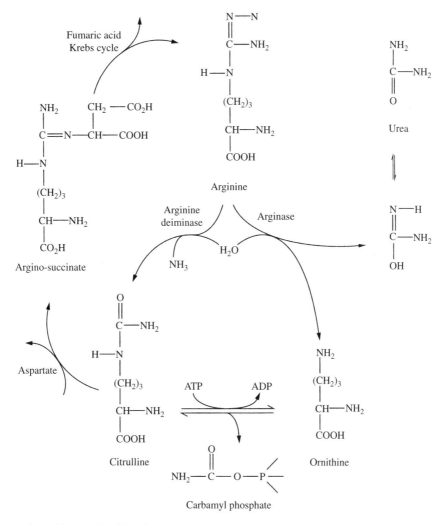

Fig. 5.8. Urea cycle and its relationship with the Krebs cycle. Comparative catabolism of arginine under the effect of arginase and arginine deiminase

significant increase in the ethyl carbamate content of the brandy. It was possible to reduce the ethyl carbamate concentration by fixing the hydrocyanic acid on resin immediately after distillation.

5.4.3 Bioamines

Since the time of Hippocrates, bioamines have been held responsible for physiological problems in humans. They are present mainly in foods and beverages produced by fermentation with lactic bacteria, including cheese, dry sausage, cider and beer. They are also observed in improperly stored foods such as fish and meat. Among these substances, histamine, responsible for allergic reactions and headaches, is the most carefully controlled. Wines containing more than 3.5 mg/l of histamine have been refused by Dutch importers. In Switzerland, a maximum level of 10 mg/l has been set to comply with health criteria that have not yet been fully defined (Bauza *et al.*, 1995). Whereas a healthy subject may be unaffected by absorbing 200–500 mg of histamine via the digestive system, 7 µg administered intravenously may produce undesirable effects (Mordelet-Dambrine and Parrot, 1970). Its structure is shown below:

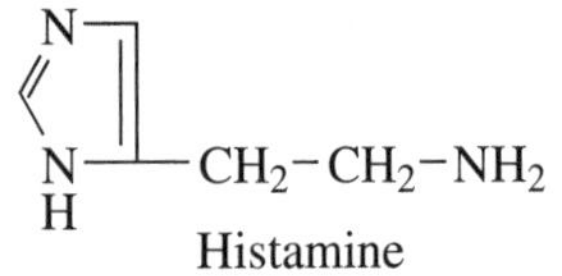

$$\underset{H}{\overset{N}{\underset{N}{\bigvee}}} - CH_2-CH_2-NH_2$$
Histamine

The automated method described by Lethonen *et al.* (1992) is capable of identifying and assaying over twenty bioamines in wine at the same time. It is thus possible to envisage a detailed study of these compounds, aimed at determining their possible responsibility for the symptoms experienced by certain subjects after ingesting wine.

From a biochemical standpoint, bioamines are mainly formed by the decarboxylation of amino acids due to decarboxylase activity. This phenomenon is assisted by pyridoxal phosphate from yeasts and bacteria (Table 5.4 and Figure 5.9). Bioamines therefore originate from fermentation.

Certain bacterial strains, such as *Pediococcus* or *Lactobacillus*, may contain this type of enzyme. They may also be capable of biosynthesizing them by induction, in the presence of an amino acid precursor (Brink *et al.*, 1990). Arginine is the amino acid precursor of several bioamines. Figure 5.9 shows that decarboxylation by bacterial arginine decarboxylase produces agmatine, a bioamine precursor of putrescine. Ornithine (Figure 5.9) may also be the direct precursor of putrescine and cause the formation of two other bioamines, spermidine and spermine. Arginine is therefore a direct or indirect precursor of four bioamines.

Besides these amines, methylamine, dimethylamine, ethylamine, hexylamine, isopentylamine, piperidine, propylamine, pyrrolidine and tryptamine are also present.

In wine, these various amines mainly originate from bacteria. Red wines have concentrations higher than those in white wines that are not biologically deacidified by malolactic fermentation. Concentrations in wine are on the order of a few tens of mg/l.

Histamine appears during fermentation, irrespective of the yeast strain used, and concentrations increase during malolactic fermentation. There is, however, no correlation between the histidine content in the must and the histamine concentration in the wine. Although wines generally only contain a few mg/l, concentrations in certain wines may exceed the 10 mg/l maximum value prescribed by legislation in some countries. The exact causes and the conditions responsible for the formation of large quantities of histamine in wine are not very well known. The most probable explanation is that this is due to the action of specific bacterial strains with a high histidine decarboxylase content (Volume 1, Section 5.4.2).

As may be expected, treating white wines with 50 g/hl of bentonite reduces bioamine

Table 5.4. The various bioamines and the corresponding amino acids

Amino acid	Bioamine
Arginine	Agmatine ⟶ Putrescine (↑ Spermidine, ↓ Spermine)
Cysteine	Mercaptoethylamine
Histidine	Histamine
Phenylalanine	Phenylethylamine
Serine	Ethanolamine
Ornithine	Putrescine
5-Hydroxytryptophan	Serotonin
Lysine	Cadaverine
Tyrosine	Tyramine

Nitrogen Compounds

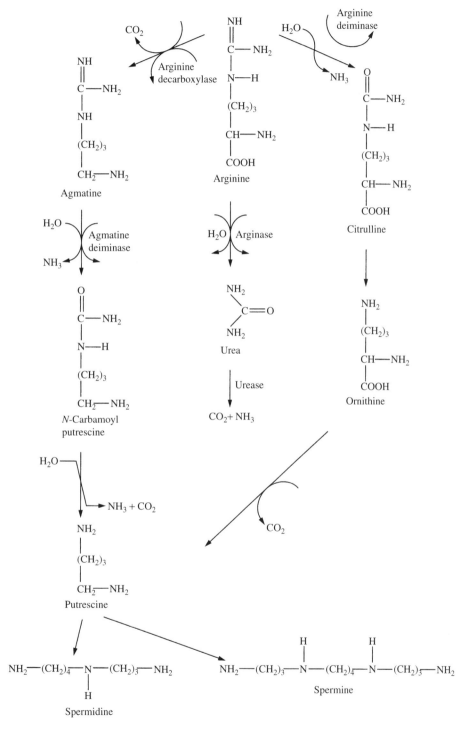

Fig. 5.9. Role of arginine in bioamine synthesis

concentrations. This treatment is particularly effective as amines are positively charged at the pH of wine, while bentonite particles have a negative charge.

5.5 PROTEINS AND PROTEIN CASSE

5.5.1 Proteins

Proteins are essential structural and functional components of all living organisms. These macromolecules, with molecular weights above 10 000, consist of clearly defined chains (or sequences) of amino acids linked by peptide bonds (Section 5.2.3). According to pH, proteins may be positively or negatively charged. They may also be at the isoelectric point (Section 9.2.4) and, therefore, neutral. The sequence of amino acids in the polypeptide chains determines proteins' three-dimensional structures, i.e. their spatial configurations.

Red wines hardly contain any free proteins, as they are precipitated by tannins. White and rosé wines, on the other hand, may have variable protein concentrations of up to a few hundred mg/l, mainly originating from grapes.

Proteins in must are a well-known cause of instability, affecting the clarity of white wines. When they precipitate, they cause 'protein casse', reported by Laborde as early as 1904. For many years, this was confused with 'white casse' or 'copper casse'. The turbidity or deposits characteristic of 'protein casse' appear in the bottle, usually when wines are stored at high temperatures. They may also occur when tannin is leached into wine from the cork. Tartrate crystallization and flocculated proteins are responsible for the main problems with clarity in bottled white wines.

Bentonite treatment has been recommended for eliminating unstable proteins from white wine for over 50 years (Saywell, 1934; Ribéreau-Gayon, 1935). However, the proteins responsible for protein casse were only studied much later, and further research is still required.

Koch and Sajak (1959) first showed that protein precipitates isolated from grape must, either by precipitation with ammonium sulfate or by heating, consisted of the same protein fractions separated by electrophoresis on paper. These protein fractions contain not only amino acids but also reducing sugars, tannins and cations. These fractions, which vary according to the grape variety, have also been identified in wine. Different fractions react differently to heat treatment. It is, however, possible to state that all of the protein compounds in must are involved to a greater or lesser extent in protein instability in wine.

According to Berg and Akioshi (1961), there is no correlation between the total protein concentration of must and the turbidity formed by heating as a result of protein instability. This indicates that proteins in must are not all equally heat sensitive. Bayly and Berg (1967) confirmed this hypothesis by purifying proteins using various techniques (dialysis, gel filtration and ion-exchange chromatography) and analyzing the fractions they had isolated by electrophoresis on polyacrylamide gel. The fractions thus purified had molecular weights between 18 000 and 23 000 Da. They had varying reactions to heat, although they were all affected.

The increase in peptide nitrogen concentrations as must turns into wine during alcoholic fermentation and barrel aging is due to peptides released by yeasts as they autolyze (Feuillat, 1974). These yeast peptides are considered to be thermostable. Heat-sensitive proteins from must generally remain at constant concentrations during fermentation. They are not assimilated by yeast as they are resistant to yeast proteases.

Hsu and Heatherbell (1987a, 1987b) studied proteins in Gewürztraminer and Riesling grapes by electrophoresis on polyacrylamide gel. They observed that the proteins which contribute to protein instability in wines are identical to those in must. These proteins have molecular weights between 12 500 and 30 000 Da, and isoelectric points (i.p.) ranging from 4.1 to 5.8. Some are glycosylated.

Paetzold *et al.* (1990) obtained similar results by fractionating proteins from Sauvignon Blanc must using electrophoresis on polyacrylamide gel, followed by chromatofocusing. Proteins in must from this grape variety have molecular weights between 13 000 and 67 000 Da (Figure 5.10) and a

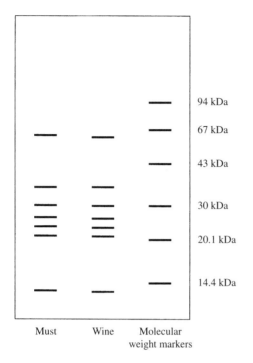

Fig. 5.10. Separation of proteins from Sauvignon Blanc must and wine by electrophoresis on polyacrylamide gel under denaturing conditions (Moine-Ledoux, 1996)

characterized by molecular weights of 24 000 and 32 000 Da. Their amino acid composition includes an unusually high proportion of aspartic acid, glycine, threonine and serine. A single fraction (34 000 Da) appears to be glycosylated.

It would seem, therefore, that the proteins responsible for instability in white wines come exclusively from grapes and have relatively low molecular weights, between 12 000 and 35 000 Da. However, the specific types of proteins, as well as their isoelectric points, degree of glycosylation and heat sensitivity, differ according to the grape variety.

Proteins in must or wine may be conveniently assayed by high-pressure liquid chromatography with molecular screening (Dubourdieu et al., 1986). Figure 5.12 shows an example of proteins from Sauvignon Blanc must separated into four fractions (A,B,C,D). Peak A contains proteins with molecular weights above 70 000 Da, as well as wine polysaccharides, as shown by the extent of the refractometric detection on this chromatogram. Peak B corresponds to proteins with molecular weights of around 65 000 Da. Peaks C and D are not completely separated. They include the group of proteins with molecular weights between 15 000 and 30 000 Da. Fractions B, C and D are likely to flocculate due to the effects of heat and tannins.

More recently, Moine-Ledoux et al. (1996) used the remarkable resolution of capillary electrophoresis (CE) to separate and assay proteins in must and wine. The wine was first dialyzed with

wide range of i.p., from 4 to over 7 (Figure 5.11). Most of the proteins separated in this way, however (5 bands out of 7 in Figure 5.10), have molecular weights in the vicinity of 20 000 and 30 000 Da.

Waters et al. (1991, 1992) showed that unstable proteins in Muscat of Alexandria wines are

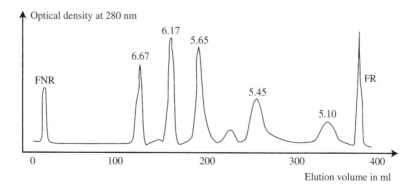

Fig. 5.11. Separating the proteins in Sauvignon Blanc must by chromatofocusing (Paetzold et al., 1990). FNR, fraction not retained, i.p. of 7 or higher; FR, fraction retained, i.p. of 4 or lower

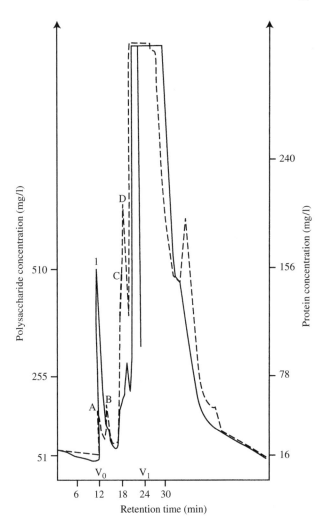

Fig. 5.12. Separating proteins from a Sauvignon Blanc wine by liquid chromatography using molecular screening (Dubourdieu et al., 1986). Proteins (- - - - -) are detected in the ultraviolet at 225 nm, polysaccharides (———) by refractometry

a 50 mM citrate buffer at pH 2.5 to eliminate the charged substances (peptides, amino acids, phenols and ions) likely to interfere with the assay. Under these conditions, the proteins in wine were stable and positively charged. They migrated from the anode toward the cathode in the silica capillary and were detected at the outlet by spectrophotometry at 200 nm. Figure 5.13a shows results of the CE separation of proteins in a Sauvignon Blanc wine at the end of alcoholic fermentation. There are six peaks, each probably corresponding to a pure protein. The quantities of each protein species (Table 5.5a) are given in equivalent bovine albumin serum (BAS). Figure 5.13b and Table 5.5b show the result of protein analyses carried out on the same wines after ten months aging on the lees. The impact of aging methods on a wine's protein composition is dealt with later, in this chapter (Section 5.6.4).

The electrophoresis profiles for a particular grape variety, e.g. Sauvignon Blanc, do not differ qualitatively according to vineyard, region or

Nitrogen Compounds

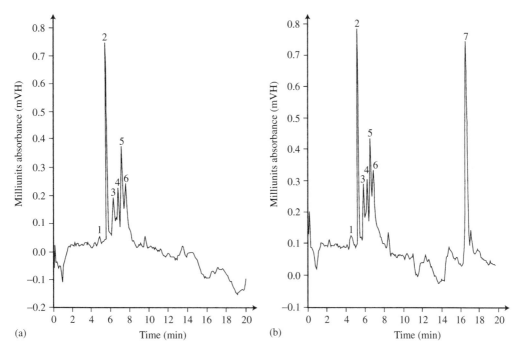

Fig. 5.13. Separating proteins in Sauvignon Blanc wine by capillary electrophoresis (Moine-Ledoux *et al.*, 1996): (a) at the end of alcoholic fermentation and (b) after 10 months aging on the lees

Table 5.5. Analysis of a Sauvignon Blanc wine by capillary electrophoresis (CE) (Moine-Ledoux *et al.*, 1996)

Peak Number	Retention time (min)	Proteins (mg/l eq. BAS)
	(a) At the end of fermentation	
1	5.01	5
2	5.73	36
3	6.44	9
4	6.9	11
5	7.3	19
6	7.67	12
	(b) After 10 months aging on the lees	
1	5.2	5
2	5.82	37
3	6.36	9
4	6.9	10
5	7.3	19
6	7.67	11
7	16.93	25

vintage. The must of a particular grape variety always contains the same protein species, but quantities naturally vary according to the origin. There is, however, considerable variation from one grape variety to another (Moine-Ledoux, 1996). The overall quantities of proteins assayed in must range from a few tens of mg to nearly 300 mg/l.

Proteins precipitated out of Sauvignon Blanc and Pinot Noir wines using ammonium sulfate have been likened to pathogenesis-related (PR) proteins. These findings are based on the fact that they have homologous sequences (Waters *et al.*, 1996; Peng *et al.*, 1997; Pocock *et al.*, 1998) and identical molecular weights (Hayasaka *et al.*, 2001) with 2 categories of plant PR proteins, chitinases, and a structural protein, thaumatin. These proteins accumulate during ripening at the same time as sugars (Salzman *et al.*, 1998) and independently of water supply (Pocock *et al.*, 1998). These PR proteins have also been reported in both grape flesh and skins (Pocock *et al.*, 2000). Their presence in grape skins was, however, not confirmed by Ferreira *et al.* (2000), who used immunodetection to locate proteins previously identified in white wine in grapes and found them only in grape flesh.

The resemblance between grape and PR proteins (chitinase and thaumatin) does not prove that they are identical molecules. In fact, none of the proteins in grapes have the extreme sweetening properties of thaumatin and thaumatin and chitinase antibodies do not react with grape proteins (Piçarra-Pereira *et al.*, 1998).

5.5.2 The Protein Casse Mechanism

The mechanism of protein casse in wine is usually (Ribéreau-Gayon *et al.*, 1976) included in the general diagram of flocculation of a hydrophilic colloid (Section 9.3.2). Flocculation requires the disappearance of two stabilizing factors: charge and hydration.

The tannin-protein complex formed in the presence of tannin is similar to the negative hydrophobic colloid that flocculates under the effect of cations. In the same way, heating a white wine to 70 or 80°C for a sufficiently long time may precipitate almost all the proteins. If the wine is heated rapidly, the turbidity only appears during cooling. Heating does not directly precipitate or coagulate proteins in wine or must, but converts them into a form that is soluble at high temperatures, becoming insoluble and flocculating at lower temperatures. Thus, heating denatures proteins by eliminating water. They then flocculate on contact with tannin and cations, which also settle out in the deposit.

In fact, the turbidity formed by proteins during heat treatment differs according to their isoelectric point (Dawes *et al.*, 1994). Proteins with an i.p. above 7 form a compact precipitate, while those with an i.p. between 5.94 and 4.65 flocculate. Proteins with an i.p. below 4.65 form turbidity in suspension. When a mixture of these fractions is heated, however, a compact precipitate is formed. Interactions must therefore occur between proteins during flocculation.

5.5.3 Vineyard Management Factors and Winemaking Techniques that Enhance the Protein Content of Must, Making it More Difficult to Stabilize the Wine

The protein concentration of must depends on the variety and ripeness of the grapes, as well as the way they are handled prior to fermentation (Paetzold *et al.*, 1990). For white Bordeaux grape varieties (Sauvignon Blanc, Semillon and Muscadelle), the concentration of unstable proteins in juice increases as the grapes ripen. Musts obtained by immediate pressing of ripe Sauvignon Blanc and Muscadelle grapes have similar protein concentrations (Figure 5.14). If the juice is extracted under the same conditions (with or without prolonged skin contact), Semillon must from the same vineyard area has a lower protein content.

In ripe Sauvignon Blanc and Semillon, prolonged skin contact causes a considerable increase in unstable protein concentrations in the must, as compared to immediate pressing. It doubles the unstable protein content of Sauvignon Blanc must and increases that of Semillon by 50%. Proteins are diffused in the must during the first few hours of skin contact (Figure 5.15). Adding sulfur to the grapes during skin contact, even at low doses, also promotes protein extraction. It is therefore preferable for grapes to be protected from oxidation by an inert CO_2 atmosphere during skin contact, rather than using SO_2.

Finally, juice produced by the immediate pressing of destemmed grapes has a much higher protein

Nitrogen Compounds

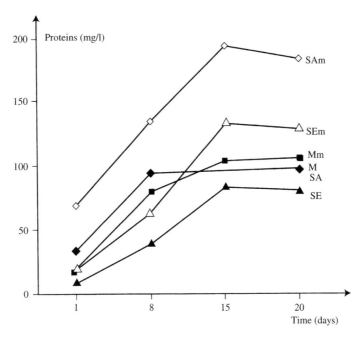

Fig. 5.14. Changes in protein concentrations in must during ripening according to the grape variety and the way the juice was extracted (Paetzold *et al.*, 1990). SE, Semillon, immediate pressing; SEm, Semillon with skin contact; SA, Sauvignon Blanc, immediate pressing; SAm, Sauvignon Blanc with skin contact; M, Muscadelle, immediate pressing; Mm, Muscadelle with skin contact

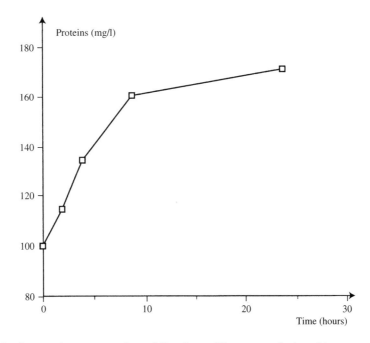

Fig. 5.15. Changes in the protein concentration of Sauvignon Blanc must during skin contact before fermentation (Paetzold *et al.*, 1990)

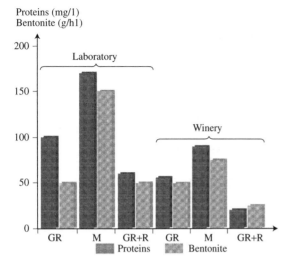

Fig. 5.16. Effect of destemming on the protein concentration of Sauvignon Blanc must and the amount of bentonite required for stabilization (Paetzold *et al.*, 1990). GR, immediate pressing of destemmed grapes; M, destemmed grapes with skin contact before fermentation; GR + R, immediate pressing of whole bunches of grapes

content than it would if the bunches were left whole. The doses of bentonite necessary for stabilization are clearly lower when whole bunches are pressed (Figure 5.16). Tannins from the stalks therefore have a particular aptitude for fixing proteins in must when the grapes are pressed (Dulau, 1990). Thus mechanical grape-harvesters that eliminate stalks may be considered one of the major factors in the protein instability of wines made from certain grape varieties.

This also explains the increase in the doses of bentonite necessary to achieve stabilization over the past couple of decades. Indeed, according to Ribéreau-Gayon *et al.* (1977), 20–40 g/hl of bentonite was generally regarded to be sufficient to prevent protein casse. Currently, doses of 80–120 g/hl are often necessary. This phenomenon is no doubt due to changes in vineyard and winemaking techniques. Riper grapes, mechanical harvesting, prolonged skin contact, particularly if the grapes are sulfured, and changes in pressing techniques, leading to the incorporation of a larger proportion of press-juice, have resulted in wines with a higher protein content and, consequently, an increase in the doses of bentonite required to stabilize them. It is also undeniable that loss of aroma occurs due to the use of excessive doses of bentonite (Simpson and Miller, 1984). White wines from all over the world have the same problems.

5.5.4 Protein Stability Tests

Various laboratory tests have been used for many years to assess the risk of protein turbidity before bottling. These tests are based on the instability of proteins under various conditions: at high temperatures, or in the presence of tannin, trichloroacetic acid, ethanol or reagents based on phosphomolybdic acid. These tests do not all produce the same results. Some of them overestimate the risk of casse during bottle aging. This may lead to the use of much higher doses of bentonite than would be strictly necessary to ensure the protein stability of the wine.

Heating wine in a water bath at 80°C for 30 min is one of the most widely used, reliable tests. Turbidity appears during cooling. This technique is the best suited for predicting protein casse during bottle aging. Wines are considered stable if the additional turbidity caused under these conditions has a turbidity level less than 2 NTU (Dubourdieu *et al.*, 1988). Heating to 90°C for 1 h is too drastic. Heating a full, hermetically sealed flask in an oven at 30 or 35°C for 10 days takes too long to be of any practical use. Furthermore, the turbidity measured when the wine is removed from the oven is generally less than that obtained using the 80°C test.

Doses of 0.5–2 g/l of tannin made from gall nut may also be added to wine. Turbidity is immediately visible. This procedure produces different amounts of turbidity from the test at 80°C, depending on the way the tannin was prepared (extracted with water, alcohol or ether). Certain tests combine adding tannin (0.5 g/l) with heating to 80°C for 30 min. They give higher turbidity values than the same heat treatment in the absence of tannin (Table 5.6).

The Bentotest (Jakob, 1962), involving the addition of a reagent based on phosphomolybdic acid

Nitrogen Compounds

Table 5.6. Turbidity levels (NTU) after different protein stability tests carried out on a Sauvignon Blanc wine during barrel aging on the lees (V. Moine-Ledoux, 1997, unpublished results)

Tests	End of fermentation	After 4 months on the lees	After 11 months on the lees
30 min at 80°C	45	34	34
30 min at 80°C with added tannins	58	64	76
Bentotest	70	103	166

Table 5.7. Doses of bentonite (g/hl) required for the protein stabilization of a dry Sauvignon Blanc wine, determined by different stability tests during the period of barrel aging on the lees (V. Moine-Ledoux, 1997, unpublished results)

Tests	End of fermentation	After 4 months on the lees	After 11 months on the lees
30 min at 80°C	140	100	60
30 min at 80°C with added tannins	140	100	60
Bentotest	180	220	300

(10%), has also been recommended for assessing the risk of protein casse. The liquid turns blue and turbidity appears instantaneously. The Bentotest does not specifically react to heat-sensitive proteins, so it systematically overestimates the risk of protein problems (Table 5.6).

The same is true of the trichloroacetic acid test (10 ml at 55% per 100 ml of wine). The mixture is heated in a waterbath at 100°C for 2 mn and turbidity is observed after 15 min at room temperature. This test is hardly more effective than the Bentotest if the addition of excessive quantities of bentonite to wine is to be avoided.

The ethanol test (Boulton *et al.*, 1996) consists of adding a volume of absolute ethanol to the same volume of wine. The turbidity formed in the presence of alcohol, measured by nephelometry shortly after the substances are mixed, does not consist only of unstable proteins, as polysaccharides and particularly mannoprotein also precipitate. It is relatively common for wines that are perfectly thermostable to become turbid when subjected to this ethanol test, particularly if they have acquired a high mannoprotein content through aging on the lees. This test leads to the use of higher doses of bentonite than are truly necessary to achieve stability.

To determine the dose of bentonite necessary for protein stabilization in a wine, laboratory tests are carried out with increasing doses of bentonite. When this has been eliminated by centrifugation or filtration, the clear wine is subjected to the protein stability test. The choice of test is of considerable importance in assessing the right dose of bentonite (Tables 5.6 and 5.7). When a white wine is aged on its lees, its turbidity, assessed by the heat test without adding tannin, clearly decreases (Table 5.6), so smaller doses of bentonite are required (Table 5.7). The turbidity of the same wine assessed by the Bentotest increases considerably during aging, as do the recommended doses of bentonite. After the same wine had been barrel aged for 10 months, it would need to be treated with 300 g/hl of bentonite to remain stable according to the Bentotest, compared to only 60 g/hl according to the heat test (with or without tannins). The latter result is much more realistic, as the spontaneous protein stability of wine is known to increase during aging on the lees (Section 5.6.4).

Thus the heat test (80°C, 30 min) without added tannin is the most effective way of assessing the protein stability of a white wine.

5.6 PREVENTING PROTEIN CASSE

5.6.1 Principle of Stabilization Treatments to Prevent Protein Casse

As early as 1904, Laborde recommended eliminating proteins by prolonged heating (15–30 min at 70–80°C) to prevent the formation of turbidity in bottled wines. As this process is obviously likely to have a negative impact on the organoleptic qualities of the wines, it is no longer used.

The addition of tannins may precipitate proteins, but very high doses are necessary to eliminate them almost completely (2 g/l) and this technique is therefore impractical. If smaller quantities of tannins are added (0.1–0.5 g/l), all of the unstable proteins are not precipitated. Furthermore, the soluble tannin-protein complex formed is markedly more heat sensitive and less easily adsorbed by bentonite. In other words, the use of low doses of tannin tends to make wine more susceptible to protein casse rather than stabilizing it. There is a parallel between these phenomena and the increase in the sensitivity of white wines to protein casse during aging in new barrels, in the absence of lees. Under these conditions, the dose of bentonite required to stabilize the wine generally increases significantly during the barrel aging period. In some cases, a wine may even be stable according to the heat test at the beginning of barrel aging and become unstable a few months later due to tannins released from the oak. In the presence of lees, on the contrary, wine evolves spontaneously toward protein stability, even in the presence of tannins (Section 5.6.4).

The effect of prolonged cooling, keeping wine around the freezing point, only causes partial precipitation of the proteins. This treatment is therefore never sufficient to ensure protein stability.

Ribéreau-Gayon demonstrated in 1932 that it was possible to adsorb unstable proteins in wine using kaolin, a negatively charged clay. However, large quantities of kaolin, on the order of 500 g/hl, are necessary to obtain protein stability. This makes kaolin treatment impracticable due to the volume of lees it produces and the amount of wine lost.

In the USA a few years later, Saywell (1934) recommended using another clay, bentonite, for clarifying and stabilizing wine. He paid no particular attention to the effect of bentonite on proteins. The following year, Ribéreau-Gayon (1935) studied the properties and effect of bentonite on proteins. Bentonite was capable of adsorbing proteins likely to precipitate in wine at doses 10 times lower than kaolin. It was also effective in preventing copper casse. No other treatment discovered since that time has proved capable of replacing bentonite in preventing protein problems in white wines.

5.6.2 Using Bentonites to Eliminate Proteins

Bentonites are hydrated aluminum silicates, mainly consisting of montmorillonites (Al_2O_3, $4SiO_2 \cdot nH_2O$) (Section 10.9.1). When bentonites are put into suspension in water or wine they form a colloidal dispersion with negatively charged particles. These are capable of fixing proteins which are positively charged at the pH of wine. Bentonite initially adsorbs proteins with higher isoelectric points (above 6). The elimination of proteins with lower isoelectric points requires higher doses of bentonite (Hsu and Heatherbell, 1987b; Paetzold et al., 1990).

Bentonites contain exchangeable cations (Mg^{2+}, Ca^{2+} and Na^+) in variable proportions according to their geographical origins. Sodium bentonites are the most widely used, as they are most effective in treating wine. They swell more in water and have a markedly higher adsorption capacity for proteins. Bentonite treatment causes a slight increase in the wine's sodium content (around 10 mg/l), but only a negligible decrease in acidity. Good quality bentonite has practically no flavor or odor. However, when bentonite is used in white wines at high doses (above 80 g/hl) it may attenuate their organoleptic characteristics. The intense aromas of Sauvignon Blanc wines are often dulled by excessive bentonite treatment, due to the elimination of part of the 4-mercapto-4-methylpentanone.

For a long time, the use of bentonite in must during fermentation was recommended for stabilizing white wines. This advice was based on several sound arguments. Bentonite treatment 'tires' the wines less during fermentation than it does at later stages. Furthermore, fewer operations are required and liquid losses are reduced, as the bentonite settles out with the yeast after fermentation is completed. The volume of the lees is not significantly increased. Bentonite treatment partially eliminates tyrosinase, thus limiting oxidation of the must. It also stimulates fermentation by acting as a support for yeast. Bentonite added to must has also been credited to a certain extent with adsorbing fungicide residues.

All these advantages depend, however, on the new wine being racked shortly after completion of alcoholic fermentation. This is indeed the case when wines are fermented in vat and intended for early bottling, within 2 or 3 months. The situation is quite different when wines are aged on the lees. Leaving the wine on bentonite for several months and stirring it into suspension at regular intervals is certainly detrimental to quality. It would, however, be a pity to have to rack barrel-fermented dry white wines at an early stage to eliminate bentonite added to the must. The wines would dry out in the new oak under these conditions, as they would be deprived of the reducing capacity of the yeast lees. As a result, bentonite treatment of the must is unsuitable for all dry white wines aged on the lees and, above all, high-quality, barrel-fermented white wines with aging potential. These wines should be treated when they have finished aging. Furthermore, contact with the lees improves their protein stability considerably and lower doses of bentonite are then required for stabilization (Section 5.6.4).

Practical considerations covering the use of bentonite and the conditions for clarifying treated wines (Ribéreau-Gayon *et al.*, 1976, 1977) are described elsewhere (Section 10.9.5).

When bentonite is used during fermentation, it is preferable to add it to clarified must, as this treatment tends to delay settling. It is not necessary to swell the bentonite in water or must a few hours before treatment. The bentonite should simply be sprinkled on the must as it drains into a bowl during pumping-over and the mixture subjected to vigorous agitation. Obviously, if this operation takes place before alcoholic fermentation, there is a risk of oxidizing the must. This is likely to be detrimental to the aromatic expression of many grape varieties. It would therefore be better to add bentonite at the beginning of alcoholic fermentation, to benefit from the presence of CO_2 and the protective effect of the yeast. At the same time, this operation provides aeration which is favorable for yeast development. Bentonite treatment of must should, however, be considered appropriate only in 'industrial' winemaking, aimed at producing very straightforward wines. Even the most modest dry white wines show an improvement in quality after aging on the lees, even for a short time.

5.6.3 Possible Substitutes for Bentonite Treatment

Several substitution treatments for bentonite have been tried, but none of them have been capable of guaranteeing stabilization without detracting from the wine's organoleptic qualities.

Tangential ultrafiltration on membranes with a cutoff at 50 000 Da does not eliminate unstable proteins from wine. In fact, a cutoff threshold on the order of 10 000 Da (Hsu and Heatherbell, 1987b) is necessary to ensure protein stability in the permeate. However, membranes this fine cause a deterioration in the aromatic qualities of the wines, probably due to the elimination of macromolecules (Feuillat *et al.*, 1987).

The addition of enological tannins only eliminates part of the proteins, and the new tannin-protein complexes in the wine are generally more heat sensitive than the original proteins.

Hyperoxygenation of must is insufficiently effective to guarantee protein stability in wine. Furthermore, oxidation of the must is clearly detrimental to the varietal aroma of wines made from certain grape varieties, such as Sauvignon Blanc.

The use of proteolytic enzymes seems, theoretically, to be the most appropriate tool for selectively eliminating proteins from wine. The first proteinases demonstrated in winemaking were initially studied in grapes. The role of these enzymes in protein breakdown in must during fermentation remains limited, however, as they are inhibited by alcohol (Cordonnier and Dugal, 1968).

Taking advantage of the proteolytic potential of winemaking yeast (Feuillat *et al.*, 1980) seemed to provide an attractive solution to the problem of protein stabilization in wine. Dulau (1990) tried to make use of the proteolytic capacity of winemaking yeast (*Saccharomyces cerevisiae*). The PEP 4 gene, coding for protease A with vacuolar targeting, was cloned in a multi-copy plasmid. These plasmids were used to convert a laboratory strain of *S. cerevisiae*. Overexpression of the PEP 4 gene in the yeast caused it to secrete protease A. This was easily visualized on a solid medium containing casein. The converted clones were used to ferment small quantities of Sauvignon Blanc must, and the results were compared with those produced by the same strain of yeast converted by the plasmid without the insert. Overexpression of the PEP 4 gene only caused minimal breakdown of proteins in the must during alcoholic fermentation. It was insufficiently effective to be of any practical use. The same strategy was also explored by Lourens and Pretorius (1996). All this research has shown that it is rather easy to modify yeast genetically, causing it to excrete protease A. The major difficulty, however, is that grape proteins are singularly resistant to protease A.

The resistance of proteins in must or wine to various types of peptidases is well known. Waters *et al.* (1992) showed that peptidases capable of hydrolyzing bovine albumin serum in a dilute alcohol medium also digested this serum when it was added to must or wine. On the other hand, these peptidases left the unstable natural grape proteins intact. It is possible to conclude that there is no specific enzyme in wine or must that inhibits protease activities, but that grape proteins have an unusually strong inherent resistance to hydrolysis. These findings indicate that there is no specific enzyme in wine or must that inhibits protease activities. However, independently of their association with tannins or degree of glycosylation, grape proteins have an unusually strong inherent resistance to hydrolysis (Waters *et al.* (1995a, 1995b)).

It has been observed that some white wines, kept on their lees for several months during traditional barrel aging, are capable of acquiring a certain stability in relation to protein casse. It therefore seems useful to start by understanding this phenomenon and then possibly finding a way to take advantage of this property to improve protein stability in wine.

5.6.4 Molecular Interpretation of the Enhanced Protein Stability of White Wines Aged on the Lees: practical Applications

The systematic improvement in the protein stability of white wines during barrel aging on the lees is easily verified by winemakers (Ledoux *et al.*, 1992). New wines kept on their lees become less and less turbid on heating as they age. Consequently, lower doses of bentonite are required to achieve protein stability. A Sauvignon Blanc wine racked directly after fermentation required treatment with 120 g/hl of bentonite to avoid protein casse. After 10 months aging on the lees, it was practically stabilized by adding 30–40 g/hl of bentonite (Figure 5.17). However, the grape proteins responsible for protein casse in white wines are

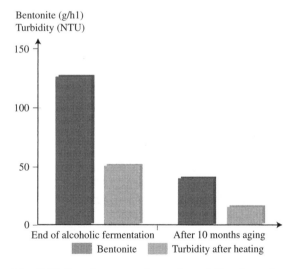

Fig. 5.17. Development of protein stability in a dry Sauvignon Blanc wine barrel aged on total lees. This was assessed by the increase in turbidity after heating and by the quantity of bentonite necessary to achieve stability (Ledoux *et al.*, 1992)

Nitrogen Compounds

neither digested nor adsorbed by the yeast lees during barrel aging. They become thermostable in the presence of certain colloids released into the wine by yeast cell walls. This observation led to the hypothesis that certain yeast parietal mannoproteins, released into wine kept in contact with the lees, were capable of decreasing the heat sensitivity of proteins. This was in agreement with the stabilizing effect of a high molecular weight mannoproteins (420 kDa) extracted from a Muscat wine by Waters *et al.* (1993). Gum arabic and arabinogalactanes purified from Chardonnay (Pellerin *et al.*, 1994) and Carignan Noir (Waters *et al.*, 1994) grapes have also been attributed a similar, though less effective, protective role.

It has already been reported (Section 5.5.1) that the proteins responsible for protein casse in a Sauvignon Blanc wine just after fermentation could be separated into six major species by capillary electrophoresis (Figure 5.13a and Table 5.5a). The same six peaks, corresponding to similar protein concentrations, were found in the same wine aged on its lees. After a few months of barrel aging, however, an additional protein fraction appeared (peak 7 on Figure 5.13b and Table 5.5b). The six peaks corresponding to proteins that were already present in the must were all relatively unstable on heating. Peak 7, however, corresponding to a protein that appeared during aging on the lees, was stable (Figure 5.18). While the proteins corresponding to peaks 1 to 6 were adsorbed by bentonite at varying doses, the protein in peak 7 was not eliminated by bentonite (Figure 5.19). Besides the protein that originated from yeast (peak 7), the protein profiles of wines aged with or without lees were identical. It was therefore perfectly logical to assume that this thermostable protein compound that was not adsorbed by bentonite had an impact on the protein stability of wines aged on the lees (Moine-Ledoux, 1996).

The compound corresponding to peak 7 may be extracted from yeast cell walls *in vitro* by digesting them with Glucanex™ (Novo-Swiss Ferment), a special enzyme preparation that includes various β-glucanases and a protease (Dubourdieu and Moine-Ledoux, 1994). This enzyme is already permitted by EU legislation for breaking

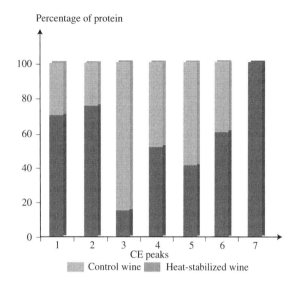

Fig. 5.18. Heat stability of various proteins in a Sauvignon Blanc wine separated by capillary electrophoresis (CE). Protein 7, which appears during aging on the lees, is perfectly stable. All of the other proteins are relatively unstable (Ledoux *et al.*, 1992)

down the (1-3:1-6)-β-D-glucane of *Botrytis cinerea* (Sections 3.7.2 and 11.5.2), a colloid responsible for difficulties in clarifying wines made from grapes affected by rot (Dubourdieu *et al.*, 1985). When the yeast cell walls are digested by Glucanex™, they release a mixture of mannoproteins including the same heat-stabilizing protein compound that appears in wine during aging on the lees (peak 7). The mannoproteins thus extracted by enzymes (MPEE) from the yeast cell walls (purified by ultrafiltration and dried) are capable, at doses of 25 g/hl, of halving the dose of bentonite required for protein stabilization of extremely heat-sensitive wines (Figure 5.20).

The active compound corresponding to peak 7 was purified in two stages by ion-exchange chromatography on DEAE Sepharose and then affinity chromatography on Concanavaline A (Moine-Ledoux, 1996). The purified product was highly effective for protein stabilization, as only a few mg/l were sufficient to make a Sauvignon Blanc wine perfectly stable. It would otherwise have required bentonite treatment at a rate of 100 g/hl.

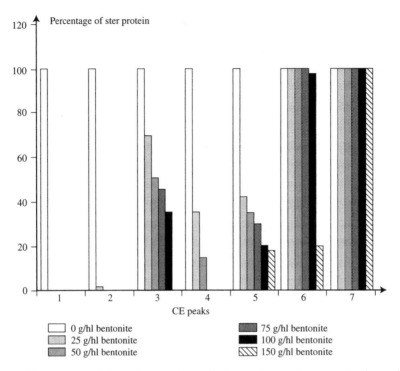

Fig. 5.19. Influence of the quantity of bentonite used to stabilize a wine on the concentrations of various proteins, separated by capillary electrophoresis (CE). Protein 6 is not properly eliminated by bentonite. Protein 7 is not eliminated at all. (Ledoux *et al.*, 1992)

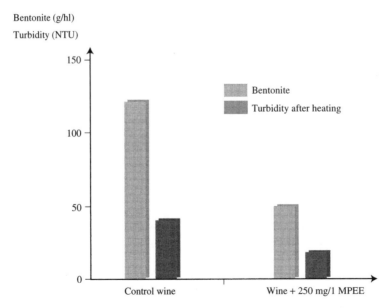

Fig. 5.20. Effect of adding (250 mg/l) mannoproteins extracted by enzymes from yeast cell walls (MPEE) on the protein stability of various white wines, measured by the quantity of bentonite necessary to stabilize them and the appearance of turbidity on heating (Dubourdieu and Moine-Ledoux, 1994)

Nitrogen Compounds

The purified heat-stabilizing product is a 31.8 kDa mannoprotein (known as MP32), consisting of 27% protein and 62% mannose. Its peptide fraction has been identified as a fragment of the parietal invertase of *Saccharomyces cerevisiae* (Dubourdieu and Moine-Ledoux, 1996). This result was obtained by microsequencing this peptide, generated by enzyme digestion of MP32, using an endoprotease. The same peptide sequence, Val–Phe–Trp–Tyr–Glu–Pro–Ser–Gln–Lys (VFWYEPSQK), is found in the parietal invertase of yeast (Figure 5.21). It corresponds to amino acids 174 to 182 in invertase. Furthermore, as the sequenced peptide results from a cut by lys C endoprotease, it is necessarily followed by a lysine residue. If the lysine is taken into account, MP32 and *S. cerevisiae* invertase have the same sequence of ten amino acids.

However, the molecular weight of MP32 (31.8 kDa) is significantly lower than that of invertase (270 kDa). In fact, MP32 is an invertase fragment, released during yeast autolysis caused by the conditions occurring during barrel aging on the lees. Two categories of yeast enzyme activity are probably involved in its release: parietal glucanases for cutting the link holding the invertase on the glucane of the cell wall, and proteases that digest the peptide part of the invertase.

From a practical standpoint, the improvement in protein stability in white wines during barrel aging is strongly affected by various parameters: the length of barrel aging, the quantity of lees, the age of the barrels and the frequency of stirring. In the example in Figure 5.22, in June, a dry Sauvignon Blanc wine barrel aged on total lees required less than 40 g/hl of bentonite for complete stabilization. A dose of 90 g/hl was required to achieve the same result in the same wine when the lees had been partially eliminated before aging. The MP32 concentration was only 10 mg/l in the wine aged

ytfteyqknp vlaanstqfr dpkvfwyeps qkwimtaaks qdykieiyss
151   174   182   200

Fig. 5.21. Sequence of amino acids 151–200 in *S. cerevisiae* invertase. The sequence 174–182 is identical to the peptide fragment sequenced from MP32 (Moine-Ledoux, *et al.*, 1996)

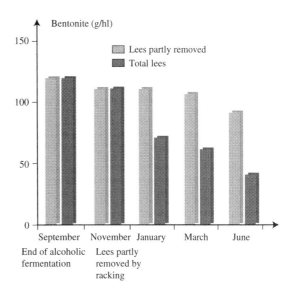

Fig. 5.22. Changes in the dose of bentonite necessary to stabilize a dry Sauvignon Blanc wine during aging on partial or total lees (Moine-Ledoux, 1996).

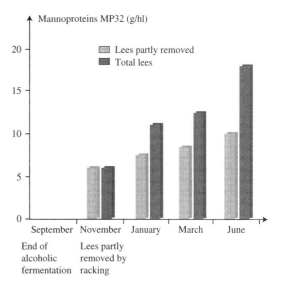

Fig. 5.23. Changes in the MP32 concentration in a dry Sauvignon Blanc wine during aging on partial or total lees (Moine-Ledoux, 1996).

on partial lees, as compared to 18 mg/l in the wine aged on total lees (Figure 5.23). This spontaneous improvement in protein stability during aging is faster in used rather than new barrels. It is also accelerated by more frequent stirring. Research is currently in progress to develop a product from yeast or invertase that would take the place of bentonite in stabilizing wine and preventing protein casse (Dubourdieu and Moine-Ledoux, 1996).

It should also be noted that the purified mannoprotein preparation obtained by digesting yeast cell walls contains another protein fraction that has a protective effect on tartrate precipitation (Moine-Ledoux *et al.*, 1997) (Section 1.7.7). It is perfectly possible to envisage using an industrial preparation of this protective colloid in the near future to stabilize white wines and prevent tartrate precipitation.

Other methods for obtaining the mannoprotein (420 kDa), other than MP32, with a protective effect on protein casse (Waters *et al.*, 1993) were tested by Dupin *et al.* (2000). The most effective extracts were obtained by treating whole yeast cells with EDTA. This mannoprotein was located in the entire cell wall by immunodetection and was linked to the other constituents by non-covalent bonds.

REFERENCES

Bauman U. and Zimmerli B. (1988) *Milt. Gebiete Lebensm. Hyg.*, 79, 175.
Bauza A., Blaise A., Teissedre P.L., Cabanis J.C., Kanny G., Monneret-Vautrin D.A. and Daumas F. (1995) *Bull. OIV*, 68 (767–768), 43.
Bayly F.C. and Berg H.W. (1967) *Am. J. Enol. Viticult.*, 18, 18.
Berg H.W. and Akioshi M. (1961) *Amr. J. Enol. Viticult.*, 12 (3), 107.
Berthier L., Marchal R., Debray H., Bonnet E. and Jeandet Maujean A. (1999) *J. Agric Food Chem.*, 47, 2193.
Bertrand A. (1993) *Les Acquisitions Récentes en Chromatographie du Vin.* Tec et Doc, Lavoisier, Paris, p. 175.
Bertrand A. and Barros P. (1988) *Conn. Vigne et Vin*, 22 (1), 39.
Bertrand A., Ingargiola M.C. and Delas J. (1991) Proceedings of the International Symposium on *Nitrogen in grapes and wine* (eds. The American Society of Enology and Viticulture), p. 215.
Bertsch K. (1992) Des eaux-de-vie d'armagnac: critères analytiques de la qualité, étude sur lo carbamate d'éthyle. Thèse Doctorat, l'Université Bordeaux II.
Boulton R.B., Singleton V.L., Bisson L. and Kunkee R.E. (1996) *Principles and Practices of Winemaking.* Chapman and Hall Enology Library, New York.
Brink B.T., Damin C., Joosten H.L.M.J. and Huis Int Velt J.H.J. (1990) *Int. J., for Microbiol.*, 11, 73.
Colagrande O. (1989) *Revue des Œnologues*, 53, 19.
Cordonnier R. and Dugal A. (1968) *Ann. Techn. Agric.*, 17, 189.
Dartiguenave C., Jeandet P. and Maujean A. (2000) *Am. J. Enol. Vitic.*, 51 (4), 347 et 51, 4, p. 352.
Dawes H., Boyes S., Keene J. and Heatherbell D. (1994) *Am. J. Enol. Viticult.*, 45 (3), 319.
De Roissart H. and Luquet F.-M. (1994) *Bactéries Lactiques. Aspects Fondamentaux et Technologiques*, Vol. I. Lorica.
Desportes C., Charpentier M., Duteurtre B., Maujean A. and Duchiron F. (2000) *J. Chromatogr.*, 893, 281.
Dubourdieu D. and Moine-Ledoux V. (1994) Brevet d'Invention Français 94-13261 and International PCT-FR95-01426.
Dubourdieu D. and Moine-Ledoux V. (1996) Demande de Brevet d'Invention Français 9608187.
Dubourdieu D., Villetaz J.C., Desplanques C. and Ribéreau-Gayon P. (1985) *Carbohyd. Res.*, 144, 277–287.
Dubourdieu D., Llauberes R.M. and Ollivier C. (1986) *Conn. Vigne Vin*, 20 (2), 119.
Dubourdieu D., Serrano M., Vannier A.C. and Ribéreau-Gayon P. (1988) *Conn. Vigne Vin*, 224, 261–273.
Dulau L. (1990) Recherches sur les protéines responsables de la casse protéique des vins blancs secs. Thèse de Doctorat, Université de Bordeaux II.
Dupin I.V.S., Stockdale V.J., Williams P.J., Jones G.P., Markides A.J. and Waters E.J. (2000) *J. Agric. Food Chem.*, 48, 1086–1095.
Ferreira R.B., Monteiro S., Piçarra-Pereira M.A., Tanganho M.C., Loureiro V. and Teixeira A. (2000) *Am. J. Enol. Vitic.*, 51 (1), 22–28.
Feuillat M. (1974) Contribution à l'étude des composés azotés du moût de raisin et du vin. Thèse Doctorat, Université de Dijon.
Feuillat M., Brillant G. and Rochard J. (1980) *Conn. Vigne Vin*, 14 (1), 37.
Feuillat M., Peyron D. and Berger J.L. (1987) *Bull. OIV*, 673, 227–244.
Hayasaka Y., Adams KS., Pocock KF., Baldock GA., Waters EJ. and Hoj PB. (2001) *J. Agric. Food Chem.*, 49, 1830–1839.
Hsu J.C. and Heatherbell D. (1987a) *Am. J. Enol. Viticult.*, 38 (1), 6.
Hsu J.C. and Heatherbell D. (1987b) *Am. J. Enol. Viticult.*, 38 (1), 11.
Jakob L. (1962) *Das Weinblatt*, 57, 805.
Koch J. and Sajak E. (1959) *Weinberg u. Keller*, 10, 35.
Laborde J. (1904), *Rev. Viticult.*, 21, 8.
Ledoux V., Dulau L. and Dubourdieu D. (1992) *J. Int. Sci. Vigne et Vin*, 26 (4), 239.
Lethonen P.K., Saarinen M., Vesanto M. and Riekkola M.L. (1992) *Z. Lebeusm. Unters. Forsch*, 194, 434–437.

Liu S.O., Pritchard G.G., Hardman M.J. and Pilone G. (1994) *Am. J. Enol. Viticult.*, 45 (2), 235.

Lourens K. and Pretorius S. (1996) Œnologie 95, Vème Symposium International d'Œnologie de Bordeaux, Tec. et Doc., Lavoisier, Paris.

Makaga E. and Maujean A. (1994) *Bull. OIV*, 67 (763–764), 753.

Millery A. (1988) Caractérisation des cépages champenois à partin de leurs acides aminés libres. Thèse Doctorat Université Reims.

Millery A., Maujean A., Duteurtre B. and Beudaille J.P. (1986) *Rev. Œnol.*, Cahiers Scientifiques, 103, 32–50.

Moine V. and Dubourdieu D. (1996) Œnologie 95, 5th Symposium International d'Œnologie de Bordeaux, Tec. et Doc., Lavoisier, Paris.

Moine-Ledoux V. (1996) Recherches sur le rôle des mannoprotéines de levure vis-à-vis de la stabilization protéique et tartrique des vins. Thèse de Doctorat, Université de Bordeaux II.

Moine-Ledoux V., Perrin A., Paladu I. and Dubourdieu D. (1997) *J. Int. Sci. Vigne Vin*, 31 (1), 23.

Moore and Stein (1951) *J. Biol. Chem.*, 192, 663.

Mordelet-Dambrine M. and Parrot J.-L. (1970) *Ann. Hyg. L. Fr. Méd. et Nut.*, 6 (3), 59.

Moutounet M. (1990) *Cah. Scientif. Rev. Fr. Œno*, 124, 27.

Paetzold M., Dulau L. and Dubourdieu D. (1990) *J. Int. Sci. Vigne et Vin*, 24 (1), 13.

Pellerin P., Waters E., Brillouet J.-M. and Moutounet M. (1994) *J. Int. Sci. Vigne Vin*, 28 (3), 213.

Peng Z., Pocock K.F., Waters E.J., Francis I.L. and Williams P.J. (1997) *J. Agric. Food Chem.*, 45, 4639–4643.

Piçarra-Pereira M.A., Monteiro S., Loureiro V., Teixeira A. and Ferreira R.B. (1998) *Pol. J. Food Nutr. Sci.*, 7/48 (2(S)), 107–111.

Pocock K.F., Hayasaka Y., McCarthy M.G. and Waters E.J. (2000) *J. Agric. Food Chem.*, 48, 1637–1643.

Pocock K.F., Hayasaka Y., Peng Z., Williams P.J. and Waters E.J. (1998) *Aust. J. Grape Wine Res.*, 4, 23–29.

Ribéreau-Gayon J. (1932) *Ann. Fasf. Fraudes*, 25, 518.

Ribéreau-Gayon J. (1935) *Rev. Viticult.*, 82, 367.

Ribéreau-Gayon J., Peynaud E., Ribéreau-Gayon P. and Sudraud P. (1976 and 1977) *Sciences et Techniques du Vin*, Vols. III and IV. Dunod, Paris.

Ribéreau-Gayon J., Peynaud E., Sudraud P. and Ribéreau-Gayon P. (1982) *Sciences et Techniques du Vin*, Vol. I: *Analyse et Contrôle du Vin.* Dunod, Paris.

Salzman R.A., Tikhonova I., Bordelon B.P., Hasegawa P.M. and Bressan R.A. (1998) *Plant Physiol.*, 117, 435–472.

Saywell L.G. (1934) *Ind. Eng. Chem.*, 26, 981.

Simpson R.F. and Miller G.C. (1984) *Vitis*, 23, 143.

Waters EJ., Pellerin P. and Brillouet P. (1994) *Biosci. Biotechnol. Biochem.*, 58 (1), 43–48.

Waters EJ., Peng Z., Pocock KF. and Williams PJ. (1995a) *Aust. J. Grape Wine Res. I*, 1, 86–93.

Waters EJ., Peng Z., Pocock KF. and Williams PJ. (1995b) *Aust. J. Grape Wine Res. I*, 1, 94–99.

Waters EJ., Shirley NJ. and Williams PJ. (1996) *J. Agric. Food Chem.*, 44, 3–5.

Waters E.J., Wallace W. and Williams P.J (1990) Proceedings of 7th Australian Wine Industry Technical Conference, 13–17 august 1989. Australian Industrial Publishers, Adelaide, p. 186.

Waters E.J., Wallace W. and Williams P.J (1991) *Am. J. Enol. Viticult.*, 42 (2), 123.

Waters E.J., Wallace W. and Williams P.J (1992) *J. Agric. Food Chem.*, 40, 1514.

Waters E., Wallace W. and William P.J. (1993) *J. Agric. Food Chem.*, 41, 724.

6

Phenolic Compounds

6.1 Introduction	141
6.2 Types of substances	142
6.3 Chemical properties of anthocyanins and tannins	152
6.4 Anthocyanin and tannin assays—organoleptic properties	172
6.5 Evolution of anthocyanins and tannins as grapes ripen	184
6.6 Extracting tannins and anthocyanins during winemaking	191
6.7 Chemical reactions occurring during barrel and bottle aging	193
6.8 Precipitation of coloring matter (color stability)	198
6.9 Origin of the color of white wines	199

6.1 INTRODUCTION

Phenolic compounds play a major role in enology. They are responsible for all the differences between red and white wines, especially the color and flavor of red wines. They have interesting, healthful properties, responsible for the 'French paradox'. They have bactericide, antioxidant and vitamin properties that apparently protect consumers from cardiovascular disease.

These molecules come from various parts of grape bunches and are extracted during winemaking. Their structure varies a great deal when wine ages in the barrel or in the tank and in the bottle, according to the conditions, but these modifications have not yet been fully explained. Indeed, even the latest chromatography techniques (HPTLC, LPLC, HPLC) still produce relatively limited results and are only capable of analyzing simple and little-polymerized molecules. Although this represents a clear advance, chromatographic analyses are still rather limited as regards the analysis and fractionation of the polymers that play a major role in all types of wine. Furthermore, physicochemical methods, focused on structural definition (NMR, mass spectrometry), are not very well-suited to the study of these types of molecules, although their applications are constantly being extended.

Further complications are due to the interference of a colloidal state that does not involve covalent bonds. This interference definitely plays a role in the structure and, consequently, the properties of phenolic compounds in wine. The colloidal state is, however, difficult to study, as it is modified by any manipulation of these substances (Section 9.3).

A great deal of work has been carried out in several research centers in recent years, building, to a large extent, on earlier findings from the Bordeaux Faculty of Enology.

6.2 TYPES OF SUBSTANCES

6.2.1 Phenolic Acids and their Derivatives

Grapes and wine contain benzoic and cinnamic acids. Concentrations are on the order of 100–200 mg/l in red wine and 10–20 mg/l in white wine. Seven benzoic acids ($C_6 - C_1$) have been identified (Figure 6.1). Two are present in trace amounts: salicylic acid (*ortho*-hydroxybenzoic acid) and gentisic acid ($2',5'$-dihydroxybenzoic acid). The various acids are differentiated by the substitution of their benzene ring. In grapes, they are mainly present as glycoside combinations, from which they are released by acid hydrolysis, and esters (gallic and ellagic tannins), from which they are released by alkaline hydrolysis. Free forms are more prevalent, mainly in red wine, due to the hydrolysis of these combinations and heat breakdown reactions of more complex molecules, especially anthocyanins (Galvin, 1993).

Several cinnamic acids ($C_6 - C_3$) are present in grapes and wines (Figure 6.1). They have been identified in small quantities in the free form, but are mainly esterified, in particular with tartaric acid (Figure 6.2) (Ribéreau-Gayon, 1965). They may also be simple glycosides of glucose (Figure 6.3). Esters with tartaric acid, especially caffeoyltartaric acid (*trans*-caftaric acid) or *p*-coumaryltartaric acid, are highly oxidizable components of grape juice, responsible for the browning of white must (Cheynier *et al.*, 1989a, 1989b). Cinnamic acids combine with anthocyanin monoglucosides (Section 6.2.3) to form acylated anthocyanins, via the esterification of caffeic acid and *p*-coumaric acid with the glucose of the glycoside.

Phenolic acids are colorless in a dilute alcohol solution, but they may become yellow due to oxidation. From an organoleptic standpoint, these compounds have no particular flavor or odor. They are, however, precursors of the volatile phenols produced by the action of certain microorganisms (yeasts in the genus *Brettanomyces* and bacteria) (Section 8.3). Ethyl phenols, with animal odors, and ethyl gaiacols are found in red wines (Figure 6.4). In white wines, vinyl phenols, with an odor reminiscent of gouache paint, are accompanied by vinyl gaiacols. It has been clearly established that these compounds result from the

(1) Benzoic acids	R_2	R_3	R_4	R_5	(2) Cinnamic acids
p-Hydroxybenzoic acid	H	H	OH	H	*p*-Coumaric acid
Protocatechic acid	H	OH	OH	H	Caffeic acid
Vanillic acid	H	OCH₃	OH	H	Ferulic acid
Gallic acid	H	OH	OH	OH	
Syringic acid	H	OCH₃	OH	OCH₃	Sinapic acid
Salicylic acid	OH	H	H	H	
Gentisic acid	OH	H	H	OH	

Fig. 6.1. Phenolic acids in grapes and wine

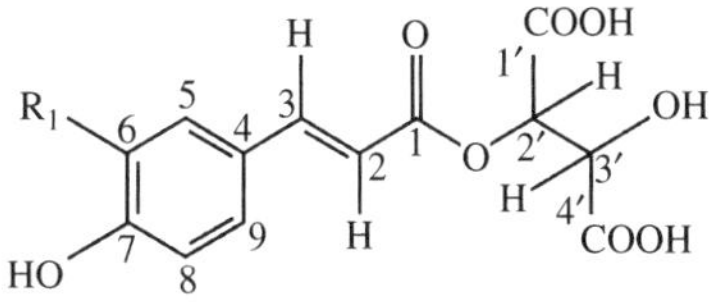

Phenolic Compounds

Fig. 6.2. Derivatives of cinnamic acids and tartaric acid. $R_1 = H$, p-coumaryl tartaric acid (coutaric acid); $R_2 = OH$, caffeoyl-tartaric acid (trans-caftaric acid) (caftaric acid)

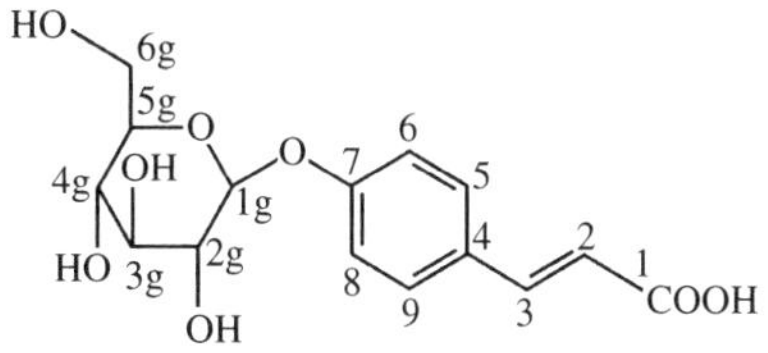

Fig. 6.3. 7-O-β-D glucosyl-p-coumaric acid (Biau, 1996)

breakdown of p-coumaric acid and ferulic acid (Chatonnet, 1995) (Section 8.4).

When wines are aged in new oak barrels, the toasting of the wood involved in barrel manufacture causes the breakdown of lignins and the formation of various components in the same family, with a variety of smoky, toasty and burnt smells (Figure 6.4): gaiacol, methyl gaiacol, propyl gaiacol, allyl gaiacol (isoeugenol), syringol and methyl syringol.

Tyrosol (Figure 6.5) or p-hydroxy-phenyl-ethyl alcohol may be included in this group of compounds (Ribéreau-Gayon and Sapis, 1965). It is always present in both red and white wine (20–30 mg/l) and is formed during alcoholic fermentation from tyrosine (p-hydroxyphenyl-alanine), in turn synthesized by yeast. This compound, which remains at relatively constant concentrations throughout aging, is accompanied by other non-phenolic alcohols like tryptophol (0–1 mg/l) and phenyl-ethyl alcohol (10–75 mg/l).

Coumarins (Figure 6.5) may be considered derivatives of cinnamic acids, formed by the intramolecular esterification of a phenol OH into the α of the carbon chain. These molecules are components of oak, either in glycosylated

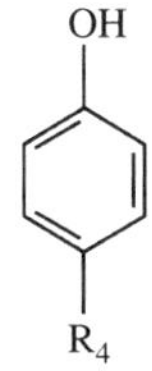

R_4	Name	Origin
$CH_2 - CH_3$	Ethyl phenol	Red wine
$CH = CH_2$	Vinyl phenol	White wine

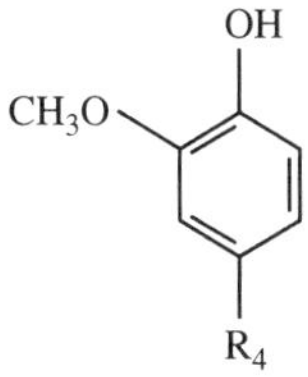

R_4	Name	Origin
H	Gaiacol	Wood
CH_3	Methyl gaiacol	Wood
$CH_2 - CH_3$	Ethyl gaiacol	Red wine
$CH = CH_2$	Vinyl gaiacol	White wine
$CH_2 - CH_2 - CH_3$	Propyl gaiacol	Wood
$CH = CH - CH_3$	Allyl gaiacol	Wood

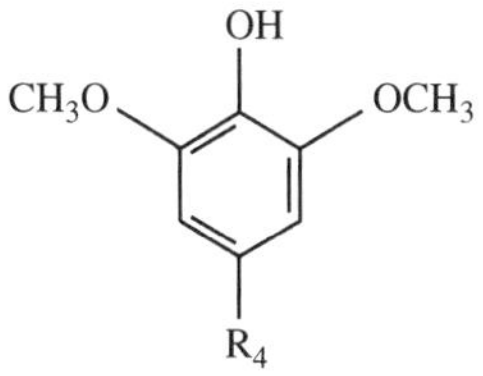

R_4	Name	Origin
H	Syringol	Wood
CH_3	Methyl Syringol	Wood

Fig. 6.4. Volatile phenols in wine

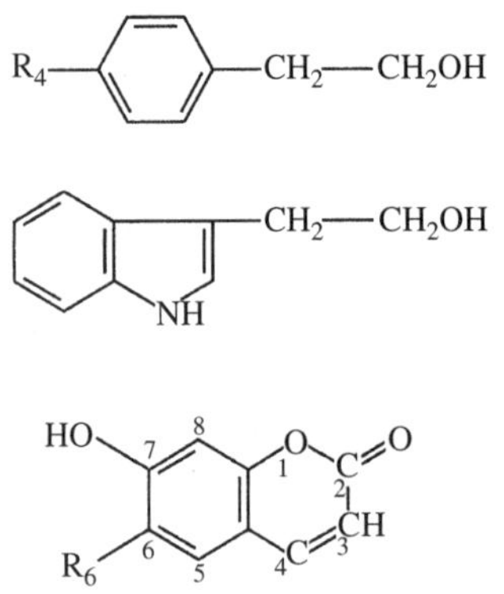

Alcohols:

R_4 = H phenyl-ethyl alcohol
R_4 = OH P-hydroxy-phenyl-ethyl alcohol (tyrosol)

Tryptophol

Coumarins:

Aglycone	Glycoside
R_6 = —OH esculetin	Esculin (glucoside 6)
R_6 = —OCH$_3$ scopoletin	Scopolin (glucoside 7)

Fig. 6.5. Phenolic alcohols and coumarins

form (esculin and scopoline) in green wood or in aglycone form (esculetin and scopoletin) in naturally seasoned wood. Although very small quantities (a few μg/l) of coumarins are found in wood-aged wine, they still affect its organoleptic characteristics, as glycosides are bitter and aglycones are acidic, with a detection threshold in red wine of 3 μg/l.

Another family of more complex polyphenols is also present in grapes, wine and oak wood. Stilbens have two benzene cycles, generally bonded by an ethane, or possibly ethylene, chain. Among these *trans*-isomer compounds, resveratrol, or 3,5,4′-trihydroxystilben (Figure 6.6), is thought to be produced by vines in response to a fungal infection (Langcake, 1981). Resveratrol, located in the skins, is mainly extracted during the fermentation of red wines and seems to have some healthful properties. Concentrations are on the order of 1–3 mg/l. Recent research (Jeandet *et al.*, 1995; Bourhis *et al.*, 1996) has identified many oligomers of resveratrol in *Vitis vinifera*.

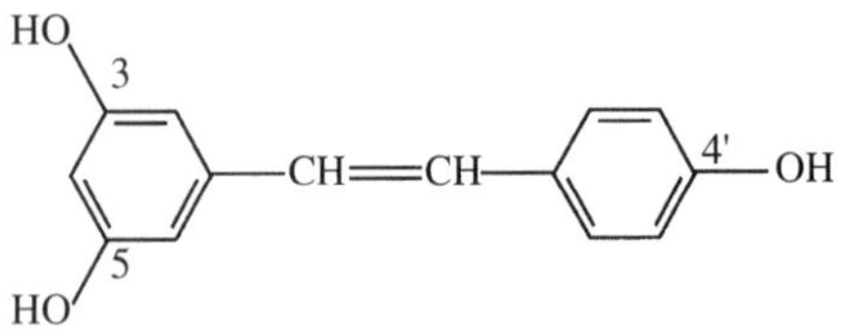

Fig. 6.6. Trihydroxy-3,5,4′-stilben (resveratrol)

6.2.2 Flavonoids

These are more-or-less intense yellow pigments, with a structure characterized by two benzene cycles bonded by an oxygenated heterocycle, derived either from the 2-phenyl chromone nucleus (flavones and flavonols) or the 2-phenyl chromanone nucleus (flavanones and flavanonols) (Figure 6.7).

The most widespread compounds are flavonols, yellow pigments in the skins of both red and white grapes and, to a lesser extent, flavanonols, which are much paler in color. In grapes, these molecules are present in glycoside form (Figure 6.8), e.g. rhamnosylquercetin. They are differentiated by substitution of the lateral nucleus, producing kaempferol (1 OH), quercetin (2 OH) and myricetin (3 OH). All three pigments are present in red wine grapes, whereas white wine grapes only have the first two (Ribéreau-Gayon, 1964).

These compounds are present in red wine in aglycone form, as the glycosides are hydrolyzed during fermentation. Concentrations are in the region of 100 mg/l. In white wine, where fermentation takes place in the absence of grape solids, typical values are from 1 to 3 mg/l according to the grape variety. Pre-fermentation maceration in the aqueous phase has less impact on this concentration than settling (Ollivier, 1987).

The flavanonol most frequently identified in grapes and wine is dihydroquercetin, also known

Phenolic Compounds

Fig. 6.7. Flavonoids: a, flavone ($R_3 = H$) and flavonol ($R_3 = OH$); b, flavanone ($R_3 = H$) and flavanonol ($R_3 = OH$)

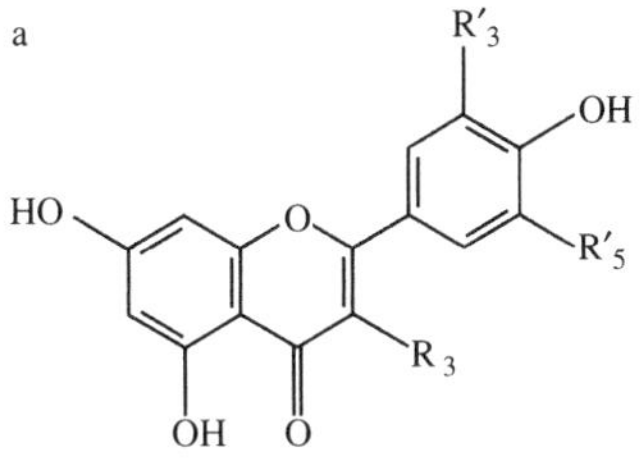

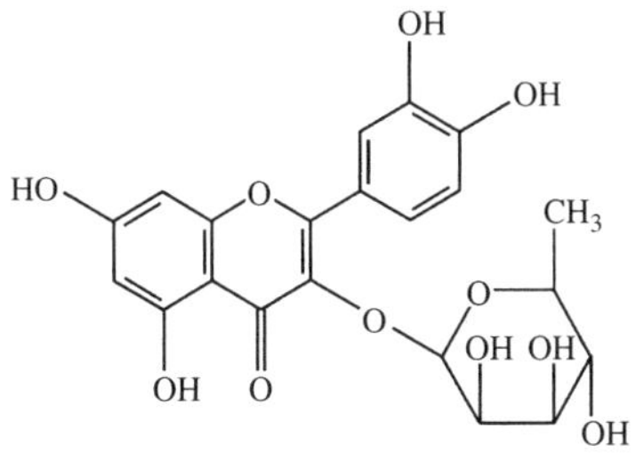

a) $R_3 = OH$

R'_3	R'_5	Name of aglycone
H	H	Kaempferol
OH	H	Quercetin
OH	OH	Myricetin

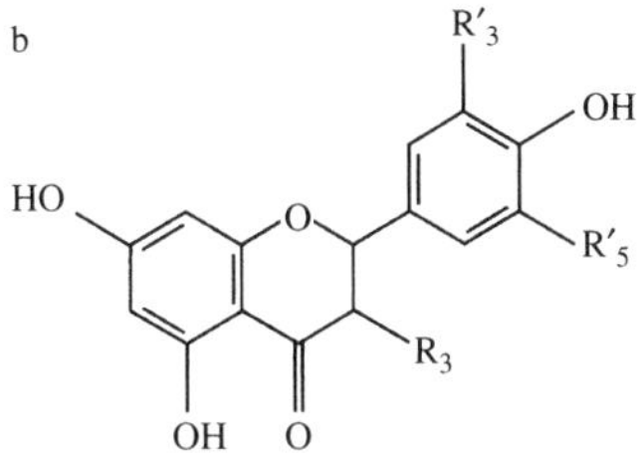

b) $R_3 = OH$

R'_3	R'_5	Name of aglycone
OH	H	Dihydroquercetin (taxifolin)

Fig. 6.8. 3-*O*-Rhamnosylquercetin

as taxifolin. The role played by these various compounds in the color of red and white wines will be discussed later in this chapter (Section 6.9).

6.2.3 Anthocyanins

Anthocyanins are the red pigments in grapes, located mainly in the skin and, more unusually, in the flesh ('teinturier' grape varieties). They are also present in large quantities in the leaves, mainly at the end of the growing season.

Their structure, flavylium cation, includes two benzene rings bonded by an unsaturated cationic oxygenated heterocycle, derived from the 2-phenyl-benzopyrylium nucleus. Five molecules have been identified in grapes and wines, with two or three substituents (OH and OCH_3) according to the substitution of the lateral nucleus (Figure 6.9).

These molecules are much more stable in glycoside (anthocyanin) than in aglycone (anthocyanidin) form. Only monoglucoside anthocyanins (Figure 6.10) and acylated monoglucoside anthocyanins have been identified in *Vitis vinifera* grapes and wines; acylation is made with *p*-coumaric (Figure 6.10), caffeic and acetic acids.

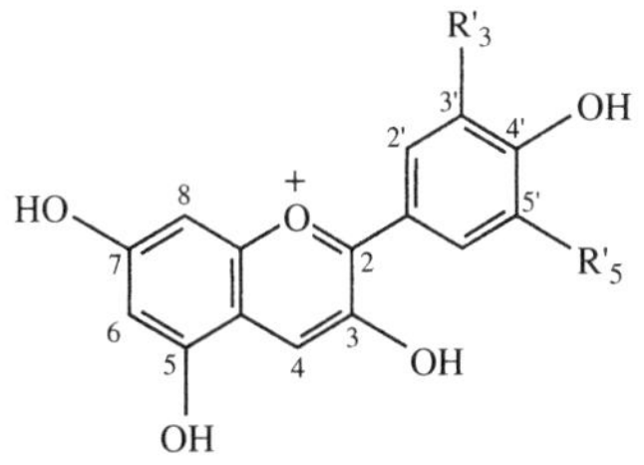

R'_3	R'_5	Name of aglycone
OH	H	Cyanidin
OCH_3	H	Peonidin
OH	OH	Delphinidin
OH	OCH_3	Petunidin
OCH_3	OCH_3	Malvidin

Fig. 6.9. Structure of anthocyanidins in grapes and wine

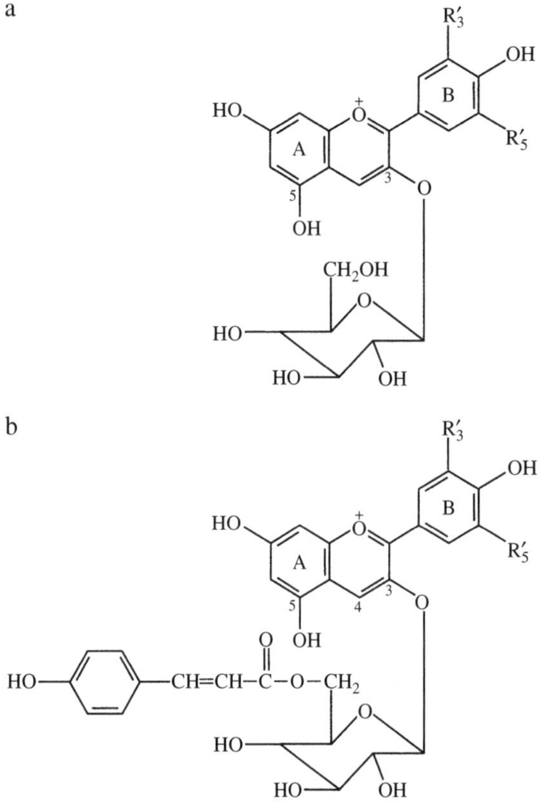

Fig. 6.10. Structure of: (a) anthocyanin 3-monoglucosides, (b) anthocyanins 3-monoglucosides acylated by *p*-coumaric acid on position 5 of the glucose (R'_3 and R'_5 see Figure 6.9)

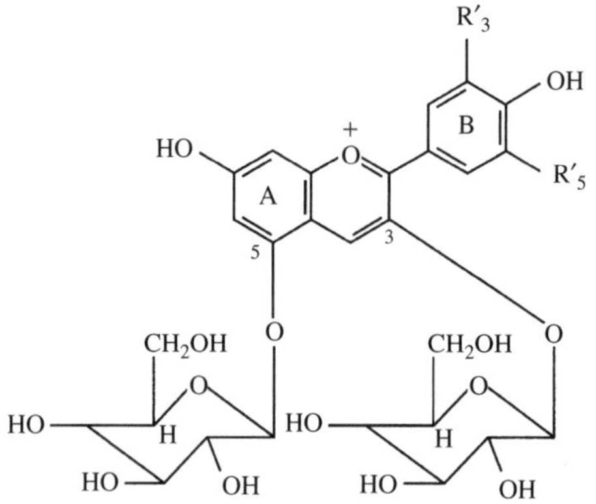

Fig. 6.11. Structure of anthocyanin 3,5-diglucosides (R'_3 and R'_5 see Figure 6.9)

The presence of diglucoside anthocyanins (Figure 6.11) in large quantities is specific to certain species in the genus *Vitis* (*V. riparia* and *V. rupestris*) (Ribéreau-Gayon, 1959). Traces have, however, been found in certain *V. vinifera* grapes (Roggero *et al.*, 1984). The 'diglucoside' character is transmitted according to the laws of genetics, as a dominant characteristic. This means that a cross between a *vinifera* grape variety and an American species (*V. riparia* or *V. rupestris*) produces a population of first-generation hybrids that have all the diglucosides. On the other hand, results obtained with a new cross between a first-generation hybrid and a *V. vinifera* vine show that the recessive 'absence of diglucoside' characteristic may be expressed in a second-generation hybrid. These findings led to the development of the method for differentiating wines by chromatographic analysis of their coloring matter (Ribéreau-Gayon, 1953, 1959). This played a major role in ensuring that traditional grape varieties were used in certain French appellations of origin, as well as in monitoring quality.

The color of these pigments depends on conditions in the medium (pH, SO_2), as well as the molecular structure and the environment. On the one hand, substitution of the lateral cycle leads to a bathochrome shift of the maximum absorption wavelength (towards violet). On the other hand, glucose fixation and acylation shift the color in the opposite direction, i.e. towards orange. These molecules are mainly located in the skin cells, with a concentration gradient from the inside towards the outside of the grape (Amrani-Joutei, 1993). Pigment molecules are in solution in the vacuolar juice in the presence of other polyphenols (phenolic acids, flavonoids, etc.) likely to affect their color. Copigmentation (Section 6.3.8) generally gives wines a violet tinge.

These factors explain the different colors of red grapes. All grape varieties have the same basic anthocyanidin structures, but there are a few small variations in composition. Indeed, among the five anthocyanins, malvidin is the dominant molecule in all grape varieties, varying from 90% (Grenache) to just under 50% (Sangiovese). Malvidin monoglucoside (malvine) may be considered

Phenolic Compounds

to form the basis of the color of red grapes and, by extension, red wine. On the other hand, the quantity of acylated monoglucosides is highly variable according to the grape variety.

In *vitis vinifera* wines, the presence of ethanol works against copigmentation (Section 6.3.8), and the acylated anthocyanins disappear rapidly a few months after fermentation, so it is not reasonable to use them to identify grape varieties. This leaves only the five monoglucosides, predominantly malvidin. Concentrations vary a great deal according to the age of the wines and the grape varieties. Starting at levels of 100 mg/l (Pinot Noir) to 1500 mg/l (Syrah, Cabernet Sauvignon, etc.) after fermentation, they decrease rapidly in the first few years, during barrel and bottle aging, until they reach a minimum value on the order of 0–50 mg/l. In fact, this concentration was determined by a free anthocyanin assay, using chemical and chromatographic methods. In fact, the majority of these pigments combine and condense with tannins in wine to form another, more stable, class of color molecules that are not detected by current assay methods (Sections 6.3.7 and 6.3.8). These complex combined anthocyanins are responsible for color in wine but cannot be identified by standard analyses. Another relatively small fraction of the anthocyanins, however, disappears (Section 6.3.3), either broken down by external factors (temperature, light, oxygen, etc.) or precipitated in colloidal coloring matter. The elimination of these pigments is particularly detrimental to the quality of the wine, as it leads to loss of color.

Another recently demonstrated property of anthocyanins (Castagnino and Vercauteren, 1996) involves their reaction with compounds containing an α-dicarbonylated group, such as diacetyl (CH_3–CO–CO–CH_3). This reaction gives rise to castavinols (Figure 6.12), not present in grapes but formed spontaneously in wine. These colorless compounds are capable of regenerating colored anthocyanins in an acid medium, by a process called the Bate–Smith reaction, which converts procyanidins into cyanidin (Section 6.3.5). However, in the case of castavinols, this reaction does not require very high temperatures and acidity as it occurs spontaneously and gradually in wine during

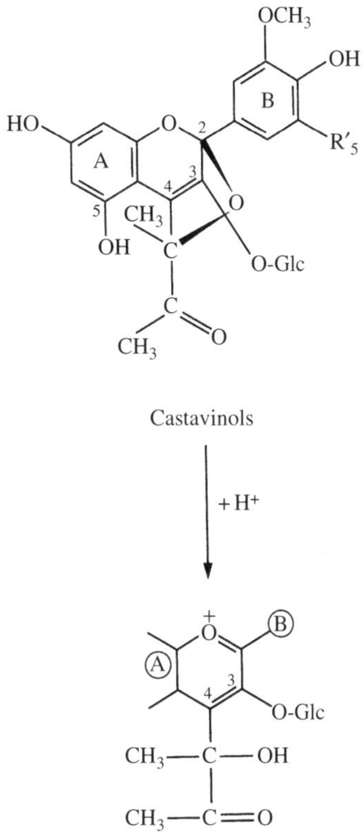

Fig. 6.12. Structure of castavinols resulting from fixing diacetyl (CH_3–CO–CO–CH_3) on carbons 2 and 4 of the anthocyanin and their transformation into flavylium substituted in 4 and colored by heating in an acid medium (R'_5 = –H, –OH, –OCH_3) (Castagnino and Vercauteren, 1996)

aging. The color of the anthocyanin is stabilized by substitution of the molecule in carbon 4. According to several authors, the concentration of castavinols in wine is on the order of a few mg/l. Nevertheless, these substances are likely to play a role as reserves of coloring matter.

6.2.4 Tannins

Tannins are, by definition, substances capable of producing stable combinations with proteins and other plant polymers such as polysaccharides. The transformation of animal skins into rotproof leather results from this property, as does astringency, fining and enzyme inhibition. Tannins react with proteins in each instance: collagen in tanning, glycoproteins in saliva and proline-rich proteins (PRP) for astringency, protein-based fining agents in fining wines and the protein fraction of enzymes.

In chemical terms, tannins are relatively bulky phenol molecules, produced by the polymerization of elementary molecules with phenolic functions. Their configuration affects their reactivity (Section 6.3.4). They must be sufficiently bulky to produce stable combinations with proteins, but if they are too bulky, they are likely to be too far from the active protein sites. The molecular weights of active tannins range approximately from 600 to 3500. Condensed or catechic tannins are distinguished from complex or mixed tannins by the type of elementary molecules.

Hydrolyzable tannins include gallotannins and ellagitannins that release gallic acid and ellagic acid, respectively (Figure 6.13a,b), after acid hydrolysis. They also contain a glucose molecule. The two main ellagitannin isomers in oak used for cooperage are vescalagin and castalagin (M = 934), as well as two less important compounds, grandinin and roburin (Figure 6.13). These molecules include a hexahydroxydiphenic

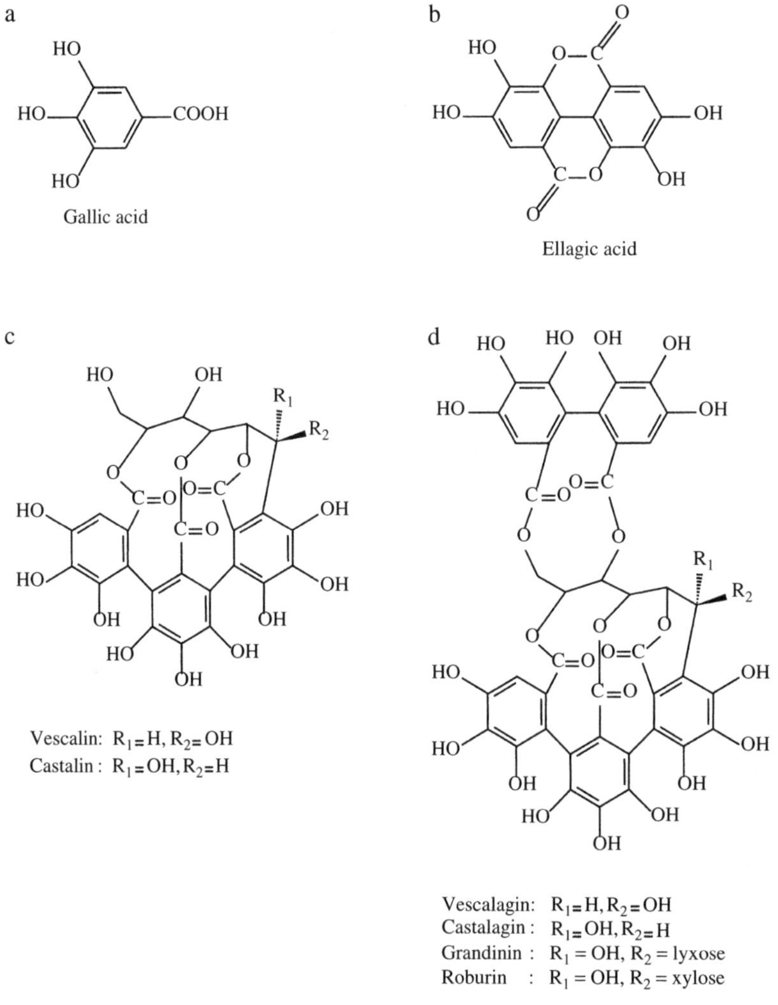

Vescalin: $R_1=H, R_2=OH$
Castalin: $R_1=OH, R_2=H$

Vescalagin: $R_1=H, R_2=OH$
Castalagin: $R_1=OH, R_2=H$
Grandinin: $R_1=OH, R_2=$ lyxose
Roburin: $R_1=OH, R_2=$ xylose

Fig. 6.13. Structure of phenolic acids (a and b) and ellagitannins (c and d) in extracts from the duramen of oak and chestnut wood (Vivas and Glories, 1996)

Phenolic Compounds

and a nonahydroxydiphenic acid, esterified by a non-cyclic glucose. The partial hydrolysis of vescalagin and castalagin, involving the loss of hexahydroxydiphenic acid, produces vescalin and castalin ($M = 632$) (Figure 6.13c).

The various molecules are water soluble and dissolve rapidly in dilute alcohol media such as wines and brandies (Moutounet et al., 1989). They play a considerable role in the aging of red and white wines in oak barrels, due to their oxidizability (Vivas and Glories, 1993, 1996) and flavor properties (Pocock et al., 1994).

The ellagitannin composition of extracts from the duramen depends on the species of oak. All four monomeric and four dimeric (roburin A, B, C, and D) ellagitannins are present in the three species of European oak, while the American species have practically no dimers.

Hydrolyzable tannins are not naturally found in grapes. On the other hand, they are the main commercial tannins legally authorized as wine additives. Ellagic acid in wine originates either from wooden containers or from the addition of enological tannins. On the other hand, gallic acid from the skins and seeds is always present in wine.

Condensed tannins in grapes and wine are more-or-less complex polymers of flavan-3-ols or catechins. The basic structural units are (+)-catechin and (−)-epicatechin (Figure 6.14). Heating these polymers in solution in an acid medium releases highly unstable carbocations that are converted into brown condensation products, mainly red cyanidin, which explains why these compounds are known as 'procyanidins', replacing the formerly used term 'leucocyanidin'.

Analysis of these molecules is particularly complex, due to the great structural diversity resulting from the number of hydroxyl groups, their position on the aromatic nuclei, the stereochemistry of the asymmetrical carbons in the pyran cycle, as well as the number and type of bonds between the basic units. In spite of the progress made in liquid chromatography, mass spectrometry and NMR, all of the structures have not been analyzed: only the procyanidin dimers and some of the trimers have been completely identified.

This diversity explains the existence of tannins with different properties, especially as regards flavor, in various types of grapes and wine. Tannin content should not be the only factor considered, as structure and colloidal status also affect the impression tannins give on tasting.

It is possible to isolate and fractionate the following constituents of grapes and wine: (+)-catechin, gallocatechin, (−)-epicatechin, epigallocatechin, and epicatechin-3-0-gallate. There are also dimeric, trimeric, oligomeric, and condensed procyanidins (Burger et al., 1990; Kondo et al., 2000). Basic 'catechin' units may not be considered as tannins, as their molecular weight is too low and they have very restricted properties in relation to proteins. They only have a high enough

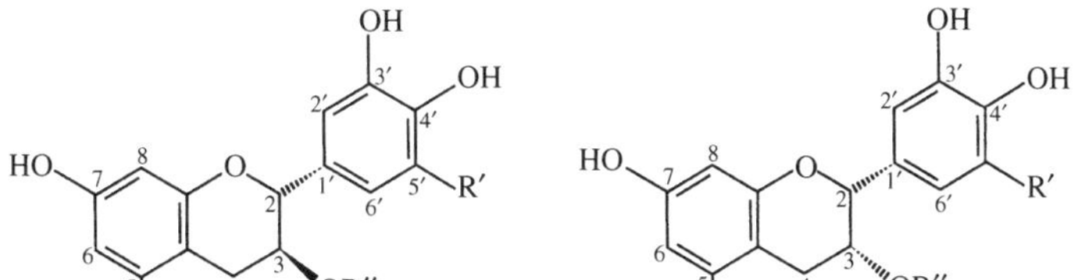

Catechin Series

R′ = H, R″ = H: (+)-catechin (2R, 3S)
R′ = H, R″ = H: (−)-catechin (2S, 3R)
R′ = OH, R″ = H: gallocatochin
R′ = H, R″ = gallic acid: galloyl catechin
 (catechin-3-0-gallate)

Epicatechin Series

R′ = H, R″ = H: (+)-epicatechin (2S, 3S)
R′ = H, R″ = H: (−)-epicatechin (2R, 3R)
R′ = OH, R″ = H: epigallocatechin
R′ = H, R″ = gallic acid: galloyl epicatechin
 (epicatechin-3-0-gallate)

Fig. 6.14. Structure of flavan-3-ol precursors of procyanidins and tannins

150   Handbook of Enology: The Chemistry of Wine

molecular weight in dimeric form to bond stably with proteins.

Catechins (Figure 6.14) have two benzene cycles bonded by a saturated oxygenated heterocycle (phenyl-2 chromane nucleus). This structure has two asymmetrical carbons (C_2 and C_3) that are the origin of the four isomers. The more stable forms are (+)-catechin and (−)-epicatechin.

Dimeric procyanidins may be divided into two categories, identified by a letter of the alphabet and a number (Weinges *et al.*, 1968; Thompson *et al.*, 1972):

1. Type-B procyanidins ($C_{30}H_{26}O_{12}$) (Figure 6.15) are dimers resulting from the condensation of two units of flavan-3-ols linked by a C_4–C_8 (B_1 to B_4) or C_4–C_6 (B_5 to B_8) bond. As there are five different types of monomers and two types of intermonomeric bonds, there may be $2 \times 5^2 = 50$ dimers in wine. The eight procyanidins presented have been identified as the most common ones in wine.

2. Type-A procyanidins ($C_{30}H_{24}O_{12}$) (Figure 6.16) are dimers that, in addition to the C_4–C_8 or C_4–C_6 interflavan bond, also have an ether bond between the C_5 or C_7 carbons of the terminal unit and the C_2 carbon of the upper unit. Procyanidin A_2 has been identified in wine (Vivas and Glories, 1996). Form B can change to form A via a radical process.

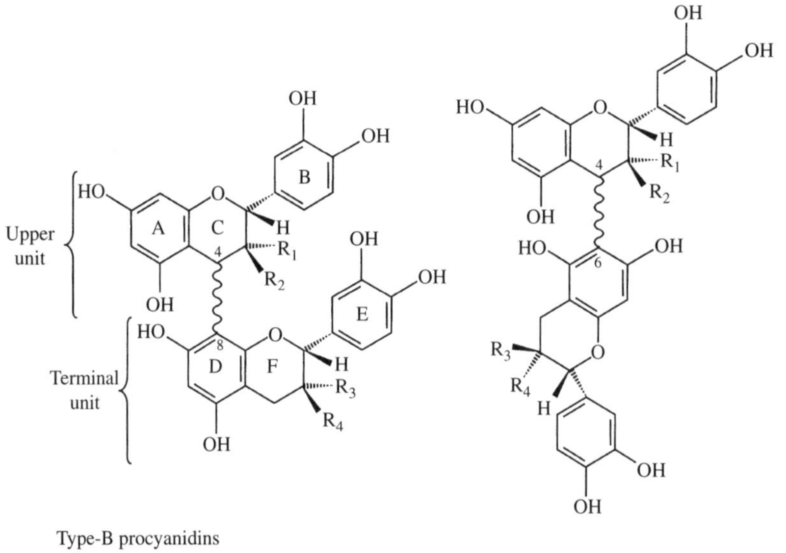

Type-B procyanidins

B_1 : R_1 = OH; R_2 = H; R_3 = H; R_4 = OH
B_2 : R_1 = OH; R_2 = H; R_3 = H; R_4 = OH
B_3 : R_1 = H; R_2 = OH; R_3 = H; R_4 = OH
B_4 : R_1 = H; R_2 = OH; R_3 = OH; R_4 = H

B_5 : R_1 = OH; R_2 = H; R_3 = OH; R_4 = H
B_6 : R_1 = H; R_2 = OH; R_3 = H; R_4 = OH
B_7 : R_1 = OH; R_2 = H; R_3 = H; R_4 = OH
B_8 : R_1 = H; R_2 = OH; R_3 = OH; R_4 = H

Schedule

B_1 : 2,3-*cis*-3,4-*trans*-2″,3″-*trans* [2R,3R,4R,2″R,3″S] epicatechin (4β → 8)-catechin
B_2 : 2,3-*cis*-3,4-*trans*-2″,3″-*cis* [2R,3R,4R,2″R,3″R] epicatechin (4β → 8)-epicatechin
B_3 : 2,3-*trans*-3,4-*trans*-2″,3″-*trans* [2R,3S,4S,2″R,3″S] catechin (4α → 8)-catechin
B_4 : 2,3-*trans*-3,4-*trans*-2″,3″-*cis* [2R,3S,4S,2″R,3″R] catechin (4α → 8)-epicatechin
B_5 : 2,3-*cis*-3,4-*trans*-2″,3″-*cis* [2R,3R,4R,2″R,3″R] epicatechin (4β → 6)-epicatechin
B_6 : 2,3-*trans*-3,4-*trans*-2″,3″-*trans* [2R,3S,4S,2″R,3″S] catechin (4α → 6)-catechin
B_7 : 2,3-*cis*-3,4-*trans*-2″,3″-*trans* [2R,3R,4R,2″R,3″S] epicatechin (4β → 6)-catechin
B_8 : 2,3-*trans*-3,4-*trans*-2″,3″-*cis* [2R,3S,4S,2″R,3″R] catechin (4α → 6)-epicatechin

Fig. 6.15. Structure and schedule of type-B dimeric procyanidins (de Freitas, 1995)

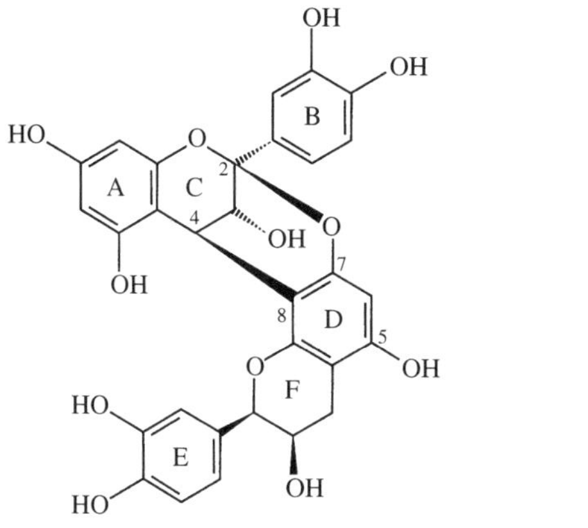

Fig. 6.16. Structure of the dimeric procyanidin A_2 (Vivas and Glories, 1996)

Trimeric procyanidins may also be divided into two categories:

1. Type-C procyanidins are trimers with two interflavan bonds corresponding to those of type-B dimers.

2. Type-D procyanidins are trimers with two interflavan bonds, one type A and one type B.

As in the case of dimers, it is possible to calculate the number of tetramers that could possibly exist in grapes and wine, i.e. $2^2 \times 5^3 = 500$. Only a few trimers have been clearly identified in grapes.

Oligomeric procyanidins correspond to polymers formed from three to ten flavanol units, linked by C_4–C_8 or C_4–C_6 bonds. An infinite number of isomers are possible, which explains why it is so difficult to separate these molecules. Condensed procyanidins (Figure 6.17) have more than ten flavan units and a molecular weight greater than 3000.

Condensed tannins, especially procyanidins and catechins, present in all of the solid parts of grape bunches (skin, seeds, stalk), are dissolved in the wine when it is left on the skins. Concentrations in red wine vary according to grape variety and, to an even greater extent, winemaking methods. Values are between 1 and 4 g/l. In dry white wine, the quality of settling determines the tannin

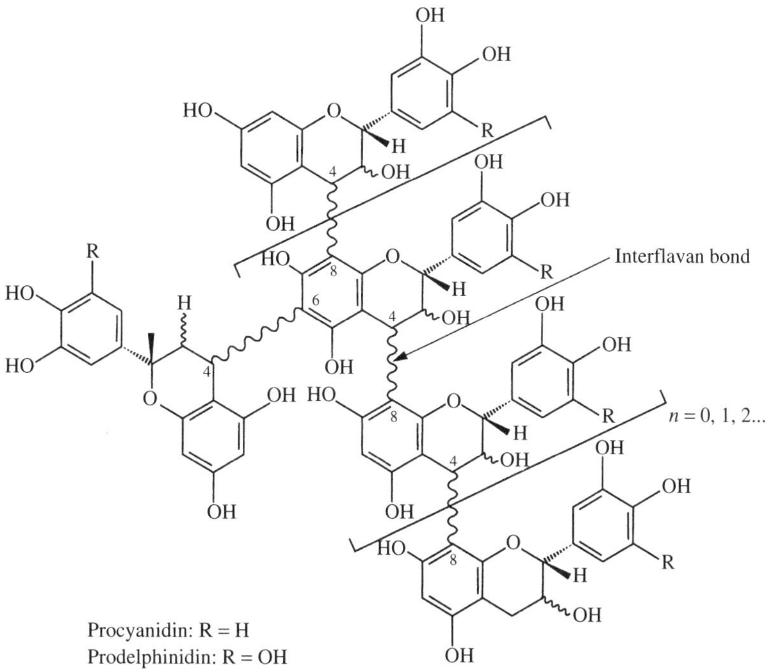

Procyanidin: R = H
Prodelphinidin: R = OH

Fig. 6.17. Structures of condensed proanthocyanidins

concentration. It ranges from 100 mg/l if the must settles properly, to 200 or 300 mg/l if fermentation takes place in the presence of lees. Sweet white wines made from botrytized grapes have a very low tannin content, as these compounds are completely broken down by the fungus.

As they are highly reactive, flavanol molecules condense together. Unlike anthocyanins and flavonols, they do not have glycosylated forms. However, they may be bonded to polysaccharides in grapes and extracted as complexes in the winemaking process.

Complex or mixed tannins are found in plants containing both hydrolyzable and condensed tannins. They consist of covalent complexes between ellagitannins and flavanols (Han *et al.*, 1994). Acutissimin A, isolated from *Castanea sativa* bark (Ampere, 1998), is a covalent complex of castalagin or vescalagin and catechin. These constituents, together with a castalagin–malvidin complex are present in very small amounts in wood-aged wines. Their organoleptic properties are currently under investigation (Jourdes *et al.*, 2003).

6.3 CHEMICAL PROPERTIES OF ANTHOCYANINS AND TANNINS

6.3.1 Properties of Phenols

Substituting a hydroxyl for a hydrogen in the benzene unit produces a low acid phenol that reacts with electrophilic reagents. In an alkaline medium, the proton is exchanged with a metal to produce a phenate.

The electron shift of the three Π doublets in the cycle and both free oxygen doublets, as well as their conjugation, cause these modifications in reactivity. They give rise to an electron shift (mesomer effect) that, in borderline cases, leads to the appearance of a positive charge on the oxygen and a negative charge on a node. Three borderline formulas are possible, with the charges located in the *ortho* (nodes 2 and 6) and *para* (node 4) position in relation to the OH:

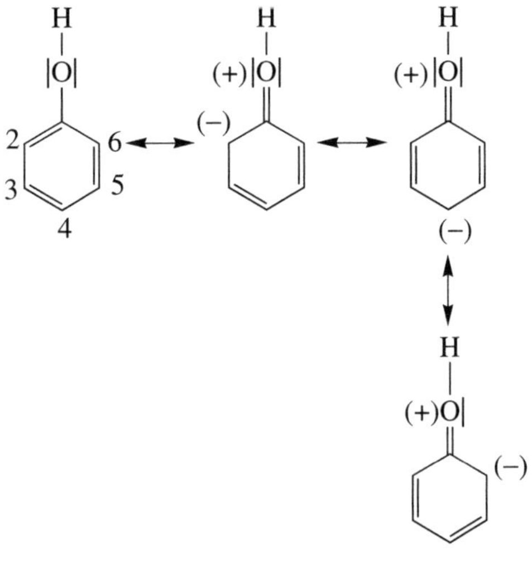

In a phloroglucinol-type nucleus, with 3 OH in the *meta* position, the mesomer effects of the 3 OH are superimposed. The nodes located in the *ortho* position of OH have a considerable excess of electrons (δ^- is close to 1). This facilitates electrophilic substitutions:

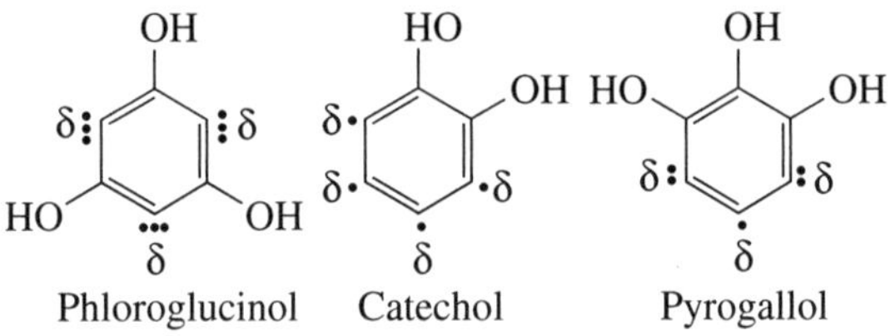

Phloroglucinol   Catechol   Pyrogallol

When the OH are located in the *ortho* position (catéchol and pyrogallol), the mesomer effects are less superimposed and the free nodes are less nucleophilic ($\delta < 1$).

Anthocyanins and condensed tannins in grapes and wine have a phloroglucinol type-A cycle and a type-B cycle substituted in the *ortho* position. Theoretically, therefore, they may react with electrophilic reagents.

6.3.2 Anthocyanin Equilibrium Depending on pH and SO_2

Anthocyanin molecules contain a flavylium nucleus with a positively charged oxygen. In view of the existence of conjugated double bonds,

Phenolic Compounds

the charge is delocalized on the entire cycle, which is stabilized by resonance. Flavylium is an example of a stable oxonium cation that can be isolated as a salt, by analogy with ammonium salts. Anthocyanins also have this property.

The colors of anthocyanin solutions are directly linked to pH. In an acid medium they are red, losing their color as the pH increases. Maximum color loss is observed at values of 3.2 to 3.5. Colors vary from mauve to blue at pH values above 4, then fade to yellow in a neutral or alkaline medium.

Brouillard *et al.* (1978, 1979) showed that these colors reflect the equilibrium between four structure groups (Figure 6.18):

1. Red flavylium cations have an electron deficit. The six limit formulae possible, according to the position of the (+) charge, are all in equilibrium:

$$A_1^+ \leftrightarrow A_2^+ \leftrightarrow A_4^+ \leftrightarrow A_5^+ \leftrightarrow A_7^+ \leftrightarrow A_{4'}^+$$

2. The blue quinonic base has an aromatic ketone, formed from phenol OH. Three limit formulae are possible, derived from the corresponding flavylium cations (AO_5, AO_7, $AO_{4'}$), but they are not in mesomeric balance (Brouillard and Cheminat, 1986).

3. The colorless carbinol base is characterized by an alcohol function, either in 2 or 4 (AOH_2 and AOH_4). Only the first form has been identified.

4. Very pale yellow-colored chalcones derive from the preceding structures when the heterocycle opens and have a ketone function in 2 or 4, C_2 being the most probable. Furthermore, these molecules may exist in two isomer forms, *cis* and *trans*.

Brouillard *et al.* (1978) showed that the flavylium cation is the location of two types of reactions, an acid–base reaction and a hydration reaction. The method used is chemical relaxation applied to jumps in pH.

Conversion from the flavylium cation (A^+) to the quinonic base (AO), corresponding to the red $\rightleftarrows$ mauve balance, occurs due to a very fast proton transfer (10^{-4} s):

$$A^+ \rightleftarrows AO + H^+$$

The equilibrium between the flavylium cation (A^+) and the carbinol base (AOH) involves the presence of a water molecule, followed by a proton transfer. This is also relatively fast (a few seconds):

$$A^+ \underset{H_2O}{\overset{H_2O}{\rightleftarrows}} AOH + H^+$$

The opening of the heterocycle and rearrangement as chalcone corresponds to a tautomerism. This balance is slow and takes a few minutes, or even a few hours according to the temperature:

$$AOH \rightleftarrows \text{chalcone } cis \rightleftarrows \text{chalcone } trans$$

The cis-trans isomers are theoretically balanced. In fact, although the cis $\rightarrow$ trans conversion raises no major problems, the reverse reaction is slow and difficult.

As the speeds of the three equilibrium reactions between these four forms are very different, each equilibrium may be considered separately. It is therefore possible to apply the mass action law to each case and calculate the equilibrium constants:

$$A^+ \overset{K_a}{\rightleftarrows} AO + H^+$$

$$K_a = \frac{[AO][H^+]}{[A^+]} \qquad pK_a = pH + \log \frac{[A^+]}{[AO]}$$

$$A^+ \overset{K_h}{\rightleftarrows} AOH + H^+$$

$$K_h = \frac{[AOH][H^+]}{[A^+]} \qquad pK_h = pH + \log \frac{[A]^+}{[AOH]}$$

$$AOH \overset{K_t}{\rightleftarrows} C$$

$$K_t = \frac{[C]}{[AOH]}$$

The following values were calculated by Brouillard *et al.* (1978) for malvidin monoglucoside

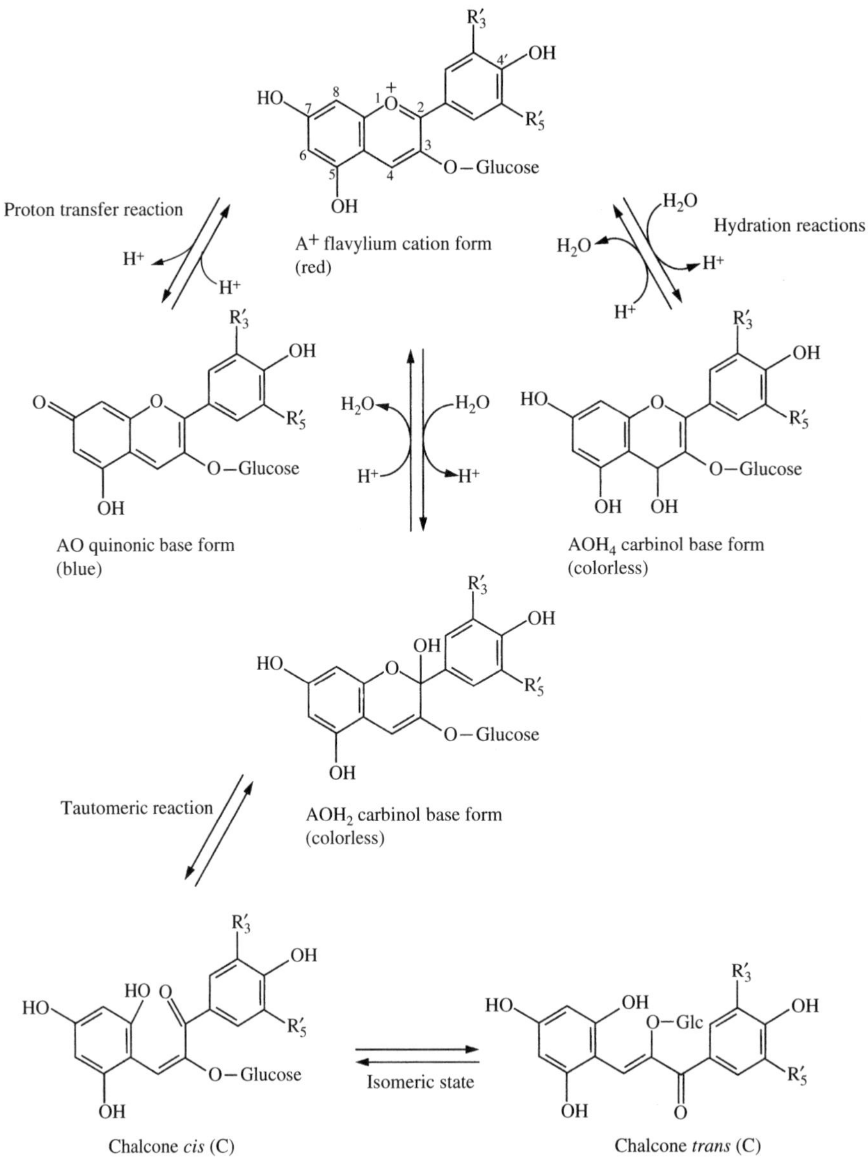

154   Handbook of Enology: The Chemistry of Wine

Fig. 6.18. The various forms of anthocyanins (R'_3 and R'_5, see Figure 6.9). (Brouillard et al., 1978)

(malvine) at 20°C:

$$pK_a = 4.25 \quad pK_h = 2.6 \quad K_t = 0.12$$

Molecular extinction coefficients at 520 nm are $\varepsilon_{(A^+)} = 27\,000$ M^{-1} cm^{-1}, $\varepsilon_{(AO)} = 14\,000$ M^{-1} cm^{-1}.

Glories (1984) also reported various values at 20°C, but with a mixture of anthocyanins extracted from Cabernet Sauvignon skins. This was made up of the five monoglucosides in grapes and wine (malvidin 45%, petunidin 25%, delphinidin 15%, peonidin 10% and cyanidin 5%):

$$pK_a = 3.41 \quad pK_h = 2.93 \quad K_t = 0.61$$

Molecular extinction coefficients at 520 nm are $\varepsilon_{(A^+)} = 18\,800$ M^{-1} cm^{-1}, $\varepsilon_{(AO)} = 7332$ M^{-1} cm^{-1}.

Phenolic Compounds

The pK_a of mixed anthocyanins in grapes had a higher acidity and pK_h than that of pure malvine. On the one hand, this facilitated the proton transfer reaction and the red–mauve balance occurred at a lower pH. On the other hand, hydration was more difficult; therefore the red forms faded at a slightly higher pH. Between pH 3 and pH 4, the color of the mixed solution was redder and more violet than that of malvine. These differences are due to the type-B cycle substitution.

It is thus possible to calculate the percentage of the various forms of anthocyanins according to pH (Figure 6.19), especially in wine with a pH between 3 and 4 (Table 6.1). The proportion of strongly colored forms is higher in mixed grape anthocyanins than in malvine alone (Table 6.2).

Anthocyanin solutions are strongly bleached in the presence of sulfur dioxide. At a pH of 3.2, 96% of the sulfuric acid ($SO_2 + H_2O$) consists of HSO_3^- (bisulfite) anions that react with the flavylium cation, most probably on carbon 2 by analogy with the hydration reaction. The product formed is colorless:

$$A^+ + HSO_3^- \underset{}{\overset{K_s}{\rightleftharpoons}} ASHO_3$$

Table 6.1. Calculated percentage coloring of anthocyanin solutions in the mixture under consideration, between pH 3 and pH 4 at 20°C (Glories, 1984)

pH	A$^+$	AO	A(OH)	C
3.0	30.6	11.9	35.7	21.8
3.1	25.8	12.6	38.2	23.3
3.2	21.7	13.4	40.3	24.6
3.3	18.0	14.0	42.2	25.8
3.4	14.9	14.5	43.9	26.7
3.5	12.2	15.0	45.2	27.6
3.6	8.9	15.4	46.4	28.3
3.7	8.0	15.7	47.4	28.9
3.8	6.5	16.0	48.1	29.4
3.9	5.2	16.2	48.2	28.8
4.0	4.2	16.3	49.4	30.1

The equilibrium constant

$$K_s = \frac{[AHSO_3]}{[A^+][HSO_3^-]}$$

was calculated using mixed grape anthocyanins.

The result ($K_s = 10^5$ M^{-1}) is similar to that obtained by Brouillard et al. (1979), with monoglucoside cyanidin ($K_s = 1.05 \times 10^5$ M^{-1}).

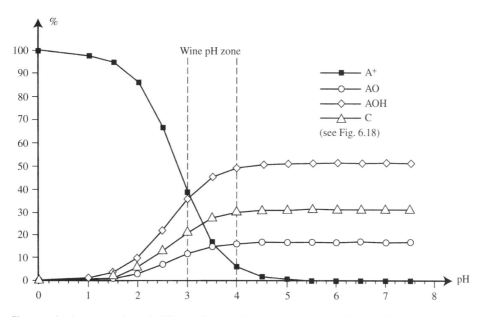

Fig. 6.19. Changes in the proportion of different forms of anthocyanins according to pH: p$K_a = 3.41$, p$K_h = 2.93$, $K_t = 0.61$ (Glories, 1984)

Table 6.2. Comparison of the percentage color at 520 nm, between pH 3 and pH 4, calculated for a grape anthocyanin solution and a malvidin monoglucoside solution. (Glories, 1984)

pH	Calculated percentages of colored forms ($A^+ + AO$)	
	Mixed grape anthocyanins[a]	Pure malvine[b]
3.0	35.2	26.6
3.1	30.7	22.5
3.2	26.9	18.9
3.3	23.5	15.7
3.4	20.6	13.1
3.5	18.0	10.8
3.6	15.9	9.0
3.7	14.1	7.4
3.8	12.7	6.2
3.9	11.5	5.2
4.0	10.6	4.3

[a] The values are calculated on the basis of the following coefficients:

$\varepsilon(A^+)520 = 18\,800$ M^{-1} cm^{-1}; $\varepsilon(AO)520 = 7332$ M^{-1} cm^{-1}

[b] The values are calculated on the basis of coefficients determined by Brouillard et al. (1978) for monoglucoside malvidin:

$\varepsilon(A^+)520 = 27\,000$ M^{-1} cm^{-1}; $\varepsilon(AO)520 = 14\,000$ M^{-1} cm^{-1}

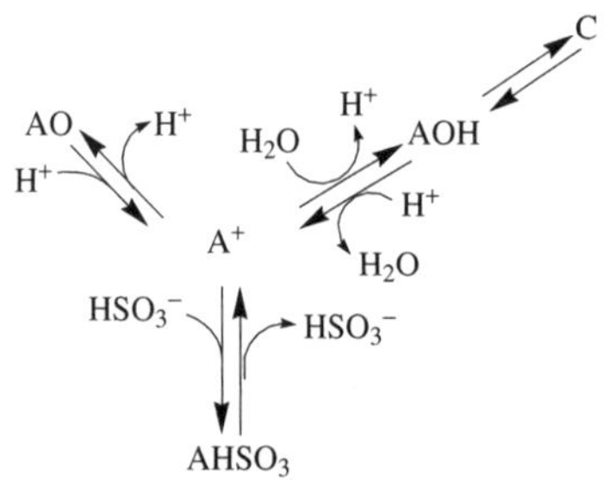

Table 6.3. Calculated percentage coloring of anthocyanin solutions according to concentration, pH, and added SO_2 as compared to the color of these solutions at pH 0 with no SO_2. Under the latter conditions, all of the anthocyanins are in flavylium form (red) (Glories, 1984)

pH	Free SO_2 (mg/l)	Anthocyanins (mg/l)			
		50	100	200	400
3.2	0	26.9	26.9	26.9	26.9
	10	9.4	13.2	18.3	22.2
	20	4.5	7.0	11.0	17.6
	30	3.0	3.8	6.4	13.3
3.4	0	20.6	20.6	20.6	20.6
	10	8.6	11.0	14.2	17.0
	20	4.7	6.0	9.2	13.6
	30	3.2	3.8	5.8	10.5
3.6	0	15.9	15.9	15.9	15.9
	10	7.9	9.4	11.4	13.3
	20	4.7	5.7	7.8	10.8
	30	3.3	4.0	5.4	8.5
3.8	0	12.7	12.7	12.7	12.7
	10	7.3	8.2	9.5	10.7
	20	4.9	5.6	6.9	8.8
	30	3.6	4.1	5.1	7.2
4.0	0	10.6	10.6	10.6	10.6
	10	7.0	7.5	8.3	9.1
	20	5.1	5.6	6.4	7.7
	30	3.9	4.3	5.1	6.4

Fig. 6.20. Bleaching of anthocyanin solutions due to pH and sulfur dioxide

Figure 6.20 summarizes the equilibria involving pH and sulfur dioxide that lead to a shift towards colorless forms.

Table 6.3 shows the percentage coloring of anthocyanin solutions, according to concentration, pH and SO_2 content, indicating that, if red wine only contained free anthocyanins, the color would not be very intense, similar to that of a rosé wine. Furthermore, at a given concentration of SO_2, the percentage coloring depends on the anthocyanin content: the higher the concentration, the more intense the color.

6.3.3 Anthocyanin Breakdown Reactions

Anthocyanin molecules are not very stable, so their concentration in wine drops sharply during the first few months of barrel aging. They disappear completely in a few years, although the wine remains red. This decrease is due to combination reactions with various other compounds in the wine, especially tannins, as well as breakdown reactions.

The stability of these pigments depends on various factors: the type of molecule, the concentration of the solution, pH, temperature, oxidation, light and the types of solvents.

The mechanisms of breakdown reactions are not very well known. It is, however, assumed that, according to prevailing conditions, they may result in:

(a) chalcones, in an alkaline medium (Ribéreau-Gayon, 1973),

(b) malvones, under the influence of peroxides (Karrer and De Meuron, 1932; Montreau, 1969; Hrazdina, 1970),

(c) phenolic acids and coumarins, in aqueous solutions with a pH between 3 and 7 (Hrazdina, 1971),

(d) dihydroflavonols, in the presence of alcohols (Glories, 1978a).

Three types of reactions were the subject of more detailed study by Glories (1978a) and Galvin (1993):

Thermal degradation of anthocyanins

Heating an anthocyanin solution to 100°C causes color fading that becomes more marked over time (Table 6.4). This result could be explained by a shift in the equilibrium towards chalcone and colorless forms. However, once it has been heated, the solution never returns to its original color, whatever the subsequent conditions (temperature, time, darkness, etc.).

HPLC and spectroscopic analysis show that a complex process occurs, involving two types of reactions:

1. Breakdown of the carbon chain of the chalcone trans and the corresponding formation of benzoic acid.

2. Glycoside hydrolysis and the formation of dihydroflavonol which may produce a cinnamic acid.

After 8 hours of heating, anthocyanin solutions contain benzoic and cinnamic acids, dihydroflavonols, catechins and a certain number of unidentified molecules. Furthermore, malvidin, the major component of wine coloring matter, has been found to be much more sensitive to thermal degradation than cyanidin (Table 6.4). The temperature factor should, therefore, be taken into account when wines are aging in barrels, vats or bottles, in order to protect their color.

Oxidative degradation of anthocyanins

Anthocyanins in an acidified alcohol solution (0.1% HCl) lose their color after a few days of exposure to light (Table 6.5). The reaction is mainly affected by alcohol concentration and type of solvent (ethanol, methanol, etc.). Oxygen and light seem to be catalysts. Dihydroflavonols (taxifolin) have been detected in the reaction medium (Glories, 1978a).

Table 6.5. Influence of storage time at room temperature in the presence of oxygen on the color of model anthocyanin solutions (500 mg/l) (Laborde, 1987)

Time (in days)	Mv-3Gl		Cy-3Gl	
	CI	Abs 520	CI	Abs 520
0	100.0	100.0	100.0	100.0
6	88.1	88.9	93.5	92.1
16	71.6	71.5	86.7	84.1
24	71.0	70.0	75.8	70.1
28	70.6	70.0	35.9	23.6
31	68.1	67.1	28.6	20.6
35	66.2	65.6	22.6	15.4
45	62.4	60.7	19.2	11.8
59	62.7	59.9	15.0	7.3
63	58.6	57.0	13.9	6.8
66	58.6	56.5	13.9	6.3
71	40.9	38.8	16.7	8.1
89	37.4	34.2	17.4	7.5
97	34.8	31.2	22.7	10.4
111	35.1	32.2	22.8	10.2
148	—	—	23.8	11.3

Model solution: 10% EtOH + 5 g/l tartaric acid at pH 3.2; CI = color intensity (Section 6.4.5); Abs 520 = absorption through 1 mm thickness at 520 nm.

Table 6.4. Influence of heating time, in a water bath at 100°C, on the color intensity of anthocyanin solutions (results expressed as percentages) (Galvin, 1993)

	Time (h)					
	0	0.5	1	2	4	8
Mv-3Gl	100	87.4	75.5	55.3	36.0	14.4
Cy-3Gl	100	76.9	88.3	79.9	67.5	46.0

Malvidin is more resistant than cyanidin to the controlled oxidation that occurs during the barrel aging of red wines (Table 6.5), due to the substitution of the B cycle (Laborde, 1987).

Orthodiphenols are oxidizable and may act as substrates for oxidation enzymes, such as polyphenol oxidase and, to a lesser extent, peroxidase. The orthoquinones thus formed are powerful oxidants and reagents. Anthocyanins react with the corresponding oxidation by-products, but less strongly than caffeoyl-tartaric acid (caftaric acid). They may be oxidized by quinones, producing highly unstable anthocyanin-quinones. Otherwise, the carbinol base form (neutral), with negative charges on nodes 6 and 8, may fix the electrophilic quinones, giving a colorless addition product that dehydrates to form a red flavylium cation. This reaction has been demonstrated in a model solution (Sami-Monchado *et al.*, 1997) and in wine (Laborde, 2000).

Degradation of anthocyanins by ketones

In an aqueous acid medium containing acetone, anthocyanins produce orange-colored compounds (Glories, 1978a). Various mechanisms have been suggested to explain the formation of these orange compounds (Section 6.3.9): hydrolysis of the anthocyanins and conversion of the anthocyanidins into dihydroflavonols, breakdown of the heterocycle with formation of benzoic acids, or a reaction between acetone and anthocyanin via polarized double bonds.

The presence of 2-oxogluconi and 5-oxogluconic acids in red wines, sometimes at fairly high concentrations in wines made from spoiled grapes (Flanzy, 1998), results in colors that shift rapidly toward orange tones. This is probably due to a reaction between these acids and anthocyanins, by fixing the polarized double bond on C4 (Section 6.3.9).

6.3.4 Reactions Involving Tannins with Protein and Polysaccharides

Polyphenols, especially tannins, are capable of forming stable combinations with proteins (Section 6.2.4) and polysaccharides. Although various types of interactions between tannins and proteins have been observed (Section 9.3.2), their mechanisms have not been completely explained. The two main types are hydrophobic interactions and hydrogen bonds (Figure 6.21), while ionic or covalent bonds are probably present but less important.

The model of interactions between tannins and proteins (Figure 6.22) described by Haslam in 1981 is still in use today. In the case of small quantities of proteins, the polyphenols spread over the surface in a single layer, thus decreasing their hydrophilic character. The proteins clump together and, eventually, precipitate. When the protein concentration increases, phenolic compounds spread over their surface act as 'ligands' or cross-linking agents between the various molecules. The superficial hydrophobic layer then recombines and causes the proteins to precipitate. Therefore, the relative concentrations of tannins and proteins affect the formation and precipitation of tannin-protein complexes.

A number of factors, including pH, reaction time, temperature, solvents and ionic strength, have an influence on the formation of tannin-protein complexes. Furthermore, the type and molecular weight of the proteins seem to play a major role in the formation of insoluble complexes. Hagerman and Butler (1980) showed that proteins with a high proline content had a great affinity for condensed tannins. This property has an impact on the organoleptic qualities of tannins in red wine and plays an important role in fining wine (Lagune, 1994), thus explaining the significance of the fining agent's protein composition.

The procyanidin and polysaccharide bonds in the skin cell walls (Amrani-Joutei, 1993) constitute another type of complex, with a less well-defined mechanism. Both acid polysaccharides (pectins), with α-D-galacturonic acid as their main monosaccharide, and arabinogalactans react strongly. In the presence of proteins, they promote the formation of tannin complexes that are one of the factors ensuring a stable head on beer and lasting bubbles in sparkling wine (Siebert *et al.*, 1996).

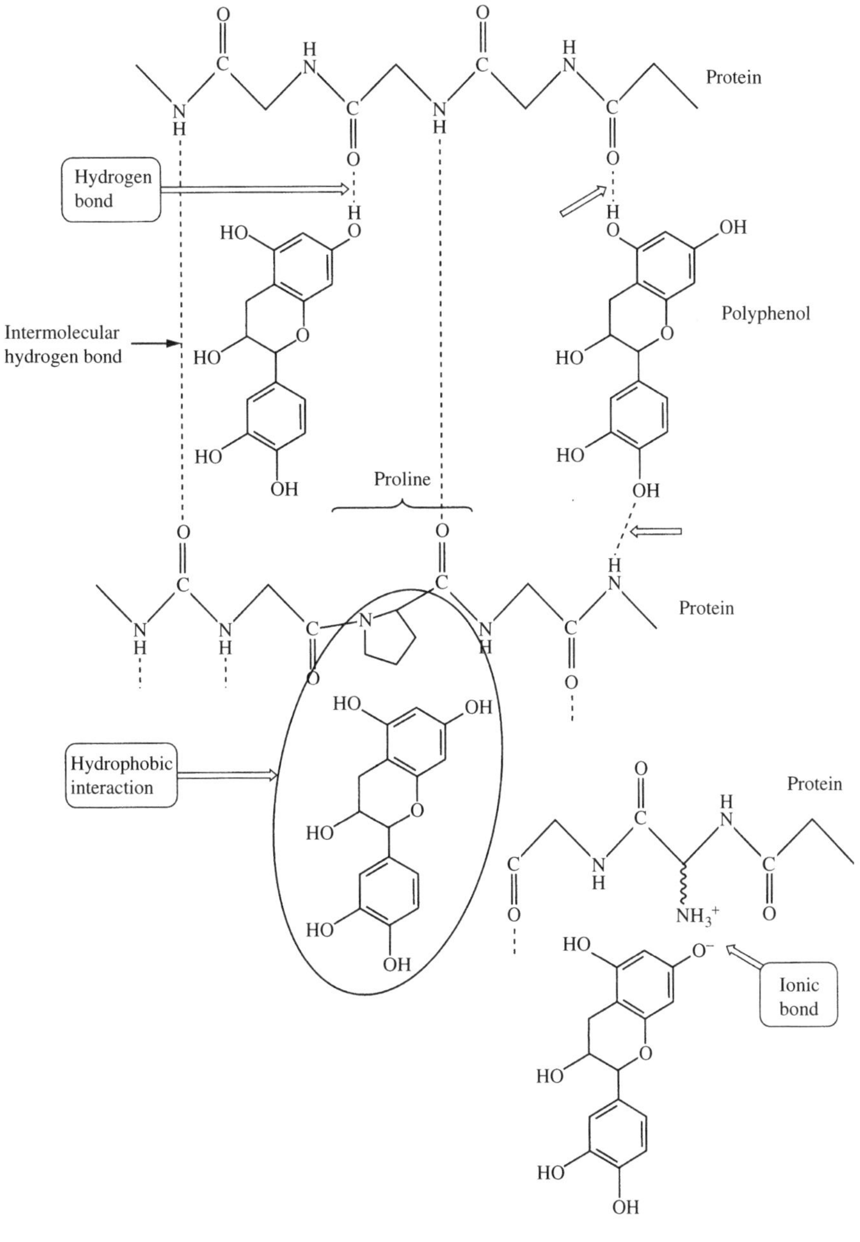

Phenolic Compounds

Fig. 6.21. Interaction between proteins and polyphenols (Asano *et al.*, 1982)

6.3.5 Formation of Carbocations from Procyanidins

Procyanidins are the building blocks of condensed tannins (Section 6.2.4). These are polymers of flavan-3-ol with a defined bond between two carbons in the flavan units: C_4 in the upper section and C_6 or C_8 in the lower part. However, this bond is relatively unstable and may be broken by acid catalysis.

A dilute alcohol solution (10% vol) with a pH similar to that of wine (Section 3.2), containing

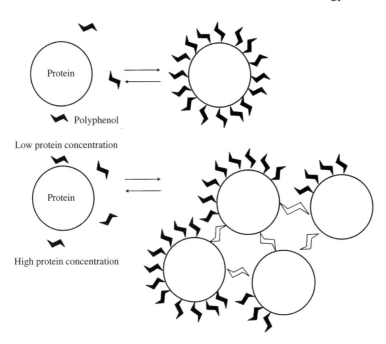

Fig. 6.22. Model of protein precipitation by polyphenols (Haslam, 1981)

procyanidin B_3 (Figure 6.15) and (−)-epicatechin (Figure 6.14), was kept at 20°C in the absence of oxygen and light and its composition monitored by HPLC. After a few weeks, the procyanidin B_3 and (−)-epicatechin concentrations decreased and two new peaks appeared, corresponding to (+)-catechin and procyanidin B_4. If the reaction was allowed to continue, polymerized forms were identified. It was as though (Figure 6.23) the acid medium caused the breakdown of the B_3 dimer, releasing the catechin (flavan-3-ol) corresponding to the lower part, which stayed in the medium, and an 'activated catechin', originating from the upper part (carbocation). The carbocation formed after protonation of the tautomeric form was in balance with the corresponding methylene quinone. It had an electrophilic center that could bond to various nucleophilic compounds (thiols, R-SH), making it highly reactive, for example, with (−)-epicatechin to form procyanidin B_4.

Heating a procyanidin solution in a strong inorganic acid medium (HCl) in the presence of oxygen produces red cyanidin. This is known as the Bate–Smith reaction (1954) and was reported in the early 20th century by Laborde (1910). It is used to detect the presence of these molecules and measure their concentrations in wine (Ribéreau-Gayon and Stonestreet, 1966). If the inorganic acid is replaced by acetic or even formic acid, the reaction is more complex, slower, requires more energy and produces derivatives with a xanthylium nucleus (Cipolli, 1975). In both instances, the first stage of the reaction is the formation of a carbocation (C_4^+). Its conversion to cyanidin (Figure 6.23) corresponds to oxidation (loss of two protons and two electrons) and requires energy. The carbocation stabilizes in the presence of organic acid.

In an acid dilute alcohol medium, comparable to wine (pH 3.2), de Freitas (1995) showed that the carbocation formed from procyanidin B_3 (Figure 6.23) could easily react with a nucleophilic compound, such as ethanethiol. The 4-α-ethylthioflavan-3-ol derived from (+)-catechin has been isolated and synthesized (Figure 6.24). Its structure has been formally established by NMR of the proton and C_{13} after acetylation. Thiolysis is a standard technique using various thiols (toluene α-thiol, benzene thiol) to study procyanidic oligomer structures. This

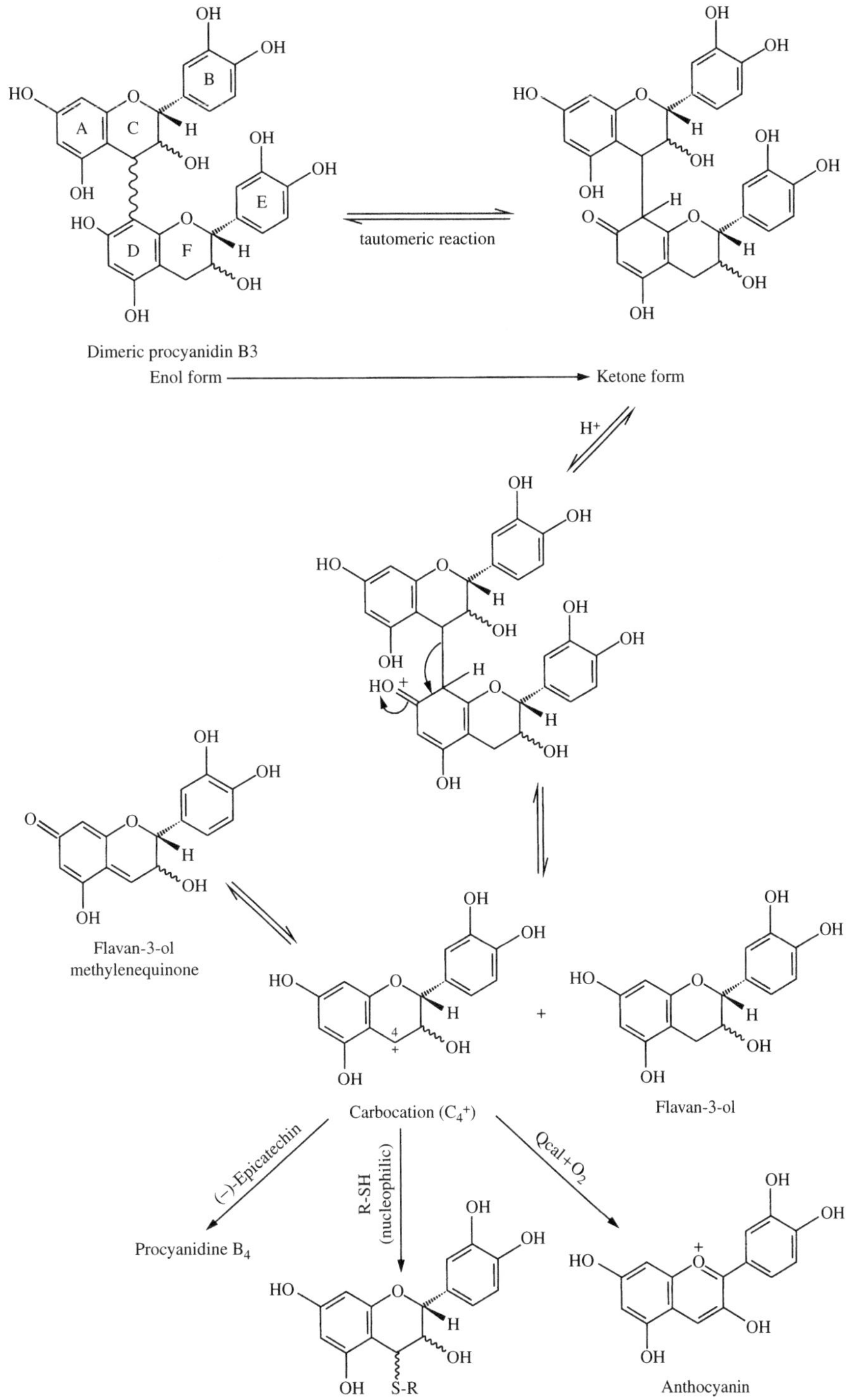

Phenolic Compounds

Fig. 6.23. Breakdown of dimeric procyanidins by acid catalysis (de Freitas, 1995)

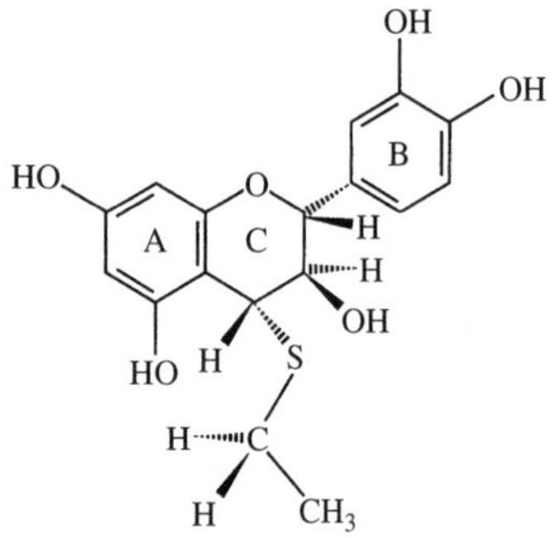

Fig. 6.24. Structure of 4-α-ethylthioflavan-3-ol derived from (+)-catechin by reacting with ethanethiol (de Freitas, 1995)

makes it possible to determine the mean degree of polymerization (MDP, Section 6.4.4) of the polymers. Accurate monitoring of the stages in this reaction is used to define the breakdown kinetics of procyanidin B_3 under acid catalysis at pH 3.2, depending on temperature. The breakdown speed of procyanidin B_3 depends on concentration and also temperature (between 5 and 37°C).

This reaction probably occurs not only in red but also in white wine. The carbocations are stabilized by reacting with nucleophilic compounds, such as the sulfur compounds present in very small quantities that are responsible either for off-flavors or characteristic aromas. The sulfur compounds may be inhibited in this way (Figure 6.24) and lose their unpleasant organoleptic characteristics (odor and flavor).

6.3.6 Procyanidin Oxidation Reactions

Oxidizability is a characteristic of the phenolic function (Ribéreau-Gayon, 1968), giving these substances a protective effect against oxidation, particularly in grapes and red wines. This reaction may be chemical or enzymatic. The phenomena, well known in enology, of oxidation of phenolic compounds in grapes by grape enzymes (tryrosinase) and *Botrytis cinerea* (laccase), are described elsewhere (Volume 1, Section 11.6.2).

Furthermore, there is considerable interest in studying the oxidative–reductive properties of polyphenols in view of their medical and nutritional implications. One of their properties is the neutralization of the oxygenated radicals responsible for tissue breakdown, which has been linked to aging and, possibly, to tumor development.

The oxidation mechanisms involved are very complex, particularly in an acid medium. Light and temperature, as well as the presence of hydroperoxides and certain metals, promote the formation of oxygenated radicals (Waters, 1964). Molecular oxygen, O_2, has a biradical structure. It may become a hydroperoxide radical $HO_2^{\bullet}$ or the superoxide anion $O_2^{-\bullet}$, responsible for creating a wide range of oxygenated free radicals. The resulting peroxides may cause the oxidative degradation of proteins and many other molecules (carbohydrates, unsaturated fats, etc.). In particular, phenolic compounds (tannins) take priority in oxidation and contribute towards eliminating free radicals.

The oxidation of tannins in an acid medium is currently thought to correspond to the hypotheses outlined in Figure 6.25, producing polymers and insoluble brown pigments known as phlobaphenes. Several parameters of this reaction are known:

1. The procyanidins polymerized during the various reactions oxidize other components in the medium, especially ethanol into ethanal.

2. Oxidation of the various flavanols, by oxygen in the presence of catalysts (Fe^{2+} and Cu^+), is completed in 20–60 days, in a medium with EtOH 10% and pH 3.2.

3. The molecular structure of the phenolic compound affects the reaction speed. Among the catechins, (−)-epicatechin is more oxidizable than (+)-catechin.

4. In the case of dimeric procyanidins (Figure 6.25) with a C_4–C_8 interflavan bond (B_1 to B_4), oxidation depends on the type of upper structural unit. The oxidizability of the C_4–C_6 bond (B_5 to B_8) originates in the lower structural unit. The presence of (+)-catechin (B_3–B_4 and B_6–B_7) makes molecules more oxidizable, as compared to (−)-epicatechin (B_1, B_2, B_6, B_8).

5. For the same basic [(−)-epicatechin] unit, oxidizability increases with the degree of

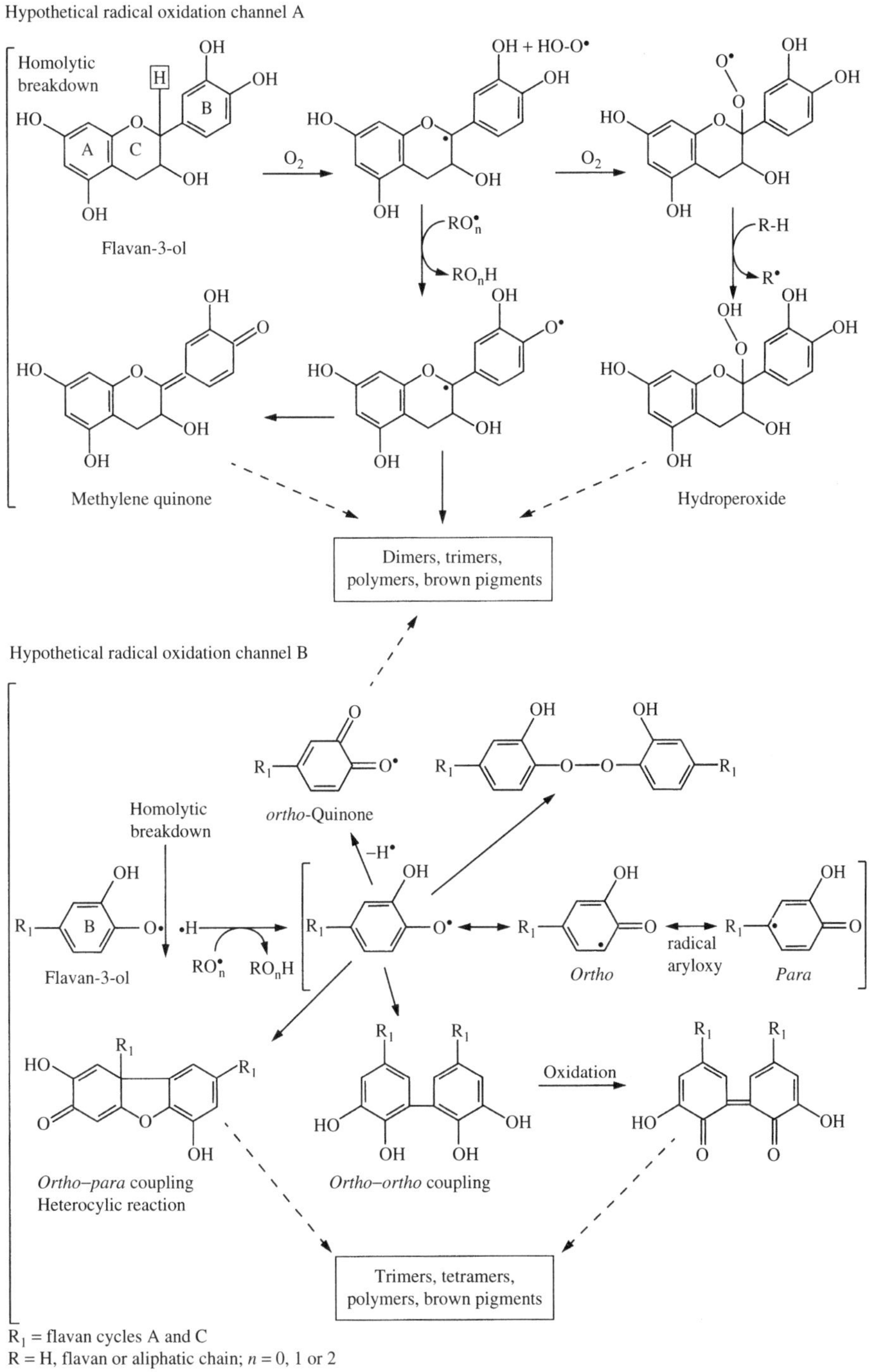

Phenolic Compounds

Fig. 6.25. Diagram of the various hypothetical channels for the oxidation of polyphenols in the presence of traces of oxygen: channel A, formation of a free radical in C_2; channel B, formation of an aryloxy free radical (de Freitas, 1995)

polymerization. The C_1 trimer is certainly more oxidizable than the B_2 dimer or the monomer, due to the number of oxidizable sites.

6. Oxidizability also depends on the type of basic unit. One significant factor is the esterification of the OH bond in position 3 by gallic acid, which increases the oxidizability of epicatechin and decreases that of the procyanidins.

In conclusion, flavanols, procyanidins and, consequently, condensed tannins react more-or-less easily with free radicals, according to their configuration. These chain reactions produce brown polymers with varying structures that precipitate. In wine, these phenomena depend on the phenol content. The oxidation kinetics are apparently slower than in a model medium, probably due to the presence of even more easily oxidizable compounds, also involved in the oxidation of procyanidins.

6.3.7 Procyanidin Polymerization Reactions

Acid solutions of dimeric, oligomeric and polymerized procyanidins are unstable. Even under nitrogen with sulfur dioxide in the absence of light, the color yellows, then browns and, after a short time, a precipitate is observed. At pH 3.2, the reaction takes about ten months at 5°C, a few months at 20°C and one to two months at 30°C. In the presence of oxygen from the air, and especially at high temperatures, conversion of the solution is more intense and the precipitates look different. It is impossible to dissolve them in any solvents other than formic acid. They can only be studied after acetylation. Results obtained by molecular screening (TSK), NMR and mass spectrometry show that these complex polymers have molecular weights above 3000.

In an acid medium without oxygen, procyanidins are capable of forming a carbocation that is likely to react with the negative nodes of another procyanidin and thus increase its degree of polymerization. When a procyanidin B_2 solution is stored, it produces C_1 trimer, several polymers and $(-)$-epicatechin. On the other hand, $(+)$-catechin and $(-)$-epicatechin solutions are perfectly stable under the same conditions. The fact that these reactions are accelerated by temperature is a clear indication that a carbocation is formed. Polymerized procyanidins, therefore, are produced by a C_4–C_8 or C_4–C_6 'organized polymerization' (Figure 6.26a).

This type of polymerization certainly occurs in wine, as catechin is always present. Even after it has been eliminated, it reappears after a few weeks of aging.

In a strongly oxidative medium, the formation of free radicals is accompanied by disorganized oxidative polymerization, as previously described (Figure 6.26b). However, with gentle,

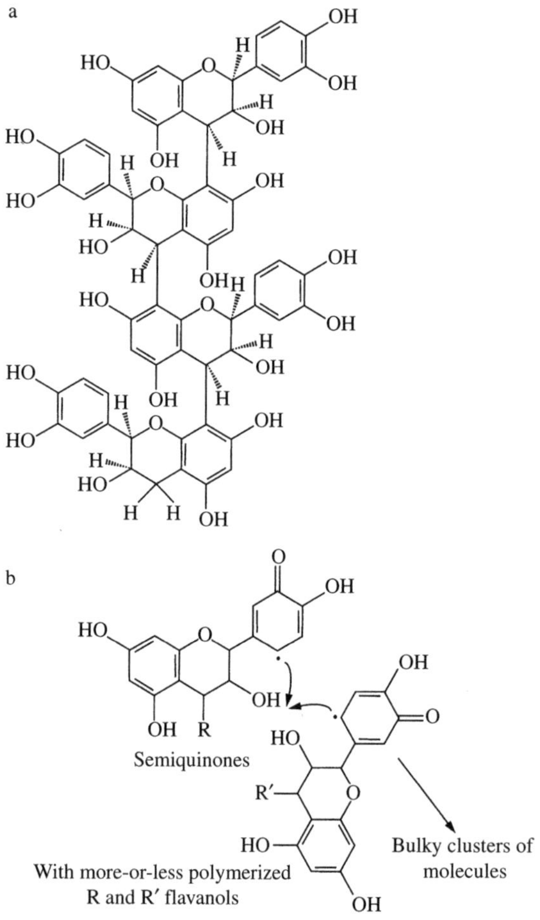

Fig. 6.26. Example of tannin polymerization: (a) 'organized polymerization' of tannins, an example of a tetrameric procyanidin; (b) disorganized polymerization of tannins (reactions with free radicals) (Glavin, 1993)

Phenolic Compounds

controlled aeration, combined oxidation of the procyanidins leads to the formation of ethanal from ethanol. This molecule is responsible for modifying the procyanidin structure. The reaction is much faster than 'organized polymerization' and produces polymers that are likely to precipitate, according to their degree of polymerization and concentration.

The aldehyde function is protonated in an acid medium (Figure 6.27). It produces an initial carbocation that stabilizes by fixing on one of the negative nodes (6 or 8) of the flavanols (P). In this way, it forms benzyl alcohol, characterized by a polarized bond that is easily broken in an acid medium, producing another carbocation that reacts with the nodes of another flavonol.

$$-\overset{|}{\underset{|}{C}}-OH$$

Benzyl alcohol

It is as though the various flavanols in the medium (catechins and procyanidins) were linked by ethyl cross-bonds

$$-\overset{|}{C}H-CH_3$$

This type of reaction produces heterogeneous polymers with very different configurations from those of homogeneous polymers. These are responsible for specific properties that are important in enology. The reaction may go as far as dodecamers (molecular mass 3600, diameter around 4 nm) and the corresponding polymers combine by hydrophobic forces to form colloidal aggregates, 400 nm in diameter, that are likely to precipitate (Saucier, 1997).

This gives an idea of the diverse structures of polymers formed in wine, according to winemaking and barrel aging conditions, and their effect on flavor and quality.

Another type of polymerization has been identified, based on aldehydes such as glyoxylic acid, furfural, methyl furfural, etc. The complexes formed are yellow, with a xanthylium structure.

Glyoxylic acid is formed by the oxidation of tartaric acid (Fulcrand et al., 1997). Furfural is present in rotten grapes and may also be released by barrels, depending on the toasting level.

The reaction is simple, starting with the formation of a cross-bond between the two flavanol units in an acid medium (Figure 6.28), followed by dehydration of the OH on 7 and cyclization. The flavene thus formed is then oxidized, producing the yellow xanthylium cycle (Es-Safi et al., 1999). This reaction contributes to the yellowing observed

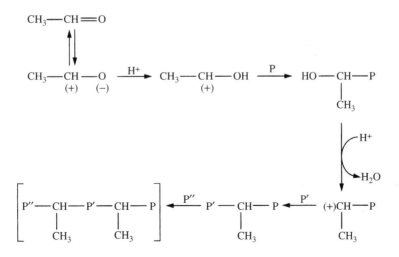

Fig. 6.27. Diagram of 'heterogeneous polymerization' of procyanidins (P) in the presence of ethanal (Glories, 1987)

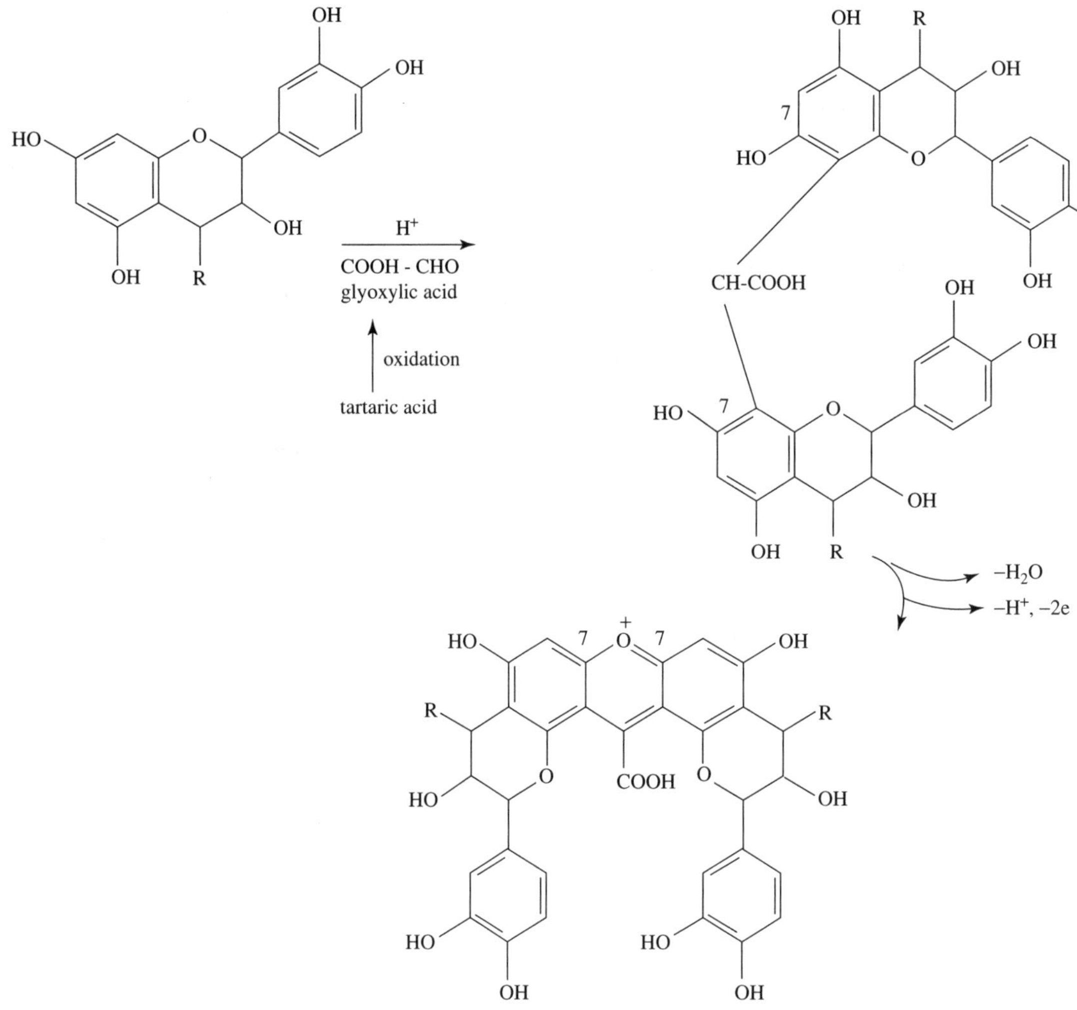

Fig. 6.28. Polymerization of flavanols in the presence of glyoxylic acid

in the color of oxidized wines, as well as those made from grapes affected by botrytis.

6.3.8 Anthocyanin Copigmentation Reactions

The color of anthocyanin solutions depends on many factors that have already been mentioned (Section 6.3.2) (concentration, pH, SO_2, temperature, etc.), but also on the presence of other components in the medium that cause both a color shift towards violet (bathochrome effect) and an increase in intensity (hyperchrome effect).

Metal cations, mainly Al^{3+}, Fe^{3+}, Cu^{2+}, Mg^{2+}, form complexes with anthocyanins that have two phenols in the *ortho* position on the B nucleus (delphinidin, petunidin and cyanidin). These are responsible for bathochrome effects of varying intensity. Two types of chelates, either directly with both phenol functions (flavylium A^+ form) or with the aromatic ketone in carbon 4' (AO quinoid base), stabilize the molecules and prevent the formation of the colorless carbinol base (AOH). The color becomes intensely blue, even at pH values around 3. This property is used to change the color of certain flowers (roses, hydrangeas, etc.). These bonds generally break in a strong acid medium.

Copigmentation involves complexation phenomena, generally at low energy (hydrogen bonds and hydrophobic interactions), either between the various forms of anthocyanins or between anthocyanins and other, mostly colorless, phenolic compounds (coumarins, phenolic acids, flavonols, flavanols, etc.) These bulky complexes modify the cation resonance and prevent the substitution of carbons 2 and 4 (Mazza and Brouillard, 1987).

Copigmentation depends on many factors: type and concentration of anthocyanins, type and concentration of copigments, pH, temperature and, in particular, the type of solvents. All of the studies to date deal with aqueous solutions comparable to those in plant cells (flowers, leaves, grapes, etc.). These bonds, however, are considerably disrupted when plant pigments are extracted in the presence of alcohols (MeOH, EtOH, etc.). Unlike grape juice, dilute alcohol solutions, especially wine, are not favorable media for this phenomenon. Somers and Evans (1979) attribute the deep purple color of the juice obtained by pressing heated red grapes to complexes involving flavylium (A^+) and quinoid base (AO), together with other compounds. Once alcohol starts to be produced during fermentation, the must changes color to bright red, due to the breakdown of the bonds formed by copigmentation.

However, another copigmentation was observed in the wine due to the presence of a high ratio of tannins (3 g/l) to anthocyanins (500 mg/l) (RM = 3) (Mirabel et al., 1999). This led (Table 6.6) to an increase in the coloring of the anthocyanins and a slight shift in λmax. toward blue. The color intensity of young wine correlated with its tannin content and ionization index (Section 6.4.5).

Table 6.6. Copigmentation of anthocyanins (A) by tannins in a dilute alcohol medium. Effect of the presence of oligomeric procyanidins (Po) on the resonance of the flavylium cation (hyperchromic effect: + d520%, bathochromic effect: + $\Delta\lambda$) (Mirabel et al., 1999)

MR = Molar Ratio (equation) (Mean molecular weight: Po = 1000; A = 500)

RM	+ d520 (%)		+$\Delta\lambda$ (nm)	
	H_2O	Et(OH)	H_2O	Et(OH)
1	33	31	2	2
3	53	80	4	3
5	61	127	7	4
10	75	215	12	3

6.3.9 Reactions between Compounds with Polarized Double Bonds and Anthocyanins

A mechanism for cycloaddition on anthocyanins involving various yeast metabolites was demonstrated by Cameira dos Santos et al. (1996). This consists of a cycloaddition between a flavylium and compounds with a polarized double bond. The new pigments formed are generally orange, stable and insensitive to variations in both pH and sulfur dioxide.

Vinyl-phenol, resulting from the decarboxylation of p-coumaric acid by yeast decarboxylases, may react with malvidin, either as a monoglucoside or in the form of an acylated monoglucoside (p-coumarylglucoside). The double bond is added between the carbon 4 of the anthocyanin and oxygen held by carbon 5, forming a new oxygen heterocycle. The resulting compound is colorless and recovers unsaturated structure and color on oxidation (Figure 6.29a).

Another group of pigments identified in wine corresponds to the addition of pyruvic acid (Bakker and Timberlake, 1997) (Figure 6.29b). The authors suggest that other compounds, such as enolic ethanol, may also be involved in producing color changes (Cheynier et al., 1997; Ben Abdeljahd et al., 2000).

Compared with other pigments in wine, these new molecules are present in very small quantities.

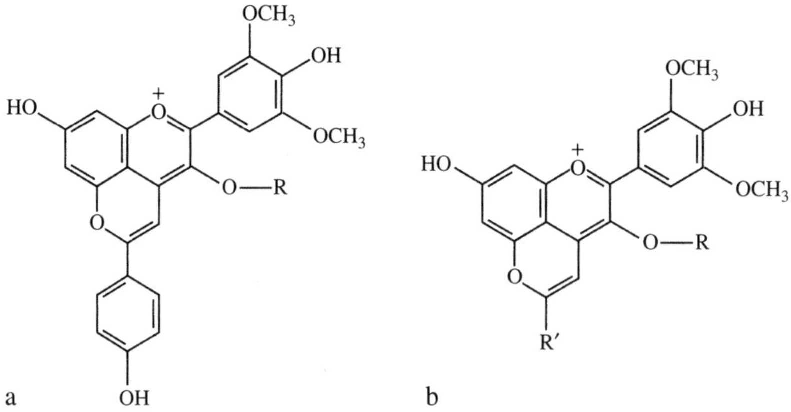

Fig. 6.29. Structure of three pigments produced by adding (a) vinyl-phenol, pyruvic acid (b, $R' =$ COOH) and ethanol (b, $R' =$ H) to malvidin-3-glucoside (R = glucose) or malvidin-3-p-coumarylglucoside (R = glucose + p-coumaric acid) (Cameria do Santos et al., 1996)

However, they are relatively stable and their concentrations change very slowly during aging, unlike those of free anthocyanins, which condense with tannins or break down under certain conditions. These molecules may represent the slightly polymerized pigments in old wines. Their structure, similar to that of the xanthylium derivatives described by Jurd in 1972, is characterized by the substitution, in positions 4 and 5, of the basic flavylium chromophore, which certainly accounts for their stability.

6.3.10 Anthocyanin and Tannin Condensation Reactions

Jurd (1972) showed that the flavylium cation could react directly with various components, such as amino acids, phloroglucinol and catechin, producing a colorless flav-2-ene substituted in carbon 4. Somers (1971) suggested that a reaction of this type was involved in the wine aging process. Indeed, the disappearance of anthocyanins has been observed while the red color remained stable, or even intensified. The complex pigments formed are not very sensitive to variations in pH and SO_2. Different mechanisms involved in condensing anthocyanins and tannins may produce compounds with various characteristics, according to the types of bonds. Colors range from orange to mauve. Three types of reactions have been identified:

1. Direct condensation reaction: anthocyanin → tannin (A–T). In this reaction, anthocyanins act as cations (A^+) on the negative nodes (6 or 8) of the procyanidins (P), forming a colorless flavene (A-P). The presence of oxygen or an oxidizing medium is necessary for the flavene to recover its color. The forms are in balance: A^+–P and AO–P (Figure 6.30).

    When anthocyanin solutions are kept in an air-free environment, in the presence of flavanols at a temperature $>20°C$, there is a decrease in color that may be reversed by aeration. A similar type of reaction occurs when wine is run off after vatting, as it 'picks up color' due to the aeration involved in this operation. Although some of these molecules have been fractionated by HPLC, their structure has not yet been formally defined. They all contribute to the red color.

2. Tannin condensation → Anthocyanin (T-A). One of the characteristics of procyanidins is that they form a carbocation after protonation of the molecule, and react with nucleophilic sites, such as nodes 6 and 8 of anthocyanin molecules as carbinol bases (neutral) (Figure 6.31).

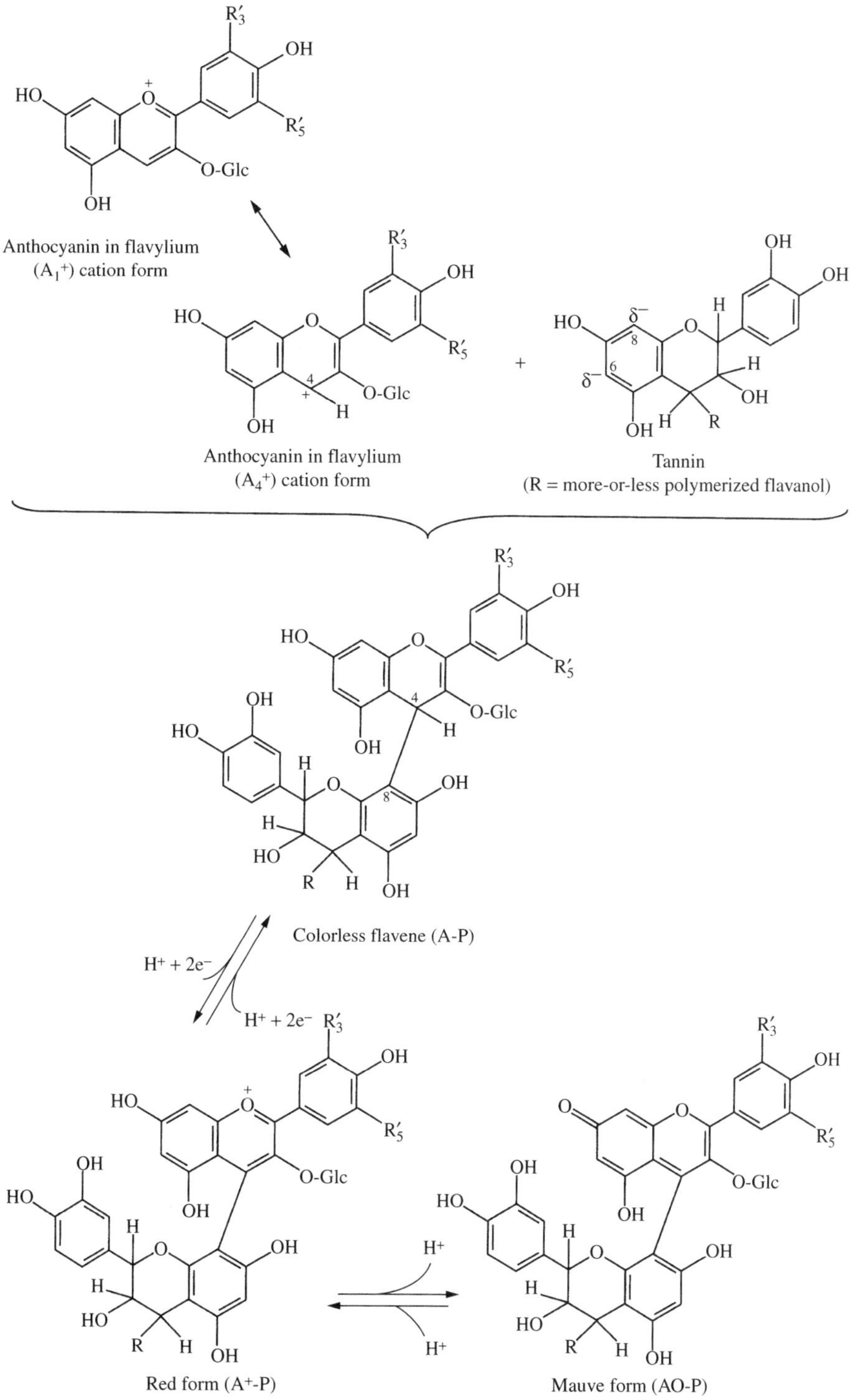

Phenolic Compounds

Fig. 6.30. Direct A-T type condensation of anthocyanins and tannins (R'_3 and R'_5, see Figure 6.9) (Galvin, 1993)

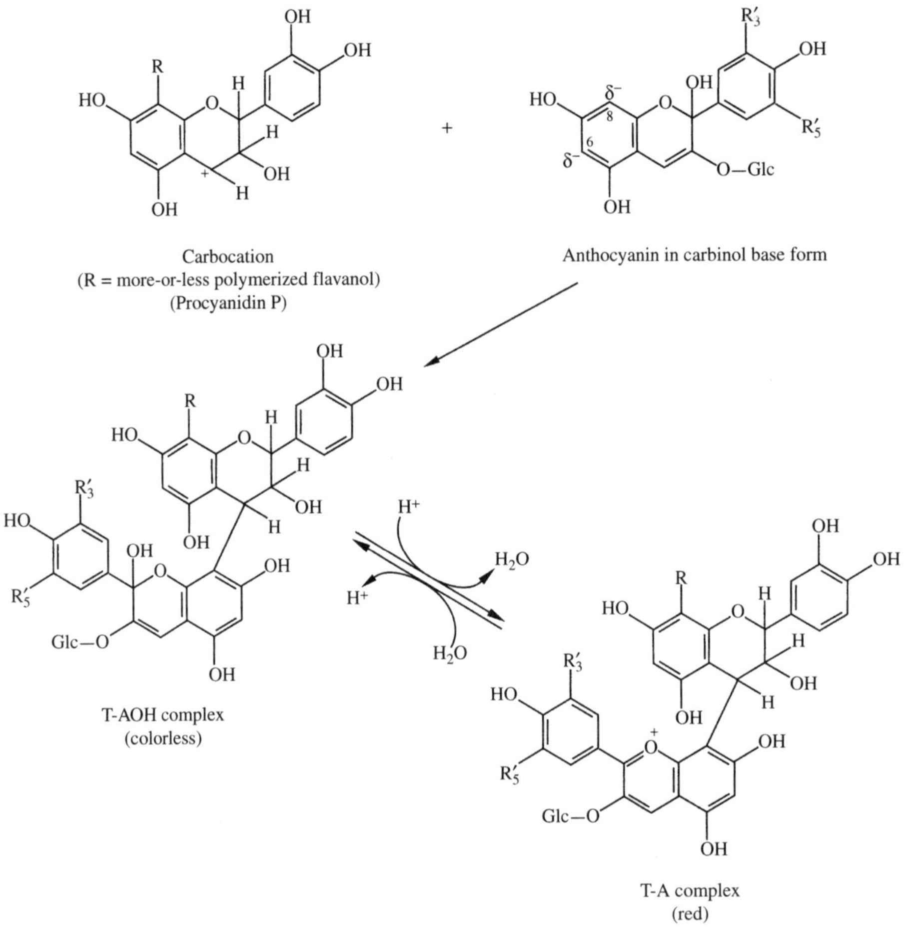

Fig. 6.31. Direct T-A type condensation of procyanidins and anthocyanins (R'_3 and R'_5, see Figure 6.9) (Galvin, 1993)

The complex thus formed is colorless and turns a reddish-orange color on dehydration. This reaction takes place in the absence of air as it requires no oxidation. It is enhanced by temperature (formation of the carbocation) and depends on the quantity of anthocyanins in the medium. Colors change according to the type of carbocation and the degree of polymerization. Keeping wine in an airtight vat or in bottle provides favorable conditions for this type of condensation.

3. Indirect reaction: condensation with an ethyl cross-bond. In an acid medium, ethanal forms a carbocation that initially reacts with the negative nodes (4 and 8) of the flavanols (procyanidin catechins), and then with the anthocyanins in neutral, i.e. carbinol base (AOH) form (Section 6.3.7).

Apparently, the bond between two C_8 takes priority (Figure 6.32), but the reaction depends on the proportion of flavanols and anthocyanins likely to react, as well as the pH of the medium. At pH 3.1 in the presence of (+)-catechin, the color ranges from reddish violet to orange, as the catechin/malvidin molar ratio increases from 1 to 10. With dimeric procyanidin B_3, the color is more orange, becoming more purplish when (−)-epicatechin reacts with malvidin monoglucoside. The reaction may continue, producing some very bulky pigments (Figure 6.33).

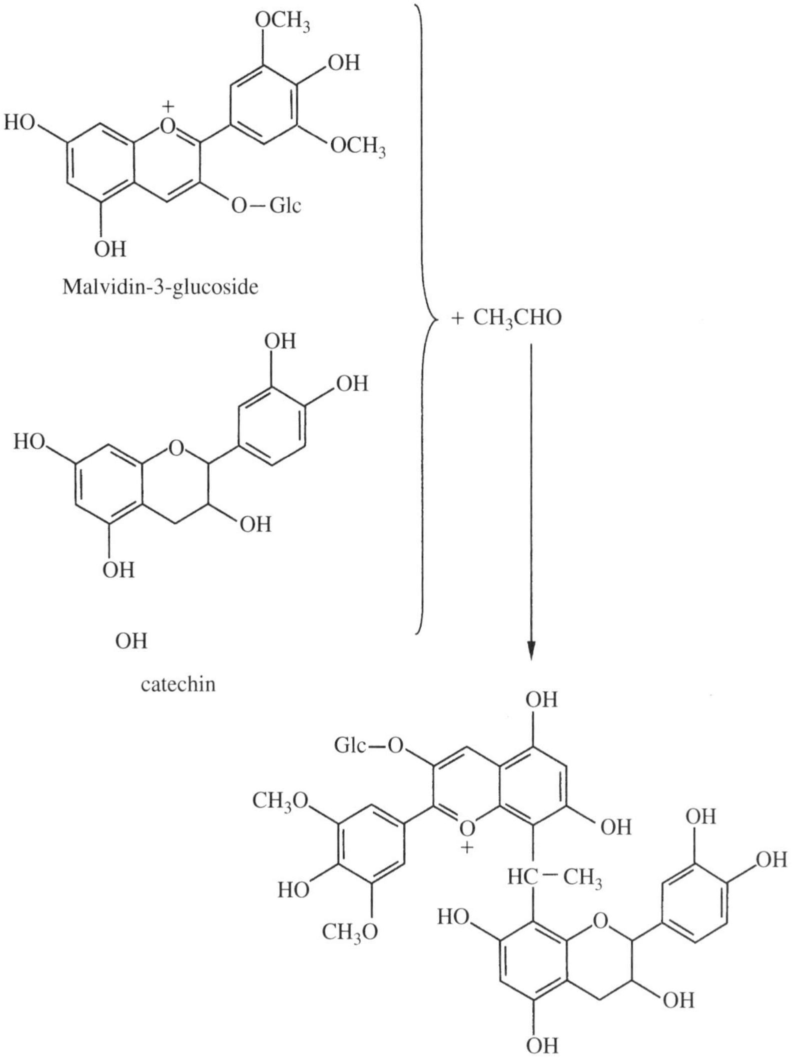

Phenolic Compounds

Fig. 6.32. Reaction between catechin and malvidin-3-glucoside in an acid medium, in the presence of ethanal (Timberlake and Bridle, 1976)

In wine, this type of reaction occurs at the same time as the heterogeneous polymerization of the procyanidins (Section 6.3.7), as a result of the controlled oxidation during barrel aging, when traces of ethanal are produced by the oxidation of ethanol. The color of the wine becomes more intense and changes tone, becoming darker after a few months in the barrel.

The presence of ethanal in a mixture of procyanidins and anthocyanins also results in orange-colored complex pyranoanthocyanin–tannin structures (Figure 6.29) (where R′ = procyanidin dimer) (Francia-Aricha *et al.*, 1997) that are highly stable over time. Complexes with ethyl cross-bonds may develop into this type of structure, thus contributing to the brick-red color of oxidized wines.

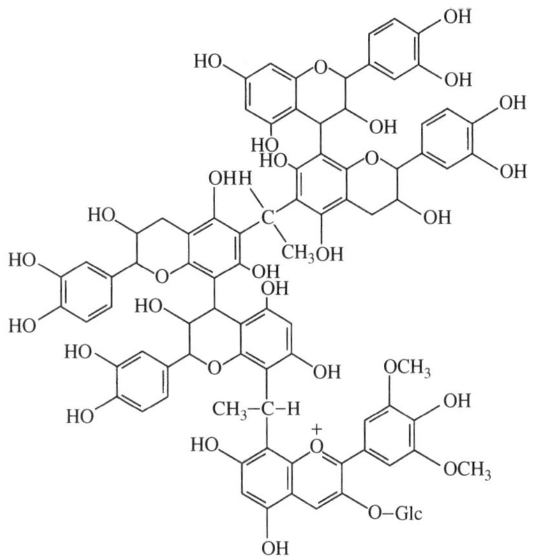

Fig. 6.33. Red-violet complex produced by the condensation of malvidin-3-glucoside and procyanidin B_3 in the presence of ethanal (Crivellaro Guerra, 1997)

6.4 ANTHOCYANIN AND TANNIN ASSAYS—ORGANOLEPTIC PROPERTIES

6.4.1 Assessing the Phenol Content of Red and White Wines

Wine contains widely varying quantities of many phenolic molecules. The ideal method for estimating the phenol content would be to define all of the compounds and assay them separately. This is not always possible, in view of the diversity of the molecules and the difficulty of analyzing them all. Furthermore, even the most effective techniques are often awkward to implement, and the results are incomplete and difficult to interpret. Although they are useful for research purposes, these techniques do not really apply to practical winemaking.

A global assessment of the phenol content of wine, expressed as a numerical value, is a rather attractive idea, along the lines of total acidity that gives winemakers a satisfactory indication of a wine's acidity. This would make it possible to classify wines according to their phenol content and measure the effects of a winemaking operation on the extraction of these compounds. However, expressing a global value in terms of the weight of certain substances (enological tannins, gallic acid, catechin, etc.) that only represent a fraction, and sometimes a minute one, of the phenols in the medium, is hardly justifiable.

The methods used for this assessment must meet three basic criteria: they must be rapid, the results obtained must be reproducible and all of the phenol molecules must be included. The various methods are based on the chemical properties of these molecules.

The permanganate value is hardly used any more (Ribéreau-Gayon et al., 1982). The Folin–Ciocalteu value (Ribéreau-Gayon, 1970) uses oxidizing agents, potassium permanganate and Folin–Ciocalteu reagent [a mixture of phosphotungstic ($H_3PW_{12}O_{40}$) and phosphomolybdic ($H_3PMO_{12}O_{40}$) acids] which act on the phenols due to their reductive properties. The first reaction, in the presence of indigo blue, produces a yellow solution, assessed visually. The second is characterized by a blue coloring, measured with a spectrophotometer. A third value is based on the characteristic absorption of the benzene cycles of the majority of phenols at 280 nm (Flanzy and Poux 1958; Ribéreau-Gayon, 1970).

Folin–Ciocalteu value

A 1 ml sample of red wine, diluted 1/10 or 1/5 with distilled water, is mixed with 5 ml of reagent, 20 ml of sodium carbonate solution (20%) and distilled water QS 100 ml. After 30 minutes, the OD at 760 nm is measured on a 10 mm optical path:

$$I_{FC} = (OD \times \text{dilution}) \times 20$$

The value is between 10 and 100.

OD 280 value

Red wine is diluted 1/100 and white wine 1/10 with distilled water. The OD is measured at 280 nm on a 10 mm optical path:

$$I_{280} = OD \times \text{dilution}$$

The value is between 6 and 120.

Measuring absorption at 280 nm seems preferable to the Folin–Ciocalteu test, as it presents a number of advantages, including speed and reproducibility. However, certain molecules, such as cinnamic acids and chalcones, have no absorption maximum at this wavelength. However, as they are present in wine at very low concentrations, any error in the value will be very small.

It is possible to define the relative contributions of phenolic acids and various non-phenolic substances in wine to this value. It is fairly constant at around 7 for both red and white wines, and corresponds to the value defined by Somers and Ziemelis (1985) using the Folin–Ciocalteu reagent. This is an important factor, especially for white and rosé wines, as it represents practically 50% of the value. In red wines, it may be considered that

$$OD\ 280 = 7 + DA + DT$$

Furthermore, the average anthocyanin and tannin coefficients in wine have been determined:

DA (anthocyanin absorption at 280 nm) = 20
  × anthocyanin concentration expressed in g/l
DT (tannin absorption at 280 nm) = 12
  × tannin concentration expressed in g/l

This value may be somewhat distorted due to an increase in the wine's gallic acid and ellagic tannin concentration during aging in new barrels. The coefficients corresponding to these molecules are certainly very high, approximately 38. This increase only lasts a limited time, due to the rapid oxidative degradation of these compounds.

6.4.2 Anthocyanin Assay

Anthocyanins (At) are present in wine in different forms: free anthocyanins (Al) and anthocyanins combined with tannins (Ac), some of which are bleached by SO_2 (TA), while the rest is unaffected (TAT):

$$At = Al + Ac = Al + TA + TAT$$

There is no accurate method for assaying At, so this value may only be estimated. On the other hand, a global value for Al + TA may be determined (As), using chemical and chromatographic methods.

The chemical methods are based on the specific properties of anthocyanins: color variation according to pH and bleaching by sulfur dioxide (Ribéreau-Gayon and Stonestreet, 1965).

The pH variation procedure requires the preparation of two samples, each containing 1 ml of wine and 1 ml of EtOH 0.1% HCl. Then 10 ml of HCl at 2% (pH = 0.7) is added to the first sample and 10 ml of buffer at pH 3.5 to the other. The difference in OD at 520 nm, Δd_1, is measured on a 10 mm optical path. In comparison with a standardized anthocyanin solution, the concentration As = Al + TA is given by the following equation:

$$C(mg/l) = \Delta d_1 \times 388$$

The SO_2 bleaching procedure requires the preparation of two samples, each containing 1 ml of wine, 1 ml of EtOH 0.1% HCl and 20 ml of HCl at 2% (pH 0.8). For the two samples, 4 ml of H_2O is added to 10 ml of the first sample, 4 ml of sodium bisulfite solution, density 1.24, is added to 10 ml of the second sample and the mixture is diluted by half. The difference, Δd_2, in OD at 520 nm is measured on a 10 mm optical path. By comparison with a standardized anthocyanin solution, the concentration As = Al + TA is given by the following equation:

$$C(mg/l) = \Delta d_2 \times 875$$

Both methods measure Al + TA. However, as the first method using the pH difference is more sensitive to the presence of free SO_2 in wine, the sulfur dioxide bleaching method is more reliable (Table 6.7).

It is possible to determine the free anthocyanin concentration (Al) after fractionation. The wine is adsorbed on a PVPP column, also known as Polyclar AT. After rinsing with water, elution with a dilute alcohol solution releases the free anthocyanins (Hrazdina, 1970), while the Ac and tannins remain adsorbed on the PVPP. After evaporation, the eluate is brought up to the initial volume and assayed using sulfur dioxide to obtain the Al concentration.

Table 6.7. Influence of the SO$_2$ content of an anthocyanin solution at pH 3.2 on the result of an assay of these pigments using two chemical methods (Glories, 1984)

Added SO$_2$ (mg/l)	Method (SO$_2$)			Method (pH)		
	d_1	d_2	Anthocyanins (mg/l)	d_1	d_2	Anthocyanins (mg/l)
0	0.300	0	262	0.682	0.144	209
10	0.301	0	263	0.746	0.140	235
20	0.307	0	269	0.769	0.136	246
30	0.305	0	267	0.784	0.123	256
50	0.308	0	269	0.800	0.123	263

On the basis of these two analyses, As and Al, Glories (1978a) defined the PVPP index used to quantify the combinations occurring in the anthocyanin assay. This value increases continuously as the wine ages:

$$I_{PVP} = \frac{As - Al}{As} \times 100$$

It is possible to analyze the anthocyanin composition of grapes and wine by means of paper chromatography (Ribéreau-Gayon, 1959, 1968; Lambri et al., 2002). These molecules may also be assayed by means of HPLC and HPTLC, using reversed-phase partition chromatography (Figure 6.34). It is not easy to identify acylated anthocyanins in new wine as their retention time corresponds to that of certain catechin–anthocyanin complexes. These molecules, however, disappear rapidly during aging. For this reason, chromatographic and chemical methods give identical results, except in the case of new wines.

6.4.3 Tannin Assay

The tannins in red wine are made up of chains of more-or-less polymerized flavanols (procyanidins). These procyanidins are either homogeneous, with regular linking, or heterogeneous, linked by different bonds (Section 6.3.7). In both cases, certain bonds are broken when these molecules are heated in an acid medium, and the resulting carbocations are partially converted into cyanidin if the medium is sufficiently conducive to oxidation. This property has been used in tannin assays (LA method) for many years (Ribéreau-Gayon and Stonestreet, 1966).

The procedure requires the preparation of two samples, each containing 4 ml wine diluted 1/50, 2 ml of H$_2$O and 6 ml of pure HCl (12 N). One of the tubes is heated to 100°C in a waterbath for 30 mn and 1 ml of EtOH at 95% is added to solubilize the red color that appears. The other sample is not heated but also receives 1 ml of EtOH. The difference in the OD at 550 nm, $\Delta D = D_2 - D_1$, is measured on a 10 mm optical path. By comparison with a standardized oligomeric procyanidin solution, the concentration is as follows:

$$LA(g/l) = 19.33 \times \Delta D$$

Although this method is highly reproducible and easy to implement, it only gives an approximate result, as it does not take into account the effect of the various structures present in wine (Porter et al., 1968), nor their degree of polymerization, nor the other components in wine that interfere with the assay. The tannin concentration in wine is often overestimated. It is not unusual to observe an increase in the results of this assay during barrel and bottle aging, which may not correspond to an increase in tannin.

On the basis of these observations, Glories (1988) developed two methods for calculating the tannin concentration in wine. The assay procedure (LA) is unchanged.

The first takes into account the various groups of molecules involved in the reaction (LA). This leads to two equations:

Young wines: $C_1(g/l) = 16.16 D_2 - 24.24 D_1$
$+ 1.71[Al]$

Phenolic Compounds

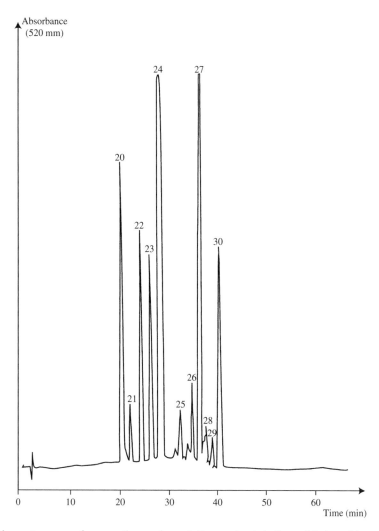

Fig. 6.34. HPLC chromatogram of an anthocyanin solution extracted from Merlot skins as monoglucosides: 20, delphinidin; 21, cyanidin; 22, petunidin; 23, peonidin; 24, malvidin; 25, 26, 27, 28, 29, 30, acylated anthocyanins (Galvin, 1993)

Older wines: $C_1(g/l) = 16.16 D_2 - 33.32 D_1 + 3.86 [Al]$

where

D_1 = optical density at 520 nm without heating, on a 10 mm optical path

D_2 = optical density at 520 nm after heating, on a 10 mm optical path

$[Al]$ = free anthocyanin concentration in g/l

The second method for calculating the tannin concentration is based on examining the visible spectrum of the reaction (LA). The following equations apply, whatever the degree of polymerization and concentration of the procyanidins. $\Delta OD\ 520$, $\Delta OD\ 470$ and $\Delta OD\ 570$ represent the difference in OD, with or without heating, for the three corresponding wavelengths:

$$\Delta OD\ 520 = 1.1 \Delta OD\ 470$$

$$\Delta OD\ 520 = 1.54 \Delta OD\ 570$$

The reaction is amplified in the presence of parasites, either towards the longer (mauve) or shorter (orange–yellow) wavelengths. The ΔOD 520 values measured and calculated from ΔOD 470 and ΔOD 570 are different, so there are three possible values for ΔOD 520. The minimum value is preferred as it is considered to provide the best estimate of procyanidin conversion alone.

The results are compared with those of an oligomeric procyanidin solution derived from grape seeds and the tannin concentration is calculated using the following equation:

$$C_2(\text{g/l}) = 15.7 \Delta\text{OD } 520$$

Both calculation methods give similar results, which are always lower than those obtained using the original (LA) method (Table 6.8).

If white wines do not contain sugar, the assay is identical to that used for red wines, without dilution or after dilution by 1/2. If residual sugar is present, it is necessary to adsorb the phenols on PVPP to separate them from the sugars. This is also true in the case of fortified wines, Port and sweet red wines.

The procedure for wines containing sugar (Voyatzis, 1984) involves the addition of 0.2 g of PVPP to 5 ml of wine mixed with 15 ml of distilled water. The mixture is agitated for 5 mn, filtered through a 0.45 nm millipore filter and the resin rinsed with water. After centrifugation, the resin is put into a test tube with 20 ml of a BuOH1–HCl (12 N) (1:1 vol) mixture, containing 150 mg/l of $FeSO_4$. After heating for 30 min in a waterbath, the optical density (d_1) at 550 nm is measured on a 1 cm optical path. A control, prepared under the same conditions but not heated, gives d_0.

The results are compared with those of a standardized oligomeric procyanidin solution and the tannin concentration is obtained by the following equation:

$$C(\text{mg/l}) = 273(d_1 - d_0)$$

Tannin concentrations range from 1 to 4 g/l in red wine and from 10 to 200 mg/l in white wine. These values depend on the type of grapes, their degree of ripeness, the effects of rot (if any) and winemaking techniques.

6.4.4 Analyzing the Characteristics of Tannins

In addition to the tannin concentration of a wine, enologists also need information on the structures that govern the properties of the various compounds. Of course, the ideal solution would be to separate and assay the various molecular units included in the concept of tannin. Thanks to high-performance techniques, these analyses are now possible, although they are still difficult to implement.

On the other hand, careful use of global values, based on the specific properties of tannins, may be of assistance in interpreting certain characteristics of a wine, including stability and organoleptic qualities (Glories, 1978a).

Polymer composition

This method is based on lysis of the polymer, by breaking the interflavane bond in a mild acid

Table 6.8. Comparison of evolutions in the results of tannin assays in wine aged under different conditions for 15 months (a, b and c are oak barrels of different origins) (Glories, 1992, unpublished)

	Tannins (g/l) (LA)	D_1 520	D_2 520	Al (mg/l)	Tannins (g/l) (C_1)	ΔOD 470	ΔOD 520	ΔOD 570	Tannins (g/l) (C_2)
Control	3.69	0.196	0.419	516	2.90	0.161	0.223	0.137	2.78
+1 month	3.03	0.187	0.354	474	2.00	0.131	0.167	0.127	2.26
+15 months a	3.30	0.130	0.323	262	2.52	0.147	0.193	0.136	2.54
b	3.36	0.128	0.323	214	2.48	0.153	0.195	0.143	2.64
c	3.34	0.114	0.311	178	2.57	0.154	0.197	0.141	2.66

Concentrations are determined by three different methods (see text): LA = standard method; C_1 = calculation involving the various pigments; C_2 = calculation on the basis of a spectrum model.

medium, leading to a reaction with toluene-α-thiol, followed by recovery of the terminal unit and blocking of the upper unit in benzyl-thio-ethanal form (Thompson *et al.*, 1972). The solution is analyzed by reversed-phase HPLC. Provided total thiolysis has occurred, it is possible to determine the composition of the monomers and the mean degree of polymerization (Rigaud *et al.*, 1991). The analysis is completed by mass spectrometry (HPLC-ESI-MS).

This complex method originally used a particularly unpleasant-smelling thiol, now replaced with phloroglucinol, which is easier to use. The results obtained are identical, as the limiting factor is the acid lysis of the polymers. Molecular screening coupled with phloroglucinolysis provides data on polymer composition. The MDP of 80 for skin tannins was not confirmed by molecular screening.

HCl index

This is based on the instability of procyanidins in a concentrated HCl medium, where the precipitation speed depends on the degree of polymerization. The procedure requires a sample consisting of 10 ml of wine, 15 ml of HCl (12 N) and 5 ml of water. After dilution to 1/30, the optical density (d_0) at 280 nm is measured immediately on a 1 cm optical path. The same measurement is made after waiting 7 h and centrifuging the mixture. A new value (d_1) is obtained.

The HCl index is given by the equation:

$$\mathbf{I}(\text{HCl}) = \frac{d_0 - d_1}{d_0} \times 100$$

The values are between 5 and 40. At values above 35–40, the tannins in wine precipitate, thus decreasing the value. At the beginning of barrel aging, very light wine has a low value, between 5 and 10. A wine suitable for aging has a value of 10–25 and a wine with a high concentration of highly polymerized phenolic compounds has a value >25.

The HCl index, therefore, reflects the state of polymerization of tannins in the wine, which, in turn, depends on the aging conditions. For example, polymerization decreases after winter cold and fining, as well as after a few years of aging in the bottle.

Dialysis index

This is related to the structure and charge of the tannins. Bulky or highly charged molecules pass through the pores of a dialysis membrane more slowly than small molecules with lower charges. The procedure consists of putting 10 ml of wine into a cellophane tube. It is dialyzed with a 100 ml model wine solution (5 g/l tartaric acid, pH 3.2, 10% EtOH) for three days and agitated manually twice a day. After dilution to 1/10 with water, the optical density (d_1) of the dialysate is measured at 280 nm on a 1 cm optical path. The control is measured in the same way (d_0).

$$\mathbf{I}(\text{dialysis}) = \frac{d_0 - d_1}{d_0} \times 100$$

Values are between 5 and 30, but there is not necessarily any direct correlation with the HCl indexes. Wines with a high concentration of procyanidins from seeds may have a high HCl index (25) and a low dialysis index (10). A high dialysis index (25) indicates that there are bulky, generally polymerized or highly charged pigments, so the HCl index will also be high (20–30). In some wines with a very high anthocyanin or oligomeric procyanidin content, the HCl index may be low (10), although the dialysis index is high.

Gelatin index

This is based on the capacity of tannins to react with proteins, forming stable combinations. Condensed tannins present in wine precipitate with gelatin in a homogeneous, reproducible manner. The gelatin index (Glories, 1974, 1978a) highlights the capacity of wine tannins to react with proteins in gelatin. This reactivity is responsible for the sensation of astringency experienced when tasting red wine. The soluble gelatin used in the assay includes a full range of proteins with different molecular weights (5000–300 000). This index reflects the reactivity of the tannins in the wine.

The procedure consists of adding 5 ml of cold-soluble gelatin solution (70 g/l) to 50 ml of red wine. After three days, the wine is centrifuged and the tannins (LA) in the supernatant diluted to 1/50 are assayed to determine the tannin concentration C_1 (g/l). C_0 is the tannin concentration of the control, prior to the addition of gelatin. The gelatin index is given by the equation:

$$I(\text{gelatin}) = \frac{C_0 - C_1}{C_0} \times 100$$

Values vary from 25 to 80, according to the origin of the wine and the winemaking methods. A high value, above 60, indicates the presence of highly reactive tannins that may be responsible for toughness, or even astringency. A low value, below 35–40, indicates a lack of body and may be the reason for an impression of flabbiness and bitterness. Average values of 40–60 show that the tannins are fairly reactive, but the wines can just as easily seem supple and full-bodied or tough and thin.

These three values are complementary and provide a satisfactory interpretation of the tannic characteristics of red wines, which are generally confirmed by sensory analysis.

6.4.5 Wine Color

The spectrum of red wines has a maximum at 520 nm, due to anthocyanins and their flavylium combinations, and a minimum in the region of 420 nm. Color intensity and hue, as defined by Sudraud (1958), only take into account the contributions of red and yellow to overall color. Of course, the results of this partial analysis cannot claim to reflect the overall visual perception of a wine's color. Application of the CIELAB universal color appreciation system, proposed by the International Commission on Illumination, certainly represents an improvement, but the results are difficult for winemakers to interpret.

The current approach to color analysis in winemaking requires optical density measurements at 420 and 520 nm, with an additional measurement at 620 nm to include the blue component in young red wines.

Probably as a result of the colloidal status of the coloring matter, there is no direct proportionality between absorption and dilution. Consequently, spectrophotometric measurements must be made on a 1 mm optical path, using undiluted wine. These measurements are used to calculate the values used to describe wine color (Glories, 1984).

Color intensity represents the amount of color. It varies a great deal from one wine and grape variety to another (0.3–1.8):

$$CI = OD\ 420 + OD\ 520 + OD\ 620$$

The *hue* indicates the development of a color towards orange. Young wines have a value on the order of 0.5–0.7 which increases throughout aging, reaching an upper limit around 1.2–1.3.

$$T = \frac{OD\ 420}{OD\ 520}$$

Color composition, i.e. the contribution (expressed as a percentage) of each of the three components in the overall color:

$$OD\ 420(\%) = \frac{OD\ 420}{CI} \times 100$$

$$OD\ 520(\%) = \frac{OD\ 520}{CI} \times 100$$

$$OD\ 620(\%) = \frac{OD\ 620}{CI} \times 100$$

The *brilliance of red* wines is linked to the shape of the spectrum. When wine is bright red, the maximum at 520 nm is narrow and well defined. On the other hand, it is flattened and relatively broad when wine is deep red or brick red. This characteristic may be shown by the expression:

$$dA(\%) = \left(1 - \frac{OD\ 420 + OD\ 620}{2\ OD\ 520}\right) \times 100$$

The results are between 40 and 60 for a young wine. The higher the value, the more dominant the red color of the wine. Furthermore, it is very interesting to assess the role played by various pigments involved in the color of a wine.

The *ionization index* (Glories 1978a), based on work by Somers and Evans (1974), is used to define the percentage of free and combined anthocyanins producing color in wine. To calculate this value, the wine is bleached by an excess of SO_2, at the normal pH of wine ($\Delta d\alpha$), on the one hand, and at pH 1 ($\Delta d\gamma$), on the other hand. The ionization value is expressed by the ratio of these two figures.

This procedure takes place in two stages. Initially, 10 ml of wine with a normal pH is mixed with 2 ml of water. The optical density value (d_1) is measured at 520 nm on a 1 mm optical path. The same operation is carried out again, replacing the water with 2 ml of sodium bisulfite solution ($d = 1.24$), waiting 5 min and measuring optical density under the same conditions to obtain the value d_2:

$$\Delta d\alpha = (d_1 - d_2)\frac{12}{10}$$

This value represents the optical density at 520 nm, including only those free (Al) and combined (TA) anthocyanins, colored at the pH of wine, that react with SO_2.

An identical measurement is made at pH 1.2, when 95% of the anthocyanins present are colored. In a mixture containing 1 ml of wine, 7 ml of HCl (N/10) and 2 ml of water, the optical density is measured at 520 nm on a 1 cm optical path, giving the value d'. A second measurement is made, replacing the water with sodium bisulfite as before, giving a value d'_2:

$$\Delta d\gamma = (d'_1 - d'_2)\frac{100}{95}$$

The ionization index is given by the expression

$$I = \frac{\Delta d\alpha}{\Delta d\gamma} \times 100$$

The ionization index for young wines varies from 10 to 30% and increases throughout aging, reaching 80 to 90% in old wines.

If the coloring matter in red wine consisted only of free anthocyanins (Al), given the usual pH (3.4–4.0) and free SO_2 concentration (10–30 mg/l), the coloring percentages would be lower (3–14%). The new pigments produced when anthocyanins combine with tannins are much less sensitive to bleaching by pH and SO_2, so the percentage of coloring increases. This phenomenon continues throughout the aging process. However, it also provides an estimate of copigmentation of the anthocyanins. A high index value for a very young wine indicates the impact of tannins on the flavylium cations.

Assessing the color of white wines is much more complex, as the spectrum has no defined maximum in the visible range. Absorption is continuous from 500 to 280 nm, with a maximum in the UV range. It is difficult to use the spectrum to translate the visual impressions corresponding to dry white wines, sweet white wines and oxidized dry white wines. As the characteristic absorption wavelength of yellow substances is 420 nm, measurements of optical density at this value provide only an approximate assessment of color.

6.4.6 Fractionation of Phenolic Compounds in Grapes and Wine

The phenol composition of grapes and wine is highly complex (Section 6.3). Methods for analyzing the structures of the various molecules in this group of compounds are currently limited to dimeric, trimeric or, possibly, tetrameric procyanidins (Ferreira and Bekker, 1996). Analysis by molecular screening (TSK gel) of extracts from seeds and wines, followed by mass spectrometry (de Freitas, 1995), shows that many compounds have molecular weights between 1400 and 3100, and thus are impossible to separate using known methods. The method currently used to interpret phenolic composition consists of rough fractionation into four or five classes of compounds with similar characteristics. The procedure involves combinations of the various standard precipitation, adsorption and solvent extraction techniques. Methylcellulose and formaldehyde were used by Montedoro and Fantozzi in 1974, polyamide by Bourzeix *et al.* in 1986 and XAD_2 resin by Di Stefano and Guidoni in 1990. The two methods proposed here use PVPP (Glories, 1978a) and Sephadex LH_{20} (Nagel and Glories, 1991) as adsorbents.

Red wine

It is possible to isolate four or five fractions (Figure 6.35) from red wine (Glories, 1978b) in a two-stage precipitation process, using EtOH and a mixture of MeOH–CHCl$_3$ (1-2), extracting the supernatant with ethyl acetate and adsorbing the residue on PVPP. This is followed by selective desorption of the various components using dilute alcohol and acid solvents. The fractionation thus obtained depends on the degree of polymerization of the tannins and their associations with polysaccharides and anthocyanins. Each of the fractions is either evaporated dry and then lyophilized to prepare it for the anthocyanin and tannin assays, or diluted to the initial volume of the sample with a dilute alcohol solution, pH 3.2, to analyze its chemical or organoleptic properties, and also for anthocyanin and tannin assays.

The following fractions have been isolated using this procedure:

1. *TP fraction*. The precipitate obtained by adding 9 volumes of ethanol to 1 volume of wine is known as the TP fraction. It consists of salts and polysaccharides, as well as tannin–polysaccharide and tannin-protein complexes with molecular weights above 5000.

2. *CT fraction*. After the first precipitation, the supernatant is evaporated dry and taken up by a volume of methanol. The addition of 2 volumes of chloroform precipitates the condensed tannins (CT), consisting of difficult-to-separate, complex, polymerized procyanidins, with molecular weights between 2000 and 5000. These are, strictly speaking, the main components of tannins.

3. *C,P fraction*. The supernatant from the previous precipitation is evaporated dry, taken up by 1 volume of water and extracted using ethyl acetate. The organic phase consists of catechins and oligomeric procyanidins (C,P), with molecular weights ranging from approximately 600 to 800.

4. *Al,TA fraction*. The aqueous residue contains free anthocyanins (Al) and anthocyanins combined with tannins (TA). It is possible to separate them by adsorption on previously activated, well-rinsed PVPP. Once the column has been rinsed with water, the free anthocyanins are desorbed with dilute alcohol solution EtOH–H$_2$O–HCl(N) (70/30/1) and the combined pigments (TA) with formic acid.

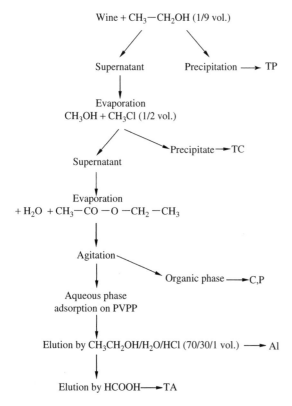

Fig. 6.35. Simplified method for fractionating various classes of phenolic compounds in wine: Al, free anthocyanins; TA, tannin–anthocyanin combinations; C,P, catechins and little-polymerized procyanidins; CT, condensed tannins; TP, tannin–polysaccharide and tannin-protein complexes (Glories, 1978)

Grapes

Starting from grapes, Nagel and Glories (1991) obtained three proanthocyanidin fractions, according to the degree of polymerization. The method consisted of a series of separations on Sephadex LH$_{20}$ using ethyl acetate and precipitations with

NaCl. It was not possible to isolate anthocyanins and their combinations with tannins.

The following fractions were separated:

1. *'Small flavanols'*: mainly monomers, dimers and trimers. The aqueous extract was deposited on Sephadex LH$_{20}$, swollen with water, then rinsed with water, phosphate buffer 0.05 M at pH 6.7 and water again. All of the flavanols were eluted with an acetone/H$_2$O (70/30) mixture. Once the acetone had been eliminated by evaporation, the 'small flavanols' were extracted from the aqueous residue using ethyl acetate.

2. *Oligomeric flavanols*: relatively uncondensed tannins.

3. *Polymerized flavanols*: condensed tannins.

6.4.7 Organoleptic Properties of the Phenolic Compounds in Red Wines

Phenolic compounds play a vital role in the flavor of red wines. They are responsible for some positive tasting characteristics, but also for some rather unpleasant, negative aspects. Body, backbone, structure, fullness and roundness are all organoleptic qualities characteristic of great red wines. On the other hand, bitterness, roughness, harshness, astringency and thinness are faults that must be avoided as they are incompatible with quality.

The overall organoleptic impression is based on a harmonious balance between these two types of sensations, directly related to the type and concentration of the various molecules, such as anthocyanins, and especially tannins. One of their properties is to react with glycoproteins in saliva (mucine) and proteins in the mouth wall, modifying their condition and lubricant properties. A study of the reaction of the B$_3$ procyanidins with synthetic, proline-rich proteins showed that three dimers were strongly bonded to the protein chains (Simon *et al.*, 2003). According to the type and concentration of tannins, they may produce a soft, balanced impression or, on the contrary, a certain aggressiveness that is either perceptible as bitterness on the end of the palate or as astringency on the aftertaste.

Making objective measurements of these basic sensations is particularly complex. However, the gelatin index provides an estimate of reactivity to proteins. The intensity of the polyphenol–gelatin reaction depends on conditions in the medium. Acidity is a favorable factor, unlike alcohol content, which inhibits the reaction and gives a sweet taste. The reaction is independent of the tannin concentration at values above 50 mg/l (Glories, 1983). Under given reaction conditions, it is thus possible to classify the various polyphenols according to their aptitude to combine.

According to Lea (1992) the reaction between tannins and proteins depends on the degree of polymerization of the procyanidins. Astringency increases up to heptamer level and then decreases, as the molecules are too bulky. Maximum bitterness occurs with tetrameric procyanidins. These findings were confirmed by Mirabel (2000), showing that the difference between bitterness and astringency varied widely from one taster to another and that the distinction was not clear.

Glories and Augustin (1994) isolated three procyanidin fractions from grapes, using the Nagel and Glories method (1991) (Section 6.4.6), as well as anthocyanins from the skins and tannins from the stalks. The following conclusions were drawn from tasting the five extracts (Figure 6.36):

1. The relatively little-polymerized catechins and procyanidins (dimers, trimers, etc.) are the least reactive with proteins. The solution tasted more acid than astringent.

2. Oligomeric and polymerized procyanidins behave in a similar way. They give an impression of body on tasting, with marked bitterness and astringency. Heterogeneous polymerization (Section 6.3.7) produces structural modifications that decrease their reactivity (gelatin index). The astringency of condensed wine tannin solutions, consisting of procyanidins, decreases with polymerization. On tasting, a combination of tannins and polysaccharides gives an impression of fullness and roundness that is highly desirable.

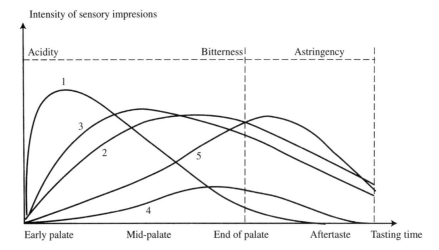

Fig. 6.36. Influence of the structure of phenolic compounds on the diversity of their organoleptic characteristics: 1, procyanidins little-polymerized procyanidins; 2, oligomeric procyanidins; 3, polymerized procyanidins; 4, anthocyanins; 5, tannins from stalks (Glories, 1994, unpublished)

3. Anthocyanins and their combinations with tannins are not very astringent, but have a marked bitterness, especially in young wines, i.e. when the molecular structures are well defined and not too complex.

4. Tannins extracted from the skins react less with the proteins in gelatin than those from seeds and stalks. The latter, exclusively procyanidins, are polymerized to varying degrees, depending on the maturity of the grapes. They do not contain any of the free anthocyanins or tannin complexes with polysaccharides or proteins that soften the tannins in the skins. The tannic balance of a young red wine comes from a good harmonization of tannins from both origins.

Seed tannins give the wine structure and body, while skin tannins provide fullness, roundness and color. However, there is a high risk of excessive astringency if seed tannins dominate, while bitterness and a herbaceous character are typical of too much extract from the skins, especially if the grapes are insufficiently ripe.

During the barrel and bottle aging of wine, many oxidative reactions modify the structures of the original procyanidins. Phenolic compounds in wine may be fractionated into four classes (Section 6.4.6) that have characteristic reactivity to gelatin (Table 6.9). According to the percentage of each of these classes in a given wine, it is

Table 6.9. Influence of the structure of different groups of phenolic compounds on the reactivity of the molecules (gelatin index) and the tasting qualities, in wines of different ages (Glories, 1992, unpublished)

	1-year-old wine	5-year-old wine	15-year-old wine	Tasting comments
Fraction I: C, P	55	58	44	Acid
Fraction II: TC	63	70	45	Tannic and astringent (good structure)
Fraction III: Al, TA	42	51	30	Bitter
Fraction IV: T-P	32	67	56	Tannic but not harsh

C,P = catechins, relatively unpolymerized procyanidins; CT = condensed tannins; Al, TA = free anthocyanins, tannin-anthocyanin combinations; TP = tannin–polysaccharide, tannin-protein combinations.

quite possible to calculate the overall reactivity of the tannins and infer the corresponding tasting characteristics.

Other methods for estimating the tannic strength of polyphenols make use of hemoglobin (Bate-Smith, 1973), PVPP (Chapon, 1993; Coupois-Paul, 1993) and, more recently, bovine albumin serum (BAS) (de Freitas, 1995). The latter consists of measuring the turbidity (expressed in NTU, Section 9.1.2) caused by associating the tannins with higher and higher doses of BAS, until a maximum is reached. The increase in turbidity is linear. The slope of the straight line is defined as 'tannic efficiency' and the maximum turbidity as 'tannic strength'. The relationship between the maximum quantity of proteins precipitated and the quantity of tannins may be considered characteristic of the latter.

The results obtained using this method confirm that maximum reactivity occurs with procyanidins that have a molecular weight around 2500, i.e. consisting of eight flavanol units. A comparison of the gelatin index and tannic strength in various wines shows that they vary in the same direction and are even complementary (Table 6.10). Indeed, the gelatin index seems mainly to be linked to astringency and tannic strength to both astringency and bitterness.

It is useful to measure the reactivity of tannins in wine, but this is not the only factor involved in assessing astringency. Other components, such as proteins, polysaccharides, ethanol, glycerol and tartaric acid, either inhibit the reaction and temper its aggressiveness or exacerbate it. The softening of the tannic feel of a wine if protein-rich cheese or polysaccharide-rich bread is eaten at the same time is quite significant in this regard. The colloidal status of the tannin molecules is dependent on their concentration and structure and plays a very important role in flavor quality, accounting for the perceived difference between great wines and more modest ones.

Furthermore, the quantity of tannins must also be taken into account. Astringency is more acceptable in a full-bodied wine with a high tannin content than in a thin wine with a low tannin content. Thus, it is quite possible to set a 'hinge value' of the relationship between the gelatin index and the tannin concentration (G/LA) above which a wine is likely to be astringent. This value is on the order of 20. For example, the gelatin index of 50 often given as standard may characterize an astringent wine if the quantity of tannins is lower than 2.5 g/l, but another wine with 3 g/l may seem well balanced.

Table 6.10. Effect of *terroir* (soil and microclimate) on the reactivity to proteins of tannins in wine: gelatin index and tannic strength (TS) (Merlot from Saint-Emilion, 1992 vintage) (de Freitas, 1995)

Vineyard	Tannins (g/l)	Gelatin index	Tannic strength (NTU)	Tasting notes
G1	1.6	31	59	Balanced, soft
P2	2.3	36	87	Balanced, soft
SP2	2.2	43	113	Full-bodied
SG1	2.5	38	123	Pleasant
SP1	3.0	37	133	Aggressive, herbaceous
SG2	2.7	47	133	Thin
P1	3.0	37	171	Tough
C2	2.5	45	173	Rustic
SP3	3.4	45	191	Good structure, Tannic
G2	2.0	43	224	Thin
C1	3.5	40	228	Herbaceous
LP1	2.8	44	268	Herbaceous, bitter, tough
LP2	2.8	45	293	Bitter, tough

6.5 EVOLUTION OF ANTHOCYANINS AND TANNINS AS GRAPES RIPEN

6.5.1 The Location of Various Phenolic Compounds in Grapes

Anthocyanins are mainly located in grape skins. In the rather rare case of grape varieties known as 'teinturiers' (Alicante, etc.), these molecules are also present in the flesh, producing grapes that are very rich in color. These molecules are also present in the leaves, mainly at the end of the growing season (leaves turning red in the fall) (Darné and Glories, 1988), although there is a different distribution of anthocyanins, with cyanidin predominating (Darné, 1991). These molecules are completely absent from the majority of white grapes, such as Sauvignon Blanc, Semillon, Chardonnay, etc., while there may be traces in other grape varieties, such as Ugni Blanc, Pinot Blanc, etc. Wines made exclusively with white grape varieties and labeled *blanc de blancs* must not contain any anthocyanins.

These pigments are located in the vacuoles of the skin cells. As the grapes ripen, they take up an increasing amount of space, to the detriment of the cytoplasm. There is a positive concentration gradient from the outside towards the inside of the grapes. The cells close to the flesh are more pigmented than those near the epidermis (Amrani-Joutei and Glories, 1995). In grapes, a distinction is made between tannins in seeds and those in skins (Souquet *et al.*, 1996). In seeds (Da Silva *et al.*, 1991), tannins are positioned in the external and internal envelopes to defend the embryo. They are only released into the outside environment if the cuticle is solubilized (Geny *et al.*, 2003).

Three types of tannins have been identified in the skins (Amrani-Joutei, 1993):

1. Tannins located in the vacuoles, forming dense clusters in the cells close to the epidermis and diffuse granulations in the internal cells of the mesocarp. The concentration gradient is reversed: the thick-walled external cells are known as 'tannin cells' (Figure 6.37).

2. Tannins bonded very strongly to the proteo-phospholipidic membrane (tonoplast) and insensitive to ultrasound (Figure 6.38) (Amrani-Joutei *et al.*, 1994).

3. Tannins integrated in the cellulose–pectin wall.

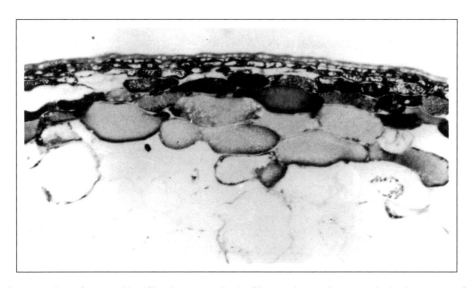

Fig. 6.37. Cross-section of grape skin (Chardonnay variety). Observation under an optical microscope after coloring with toluidine blue (× 207). Condensation of tannins at different levels in the skin. The darker-colored cells located near the surface of the skin correspond to the maximum condensation of tannins (Amrani-Joutei, 1993)

Phenolic Compounds

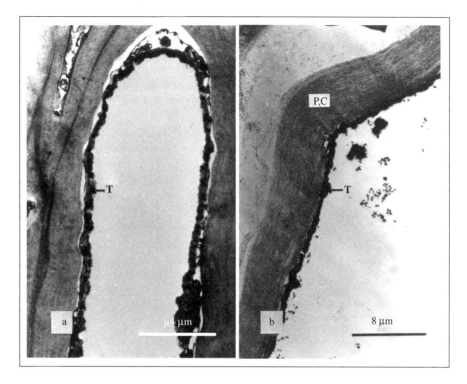

Fig. 6.38. (a) Presence of tannins (T) as a continuous layer on the internal surface of the tonoplast. (b) Effects of ultrasound on the cell wall (P,C). The tannin–tonoplast bond remains intact (Amrani-Joutei, 1993)

The distribution of these molecules is perfectly consistent with their antifungal properties, as they stop the mycelial development of fungi lacking in laccase, the only enzyme capable of breaking them down without being deactivated. The skin also contains phenolic acids and flavanols in the cell vacuoles. Phenolic acids are the main phenol components of the flesh.

6.5.2 Evolution of Anthocyanins and Tannins as Grapes Ripen

Changes in concentration

From color change to full ripeness, as defined by the ratio sugar/total acidity, the phenolic compound content in skin extract increases (Figure 6.39). Anthocyanins appear as the color changes and accumulate throughout the ripening process, reaching a maximum at full maturity. They are then broken down if the grapes become overripe. The

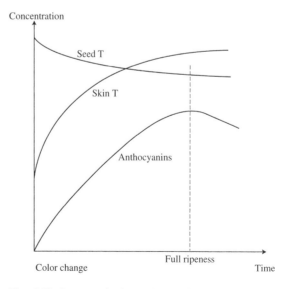

Fig. 6.39. Increase in the anthocyanin and tannin concentration in the skins and seeds as the grape ripens (Glories, 1986)

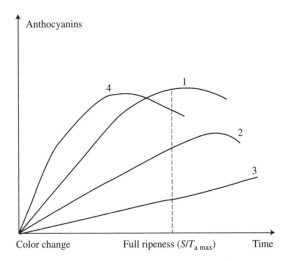

Fig. 6.40. Variations in the accumulation of anthocyanins in grape skins during ripening, according to vintages and vineyards. For the same vintage: 1, ideal situation, good grape–vineyard match; 2, late-ripening vineyard requiring slight overripeness; 3, very late-ripening vineyard, where the grapes are unlikely to produce a high-quality red wine; 4, Vineyard not very well-suited to this grape variety, as phenolic maturity occurs too early (Glories, 1986)

tannin concentration increases in a comparable manner, although it is already fairly high at color change (Guilloux, 1981). Although this pattern is valid for all grape varieties and most vineyard conditions, the accumulation of anthocyanins and the maximum values vary widely according to the environment and climate (Figure 6.40). Indeed, depending on the environment, the maximum may coincide with that of the S/T_a ratio, but it may also occur earlier, later or not at all. Furthermore, the total quantity of anthocyanins may vary by a factor of three. Even data that are valid for a particular vineyard fluctuate from one vintage to the next, depending on the weather conditions.

In seed extract, the tannin concentration generally decreases after color change, as the grapes ripen. It decreases to a greater or lesser extent according to ripening conditions, and is apparently related to the accumulation of anthocyanins in the skins (Darné, 1991). However, in certain cases, the decrease occurs at an earlier stage, before color change, and the concentration then remains relatively constant throughout the ripening period.

The decrease in seed tannins also varies from one grape variety to another. Some have a naturally low concentration (e.g. Cabernet Sauvignon), while others have much higher levels (Cabernet Franc, Pinot Noir, etc.). Tannin concentrations in the stalks are very high at color change, and vary little during ripening. The same phenomena occur in white grapes (Voyatzis, 1984): tannins accumulate in the skins, whereas concentrations in the seeds decrease regularly.

HPLC analysis of the development of dimeric and trimeric procyanidins, as well as simple flavanols, extracted from the skins and seeds of red and white grapes, shows that concentrations decrease to a greater or lesser extent, but never increase (de Freitas, 1995). It has been observed that procyanidin B_2 is the most common dimer in ripe Merlot and Cabernet Sauvignon (Table 6.11), followed by trimeric C_1. All the dimers are present in the seed extract, whereas procyanidins B_4,

Table 6.11. Concentrations of different dimeric procyanidins, trimer C_1, (−)-monogalloylated epicatechin (epigal) and catechins [(+)-catechin and (−)-epicatechin] in dilute alcohol extracts from seeds and skins of ripe Merlot and Cabernet Sauvignon grapes (1994). (Results are expressed in mg/l of equivalent (+)-catechin) (de Freitas, 1995)

Grape variety		Dimeric procyanidins (mg/l)								Trimer C_1 (mg/l)	Epigal (mg/l)	Catechins (mg/l)
		B_1	B_2	B_3	B_4	B_5	B_6	B_7	B_8			
Merlot	Seeds	29.6	120.0	36.5	61.2	22.0	11.2	2.6	7.6	32.7	22.1	146.5
	Skins	0.82	9.2	0.024	—	1.4	Traces	—	—	3.1	2.3	7.2
Cabernet Sauvignon	Seeds	26.6	150.0	24.9	40.1	22.6	8.4	9.9	0.99	62.7	21.9	210.1
	Skins	1.06	8.4	0.46	—	1.15	Traces	—	—	0.98	0.82	4.0

B_7 and B_8 are absent from the skins. Red and white grape seeds have similar distributions of procyanidins. On the other hand, dimer B_2 is in the majority in red grape skins, but practically absent from white grapes, where it is replaced by B_1. These results show that dimeric and trimeric procyanidins, present in very low concentrations, are not the most important phenolic compounds in grapes.

Development of molecular structures

The order in which anthocyanins are synthesized or, more precisely, the substitution of the lateral nucleus and the acylation of glucose have not yet been fully explained. As early as 1970, Ribéreau-Gayon identified the different types of tannins in skins, seeds and stalks. The following conclusions can be drawn on the basis of all the research published on this subject (Tables 6.12 and 6.13):

1. Tannins in seeds are procyanidins, with a relatively low degree of polymerization at color change that increases during ripening. The quantity of dimers and trimers decreases by 90% (de Freitas, 1995) and the HCl index increases considerably, from 12 to 40. These free, non-colloidal molecules are highly reactive with both proteins (gelatin index on the order of 80) and cellophane (the dialysis index, already high at color change, increases until maturity, from 35 to 45). All these characteristics give procyanidins markedly tannic properties (astringency). From an analytical standpoint, seed flavanols contain (+)-catechin,

Table 6.12. Development of anthocyanins and tannins from skins and seeds at three stages in the ripening of three grape varieties (800 berries) (Glories, 1980)

		Anthocyanins (m/g)	Skin tannins (g)	Seed tannins (g)
Merlot	Half color change	310	1.55	3.75
	Intermediate stage	881	2.40	2.18
	Full ripeness	784	2.14	1.54
Cabernet Sauvignon	Half color change	350	2.10	1.95
	Intermediate stage	822	2.10	1.00
	Full ripeness	950	2.05	1.00
Cabernet Franc	Half color change	291	1.66	2.75
	Intermediate stage	665	2.00	2.60
	Full ripeness	722	1.85	2.10

Table 6.13. Development of values characteristic of the structure of the tannin molecules at three stages in the ripening of three grape varieties (Glories, 1980)

		HCl index		Dialysis index		Gelatin index	
		Skins	Seeds	Skins	Seeds	Skins	Seeds
Merlot	Half color change	12	13	32	34	69	80
	Intermediate stage	13	13	28	35	67	80
	Full ripeness	13	31	25	43	54	84
Cabernet Sauvignon	Half color change	13	13	33	31	69	80
	Intermediate stage	13	23	29	32	67	80
	Full ripeness	15	48	22	53	42	86
Cabernet Franc	Half color change	13	13	32	31	70	81
	Intermediate stage	13	23	29	33	69	82
	Full ripeness	14	45	23	52	58	86

(−)-epicatechin, (−)-epicatechin gallate, and oligomeric procyanidins P2–P9, with 1 to 3 galloylated units. On the one hand, these large numbers of galloylated units absorb UV intensely, leading to an increase in the absorbance of these molecules at 280 nm. On the other hand, they are responsible for the adsorption phenomena that disturb the dialysis results. The elementary molecules and the dimers represent a value on the order of 500 mg/l. The polymer molecules have an MDP of at least 10.

2. Tannins in the skins have more complex structures and there is little variation in their degree of polymerization. The quantity of dimers and trimers is already very low at color change and hardly decreases during ripening, while the HCl index is relatively constant (12–15). These molecules have colloidal properties and, towards maturity, they become less and less reactive with gelatin proteins (the value may decrease by up to 40%) and cellophane (the dialysis index decreases from 35 to 25). It is as though the tannins in the skins are gradually deactivated during ripening, consequently losing their aggressiveness and astringency. Analysis of flavanols extracted from skins revealed the presence of (+)-catechin, (−)-epicatechin, (+)-gallocatechin, (−)-epigallocatechin, and oligomeric procyanidins that are not highly galloylated. The total only amounts to approximately 20 mg/l. The remainder consists of more complex molecules that are polymerized and condensed with natural macromolecules (polysaccharides and polyphenols), with an MDP of approximately 30.

3. Tannins in the stalks are polymerized, non-colloidal procyanidins (HCl index = 35–40) with a reactivity similar to that of tannins in seeds.

The various parts of grape bunches (stalks, seeds, skins) contain phenolic compounds that may be fractionated (Section 6.4.6) into four groups. The skins have a particularly high concentration of tannin–polysaccharide and tannin–protein complexes that give a nicely rounded impression. On the other hand, the stalks and seeds have high concentrations of polymerized procyanidins and condensed tannins which produce a more marked tannic astringency. There is a similar proportion of catechins and relatively little-polymerized procyanidins in all parts of the grape bunch.

In conclusion, it is possible to define the phenol composition of grapes according to the stage of maturity, taking into account the high concentration of phenolic compounds and the properties of tannins, especially their reactivity with proteins. A ripe grape is characterized by skins rich in anthocyanins and complex, relatively inactive tannins and seeds with a low content of polymerized tannins that react strongly with proteins. Unripe grapes, on the other hand, have skins with low concentrations of anthocyanins and relatively simple tannins that have not lost their reactivity, and seeds with a high content of little-polymerized and therefore highly reactive tannins.

6.5.3 The Concept of Phenolic Maturity

Grapes reach enological maturity when various factors are in balance, giving the potential to produce the highest quality wine. Technological (sugar/acid ratio), aromatic (greatest aroma potential), and phenolic maturity are independent variables that must all be taken into account in assessing enological maturity and deciding when the grapes should be harvested.

Phenolic maturity covers not only the overall concentration of substances in this family, but also their structure and capacity to be extracted from grapes during vinification.

Analysis of the anthocyanin and tannin content of grapes during the ripening period is used to monitor the development of these molecules and classify vineyards, or even individual plots, according to their phenol content. Theoretically, under comparable winemaking conditions, grapes with a higher anthocyanin content should produce wines with more color, but this is not always the case (Table 6.14). Grapes, therefore, have a variable extraction potential or 'extractability',

Phenolic Compounds

Table 6.14. Variations in the anthocyanin extraction coefficient in wine (α) according to the origin of the grapes (grape varieties, vineyards, vintages) (Glories, 1997, unpublished)

	Anthocyanins		α (%)	Color intensity of the wine
	Skins (mg/l must)	Wine (mg/l)		
CS MC 1995	1600	1230	77	1.12
M SE 1993	1450	925	64	0.73
M SE 1992	1150	977	85	1.24
M Bx 1993	1012	573	57	0.67
CS Bx 1994	780	610	78	0.60

CS Cabernet Sauvignon, M Merlot, MC Medoc, SE Saint-Emilion, Bx Bordeaux

according to differences in ripening conditions and grape varieties.

This notion of anthocyanin extractability depends on the state of maturity that controls the breakdown of skin cells (Table 6.15). All other conditions being equal, when grapes are perfectly ripe or slightly overripe, the anthocyanin content in the wine is higher than it would have been prior to maturity, although these pigments tend to decrease in grapes. Both color and total phenol content are at a maximum.

In the vineyard, it is possible to obtain an approximate idea of the potential breakdown of the skin cells by squashing a grape between the thumb and forefinger and assessing the color.

Although a high concentration of anthocyanins in the skins is necessary to obtain a deep-colored wine, it is not the only condition. The cells must also be sufficiently decayed to make these molecules easily extractable by non-aggressive technology. At phenolic maturity, grapes have both a high pigmentation potential and a good capacity for releasing these substances into wine.

6.5.4 Methods for Measuring Phenolic Maturity

Various methods for assessing the total phenol content in grapes have been suggested, but they are incapable of providing an accurate prediction of the phenol content of the corresponding wine. Nor do they provide an assessment of phenolic maturity that would be helpful in setting the date for the grape harvest. One method that has been suggested for assessing phenolic maturity and setting the date for picking the grapes is to note when the total concentration of phenolic compounds has reached its maximum value and has just started to decrease. This system is fairly easy, provided that extraction always takes place under the same conditions and the maximum value is visible.

Another fairly simple method, giving results that are both more comprehensive and easier to interpret, has been suggested by Glories (1990).

The principle

The principle of the method consists of rapidly extracting the anthocyanins from the skins, gently at first and then under more extreme conditions, where the diffusion barriers are broken down. Acidity is used as a vector to facilitate extraction. In addition, it is recommended that the grapes should be roughly crushed and the resulting flesh diluted by half. Crushing the seeds also results in partial extraction of their tannins, which is necessary to assess the characteristics of the grapes. The solutions are aqueous, pH 1 (HCl N/10) and pH 3.2 (solution with 5 g/l tartaric acid neutralized by 1/3).

The acid medium ruptures the proteophospholipid membrane, breaking the protein bonds and, consequently, releasing the contents of the vacuoles. All of the anthocyanins are then extractable and solubilized in the solution at pH 1. At pH 3.2, extraction is approximately comparable to that occurring in fermentation vats. If the membrane is not porous, anthocyanins circulate very little, but if it is broken down by grape enzymes, the pigments

Table 6.15. Effect of the harvest date on the anthocyanin extraction coefficient in wine (α) (Cabernet Sauvignon, 1995) (Glories, 1997, unpublished)

Dates	Anthocyanins		α (%)	Color intensity of the wine
	Skins (mg/l must)	Wine (mg/l)		
September 13	1550	930	61	0.686
September 20	1743	1046	59	0.812
September 28	1610	1207	75	0.915

are released from the vacuoles and extraction tends towards the same levels as in the previous case. The difference between the results obtained at both pH levels therefore reflects the fragility of the membrane, as it relates to the pigment extraction potential. This indicates the level of ripeness of the grapes.

It is interesting to note that anthocyanins and tannins are extracted from the skins under similar conditions. An extract with a high anthocyanin content also has a high level of tannins. Anthocyanins may, therefore, be considered as markers for tannins in the skins. As the OD 280/anthocyanin ratio of extracts at pH 3.2 is between 35 and 45 for ripe grapes from all varieties investigated (Merlot, Cabernet Franc, Cabernet Sauvignon, Syrah, Grenache, Tempranillo, etc.), an average value of 40 is used. Once the concentration of total phenolic compounds (OD 280) and anthocyanins (A) in the extract at pH 3.2 is known, it is possible to calculate the proportion of the phenolic compounds derived from the skins (OD 280 = A pH 3.2 × 40). The remainder, therefore, originated from the seeds.

The method also requires that a sample of 200 grapes be weighed. They are pressed manually using a nylon gauze and the juice is weighed. Density, sugar content and total acidity are also determined. The density is used to define the dilution volume corresponding to 50 g of crushed grapes.

The procedure

The procedure requires two samples of 200 grapes each. The first is pressed in order to obtain the weight of the pomace, as well as the weight and volume of the must, sugar content and acidity. The second is crushed. One 50 g of sample is added to its own volume of HCl N/10. Another 50 g of sample is added to its own volume of a solution at pH 3.2. Both samples are stirred manually and left for 4 h. The samples are then filtered through glass wool, producing two solutions identified as 'pH1' and 'pH 3.2'. An anthocyanin assay is carried out on both solutions using the SO_2 bleaching method (Section 6.4.2). The anthocyanin content A pH 1 and A pH 3.2 is expressed in mg/l of grape juice. The total phenol content is also determined for the second solution, using OD 280 (Section 6.4.1).

These simple analyses produce the following results:

1. The anthocyanin potential is given by A pH 1. It varies from 500 to over 2000 mg/l, according to the grape variety.

2. Anthocyanin extractability (AE) is expressed by the equation:

$$AE(\%) = \frac{A\ pH\ 1 - A\ pH\ 3.2}{A\ pH\ 1} \times 100$$

The lower this value, i.e. the smaller the difference between these two measurements, the more easily extractable the anthocyanins.

AE is between 70 and 20, depending on the grape variety and ripeness. Cabernet Sauvignon, with its thick, tough skin, always gives higher values than Merlot. AE decreases during ripening (Table 6.16). Certain vineyard operations, such as thinning the bunches of grapes or late removal of the leaves, facilitate ripening and decrease the AE value. The anthocyanin content and color of the wine increase, even if the anthocyanin content of the grapes does not.

3. The contribution of tannins from seeds to the total phenol content of the extract is obtained by the equation:

$$MP(\%) = \frac{OD\ 280 - (A\ pH\ 3.2 \times 40)}{OD\ 280} \times 100$$

It has already been mentioned that the OD 280 of skin extracts is correlated with the anthocyanin concentration by the approximate relation OD 280 = A pH 3.2 × 40, where A pH 3.2 is expressed in g/l. The higher the MP value, the higher the tannin content of the seeds and the greater the risk that it may have a negative effect on the flavor of the wine. MP decreases during ripening (Table 6.16).

MP varies widely, from 60 to 0, according to the grape variety, the number of seeds in the grapes and their ripeness. For example, Pinot

Phenolic Compounds

Table 6.16. Evolution of the phenolic maturity of Cabernet Sauvignon grapes from different vineyards in 1993 (Glories, 1994, unpublished)

Vineyard	Date	S/T^a	A pH1 (mg/l)	AE%	MP	Pomace (g/l of juice)
Premier Cotes de Bordeaux	6/9	24.5	1576	50	36	184
	20/9	31	1961	47	19.5	186
	4/10	33	1587	45	17	191
Saint-Emilion	30/8	21	1318	43.5	34	200
	13/9	30.5	1405	43	30	207
	27/9	34.5	1881	41.5	14.5	235
	4/10	35.5	1982	40.5	13.5	220
Médoc	30/8	21.5	1185	47	39	204
	13/9	27.5	1345	42	31	212
	27/9	32.5	1590	41	14	231
	4/10	33.5	1758	41	13	214
Graves	22/8	21	1472	44	33.5	202
	30/8	25.5	1708	44	32	212
	6/9	31	1727	41	20	208
	13/9	35	1550	37.5	19	216
	20/9	39	1743	36	15.5	225
	23/9	41	1745	35	13.5	201

$^a S/T_a$ = ratio of sugar (in g/l) over total acidity (in g/l expressed in H_2SO_4)

Noir, Grenache and Tempranillo have a high concentration of tannins in the seeds. Perfectly ripe Cabernet Sauvignon, on the other hand, has very little—less than Merlot.

4. The pomace/juice ratio, expressed in g/l, reflects the dilution of the grapes, i.e. an excess of juice in relation to the solid matter that provides the phenol content. It is possible to assess whether this ratio should be modified, either by running off juice or by eliminating water (vacuum evaporation or reverse osmosis; see Volume 1, Section 11.5.1). This ratio varies between 100 and 300 g/l, according to the grape variety, microclimate and growing conditions.

These simple data may be obtained within one day after the sample grapes are picked, so they may be used to assist in setting the date the grapes are to be harvested, as well as adapting winemaking techniques to their specific characteristics.

6.6 EXTRACTING TANNINS AND ANTHOCYANINS DURING WINEMAKING

6.6.1 Extracting Pigments During Vatting

Vatting is the period between the time the vat is filled with crushed grapes and the wine is run off. Its duration varies from a few days to three weeks, or even longer, depending on the type of wine. It may be divided into three distinct parts (Figure 6.41):

1. Maceration prior to fermentation (MpF), generally relatively short (a few hours to a few days), is the first period before the start of alcoholic fermentation.

2. Maceration during alcoholic fermentation lasts a few days, generally from 2 to 7, depending on the conditions.

3. Post-fermentation maceration continues after the completion of alcoholic fermentation and is specific to wines with aging potential. Its length is highly variable, from a few days to a few weeks.

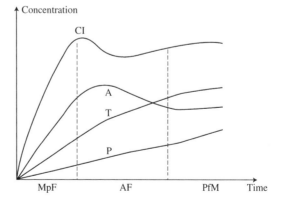

Fig. 6.41. Influence of vatting on the extraction of various compounds from grapes. A, anthocyanins; T, tannins; P, Polysaccharides; CI, color intensity; MpF, maceration prior to fermentation; AF, alcoholic fermentation; PfM, Post-fermentation maceration

Anthocyanins are extracted at the beginning of vatting, mainly in the aqueous phase, during maceration prior to fermentation and at the beginning of alcoholic fermentation. When the alcohol content reaches a certain level, a decrease is observed in the results of assays for these molecules. At this stage, extraction of anthocyanins from the grapes is almost completed and several mechanisms intervene to decrease concentrations. These include adsorption of anthocyanins on solids (yeast, pomace), modifications in their structure (formation of tannin-anthocyanin complexes) and, possibly, breakdown reactions (Volume 1, Section 12.5.2).

Tannins from the skins are extracted with the anthocyanins at the beginning of vatting, but extraction continues for a longer period due to the location of the tannins in the skin cells. Tannins from the seeds are solubilized when the cuticle is dissolved by ethanol, i.e. towards the mid-point of alcoholic fermentation. This continues during the post-fermentation phase.

Polysaccharides are mainly extracted from the grapes at the beginning of vatting and partially precipitate in the presence of alcohol. Yeast mannoproteins may be solubilized during post-fermentation maceration.

Color intensity reaches a maximum at the beginning of vatting (Volume 1, Section 12.5.3) and increases again later in some cases. During the first phase, corresponding to extraction of coloring matter from the grapes, the anthocyanins are copigmented to a certain extent with simple phenols. Color intensity may increase again in the third phase, due to the formation of new tannin–anthocyanin complexes as well as new anthocyanin–tannin copigments, if these substances are present in large enough quantities. The alcohol produced in the second phase breaks down these copigmentations. In the third phase, color intensity may increase again due to the formation of tannin–anthocyanin complexes.

The end of alcoholic fermentation and post-fermentation maceration are characterized both by in-depth extraction from the plant matter and modifications in pigment structures (polymerization of tannins and formation of tannin–anthocyanin complexes). These phenomena have varying effects on the organoleptic impression according to the type of grapes and their composition.

6.6.2 Adapting Winemaking to Various Factors

Winemakers have techniques for facilitating extraction of pigments from grapes that may have positive or negative effects, according to grape quality and ripeness (Volume 1, Section 12.5). If the grapes have perfectly ripe must (high S/T_a ratio) and good phenolic maturity, the processes for extracting phenolic compounds are relatively simple to implement. However, if conditions are not ideal, it may be possible to try to compensate for certain faults. Many techniques are available: adjusting maceration time, reducing the volume of juice by running off or eliminating water, sulfuring or oxygenation, treatment with color-extraction enzymes, selected yeast, temperature control, crushing, flash expansion, pumping-over, punching down the cap and agitation. In any case, a major characteristic of great wines is the balance between seed and skin tannins.

In red winemaking, maceration (Volume 1, Section 12.5) must be modulated according to the type of grapes. Research into Bordeaux grape varieties indicates that a distinction should be made between the extraction of anthocyanins and tannins:

1. The grapes are healthy and have a high anthocyanin content (A pH 1 > 1200 mg/l juice). If the anthocyanins are highly extractable (AE ≤ 30), slight sulfuring (3 g/hl) is necessary for maceration prior to fermentation. There is plenty of color in the wine when it is first racked, indicating that there will be no color deficit at a later stage.

   If the anthocyanins are not very extractable (AE 50 to 60), the pigments are released slowly, so sulfuring at 5–6 g/hl, combined with low temperatures (<10°C), perforates the membrane, delaying the start of alcoholic fermentation by 1 to 4 days, and promoting color release. Enzymes noticeably increase the rate of extraction, but have little effect on the final level. They are more effective at extracting tannins from the skins and are recommended when the skins have a herbaceous character.

2. The grapes are healthy but have a low anthocyanin content (A pH 1 < 1000 mg/l juice). Under these conditions, anthocyanins are generally difficult to extract. The previous technique may also be used, with the addition of more frequent pumping-over. Certain sophisticated technologies may also be useful: flash expansion, high-temperature fermentation, cryoextraction, and initial maceration at low temperatures using liquid CO_2. All these techniques burst the cells and release their contents, so color is likely to be improved, but they may also have a negative impact on quality if they are not carefully controlled.

3. If the grapes are insufficiently ripe and affected by rot. This type of situation is dangerous and must be avoided, as far as possible. The anthocyanins have necessarily suffered some deterioration. Laccase is always present, leading to a risk of oxydasic casse. Treatment with SO_2 (6 to 8 g/hl) is indispensable, and contact with air must be avoided in the pre-fermentation phase. Heating the must is an acceptable solution, but there is a major risk of obtaining unstable colloidal coloring matter. The use of dried, activated yeasts is recommended to ensure that fermentation starts rapidly.

4. If the grape seeds have a high tannin content (MP > 50). Great care should be taken with pumping-over from the middle of fermentation and especially at the end, to avoid extracting too much aggressive tannin and making the wine unbalanced.

5. If the grape seeds have a low tannin content (MP $\leq$ 15). There is no risk of excess tannin from the seeds affecting the quality of the wine. On the contrary, every effort should be made to achieve maximum extraction, as these tannins are indispensable for good balance and structure. Pumping-over and high temperatures are recommended from the middle of alcoholic fermentation until the end, possibly with extra pumping-over during post-fermentation maceration.

In general, if the aim is to produce a wine with good color, balance, softness and distinctive grape aromas, but without aggressiveness, it is important to promote extraction of tannins from the skins. Excessive extraction, however, should be avoided, as it gives a herbaceous character. The emphasis should be on maceration prior to fermentation, pumping-over should be kept to a minimum, concentrated mainly at the beginning of maceration, while vatting should be short and the temperature should be kept below 30°C.

If the objective is to make a wine with aging potential, good tannic structure is indispensable. It must, however, be adapted to the grapes in order to avoid developing a rustic character. Tannins from the seeds are just as necessary as those from the skins. Furthermore, the molecular structures of the phenolic compounds must be modified to soften them. This requires a certain aeration and a relatively high temperature at the end of alcoholic fermentation, as well as during post-fermentation maceration. Vatting time depends on the origin of the grapes, as long vatting, 3–4 weeks, is only beneficial if the grapes are of high quality and perfectly ripe. It must be reduced if the grapes are underripe, to avoid herbaceous character, or markedly overripe, to minimize the harshness of tannins from the seeds.

6.7 CHEMICAL REACTIONS OCCURRING DURING BARREL AND BOTTLE AGING

6.7.1 Reactions Essentially Involving Anthocyanins and Their Effect on Color

If the composition of a wine is monitored regularly from the end of malolactic fermentation until after bottling, the results of the anthocyanin assay decrease regularly. Free anthocyanins disappear completely after a few years, although the wine remains red. Indeed, these molecules are unstable and must combine with tannins to form the stable pigments responsible for the color of older wines. Tannin levels are the most important factor in color stabilization. Results of tannin assays vary as their structure changes (Ribéreau-Gayon and Stonestreet, 1966), and old wine contains only tannins.

The decrease in anthocyanin content is due to degradation and stabilization reactions, as well as changes in their structure.

The decrease in the anthocyanin concentration results from both breakdown reactions and stabilization reactions. In breakdown reactions (Section 6.3.3), free anthocyanins are broken down by heat into phenolic acids (mainly malvidin) and by violent oxidation, mainly delphinidin, petunidin and cyanidin. They are highly sensitive to quinones and the action of oxidases, either directly or in combination with caftaric acid. This acid may even react in the (nucleophilic) quinone form and bond to anthocyanin's (electrophilic) node 8 as a carbinol base.

In reactions leading to structural changes, anthocyanins react with compounds that have a polarized double bond and form new, orange-colored pigments that are relatively insensitive to SO_2 and variations in pH. These are no longer taken into account by the assay and contribute to the apparent decrease in anthocyanin content. These compounds have a number of different origins. They may be formed by oxidation (CH_3CHO...), yeast (pyruvic acid), or bacterial (diacetyl) metabolism, or they may be present due to the presence of *Botrytis cinerea* on the grapes (furfural).

In stabilization reactions, several mechanisms lead to the formation of tannin-anthocyanin combinations, depending not only on the conditions in the medium (temperature, oxidation) but also on the type of tannins and the tannin/anthocyanin ratio. The color of the new pigments ranges from mauve to orange and is more intense at the pH of wine than that of free anthocyanins. It depends relatively little on the SO_2 content of the medium. These molecules are not detected by chemical anthocyanin assays and are only very partially taken into account in the results.

Two main types of reactions are involved (Section 6.3.10):

1. *Direct reactions*: Reactions between anthocyanins (+) and tannins (−) ($A^+ \rightarrow T^-$). The molecules formed are colorless and turn red when the medium oxidates. They also evolve towards orange due to the appearance of xanthylium structures.

    Reactions between tannins (+) and anthocyanins (−) ($T^+ \rightarrow A^-$). The formation of carbocations (+) from procyanidins is promoted by higher temperatures and requires an acid medium (wine). Anthocyanins (−) correspond to the carbinol base. The molecules formed are theoretically colorless, but are rapidly dehydrated into a stable, reddish-orange form. This reaction is completely independent of the oxidation conditions in the medium.

2. *Indirect reaction:* Reactions of tannins and anthocyanins with ethanal, formed from ethanol by oxidation of the medium. The ethyl cross-bond acts as a bonding agent between the two groups of molecules. The pigments formed are mauve in color, with very variable structures (dimers, trimers, etc.).

These oxidative phenomena not only lead to the formation of ethanal, but also oxidize tartaric acid to form glyoxylic acid. Like ethanal, this aldehyde acts as a cross-bond between two flavanol units, which are then dehydrated and oxidized to produce a yellow xanthylium pigment. Similar reactions convert furfural and hydroxymethylfurfural.

All these reactions produce colors ranging from red to mauve, to brick red and then to brown−orange, through the following stages:

During vinification: If the must is not protected during pre-fermentation skin contact, enzymatic breakdown and oxidation reactions may cause discoloration.

Fermenting must is a reducing medium, so no oxidation can occur, but reactions with yeast metabolites may give the color an orange tinge. Direct $A^+ \rightarrow T^-$ reactions also produce colorless compounds that react when the wine is run off or after malolactic fermentation. During post-fermentation skin contact, the medium is still saturated with CO_2 and the other type of direct $T^+ \rightarrow A^-$ reaction forms colorless or red compounds. This does not occur in the case of brutal aeration due to micro-oxygenation, which leads rather to the formation of a mauve T-ethyl-A complex. In this case, depending on the amount of ethanal formed and the type of tannins extracted (skin or seed tannins and degree of polymerization), these complexes may be precipitated and red pigments formed. Micro-oxygenation also eliminates reduction odors and enhances the fruity aroma of the wine, perceptible as soon as it is run off.

During aging: Reactions with ethanal should be promoted by: 1) using oak barrels that allow oxygen to penetrate and oxidize the wine, as well as release ellagitannins, which act as oxidation cofactors, 2) micro-oxygenation, or 3) racking with aeration for wines aged in vat.

However, if the wine oxidizes too rapidly, anthocyanin breakdown reactions may also occur, causing loss of color, possibly accompanied by the formation of glyoxylic acid and yellow xanthylium. The end result depends on the relative quantities of anthocyanins and tannins in the wine.

An increase in color intensity is observed in well-balanced, properly aged wines; they become deeper and denser and the 'wine takes on color'. Direct $T^+ \rightarrow A^-$ reactions do occur when aging takes place in airtight vats, with little aeration through racking, leading to insufficient oxidation of the coloring matter, although they are very slow, unless the temperature is relatively high ($>20°C$). Color increases little and is even likely to take on yellow tinges during summer. Wine may also yellow in the barrel if the temperature is over $20°C$. This reaction involves the thermal degradation of malvidin, which facilitates the production of red pigments.

Throughout aging in bottle, characterized by the absence of oxidation, the color evolves fairly rapidly towards brick red and orange, due to the second type of reaction. This development depends on the phenol content of the wine and the combinations produced during barrel aging.

The color of a bottled wine with a high seed-tannin content is likely to develop rapidly, as these molecules are highly reactive. Pigments must be stabilized by oxidation mechanisms during aging to avoid this color loss. However, color development is slower in wines with a high concentration of relatively non-reactive skin tannins, as is generally the case with Cabernet Sauvignon. Some yellowing may be observed if the temperature is too high. Furthermore, these wines are liable to precipitate colloidal coloring matter.

The stable tannin-ethyl-anthocyanin structures are apparently transformed at varying rates into orange compounds, via the fixation of the polarized double bond of the vinyl-procyanidins on the anthocyanins, to form procyanidin–pyranoanthocyanin complexes (Francia-Aricha *et al.*, 1997). The rate of conversion depends on the wine's phenol content, the origin of the tannins (skins or seeds), and the phenolic structures (tannin–anthocyanin combinations) present at the end of the aging period.

6.7.2 Reactions Essentially Involving Tannins and their Effects on Flavor

If the composition of a wine is monitored regularly from the end of malolactic fermentation, the results of the tannin assay (LA method) decrease, or change little, during barrel aging and then increase regularly after bottling. At the same time, the values characteristic of structure (HCl index) and aptitude to react with proteins (gelatin index) vary considerably, either increasing or decreasing, thereby indicating structural modifications.

The procyanidin molecules from the grapes tend to polymerize, condense with anthocyanins and combine with plant polymers such as proteins and polysaccharides. Several reactions are involved (Section 6.3):

1. *Polymerization reactions* producing 'homogeneous polymers' (Section 6.3.7), i.e.

procyanidins polymerized by organized C_4–C_8 or C_4–C_6 bonds, are likely to occur in wine as it is an acid medium. These reactions are promoted by warm temperatures, but are independent of the oxidation levels.

Several reactions are possible in the presence of oxygen. They involve bonds between various procyanidins mediated by ethanal and, possibly, bonds between quinone functions. The molecules formed have fairly bulky structures, as well as properties different from those of procyanidins, especially their stability and capacity to react with proteins.

Polymerization is limited by the precipitation of compounds that have become excessively bulky, hydrophobic and insoluble. This corresponds to a form of stripping, showing that the wine is highly evolved. Aging must be modulated and adapted to promote certain reactions and stabilize the wine, while inhibiting or slowing down its development. These transformations have a major effect on flavor. The drop in polymerized tannins is not always accompanied by a decrease in the astringency of the wine (Haslam, 1980), but frequently gives an impression of thinness.

2. *Condensation reactions* involve other compounds such as anthocyanins, polysaccharides and proteins. Combinations with anthocyanins (Section 6.3.10) increase and stabilize color. Combinations with polysaccharides and proteins, however, are less well known. They depend on the type of polymer and are affected by temperature. Various types of polysaccharides, from grapes, yeast and fungi, are likely to be present in wine, including neutral polysaccharides (glucane, dextrane, mannane, cellulose, etc.) acid polysaccharides (pectins, etc.) and glycoproteins (mannoproteins). Red wine may also contain proteins from fining agents added during barrel aging. Apparently, tannins that are partially bound to polysaccharides and polypeptides react less strongly with proteins, particularly those in saliva.

These transformations also have a major effect on flavor. Organized polymerization produces polymerized procyanidins that are increasingly reactive with proteins and, therefore, have an increasingly pronounced tannic character. This development continues up to a limit of 8 or 10 flavan units (Figure 6.42). On the contrary, polymerization mediated by ethanal softens the flavor. Although they have the same quantity of flavanols, molecules of this type are less reactive than procyanidins. Combinations with other components such as anthocyanins, neutral polysaccharides and proteins decrease their reactivity. The reverse is true in the case of acid polysaccharides.

It is difficult for tasters to distinguish between astringency and bitterness. Mirabel (2000) demonstrated that the gelatin index was preferentially

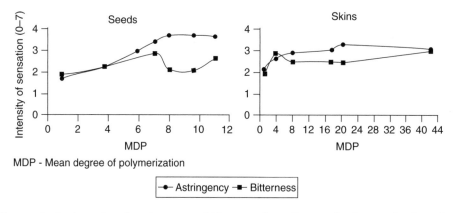

Fig. 6.42. Changes in the intensity of astringency and bitterness depending on the degree of polymerization of skin and seed tannins (Mirabel, 2000)

correlated with bitterness in wine and provided an estimate of this flavor characteristic. However, astringency, a more tactile sensation, is apparently more closely related to the tannin content. Tasters perceive an overall sensation that they describe as "aggressiveness".

6.7.3 Reactions During Barrel and Bottle Aging

During aging, winemakers may take action to affect oxidation, temperature, time and fining.

Oxidation is promoted by aerating wine (aging in barrel, racking, and the controlled introduction of oxygen) and by the presence of catalysts from the wood (ellagic tannins). This is desirable as it intensifies and stabilizes the color and softens the flavor, but must be carefully controlled, otherwise it may cause irreversible deterioration. Indeed, an excess of oxygen can lead to: a) oxidative breakdown of the anthocyanins, b) partial stabilization of the anthocyanins by the formation of mauve complexes with an ethyl cross-bond (Guerra and Glories, 1996), c) the development of orange-colored ethanal addition compounds, and d) oxidation of tartaric acid to form yellow xanthylium.

The reactions that actually take place depend on the relative concentrations of tannins and anthocyanins, as well as on the type of tannins present. The kinetics have not yet been described in detail, but the first two reactions always represent a majority of the changes that occur, and the third reaction is very slow.

The temperature depends on the winery. Low temperatures are useful for precipitating unstable colloids. On the one hand, temperatures above 20°C promote the formation of carbocations from procyanidins, and therefore the TA complex (red, orange), as well as homogeneous polymerization. On the other hand, they also facilitate combinations with polysaccharides as well as color breakdown reactions. Furthermore, it promotes the thermal degradation of some anthocyanins, particularly malvidin. Alternating a low temperature with a temperature around 20°C promotes development, while maintaining it within certain limits.

The length of barrel aging necessary to obtain the desired quality depends on the type of wine and the modifications required. A wine with a balanced tannic structure that already has a certain finesse after malolactic fermentation is likely to 'dry out' unnecessarily if aging is prolonged. Conversely, a wine with a high concentration of phenolic compounds requires longer aging to soften the tannins.

In many European wine growing areas where large quantities of tannins are extracted during the winemaking process, the wines are turbid, unrefined, and aggressive, requiring several years' aging to clarify and acquire finesse. This is the only case in which precipitation of polymerized tannins makes the wine softer.

Throughout the time it ages in an airtight bottle, wine is initially subject to a slight oxidative reaction. It is then mainly affected by transformation reactions independent of oxidation. These reactions involve the carbocations formed from procyanidins, with condensation of anthocyanins (reddish orange) and polymerization of homogeneous tannins. Temperatures that are slightly too high promote these reactions and are responsible for accelerated aging.

Changes in coloring matter: the (mauve) complexes with ethyl cross-bonds develop into orange pyrano-vinyl procyanidins. The reaction kinetics are disturbed by the presence of polysaccharides and, are apparently temperature dependent. This explains the difference between great wines that remain truly red for many years and more modest wines that rapidly take on a more yellow hue, as well as those that are saturated with oxygen during the winemaking process.

A few years after bottling (1 to 3), however, a modification is observed in the flavor of certain wines, in particular their tannic character. These wines seem temporarily thinner, with less body, although their color is still strong and not very evolved. Analyses show that there has been a rearrangement of the structural tannins in a non-oxidizing medium, leading to depolymerization (changes in the result of the tannin assay, decreases in the HCl and dialysis indexes). Part of the heterogeneous polymers may be destroyed, prior to a later homogeneous repolymerization. This is a feature of high-quality wines that form a wide range of tannin molecules during aging. These reactions slow down the wine's development and add complexity to the color and flavor. This complex tannin chemistry does not occur during the development of more modest wines, which tend to evolve continuously and rapidly towards a 'mature' wine character.

In view of their influence on the character, flavor and development of wine, it is understandable that there has been considerable interest in analyzing the chemical composition of the tannins in great wines, in order to try to copy these characteristics. Some scientists thought this objective could be reached by NMR. Until now, this highly publicized approach has not met with any significant success.

6.8 PRECIPITATION OF COLORING MATTER (COLOR STABILITY)

6.8.1 Precipitation of Coloring Matter in Young Wines

If a young wine is placed in a refrigerator just after malolactic fermentation, it rapidly becomes turbid and a precipitate is deposited at the bottom of the container. The appearance of this deposit is very different from that found in bottles of old wine. It is fairly gelatinous and very bright red, with a pearly sheen. It is similar to the lees removed from barrels and vats after the first racking.

The composition of these precipitates is relatively constant: tartrates, anthocyanins, tannins and polysaccharides. The precipitation of potassium hydrogen tartrate is a well-known phenomenon. The behavior of the phenolic compounds is related to their colloidal state, as demonstrated by Ribéreau-Gayon as early as 1931. It is possible to eliminate this fraction and avoid precipitation by dialyzing the wine with a cellophane membrane. 'The physicochemical stripping mechanism that operates in wine during barrel and bottle aging consists of the formation of this colloidal coloring matter, mainly in summer, and its precipitation, mainly during winter' (Ribéreau-Gayon et al., 1976).

All wines have this characteristic, but some more than others, especially when winemaking methods promoted high extraction. Wines also have more colloidal coloring matter if the grapes are damaged due to disease (rot), overheating (high-temperature fermentation) or mechanical operations (rough crushing, pumping, excessive pumping-over, stirring the lees, etc.). All of these lead to forced extraction, either of non-hydrolyzed polysaccharides from the grape skins or exocellular polysaccharides from fungi. All of these colloids are relatively unstable, according to their molecular size. They form a colloid base for coloring matter and are pigmented by phenolic compounds during precipitation. The degree of precipitation also depends on the wine's alcohol content and storage temperature.

The results in Table 6.17 show the high precipitation of phenolic compounds (OD 280) in a wine made in a rotary vat, as compared to a control wine. Furthermore, after 12 months, the wines did not have the same balance. The gelatin

Table 6.17. Influence of the type of winemaking methods on changes in the phenol content of different wines during aging (Merlot, 1985)

	OD 280	Tannins (g/l)	Anthocyanins (mg/l)	Polysaccharides (mg/l)	Dialysis index	Gelatin index
C						
$t = 0$	56	2.9	600	650	8	50
$t = 12$ months	55	3.0	400	500	18	48
RV						
$t = 0$	65	3.5	900	950	25	55
$t = 12$ months	52	2.7	550	620	15	60

C = control, normal winemaking methods; RV = rotary vat, with unusually high extraction of the pomace.

index increased in sample RV, showing that the tannins that precipitated were less reactive with proteins and therefore less aggressive. There was less precipitation in the control, as well as a change in structure, indicating a softening of the tannins (gelatin index decreased).

Wines do not necessarily stabilize in the first year. If warm temperatures promote the combination of tannins and polysaccharides and the medium still has a high enough concentration of colloidal molecules of the same type, more colloidal coloring matter may be formed. To avoid this repeated precipitation, it is possible to use 'protective colloids' (Section 9.4) and eliminate all the colloids by fining (Section 10.4). Protective colloids, such as gum arabic (Section 9.4.3) and mannoproteins (Section 5.6.3), prevent the flocculation of unstable colloids, maintaining the particles in suspension rather than eliminating them.

Bentonite has a negative charge that fixes the positive unstable colloids and pulls them down. It is more efficient than cold flocculation of the colloids. The problem is different in the case of protein-based fining agents. Some of the colloids are pulled down by the flakes of tannin-protein complexes, while the rest are stabilized by residual proteins that are also part of the wine's colloidal structure.

6.8.2 Precipitation of Coloring Matter in Old Wines

A long period of bottle aging involves a set of stripping reactions that continue until the wine has finished developing. These reactions cause the polymerization of tannins and anthocyanins (Section 6.3). It is also possible to envisage associations in the form of micelles that become hydrophobic and precipitate, even if the polymerized tannin molecules are smaller than 100 Å.

These unstable colloids are deposited in layers, coating the sides of the bottles. Solubilization tests on these particles show that they are very different from colloidal coloring matter, as formic acid mixed with methanol is only capable of dissolving a small fraction of the deposit. Besides tannins and anthocyanins, potassium and iron are also present, as well as nitrogen and, sometimes, small quantities of polysaccharides.

Under similar temperature conditions, however, these precipitations develop at varying speeds in different wines. Great wines develop more slowly than more modest ones, even those with a similar phenol content. In the former, precipitation occurs after about twenty years, whereas, in the latter, it may occur after only a few years in the bottle. The specific phenolic composition of great wines continues to affect their character and development throughout the aging process.

6.9 ORIGIN OF THE COLOR OF WHITE WINES

6.9.1 Phenolic Compounds in White Wines

Sweet and dry white wines result from the alcoholic fermentation of pure grape juice, clarified by settling. In comparison to red wines, only very small amounts of phenolic compounds are dissolved, as contact with grape solids only occurs during maceration prior to fermentation.

White musts and wines contain benzoic and cinnamic acids, catechins, procyanidins and flavonols (Ribéreau-Gayon, 1964; Weinges and Piretti, 1972). A recently discovered class of protein–tannin complexes has been shown to contribute towards the phenol content of white wine (Lea et al., 1979; Singleton et al., 1979). This is the reason for the excessively high result of the Folin–Ciocalteu test and the optical density value of 280 nm (Section 6.4.1)

The 13 molecules present in sweet white wines were formally identified (Biau, 1996) by fractionation and analysis of the molecules using HPLC (Voyatzis, 1984; Kovac et al., 1990) and then capillary electrophoresis (Biau et al., 1995), HPLC and NMR. It is thus possible to define the phenolic composition of various white wines, although concentrations are still somewhat approximate (Table 6.18).

Tyrosol (p-hydroxyphenylethyl alcohol) is the main phenolic substance in various types of white

Table 6.18. Influence of the origin of a wine on its phenol content (concentrations expressed in mg/l) (Biau, 1996)

	Dry white wine		Sweet white wine (botrytized)	
	Content	%	Content	%
Tyrosol	25	22.2	29	40.5
Gallic acid	1.4	1.2	1.0	1.4
Caffeic acid				
Total	13.5	12	0.2	0.3
Free	3.5		ε	
Paracoumaric acid				
Total	2.6	2.3	ε	
Free	1.0		ε	
Quercetin	0.2	0.2	ε	
Catechins	10	8.9	12.3	17.2
(+)-catechin	3.4		10.1	
(−)-epicatechin	6.6		2.2	
Tannins				
(procyanidins)	60	53.2	29	40.5
Total phenols	112.7	100	71.5	100

wines. It is formed from tyrosine by yeast and is present at concentrations of 6–25 mg/l in all fermented media (Sapis, 1967). The presence of benzoic acids and gallic acid as well as protocatechuic acid, *p*-hydroxybenzoic acid and vanillic acid has been noted.

White wines contain cinnamic acids, *p*-coumaric acid and caffeic acid, with traces of ferulic acid. These are present in free form and in combination with tartaric acid (coutaric and caftaric acids) (Ribéreau-Gayon, 1965). They are involved in the browning of white grape must (Cheynier *et al.*, 1995). White wines also contain quercetin derivatives, catechins and procyanidins.

In dry white wines, the total phenol content is between 50 and 250 mg/l, or less than 10% of the value in red wines. It is even lower in sweet wines, made from grapes affected by noble rot. Indeed, the development of *Botrytis cinerea* is accompanied by a large-scale breakdown of the phenolic compounds in the skins. Only tyrosol remains, as well as the components from seeds (gallic acid, catechin and procyanidins). Although their deep yellow color might seem to indicate the contrary, these wines always have a very low tannin content.

If sweet wines are made from grapes concentrated by drying, their phenol content is similar to that of dry white wines.

6.9.2 Contribution of the Various Components to the Color of White Wines

The chemical interpretation of the yellow color in white wines has always been a little-known field. Phenolic compounds are certainly involved, but concentrations are low and their contribution has never really been established. Many studies have investigated the oxidative browning of wines, independently of enzyme mechanisms. Other molecules are involved besides tannins (Sapis and Ribéreau-Gayon, 1968), especially compounds that have a high absorption in the visible—and especially ultraviolet—spectrum (Somers and Ziemelis, 1972). Cafeic and coutaric acids are responsible for browning in white wines (Cheynier, 2001).

Besides the phenolic fraction, Myers and Singleton (1979) and Voyatzis (1984) identified a 'non-phenolic' fraction in all types of wines. It consists mainly of polysaccharides and protein compounds, but also contains tyrosol and traces of catechins. The 'non-phenolic' fraction represents 50% of the ultraviolet absorption of dry white wines and, therefore, affects the optical density at 280 nm so that this value cannot be considered to express the phenolic composition alone. In sweet white wines, this fraction has a high concentration of nitrogen compounds and represents more than 50% of absorption at 280 nm.

The yellow color of wine is measured at 420 nm, although the spectrum has no maximum for this value. The respective participation of the two preceding fractions in this color is around 50% for dry white wine, but changes a great deal when the wine is oxidized, either by chemical or enzymatic means (laccase). The phenolic fraction is then responsible for most of the color.

Among the phenolic components identified, derivatives of quercetin, caffeic acid and *p*-coumaric acid are all more-or-less intensely yellow-colored. The maximum absorption wavelengths are between 310 and 350 nm. Tannins, consisting of procyanidins, are also yellow and their color varies according to the oxidation level of the medium. Oxidation of dry white wine produces browning, due to modifications in tannins and highly oxidizable caffeic acid derivatives (Cheynier *et al.*, 1990). The other compounds are relatively unaffected by oxidation, especially the non-phenolic protein and glucide fractions.

The particularly intense yellow color of sweet white wine is different from that of (even oxidized) dry white wine. Its adsorption spectrum is continuous, with a high maximum at 270 nm and no shoulder at 320 nm, as the hydroxycinnamic acids have been broken down by *Botrytis cinerea*. Concentration by evaporation of water from grapes affected by noble rot and compounds produced by the action of oxydases are responsible for high absorption in the ultraviolet range.

The color of white wines therefore involves the oxidation of phenolic compounds. However, the consequences of enzymatic and chemical oxidation are not the same. Chemical oxidation of a catechin solution produces maximum absorption at 400 nm, with a more intensely yellow color than in solutions where oxidation is catalyzed by laccase.

REFERENCES

Ampere O. (1998) *Phytochemistry*, 49, 623.
Amrani-Joutei K. (1993) Thèse de Doctorat Œnologie-Ampélologie, Université de Bordeaux II.
Amrani-Joutei K., Glories Y. and Mercier M. (1994) *Vitis*, 33, 133.
Amrani-Joutei K. and Glories Y. (1995) *Rev. Fr. Œnol.*, 153, 28.
Arnold G.M. (1978) *J. Sci. Food. Agric.* 29, 478.
Asano K., Shinagawa K. and Hashimoto N. (1982) *am. Soc., Brew. Chem.*, 40 (4), 147.
Bate-Smith E.C. (1954) *Food*, 23, 124.
Bate-Smith E.C. (1973) *Phytochemistry*, 12, 907.
Bakker J. and Timberlake C.F. (1997) *J. Agric. Food. Chem.*, 45, 35.
Ben Abdeljahil C., Cheynier V., Fulcrand H., Hakiki A., Mosaddak M. and Moutounet M. (2000) *Sci. Aliments*, 20, 2003.
Biau S.(1996) Thèse de Doctorat Œnologie-Ampélologie, Université de Bordeaux II.
Biau S., Dumon M.C., Glories Y. and Vercauteren J. (1995) Actualités Œnologiques 95, 5th International Symposium Œnologie, Bordeaux, p. 649.
Bourhis M., Théodore N., Weber J.F. and Vercauteren J. (1996) *Polyphenols Communications 1996*, Bordeaux, France, July, p. 43.
Bourzeix M., Clarens M. and Hérédia N. (1986) *C.R. des JEP 86*, Montpellier, Groupe Polyphénols, Narbonne, 13, 123.
Brouillard R. and Cheminat A. (1986) *Tetrahedon Lett.*, 27, 4457.
Brouillard R. El Hage Chahine J.-M. (1979) *C.R. de l'Assemblée Générale du Groupe Polyphénols*, Vol. 9, Narbonne.
Brouillard R., Delaporte B. and Dubois J.-E. (1978) *J. Am. Chem. Soc.*, 100, 6202.
Burger J.F.W., Kolodzici H., Hemingway R.W., Steynberg J.P., Young D. and Ferreira D. (1990) *Tetrahedron lett.*, 46 (16), 5733.
Cameira dos Santos J.P., Brillouet J.M., Cheynier V. and Moutounet M. (1996) *J. Sci. Food Agric.*, 70, 204.
Castagnino C. and Vercauteren J. (1996) *Tetrahedron Lett.*, 37, 7739.
Chapon L. (1993) *J. Inst. Brew.*, 99, 49.
Chatonnet P. (1995) Thèse de Doctorat Œnologie-Ampélologie, Université de Bordeaux II.
Cheynier V. (2001) Groupe polyphénols and evolutions reactions in wine, *Polyphénols actualités*.
Cheynier V., Basire N. and Rigaud J. (1989a) *J. Agric. Food Chem.*, 37, 1069.
Cheynier V., Rigaud J. and Moutounet M. (1989b) *J. Chromatography*, 472, 428.
Cheynier V., Rigaud J. and Moutounet M. (1990) *Phytochemistry*, 29, 1751.
Cheynier V., Fulcrand H.H., Guyot S., Souquet J.M. and Moutounet M. (1995) 4th International Symposium on *Innovation in Enology*, Messe, Stuttgart, Killesberg, p. 50.
Cheynier V., Doco T., Fulcrand H., Guyot S., Le Roux E., Souquet J.M., Rigaud J. and Moutounet M. (1997) *Analusis*, 25, 132.
Cipolli G. (1975) Thèse Doctorat Pharmacie, Université de Bordeaux II.
Coupois-Paul I. (1993) Thèse de Doctorat en Biotechnogies et Industries Alimentaires, INPL, Nancy, France.
Crivellaro Guerra C. (1997) Thèse de Doctorat Œnologie-Ampélologie. Université de Bordeaux II.
Darné G. (1991) Thèse de Doctorat ès Sciences, Université de Bordeaux I.
Darné G. and Glories Y. (1988) *Vitis*, 27, 71.
de Freitas V. (1995) Thèse de Doctorat Œnologie-Ampélologie, Université de Bordeaux II.
Da Silva J.M.R., Cheynier V., Rigaud J., Cheminat A. and Moutounet M. (1991), *Phytochemistry*, 30, 1259.
Di Stefano R. and Guidoni S. (1990) *Vitis*, 29, 71.
Es-Safi N., Le Gueneve C., Laborde B., Fulcrand H., Cheynier V. and Moutounet M. (1999) *Tetrahedron Lett.*, 40, 5869.
Ferreira D. and Bekker R. (1996) Natural Products Reports, p. 411.

Flanzy C. (1998) *Œnologie. Fondements Scientifiques et Technologiques*. Tec et Doc, Lavoisie, Paris.

Flanzy M. and Poux C. (1958) *Ann. Tech. Agric.*, 7, 377.

Francia-Aricha E.M., Guerra M.T., Rivas-Gonzalo J.C. and Santos-Buelga C. (1997) *J. Agric. Food Chem.*, 45, 2262.

Fulcrand H., Cheynier V., Osymianski J. and Moutounet M. (1997) *Phytochemistry*, 46, 223

Galvin C. (1993) Thèse de Doctorat Œnologie-Ampélologie, Université de Bordeaux II.

Geny L., Saucier C., Bracco S., Daviaud F. and Glories Y. (2003) *J. Agric. Food Chem.*, 51, 8051.

Glories Y. (1974) *Conn. Vigne Vin*, 8, 57 and 3, 75.

Glories Y. (1978a) Recherches sur la matière colorante des vins rouges. Thèse de Doctoratès Sciences, Université de Bordeaux II.

Glories Y. (1978b) *Annales de la Nutrition et de l'Alimentation*, 32 (5), 1163.

Glories Y. (1980) *Bull. de Liaison Groupe Polypéhnols*, 9, 129.

Glories Y. (1983) *Bull. de Liaison Groupe Polyphénols*, 11, 577.

Glories Y. (1984) *Conn. Vigne Vin*, 18, 253.

Glories Y. (1986) Symposium sur les critères Modernes de Macèration pour la production des vins Blancs et Rouges, Association deo Œnologues Italiens, Vignale Monferrato, Italy.

Glories Y. (1988) *CR Activités de Recherches Institut d'Œnologie*, 1986–1988, p. 81.

Glories Y. and Augustin M. (1994) In INRA-Viti, Bordeaux, 37.

Guerra C., Glories Y. (1996) Actualités Œnologiques 1995, *Vème Symposium International D'œnologie, Bordeaux*. Tec et Doc, Lavoisier.

Guilloux M. (1981) Thèse de 3ème Cycle, Université de Bordeaux II.

Hagerman A.E. and Butler L.G. (1980) *J. Agric. Food Chem.*, 28, 944.

Han L., Hatano T., Yoshida T. and Okuda T. (1994) *Chem. pharm. Bull.*, 42, 1399.

Haslam E. (1980) *Phytochemistry*, 25, 77.

Haslam E. (1981) *J. Chem. Soc. Chem. Comm.*, 309.

Hrazdina G. (1970) *Phytochemistry*, 9, 1647.

Hrazdina G. (1971) *Phytochemistry*, 10, 1125.

Jeandet P., Bessis R., Maume B.F., Meunier P., Peyron D. and Trollat P. (1995) *J. Agric. Food Chem.*, 43, 316.

Jourdes M., Quideau S., Saucier C. and Glories Y. (2003) Actualités Œnologiques 2003, *VIIème Symposium International d'œnologie*, Bordeaux.

Jurd L. (1972) *Adv. Food Res., Suppl.*, 3, 123.

Karrer P. and De Meuron G. (1932) *Helv. Chem.*, 15, 507.

Kondo K., Kurihara M., Fukuhara K., Tonaka T., Suzuki T., Miyata N. and Toyoda M. (2000) *Tetrahedron lett.*, 41, 485.

Kovac V., Bourzeix M., Heredia N. and Ramos T. (1990) *Rev. Fr. Œnol.*, 125, 7.

Laborde J. (1910), *Rev. Viticult.*, 206.

Laborde J.-L. (1987) Thèse de Doctorat Œnologie-Ampélologie, Université de Bordeaux II.

Laborde B. (2000) Thèse Doctorat ENSAM UMI, Montpellier, France.

Lagune L. (1994) Thèse de Doctorat Œnologie-Ampélologie, Université de Bordeaux II.

Lambri M., Jourdes M., Glories Y. and Saucier C. (2003) *J. Planar Chromatogr.*, 16, 88.

Langcake P. (1981) *Physiol. Plant Pathol.*, 18, 213.

Lea A.G.H. (1992) In *Plant Polyphenol* (ed. Hemingway).

Lea A.G.H., Bridle P., Tumberlake C.F. and Singleton V.L. (1979) *Am. J. Enol. Viticult.*, 30, 289.

Mazza G. and Brouillard R. (1987) *Food Chem.*, 25, 207.

Mirabel M. (2000) Caractéristiques chimiques et organoleptiques des tanins des raisins de Vitis vinifera Var. Merlot et Cabernet Sauvignon issus de différents terroirs bordelais. Thèse de Doctorat, OEnologie-Ampélologie, Université Victor Segalen Bordeaux 2.

Mirabel M., Saucier C., Guerra C. and Glories Y. (1999) *Am. J. Enol. Vitic.*, 50, 211.

Montedoro G. and Fantozzi P. (1974) *Lebensm-Wiss-u-Technol.*, 7, 155.

Montreau F.R. (1969) Thèse Docteur Ingénieur, Université de Toulouse.

Moutounet M., Rabier Ph., Puech J.L., Verette E. and Barillère J.M. (1989) *Sci. Aliments*, 9 (1), 35.

Myers T.E. and Singleton V.L. (1979) *Am. J. Enol. Viticult.*, 30, 98.

Nagel C.W. and Glories Y. (1991) *Am. J. Enol. Viticult.*, 42, 364.

Ollivier C. (1987) Diplôme Études et Recherches, l'Université de Bordeaux II.

Pocock K.F., Seftlon M.A. and Williams P.J. (1994) *Am. J. Enol. Viticult.*, 45 (4), 429.

Ribéreau-Gayon J. (1931) Thèse Doctorat ès Sciences Physiques, Bordeaux.

Ribéreau-Gayon J., Peynaud E., Ribéreau-Gayon P. and Sudraud P. (1976) *Sciences et Techniques du Vin*, Vol. III. Dunod, Paris.

Ribéreau-Gayon J., Peynaud E., Ribéreau-Gayon P. and Sudraud P. (1982) *Sciences et Techniques du Vin*, Vol. I, 2nd edn. Dunod, Paris.

Ribéreau-Gayon P. (1953) *CR Acad. Agric.*, 39, 800.

Ribéreau-Gayon P. (1954) *Ann. Falsif. Fraudes*, 47, 431.

Ribéreau-Gayon P. (1959) *Recherches sur les Anthocyanes des Végétaux*. Libr. Générale de l'Enseignement, Paris.

Ribéreau-Gayon P. (1964) *Les Composés Phénoliques du Raisin et du Vin*. Institut National de la Recherche Agronomique, Paris.

Ribéreau-Gayon P. (1965) *C.R. Acad. Sciences*, 260, 341.
Ribéreau-Gayon P. (1968) *Les Composés Phénoliques des Végétaux*. Dunod, Paris.
Ribéreau-Gayon P. (1970) *Chimie Anal.*, 52, 627.
Ribéreau-Gayon P. (1973) *Vitis*, 12, 119.
Ribéreau-Gayon P. and Sapis J.C. (1965) *C.R. Acad. Sciences*, 261, 1915.
Ribéreau-Gayon P. and Stonestreet E. (1965) *Bull. Soc. Chim.*, 9, 2649.
Ribéreau-Gayon P. and Stonestreet E. (1966) *Chimie Anal.*, 48 (4), 188.
Rigaud J., Perez-Uzarbe J., Ricardo Da Silva J.M. and Cheynier V. (1991), *J. Chromatogr.*, 540, 401.
Roggero J.P., Ragonnet B. and Coen S. (1984) *Bull. Liaison Groupe Polyphénols*, 12, 594.
Sami-Monchado P., Cheynier V. and Moutounet M. (1997) *Phytochemistry*, 45, 1365.
Sapis J.-C. (1967) Thèse de Docteur en Œnologie, Université de Bordeaux II.
Sapis J.-C. and Ribéreau-Gayon P. (1968) *Conn. Vigne Vin*, 2, 323.
Saucier C. (1997) Les tanins du vin. Etude de la stabilité colloïdale. Thèse de Doctorat ŒEnologie-Ampélologie, Université Victor Segalen Bordeaux 2.
Siebert K.J., Carrasco A. and Lynn P.Y. (1996) *J. Agric. Food Chem.*, 44, 1997.
Simon C., Barathien K., Lagune M., Schmitter J.M., Fouquet E., Pianet I. and Dufranc E. (2003) *Actualités œnologiques 2003, VIIème Symposium International d'œnologie, Bordeaux*.
Singleton V.L., Trousdale E. and Zaya J. (1979) *Am. J. Enol. Viticult.*, 30, 49.
Somers T.C. (1971) *Phytochemistry*, 10, 2175.
Somers T.C. and Evans M.E. (1974) *J. Sci., Food Agric.*, 25, 1369.
Somers T.C. and Evans M.E. (1979) *J. Sci., Food Agric.*, 30, 623.
Somers T.C. and Ziemelis G. (1972) *J. Sci., Food Agric.*, 23, 441.
Somers T.C. and Ziemelis G. (1985) *J. Sci., Food Agric.*, 36, 1275.
Souquet J.M., Cheynier V., Brossaud F. and Moutounet M. (1996) *Phytochemistry*, 43, 509.
Sudraud P. (1958) *Ann. Technol. Agric.*, 7, 203.
Thompson R.S., Jacques O., Haslam E. and Tanner D.J.N. (1972) *J. Chem. Soc., Perkins Trans. I*, 1387.
Jimberlake C.F. and Bridle P. (1976) *Am. J. Enol. Vitic.*, 27 (3), 97.
Vivas N. and Glories Y. (1993) *Rev. Fr. Œnol.*, 33 (142), 33.
Vivas N. and Glories Y. (1996) *Am. J. Enol. Vitic.*, 47, 103.
Voyatzis I. (1984) Thèse de Doctorat Œnologie-Ampélologie, Université de Bordeaux II.
Waters W.A. (1964) *Mechanisms of Oxidation of Organic Compounds*, Vol. 6, John Wiley and Sons Inc., New York, pp. 28–138.
Weinges K. and Piretti M.V. (1972) *Annali Chimica*, 62 (29), 45–6.
Weinges K., Kaltenhausser W., Marx H.D., Nader E., Nader F., Perner J. and Seler D. (1968) *Liebigs Ann. Chem.*, 711, 184.

7

Varietal Aroma

7.1 The general concept of varietal aroma	205
7.2 Terpene compounds	206
7.3 C_{13}-norisoprenoid derivatives	211
7.4 Methoxypyrazines	214
7.5 Sulfur compounds with a thiol function	216
7.6 Aromas of American vine species	222
7.7 Development of grape aromas during ripening and the impact of vineyard factors	223

7.1 THE GENERAL CONCEPT OF VARIETAL AROMA

Wine aromas are made up of several hundreds of volatile compounds, at concentrations ranging from several mg/l to a few ng/l, or even less. The olfactory perception thresholds of these compounds also vary quite considerably. Consequently, the olfactory impact of the volatile compounds in wine depends both on concentration and type. Certain compounds, present in trace amounts, on the order of ng/l, may play a major role in aroma, whereas other, much more plentiful, compounds may make only a slight contribution. Furthermore, the impact of each component on the attractiveness of a wine's aroma depends on its specific properties.

The concept of 'thresholds', always applied in a given medium (water, dilute alcohol solution, white wine or red wine), is used to indicate the characteristics of various aromatic substances:

1. *Perception threshold*. This is the minimum concentration at which the presence of an odoriferous substance is detected by 50% of tasters in a triangular test, although they are not necessarily capable of identifying the smell.

2. *Recognition threshold*. This is the threshold for the perception and identification of a specific odoriferous compound.

Handbook of Enology Volume 2: The Chemistry of Wine and Stabilization and Treatments P. Ribéreau-Gayon, Y. Glories, A. Maujean and D. Dubourdieu © 2006 John Wiley & Sons, Ltd

3. *Preference threshold*. This is the maximum concentration at which a compound may be present without giving rise to a negative judgment.

The complexity of wine aromas, which makes them particularly difficult to study, is due to the diversity of the mechanisms involved in their development:

1. Grape metabolism, depending on the variety, as well as soil, climate, and vineyard management techniques.

2. Biochemical phenomena (oxidation and hydrolysis) occurring prior to fermentation, triggered during extraction of the juice and maceration.

3. The fermentation metabolisms of the microorganisms responsible for alcoholic and malolactic fermentations.

4. Chemical or enzymic reactions occurring after fermentation, during aging of the wine in vat, barrel and bottle.

The many odoriferous compounds released into barrel-aged wine by the oak also have an impact on aroma.

However, odoriferous compounds from grapes (reflecting the particular variety, climate and soil) play a more decisive role in the quality and regional character of wines than any other aroma component. These compounds are responsible for the varietal aromas of wines. Paradoxically, these may differ from those found in the free state in grapes. The so-called aromatic varieties, such as the Muscats, produce odoriferous must with similar aromas to those of the resulting wines. However, the musts of many 'simple-flavored' grape varieties are practically odorless. Nevertheless, they produce wines with characteristic aromas that are relatively specific to the grape variety from which they were made. This is true of most of the major grape varieties: Merlot, Cabernet Sauvignon, Cabernet Franc, Sauvignon Blanc, Semillon, the different Pinot varieties, Gamay, Chardonnay, Chenin Blanc, etc. The concept of varietal aroma precursors, odorless forms of the substances that produce varietal aromas in wines, is, therefore, very important in winemaking

The term 'varietal aroma' should not, however, be taken to imply that each grape variety has specific volatile compounds. In fact, the same odoriferous compounds and their precursors are found in the musts and wines of several grape varieties in the same family, as well as other fruits or plants. The individual aromatic personality of wines made from each grape variety is due to the infinitely varied combinations and concentrations of the various compounds.

The odoriferous compounds in *Vitis vinifera* grapes which have been studied in the greatest detail belong to the terpene family. These compounds are responsible for the characteristic aroma in Muscat grapes and wines, although they are also present (at low concentrations) in simple-flavored varieties. Both free forms and odorless, mainly glycosylated, precursors have been identified in wine and grapes.

Other compounds also contribute to varietal aroma. Norisoprenoids, not strictly considered terpenes, are produced by the chemical or enzymic breakdown of carotenoids in grapes. They also occur in the form of glycosylated precursors.

The role of methoxypyrazines in the herbaceous aroma of certain grape varieties, such as Cabernet Sauvignon, is now well-established. These compounds exist in a free state in grapes and no precursor forms have been identified.

More recently, some highly odoriferous sulfur compounds with thiol functions have been shown to participate in the aromas of certain grape varieties, especially Sauvignon Blanc. These compounds occur in grapes in *S*-cysteine conjugate form.

7.2 TERPENE COMPOUNDS

7.2.1 Odoriferous Terpenes

The large family of terpene compounds (approximately 4000) are very widespread in the plant kingdom. Compounds within this family likely to be odoriferous are monoterpenes (compounds with 10 carbon atoms) and sesquiterpenes (15 carbon atoms), formed from two and three isoprene units, respectively. Monoterpenes occur in the form of simple hydrocarbons (limonene, myrcene, etc.), aldehydes (linalal, geranial, etc.), alcohols (linalol, geraniol, etc.), acids (linalic and geranic acid, etc.), and even esters (linalyl acetate, etc.).

In 1946, Austerweil first put forward the hypothesis that terpene compounds were involved in the aroma of Muscat. The presence of three monoterpene alcohols (linalol, α-terpineol and geraniol) in Muscat grapes was first suspected by Cordonnier as early as 1956. Since then, a great deal of research has been devoted to terpene compounds in grapes and wine (Ribéreau-Gayon et al., 1975; Marais, 1983; Strauss et al., 1986; Rapp, 1987; Bayonove, 1993).

About forty terpene compounds have been identified in grapes. Some of the monoterpene alcohols are among the most odoriferous, especially linalol, α-terpineol nerol, geraniol, citronellol and ho-trienol, which has a floral aroma reminiscent of rose essence (Figure 7.1). The olfactory perception thresholds of these compounds are rather low, as little as a few hundred micrograms per liter (Table 7.1). The most odoriferous are citronellol and linalol. Furthermore, the olfactory impact of terpene compounds is synergistic. They play a major role in the aromas of grapes and wines from the Muscat family (Muscat á Petits Grains, Muscat of Alexandria, Muscat of Ottonel and Muscat d'Alsace), as concentrations are often well above the olfactory perception thresholds (Table 7.2).

These compounds also play a role in the 'Muscat' aroma of some Alsatian and German grape varieties: Gewürztraminer, Pinot Gris, Riesling, Auxerrois, Scheurebe, Muller-Thurgau, etc. However, terpenes are only partially responsible for the varietal aromas of these wines and do not explain all of the nuances. Monoterpenes also give a "Muscat" character to Viognier, Albariño, and Muscadelle.

Terpenol concentrations in wines made from grape varieties with simple flavors (Sauvignon Blanc, Syrah, Cabernet Sauvignon, Cabernet Franc, Merlot, etc.) are generally below the perception threshold. There are, however, Chardonnay clones with the Muscat character. These are normally eliminated from clonal selections of vines, as their wines do not have typical varietal character.

About fifteen oxidized and hydroxylated forms (Figure 7.1) of the main monoterpene alcohols have been identified in grape varieties with the Muscat character (Schreier et al., 1976; Strauss et al., 1986, 1988; Rapp, 1987).

Table 7.1. Characteristics of the main monoterpenes and examples of concentrations in wines made from different varieties

Monoterpene	Olfactory description	Olfactory perception threshold (μg/l[a])	Concentration (μg/l) in wines made from:						
			Muscat of Alexandria[d]	Muscat de Frontignan[d]	Gewürztraminer[e]	Albariño[f]	Riesling[e]	Muscadelle[g]	Sauvignon Blanc[g]
Linalol	Rose	50[b]	455	473	6	80	40	50	17
α-Terpineol	Lily of the Valley	400[b]	78	87	3	37	25	12	9
Citronellol	Citronella	18[b]	ND[h]	ND	12	ND	4	3	2
Nerol	Rose	400[b]	94	135	43	97	23	4	5
Geraniol	Rose	130[b]	506	327	218	58	35	16	5
Ho-trienol	Linden	110[c]	ND	ND	ND	127	25	ND	ND

[a]Olfactory perception thresholds have been determined in wine by the following.
[b]Boidron (unpublished work).
[c]Simpson (1978).
[d]Ribéreau-Gayon et al. (1975).
[e]Günata, (1984).
[f]Falqué-Lopez et al. (1994).
[g]Darriet (1993).
[h]ND = not detected.

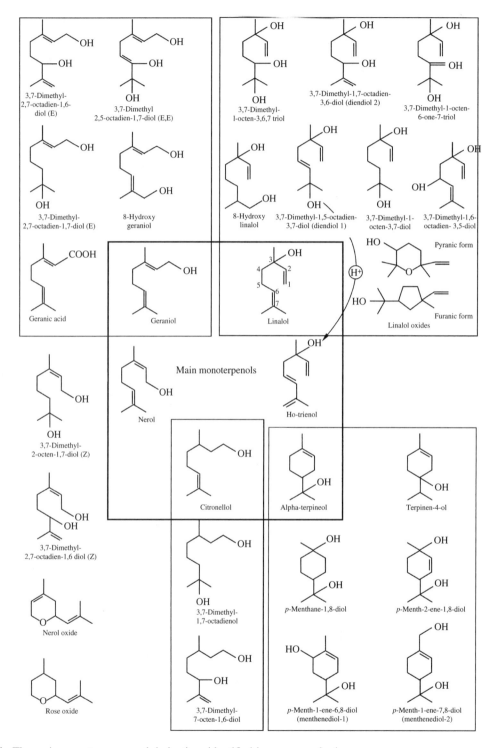

Fig. 7.1. The main monoterpenes and derivatives identified in grapes and wine

Table 7.2. Example of the distribution of free and bonded forms of the main monoterpenols and several C_{13}-norisoprenoid derivatives in ripe grapes

Grape variety	Free terpenols (µg/l)	Terpene glycosides (µg/l)	C_{13}-norisoprenoid glycosides[a] (µg/l)	Reference
Muscats:				
Alexandria	1513	4040	ND[b]	Günata (1984)
Frontignan	1640	1398	ND	Günata (1984)
Hamburg	594	1047	ND	Günata (1984)
Ottonel	1679	2873	ND	Günata (1984)
Gewürztraminer	282	4325	ND	Günata (1984)
Riesling	73	262	182	Razungles et al. (1993)
Sauvignon Blanc	5	107	104	Razungles et al. (1993)
Sémillon	17	91	265	Razungles et al. (1993)
Syrah	13	65	84	Razungles et al. (1993)
Chardonnay	41	12	140	Razungles et al. (1993)
Cabernet Sauvignon	0	13	100	Razungles et al. (1993)

[a] C_{13}-norisoprenoids analyzed: hydroxy-3-β-D-damascone, oxo-3-α-ionol, oxo-4-β-ionol, hydroxy-3-β-ionol and hydroxy-3-dihydro-7,8-β-ionol.
[b] ND = not detected.

In view of their high perception thresholds (1–5 mg/l), linalol and nerol oxides have very little olfactory impact on wines. Rose oxide is a more odoriferous compound. According to Guth (1997), it is partly responsible for the floral aroma of Gewürztraminer wines.

Monoterpene polyols (diols and triols), present in grapes at concentrations up to one milligram per liter, or even more, are not highly odoriferous. They may, however, form other monoterpenes by hydrolysis at acid pH, some of which are odoriferous. Thus, acid hydrolysis of 3,7-dimethylocta-1,5-dien-3,7-diol produces ho-trienol (Figure 7.1) (Strauss et al., 1986).

A number of monoterpene and sesquiterpene hydrocarbons with resin-like odors have been identified, including limonene, α-terpinene, p-cimene and myrcene, as well as sesquiterpene alcohols such as farnesol. The organoleptic role of these compounds in wine has not been clearly established (Schreier et al., 1976; Bayonove, 1993).

Aldehydes (geranial and linalal), acids (trans-geranic acid) and monoterpene esters (geranyl and neryl acetate) have been identified in grapes (Schreier et al., 1976; Etiévant et al., 1983; Di Stefano and Maggiorotto, 1993). The aldehydes are reduced to alcohols during fermentation. More recent research has also investigated certain menthenediols, derived from α-terpineol (Bitteur et al., 1990; Versini et al., 1992), but these compounds are not highly odoriferous (Sefton et al., 1994).

Terpenols may also be rearranged in an acid medium to produce other monoterpene alcohols (Voirin et al., 1990). The development of Botrytis cinerea on grapes may also modify the monoterpene composition to a considerable extent, by breaking down the main monoterpenols (Boidron, 1978) and converting them into generally less odoriferous compounds (Rapp, 1987) (Volume 1, Section 13.2.1). The enzymic oxidation of linalol by Botrytis cinerea produces 8-hydroxylinalol, but this reaction also occurs naturally in musts made from non-botrytized grapes.

7.2.2 Glycosylated Forms of Volatile Terpenols

The existence of a non-volatile, odorless fraction of the terpene aroma in Muscat grapes which could be revealed by chemical or enzymic means was demonstrated for the first time by Cordonnier and Bayonove (1974). Several teams (Williams et al., 1982; Günata, 1984; Voirin et al., 1990) later established that the main monoterpenols and terpene polyols were present in grapes in glycoside form, including the basic 'oses': glucose, arabinose, rhamnose and apiose.

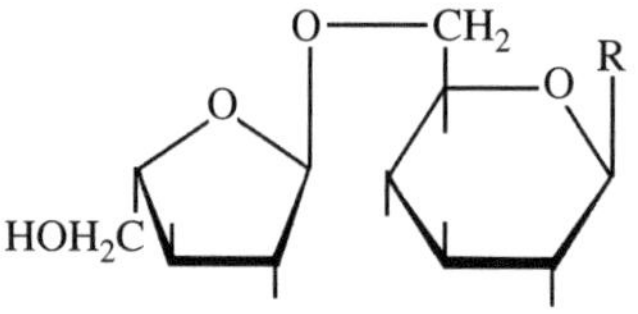

6-*O*-α-L-Arabinofuranosyl-β-D-glucopyranoside

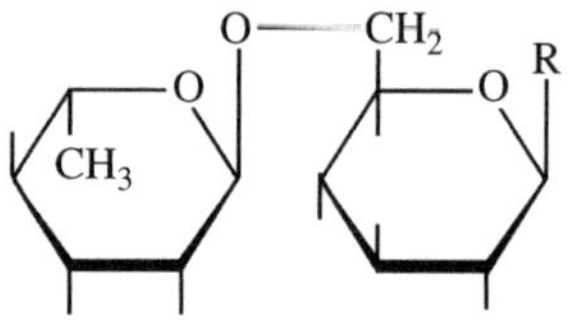

6-*O*-α-L-Rhamnopyranosyl-β-D-glucopyranoside

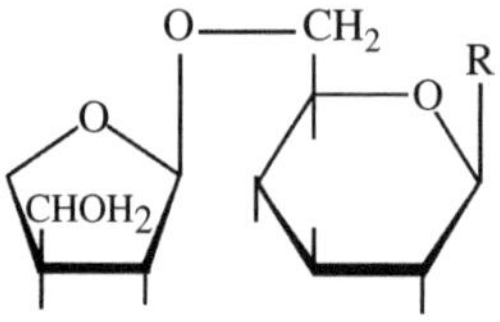

6-*O*-β-Apiofuranosyl-β-D-glucopyranoside

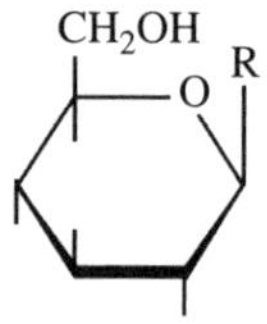

β-D-Glucopyranoside

Fig. 7.2. The various forms of terpene glycosides (or norisoprenoids) identified in grapes: R = terpenol or C_{13}-norisoprenoid

Four types of glycosides have thus been identified (Figure 7.2): three diglycosides (6-*O*-α-L-arabinofuranosyl-β-D-glucopyranoside, 6-*O*-α-L-rhamnosyl-β-D-glucopyranoside or rutinoside, 6-*O*-β-D-apiosyl-β-D-glucopyranoside) and one monoglucoside (β-D-glucopyranoside).

All grape varieties contain similar glycosides, but the Muscat-flavored grape varieties have the highest concentrations. Glycosylated forms are frequently more common than free aromas (Table 7.2). Among the glycosides corresponding to the most odoriferous aglycones, apiosylglucosides and arabinosylglucosides are the most widespread, followed by rutinosides and then β-glucosides. Terpene glycosides are very common in plants. However, in vines, unlike other plants, monoglucosides are in the minority as compared to diglycosides.

Grape skins have a higher concentration of free and glycosylated monoterpenes than the flesh or juice. The free terpenol composition varies a great deal in the different parts of grapes. Thus, geraniol and nerol are more common in the skin than in the flesh and juice. The proportions of the various bonded terpenols are largely the same throughout the grape. The relative proportions of free and bonded compounds depend on the grape variety. Muscat of Alexandria juice contains more bonded terpenols, while the skins have almost equal amounts of both bonded and free compounds. In Muscat de Frontignan, the proportion of free and glycosylated terpenols is approximately the same in the juice and skins (Table 7.3).

As glycosides are much more water soluble than aglycones, they are considered to be vectors for the transport and accumulation of monoterpenes in

Table 7.3. Locations of the free and bonded terpenols in grapes (Günata, 1984)

	Muscat de Frontignan (monoterpenes) (μg/kg)		Muscat of Alexandria (monoterpenes) (μg/kg)		Cabernet Sauvignon (isobutylmethoxypyrazine) (%)
	Free form	Bonded form	Free form	Bonded form	
Pulp	444	457	212	577	10
Juice	485	1691	291	2126	40
Skin	2237	6311	2904	3571	50

plants (Stahl-Biskup, 1987). Glycosides have been identified primarily in vine leaves and leaf stems (Di Stefano and Maggiorotto, 1993).

In these glycosylated derivatives, the aglycones are not exclusively alcohols or terpene polyols. Linear or cyclic alcohols (hexanol, phenylethanol and benzyl alcohol) and some C_{13}-norisoprenoids, as well as, probably, volatile phenols such as vanillin, may also be present (Section 7.3.1).

7.2.3 Enhancing the Glycosylated Aromatic Potential of Grapes

Grapes contain β-glycosidases capable of releasing certain free, odoriferous terpenols from their non-odoriferous glycosides (Bayonove *et al.*, 1984; Ayran *et al.*, 1987; Biron *et al.*, 1988; Günata *et al.*, 1989). Under normal winemaking conditions, these endogeneous enzymes have a limited effect on the development of the must's aroma, for several reasons. Firstly, the activity of these enzymes is optimum at pH 5 and low in must. Secondly, grape glycosidases are not capable of hydrolyzing the glycosides of tertiary alcohols such as linalol due to a lack of specific reaction to certain aglycones. Thirdly, the clarification of must inhibits its glycosidase activity (Grossmann *et al.*, 1990).

In the same way, alcoholic fermentation has little effect on the glycosylated potential of grapes. The glycoside concentration in wine is similar to that of grapes. Winemaking yeast certainly has periplasmic glycosidases (β-glucosidase, α-arabinosidase and α-rhamnosidase) that have been found to act on must glycosides *in vitro*, but their optimum activity occurs around pH 5.

The application of exogenous enzyme activity to enhance the aromatic potential has, therefore, been envisaged. These enzymes are present as contaminant activity in industrial pectinase preparations made from *Aspergillus niger* cultures. Several enzyme systems are involved in a two-stage process (Günata *et al.*, 1988). Firstly, an α-L-rhamnosidase, α-L-arabinosidase or β-D-apiosidase splits the disaccharide. Then, a β-D-glucosidase releases the corresponding odoriferous aglycone. These types of preparations (Grossmann and Rapp, 1988; Cordonnier *et al.*, 1989; Günata *et al.*, 1990; Günata, 1993) are only effective in dry wines, as fungal β-glucosidases are inhibited by glucose. They undoubtedly bring out the aromas of young wines made from 'Muscat-flavored' grape varieties. Glycosidase preparations have less effect on simple-flavored grape varieties for several reasons. First of all, the varietal aroma precursors are not necessarily all glycosylated and not all non-terpene aglycones are odoriferous. Furthermore, it is not desirable for all grape varieties to acquire a terpene background aroma, if varietal character is to be preserved.

7.3 C_{13}-NORISOPRENOID DERIVATIVES

7.3.1 Odoriferous C_{13}-Norisoprenoid Derivatives

The oxidative degradation of carotenoids (Figure 7.3), terpenes with 40 carbon atoms (tetraterpenes), produces derivatives with 9, 10, 11 or 13 carbon atoms (Enzel, 1985). Among these compounds, norisoprenoid derivatives with 13 carbon atoms (C_{13}-norisoprenoids) have interesting odoriferous properties. These compounds are common in tobacco, where they were initially studied (Demole *et al.*, 1970; Demole and Berthet, 1972), but they have also been studied in grapes (Schreier *et al.*, 1976; Simpson *et al.*, 1977; Simpson, 1978; Sefton *et al.*, 1989; Winterhalter, 1993).

From a chemical point of view these norisoprenoid derivatives are divided into two main forms: megastigmane and non-megastigmane. Each of these includes a large number of volatile compounds (Figure 7.4). The megastigmane skeleton is characterized by a benzene cycle substituted on carbons 1, 5 and 6, and an unsaturated aliphatic chain with four carbon atoms attached to C_6.

Megastigmanes (Figure 7.4) are oxygenated C_{13}-norisoprenoids, with skeletons oxygenated on carbon 7 (damascone series) or carbon 9 (ionone series). Among these compounds, β-damascenone, with a complex smell of flowers, tropical fruit and stewed apple, has a very low olfactory perception threshold in water (3–4 ng/l) and a relatively

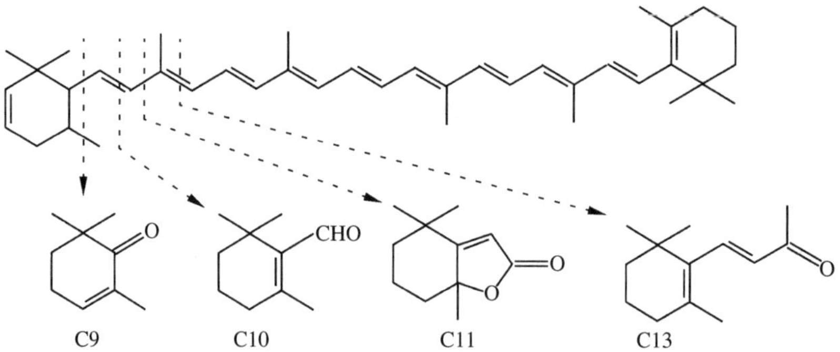
O
CHO
O
O
O
C9
C10
C11
C13

Fig. 7.3. Breakdown of carotenoids leading to the formation of C_9, C_{10}, C_{11} and C_{13}-norisoprenoids in grapes (Enzel, 1985)

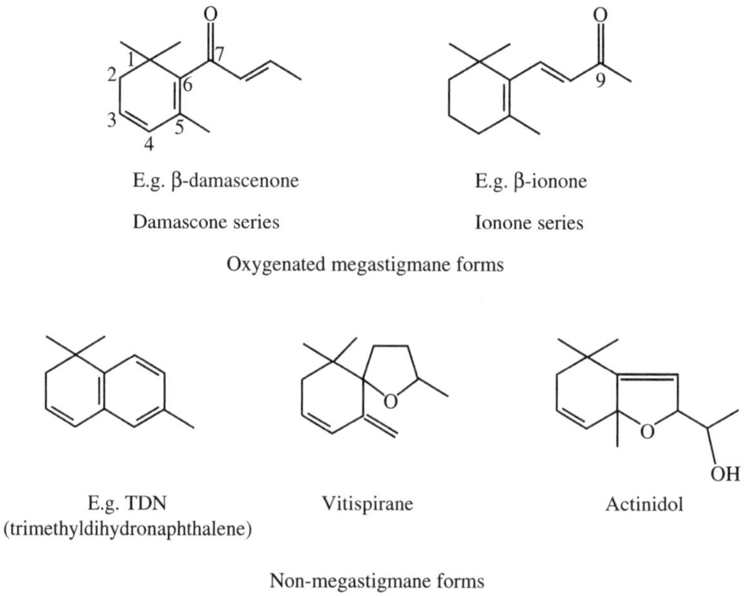
O
1
7
2
6
3
5
4
O
9
E.g. β-damascenone
Damascone series
E.g. β-ionone
Ionone series
Oxygenated megastigmane forms
O
O
OH
E.g. TDN
(trimethyldihydronaphthalene)
Vitispirane
Actinidol
Non-megastigmane forms

Fig. 7.4. Main families of C_{13}-norisoprenoid derivatives in grapes

low threshold in model dilute alcohol solution (40–60 ng/l). This compound was first identified in Riesling and Scheurebe grape juice (Schreier *et al.*, 1976) and Muscat (Etiévant *et al.*, 1983), but is probably present in all varieties of grapes (Baumes *et al.*, 1986; Sefton *et al.*, 1993). Assay results (Table 7.4), show that β-damascenone concentrations in white and red wines are extremely variable and that this compound has a major olfactory impact on certain wines. Higher values are found in red wines than in dry whites and concentrations are especially high in *vins doux naturels* made from Muscat. Average concentrations in Merlot, Cabernet Sauvignon and Cabernet Franc wines do not vary to any significant extent.

With its characteristic aroma of violets, β-ionone has a perception threshold of 120 ng/l in water and 800 ng/l in model dilute alcohol solution, and it has been identified in various white grape varieties (Schreier *et al.*, 1976), as well as Muscat (Etiévant *et al.*, 1983). Like β-damascenone, it is present in all grape varieties. The contribution of

Table 7.4. Concentrations (ng/l) of β-damascenone and β-ionone in various wines (Chatonnet and Dubourdieu, 1997)

	β-Damascenone	β-Ionone
Dry white wines		
(12 samples)		
Mean	709	13
Standard deviation	561	19
Maximum spread	89–1505	0–59
Red wines		
(64 samples)		
Mean	2160	381
Standard deviation	1561	396
Maximum spread	5–6460	0–2451
Muscat VDN (*vin doux naturel*)		
(1 sample)	11 900	72
Perception threshold in water	3–4	120
Perception threshold in model solution	40–50	800

β-ionone to the aroma of white wines is negligible (Table 7.4). It may, however, play a significant role in the aroma of red wines. Concentrations are more variable than those of β-damascenone and the grape variety does not seem to be a significant factor in these variations.

The other oxygenated C_{13}-norisoprenoids identified in wine are 3-oxo-α-ionol (tobacco), 3-hydroxy-β-damascone (tea and tobacco) and β-damascone (tobacco and fruit). Their perception thresholds are much higher and their olfactory impact in wine negligible, in spite of relatively high concentrations in some cases.

Non-megastigmane C_{13}-norisoprenoid derivatives have also been identified, including a few rather odoriferous compounds. The most important of these is TDN (1,1,6-trimethyl-1,2-dihydronaphtalene), which has a distinctive kerosene odor. It plays a major role in the 'petroleum' smell of old Riesling wines (Simpson, 1978). TDN is generally absent in grapes and young wine, but may appear during bottle aging, reaching concentrations of 200 µg/l, whereas its perception threshold is on the order of 20 µg/l.

Actinidols and vitispirane (Figure 7.4), also in the same family, have odors reminiscent of camphor. Some of the non-megastigmane C_{13}-norisoprenoids are derived from megastigmanes by chemical modifications in an acid medium (Sefton *et al.*, 1989). It is possible that vitispirane, formed during bottle aging, contributes towards the camphorated odor defects in wines that seem prematurely aged.

7.3.2 Precursors of Odoriferous C_{13}-Norisoprenoid Derivatives

In an acid medium, several not very odoriferous oxygenated C_{13}-norisoprenoids undergo chemical modifications that may result in the formation of odoriferous β-damascenone (Skouroumounis *et al.*, 1992; Winterhalter, 1993) (Figure 7.5). Certain non-megastigmane C_{13}-norisoprenoids, in particular TDN, are also derived from megastigmanes by chemical modifications in an acid medium (Winterhalter, 1993). However, C_{13}-norisoprenoids are mainly present in grapes in the form of non-volatile precursors (carotenoids and glucosides).

Like monoterpenes, certain C_{13}-norisoprenoids (vomifoliol, 3-oxo-α-ionol, 3-hydroxydamascone) exist in glycosylated form (Table 7.2) (Günata, 1984; Razungles *et al.*, 1993; Skouroumounis and Winterhalter, 1994). The currently identified glycosides of C_{13}-norisoprenoids are all monoglucosides. They are not hydrolyzed by grape and yeast glycosidases but they may be revealed by exogeneous fungal glycosidases. However, the volatile compounds thus released are not highly odoriferous. Theoretically, in an acid medium, some

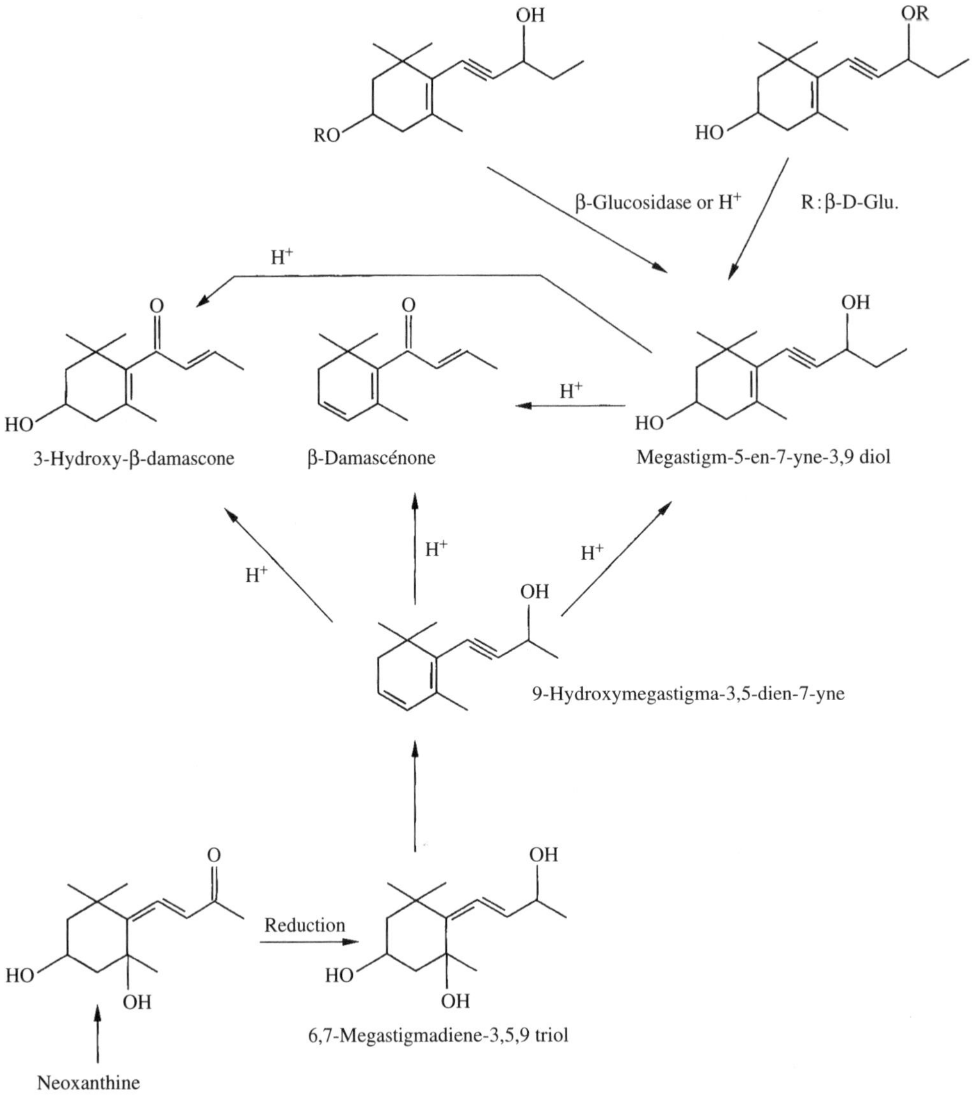

Fig. 7.5. How β-damascenone is formed in grapes and wine (Skouroumounis *et al.*, 1992; Winterhalter, 1993; Puglisi *et al.*, 2001)

of them, especially 3-hydroxydamascenone, could produce β-damascenone. The practical importance of these reactions in winemaking has not been demonstrated.

7.4 METHOXYPYRAZINES

Methoxypyrazines are nitrogenated heterocycles produced by the metabolism of amino acids. The compounds shown in Figure 7.6, 2-methoxy-3-isopropylpyrazine, 2-methoxy-3-*sec*-butylpyrazine and 2-methoxy-3-isobutylpyrazine, have odors reminiscent of green pepper and asparagus, or even earthy overtones. These highly odoriferous compounds have extremely low perception thresholds in water, on the order of 1 ng/l (Table 7.5). Many plants, including green peppers and peas (Buttery *et al.*, 1969; Murray *et al.*, 1970), as well as

Varietal Aroma

Table 7.5. Descriptions and olfactory perception thresholds of the main methoxypyrazines

Pyrazine	Olfactory perception threshold in water (ng/l)	Description
2-Methoxy-3-isobutyl	2	Green pepper
2-Methoxy-3-isopropyl	2	Green pepper, earthy
2-Methoxy-3-sec-butyl	1	Green pepper
2-Methoxy-3-ethyl	400	Green pepper, earthy

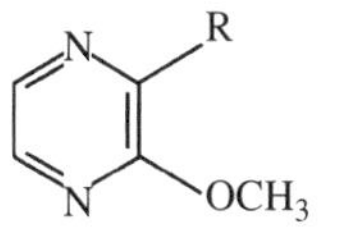

R: $CH_2CH(CH_3)_2$ — 2-Methoxy-3-isobutylpyrazine
R: $CH(CH_3)_2$ — 2-Methoxy-3-isopropylpyrazine
R: $CH(CH_3)CH_2CH_3$ — 2-Methoxy-3-sec-butylpyrazine

Fig. 7.6. The main methoxypyrazines

potatoes (Maga, 1989), have been shown to contain 2-methoxy-3-isobutylpyrazine. This was first identified in grapes (Cabernet Sauvignon) by Bayonove et al. (1975).

Since then, 2-methoxy-3-isobutylpyrazine and the other pyrazines have been identified in many grape varieties and their wines (Sauvignon Blanc, Cabernet Franc, Merlot, Pinot Noir, Gewürztraminer, Chardonnay, Riesling, etc.) (Augustyn et al., 1982; Harris et al., 1987; Calo et al., 1991; Allen et al., 1994). However, concentrations of these compounds are only significantly above the recognition threshold in Sauvignon Blanc, Cabernet Sauvignon and Cabernet Franc grapes and wines, and sometimes Merlot. This herbaceous methoxypyrazine aroma, usually most apparent when the grapes are underripe, is not appreciated in red Bordeaux wines.

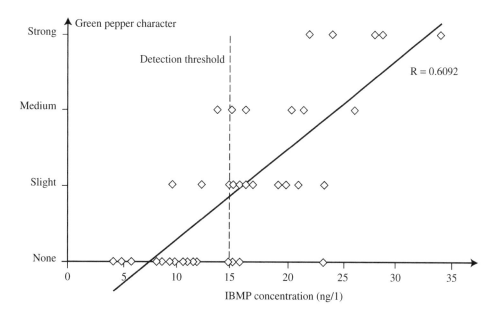

Fig. 7.7. Correlation between the 'green pepper' character identified on tasting and the concentration of 2-methoxy-3-isobutylpyrazine (IBMP) in various red Bordeaux wines (Roujou de Boubée, 1996)

Concentrations of 2-methoxy-3-isobutylpyrazine in Sauvignon Blanc and Cabernet Sauvignon must and wine range from 0.5 to 50 ng/l (Lacey *et al.*, 1991; Allen and Lacey, 1993; Kotseridis, 1999; Roujou de Boubée, 2000). In red Bordeaux wines, the recognition threshold of 2-methoxy-3-isobutylpyrazine is in the order of 15 ng/l (Figure 7.7). At higher concentrations, the herbaceous character of IBMP is clearly perceptible and spoils wine aroma.

The distribution of IBMP in Cabernet Sauvignon grape bunches has been described by Roujou de Boubée *et al.* (2002). The stems contain over half (53%) of the IBMP, while the highest concentration of IBMP in the grapes themselves is in the skins (67%). Less than 1% of the IBMP in the grapes is in the flesh and the remainder is located in the seeds (Figure 7.8).

Concentrations of 2-methoxy-3-isopropylpyrazine and 2-methoxy-3-*sec*-butylpyrazine in Sauvignon Blanc and Cabernet Sauvignon wines are systematically lower than those of 2-methoxy-3-isobutylpyrazine. They have no influence on taste.

The following methoxypyrazines have also been identified in grapes and wines: 2-methoxy-3-methylpyrazine (Harris *et al.*, 1987) and 2-methoxy-3-ethylpyrazine (Augustyn *et al.*, 1982). They are much less odoriferous than 2-methoxy-3-isobutylpyrazine. Allen *et al.* (1995a, 1995b) also suggested that some methoxypyrazines in wine might be microbial in origin.

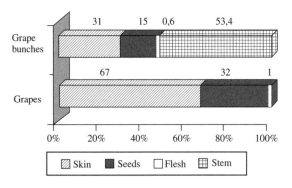

Fig. 7.8. Distribution (%) of IBMP in the various parts of Cabernet Sauvignon grape bunches at harvest time (Roujou de Boubée *et al.* 2002)

7.5 SULFUR COMPOUNDS WITH A THIOL FUNCTION

7.5.1 Odoriferous Volatile Thiols Involved in the Varietal Aromas of Wines

Sulfur compounds in the thiol family (or mercaptans) are generally held responsible for olfactory defects (Section 8.2.2). However, their major contribution to the aromas of certain fruits and aromatic plants has been clearly established. Thus, specific thiols are involved in the characteristic aromas of fruits such as blackcurrant (Rigaud *et al.*, 1986), grapefruit (Demole *et al.*, 1982), passion fruit (Engel and Tressel, 1991) and guava (Idstein and Schreier, 1985; Bassols and Demole, 1994). Two mercaptans, ethyl-3-mercaptopropionate and ethyl-2-mercaptopropionate, have been identified as components in the aroma of *Vitis labrusca* grapes (variety Concord) (Kolor, 1983; Winter *et al.*, 1990).

Since the early 1990s, a number of highly odoriferous thiols have been identified in Sauvignon Blanc wines. These wines have marked, characteristic aromas, featuring various herbaceous, fruity, and empyreumatic nuances. The first and second groups include green pepper, boxwood, broom, eucalyptus, blackcurrant buds, rhubarb, tomato leaves, nettles, grapefruit, passion fruit, white peaches, gooseberries, and asparagus broth, as well as acacia wood and blossoms. After a few years' bottle aging, some wines develop aromas reminiscent of smoke, roast meats, and even truffles. Until recently, the compounds responsible for the characteristic aromas of Sauvignon Blanc wines had not been identified, with the exception of methoxypyrazines (Augustyn *et al.*, 1982; Allen *et al.*, 1989), whose green pepper odors were more pronounced in immature grapes.

The first molecule found to be a characteristic component of the aroma of Sauvignon Blanc wines was 4-mercapto-4-methyl-pentan-2-one (Figure 7.9a) (Darriet *et al.*, 1993; Darriet, 1993; Darriet *et al.*, 1995). This extremely odoriferous mercaptopentanone has a marked smell of

Varietal Aroma

boxwood and broom. Its perception threshold in a model solution is 0.8 ng/l. It has an undeniable organoleptic impact, as concentrations may even exceed a hundred mg/l in Sauvignon Blanc wines with strong varietal character (Bouchilloux et al., 1996, Tominaga et al. (1998b)). This compound occurs in box leaves and leafy broom twigs at concentrations ranging from a few ng to a few tens of ng/g by fresh weight (Tominaga and Dubourdieu, 1997). Thus, the 'boxwood' and 'broom' descriptions that have long been used by wine tasters to define the aroma of Sauvignon Blanc actually correspond to a chemical reality.

Several other odoriferous volatile thiols have also been identified in Sauvignon Blanc wine (Figure 7.9): 3-mercaptohexan-1-ol acetate (Tominaga et al., 1996), 4-mercapto-4-methylpentan-1-ol, 3-mercaptohexan-1-ol and 3-mercapto-3-methylbutan-1-ol (Tominaga et al., 1998a). Tables 7.6 and 7.7 specify their organoleptic roles (Tominaga et al., 1998b).

The complex odor of 3-mercaptohexyl acetate is reminiscent of boxwood, as well as grapefruit zest and passion fruit. This compound was previously identified in passion fruit by other authors (Engel and Tressel, 1991). Its perception threshold is 4 ng/l and some Sauvignon Blanc wines may contain several hundreds of ng/l. Concentrations decrease as the wine ages and 3-mercaptohexanol is formed.

The aroma of 3-mercaptohexanol is redolent of grapefruit and passion fruit, in which it has also been identified. The perception threshold is on the order of 60 ng/l. It is always present in Sauvignon Blanc wine at concentrations of several hundred ng/l, and there may be as much as a few μg/l.

The organoleptic role of 4-mercapto-4-methyl-pentan-1-ol, which smells of citrus zest, is more limited. Concentrations in wine are rarely over the perception threshold (55 ng/l), but this value may be reached in a few wines.

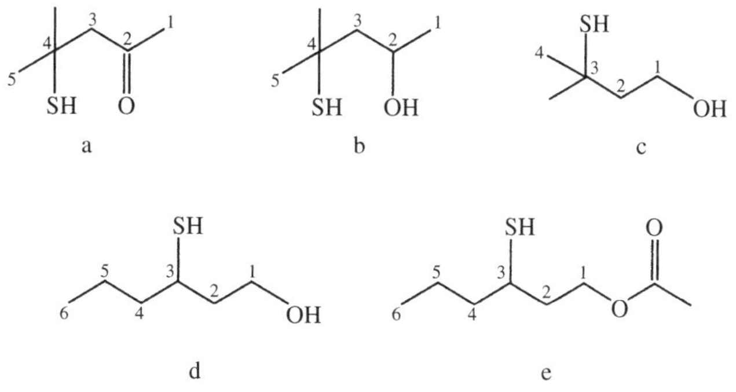

Fig. 7.9. Odoriferous volatile thiols identified in Sauvignon Blanc wine: (a) 4-mercapto-4-methyl-pentan-2-one (4-MMP), (b) 4-mercapto-4-methyl-pentan-1-ol (4-MMPOH), (c) 3-mercapto-3-methyl-butan-1-ol (3-MMB) (d) 3-mercaptohexan-1-ol (3-MH), (e) 3-mercaptohexyl acetate (A3-MH)

Table 7.6. Organoleptic impact of volatile thiols identified in Sauvignon blanc wines (Bordeaux and Loire)

Compound identified	Description	Perception threshold* ng/l	Content (ng/l)
4-mercapto-4-methyl-pentan-2-one	Boxwood, broom	0.8	0–120
3-mercaptohexyl acetate	Boxwood, passion fruit	4	0–500
3-mercaptohexanol	Passion fruit, grapefruit zest	60	150–3500
4-mercapto-4-methyl-pentan-2-ol	Citrus zest	55	15–150
3-mercapto-3-methyl-butan-1-ol	Cooked leeks	1500	20–150
benzenemethanethiol	Gunflint, smoke	0.3	5–20

*in model dilute alcohol solution.

Table 7.7. Assay of volatile thiols (ng/l) in Sauvignon Blanc wines from the same Bordeaux estate in several vintages (Tominaga et al., 1998b)

Compounds	Samples			
	1992	1993	1994	1995
4-Mercapto-4-methyl-pentan-2-one (4-MMP)	7 (9)	45 (50)	10 (13)	44 (55)
3-Mercaptohexyl acetate (A3-MH)	0 (0)	0 (0)	0.4 (0.08)	2.8
3-Mercaptohexan-1-ol (3-MH)	871 (15)	1178 (20)	600 (10)	1686 (28)
4-Mercapto-4-methyl-pentan-2-ol (4-MMPOH)	46 (0.8)	111 (2)	25 (0.5)	33 (0.6)
3-Mercapto-3-methyl-butan-1-ol (3-MMB)	128 (0.08)	89 (0.06)	97 (0.06)	104 (0.07)

Entries in parentheses are the aromatic intensities of each compound: concentration/perception threshold. The wines found on tasting to have typical Sauvignon Blanc character were the 1993 and 1995 vintages.

The much less odoriferous 3-mercapto-3-methyl-butan-1-ol smells of cooked leeks. It never reaches the perception threshold of 1500 ng/l in wine.

These volatile thiols have also been identified in white wines made from other grape varieties and contribute to their varietal aromas.

As early as 1981, Du Plessis and Augustyn had deduced by an olfactory analogy that 4-MMP was involved in the aroma of guava, Chenin Blanc, and Colombard, but they had no formal proof. Similarly, Rapp and Pretorius (1990) suggested that unidentified sulfur compounds smelling of blackcurrant were present in Scheurebe, Kerner, Bacchus, and Müller-Thurgau wines. In 1997, Guth demonstrated the major impact of 4-MMP on the aroma of Scheurebe wines, which can contain up to 400 ng/l, a concentration that is much higher than that in any Sauvignon Blanc wine.

Research by Tominaga et al. (2000) (Table 7.8) showed that the thiols identified in Sauvignon Blanc wines also contributed to the aroma of several Alsace grape varieties, as well as Colombard, Manseng, and botrytized Sémillon.

Thus, 4-MMP plays an important role in the characteristic boxwood aroma of Sauvignon Blanc, also found in Muscat d'Alsace and sometimes in Riesling. The same aroma is produced by 3-mercaptohexyl acetate (A3-MH), which makes a major contribution to the bouquet of young Colombard and Manseng wines. The grapefruit and tropical fruit nuances of 3-MH contribute markedly to the bouquet of Gewürztraminer (which may contain up to 3000 ng/l), Muscat d'Alsace, Pinot Gris, Riesling, Manseng, and botrytized Sémillon. The 3-MH content of some great Sauternes is remarkably stable, remaining as high as 5000 ng/l after several decades in bottle.

Table 7.8. Volatile thiol composition (ng/l) of several Alsace grape varieties, as well as Colombard, Manseng, and Sémillon from Sauternes (Tominaga et al., 2000)

	4-MMP	4-MMPOH	3-MMB	A3-MH	3-MH
Gewürztraminer	0.7–15	0–14	137–1322	0–6	40–3300
Riesling	0–9	0–3	26–190	0–15	123–1234
Muscat d'Alsace	9–73	0–45	19–326	0–1	100–1800
Pinot Gris	0–3	0–0.5	21–170	0–51	312–1042
Pinot Blanc	0–1	0	2–83	0	88–248
Sylvaner	0–0.5	0	1–99	0	59–554
Colombard	0	0	0	20–60	400–1000
Petit Manseng (Jurançon)	0	0	40–140	0–100	800–4500
Sémillon (Sauternes)	0	0	100–500	0	1000–6000

3-MH is also present in small quantities, but above the perception threshold, in wines made from Melon de Bourgogne (Schneider, 2001) and Chardonnay (Tominaga, unpublished results). It also contributes to the aroma of Chenin Blanc and Petite Arvine as well as, probably, to that of many other white grape varieties.

Several of the volatile thiols mentioned above have also been identified in red Bordeaux grape varieties (Bouchilloux *et al.*, 1998b). However, only 3-MH, detectable at concentrations above the perception threshold, has a real impact on aroma, contributing sulfur nuances reminiscent of black-currant (Blanchard, 2000). The 3-MH content of red Bordeaux wines decreases considerably during aging, especially in barrel, dropping from several µg/l at the end of alcoholic fermentation to 300–600 ng/l after 12 months in barrel. The controlled oxidation conditions during barrel aging are no doubt responsible for this decrease in 3-MH, as it is both highly oxidizable and extremely reactive with the quinones produced by the oxidation of phenolic compounds.

3-MH and its acetate make a decisive contribution to the aroma of rosé wines made from Merlot and Cabernet Sauvignon (Murat *et al.*, 2001a). The fact that rosé wines are protected from oxidation during aging (short aging period in vat on fine lees, with limited racking) preserves their high thiol content. Furthermore, the protective effect of anthocyanins on the volatile thiols in rosé wines has been clearly demonstrated in a model medium (Murat *et al.*, 2003). This is the likely scientific explanation for the empirical observation that deeper-colored rosé wines keep their aroma better.

Benzenemethanethiol, an extremely odoriferous mercaptan with a perception threshold in the vicinity of 0.3 ng/l in model dilute alcohol solution, was recently identified and assayed in several white (Chardonnay, Sauvignon Blanc and Sémillon) and red (Merlot and Cabernet) wines (Tominaga *et al.*, 2003). It has a smoky odor, reminiscent of gunflint. Concentrations in Chardonnay wines (Burgundy and Limoux) are around 30–40 ng/l, while those in Sauvignon Blanc (Bordeaux and Sancerre) are 10–20 ng/l. This compound certainly contributes to the empyreumatic nuances of these wines.

The presence of 2-methyl-furanthiol, an odoriferous compound smelling of cooked meat with a perception threshold in model solution of approximately 5 ng/l, has been reported in red Bordeaux (Bouchilloux *et al.*, 1998a; Kotseridis and Baumes, 2000) and Rioja wines (Aznar *et al.*, 2001). However, as it has not yet been assayed in wine, its impact on aroma has not been verified. Furthermore, it is not certain that this compound contributes to varietal aroma, as it could be formed from furfural released by the staves of the barrels used to age these wines.

7.5.2 Precursors of Volatile Thiols Derived from Cysteine

Sauvignon Blanc musts, like those of many grape varieties with relatively simple aromas, are not highly odoriferous. The characteristic aroma of the grape variety appears during alcoholic fermentation.

Peynaud (1980) had a remarkable intuition of the existence of aroma precursors in Sauvignon Blanc must, which he described as follows:

'When you taste a thick-skinned, golden Sauvignon Blanc grape, you can detect its characteristic flavor, although it is not very intense. In the same way, freshly pressed juice is not highly odoriferous, and the initial flavor is quite discreet. Twenty or thirty seconds later, after you have swallowed it, an intense aromatic Sauvignon Blanc aftertaste suddenly appears in the rear nasal cavity. Fermentation brings out the primary aroma hidden in the fruit. Wine has more fruit aroma than grapes, etc. Fermentation reveals the aroma and releases the odoriferous substances from the grapes.'

The precursors to Sauvignon Blanc aroma compounds were identified at Bordeaux Faculty of Enology in the 1990s. First of all, Darriet *et al.* (1993) demonstrated that 4-MMP was released from an odorless must extract, either because of bioconversion by yeast during alcoholic fermentation or chemically, *in vitro*, because of the action of ascorbic acid. A precursor of 4-MMP had been shown to exist in grapes, but its chemical composition was still not known. The fact that it was impossible to release 4-MMP from its precursor by acid hydrolysis or using exogenous β-glucosidases meant that the precursor of this sulfur-based Sauvignon Blanc aroma compound could not be a glycoside.

Later in the decade, a β-lyase specific to S-cysteine conjugates was used to release 4-MMP, 4-MMPOH, and 3-MH from a non-volatile extract of Sauvignon Blanc aroma precursors, suggesting that these three thiols were present in grapes in cysteinylated form (Figure 7.10) (Tominaga et al., 1995; Tominaga et al., 1998c).

It has now been established by gas-phase chromatography/mass spectrometry of the precursors in trimethylsilylated form that 3-MH, 4-MMP, and 4-MMPOH are present in must in S-cysteine conjugate form (Figure 7.11): S-3-(hexan-1-ol)-L-cysteine, S-4-(4-methylpentan-2-one)-L-cysteine, and S-4-(4-methylpentan-2-ol)-L-cysteine (Tominaga et al., 1998c). This category of aroma precursors had not previously been identified in grapes or other fruit. Since then, S-3-(hexan-1-ol)-L-cysteine has been identified in passion fruit juice (Tominaga and Dubourdieu, 2000).

From an organoleptic standpoint, S-cysteine conjugates are responsible for the unusual sensation known as "Sauvignon Blanc aftertaste," perceived on tasting Sauvignon Blanc grapes or must. Enzymes in the mouth probably have a β-lyase activity capable of releasing volatile thiols from their cysteinylated precursors in just a few seconds.

Tominaga (1998) demonstrated experimentally that this reaction could be catalyzed by a protein extract from beef tongue mucous membrane.

The cysteinylated precursor content of Sauvignon Blanc must (Peyrot des Gachons et al., 2000) can be measured indirectly. The method consists of percolating the must through an immobilized β-lyase column that catalyzes an α,β-elimination reaction on the S-cysteine conjugates to release the corresponding volatile thiols, which are then assayed by gas-phase chromatography coupled with mass spectrometry. The precursor of 3-mercaptohexan-1-ol is present in larger quantities and has also been assayed directly in trimethylsilylated form by gas-phase chromatography coupled with mass spectrometry (Murat et al., 2001b).

These assay methods have made it possible to determine the location of cysteinylated thiol precursors in Sauvignon Blanc grapes (Figure 7.12). The 4-MMP and 4-MMPOH precursors are mainly located in the flesh (approximately 80%), while the skin and flesh contain equal amounts of 3-MH precursor (Peyrot des Gachons et al., 2002a). Similarly, a majority (60%) of the 3-mercaptohexan-1-ol is located in the skins of Cabernet Sauvignon and Merlot grapes (Murat et al., 2001b). This distribution of aroma precursors explains why skin contact enhances the aromatic potential of Sauvignon Blanc musts (and that of rosés made from

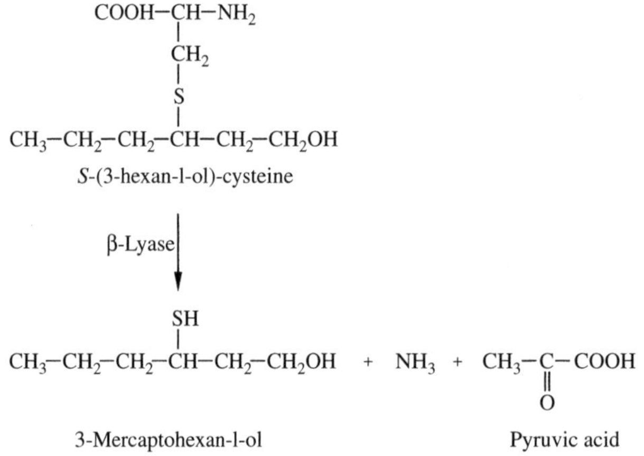

$$COOH-CH-NH_2$$
$$|$$
$$CH_2$$
$$|$$
$$S$$
$$|$$
$$CH_3-CH_2-CH_2-CH-CH_2-CH_2OH$$

S-(3-hexan-1-ol)-cysteine

↓ β-Lyase

$$SH$$
$$|$$
$$CH_3-CH_2-CH_2-CH-CH_2-CH_2OH \quad + \quad NH_3 \quad + \quad CH_3-\underset{\underset{O}{\|}}{C}-COOH$$

3-Mercaptohexan-1-ol   Pyruvic acid

Fig. 7.10. The cysteine conjugate form of 3-mercaptohexanol, revealed by a specific β-lyase

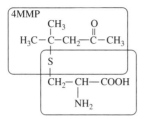
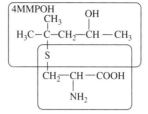

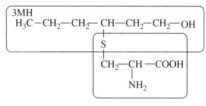

Fig. 7.11. S-cysteine conjugate precursors of volatile thiols identified in Sauvignon Blanc must

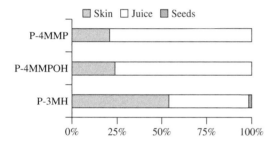

Fig. 7.12. Distribution of cysteinylated precursors of 4-MMP, 4-MMPOH, and 3-MH in Sauvignon Blanc grapes

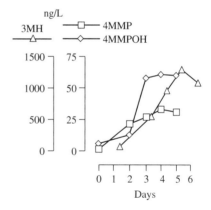

Fig. 7.13. Formation of volatile thiols from their cysteinylated precursors in Sauvignon Blanc must during alcoholic fermentation

Merlot and Cabernet Sauvignon), because of the increase in the 3-MH precursor content.

A clear increase in 4-MMP, 4-MMPOH, and 3-MH content is observed during alcoholic fermentation (Figure 7.13). The varietal aroma is amplified by the fermentation yeast (*S. cerevisiae*) metabolism because of the conversion of cysteinylated aroma precursors in the grapes. A3-MH is generally formed when acetic acid esterifies the 3-MH that has been released. The yeast genes and enzymes involved in these reactions have not yet been identified.

Much larger quantities of cysteinylated precursors of volatile thiols are present in the must than are accounted for by the aromas they generate in the wine (Peyrot des Gachons 2000; Murat et al., 2001b). The molar concentrations of 4-MMP, 4-MMPOH, and 3-MH formed in Sauvignon Blanc must (or a model medium supplemented with precursors) during alcoholic fermentation only account for approximately 10% of the precursors degraded (Figure 7.14).

Winemaking yeasts have variable aptitudes to reveal sulfur-based Sauvignon Blanc aromas. This

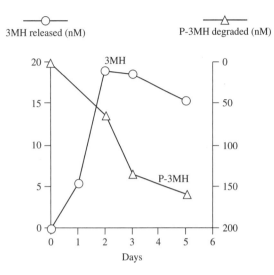

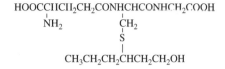

Fig. 7.15. S-3-(Hexan-1-ol)-Glutathion

Fig. 7.14. 3-MH release and degradation of its precursor (P-3-MH) in a Sauvignon Blanc must during alcoholic fermentation

property is discussed in the chapter concerning winemaking yeasts for use in white must (Volume 1, Section 13.7.2).

Finally, S-3-(hexan-ol)-glutathion (Figure 7.15) has recently been identified in Sauvignon Blanc must (Peyrot des Gachons *et al.*, 2002b). The presence of this compound, which may be considered a "pro-precursor", indicates that the S-3-(hexan-ol)-L-cysteine in grapes results from the catabolism of S-3-(hexan-ol)-glutathion, as shown in Figure 7.16. S-glutathion conjugates may be involved in detoxification processes in vines, as is the case in other plant and animal organisms. The contribution of S-3-(hexan-ol)-glutathion to the aroma potential of must and, hence, its significance in winemaking has not yet been determined. It is, however, known that the conversion rate of this compound into 3-MH during alcoholic fermentation in a model medium is 20–30 times lower than that of S-3-(hexan-ol)-L-cysteine. The possibility that S-3-(hexan-ol)-glutathion may be partially converted into S-3-(hexan-ol)-L-cysteine by grape enzymes during pre-fermentation operations cannot be excluded.

7.6 AROMAS OF AMERICAN VINE SPECIES

Methyl anthranilate was long considered to be solely responsible for the 'foxy' smell of *Vitis labrusca* and *Vitis rotundifolia* grapes (Power and Chesnut, 1921). It is now known that other compounds are involved in the aromas of these varieties (Figure 7.17).

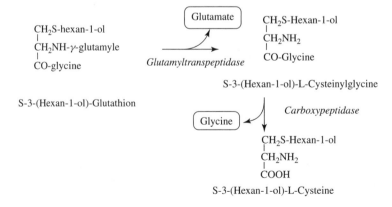

Fig. 7.16. Proposed pathway for the conversion of glutathionylated "pro-precursor" into the cysteinylated precursor of 3-MH

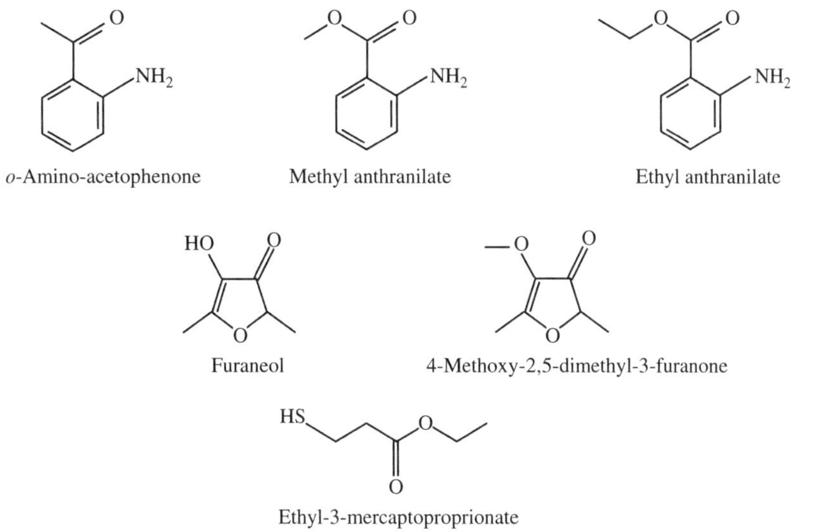

Varietal Aroma 223

o-Amino-acetophenone   Methyl anthranilate   Ethyl anthranilate

Furaneol   4-Methoxy-2,5-dimethyl-3-furanone

Ethyl-3-mercaptoprionate

Fig. 7.17. Various compounds identified in *Vitis labrusca* and *Vitis rotundifolia* grapes and wines

Low concentrations of ethyl-2-and 3-mercaptopropionate, mentioned earlier in this chapter, have a fruity aroma, whereas in larger quantities they have sulfurous smells. Acree *et al.* (1990) also demonstrated the presence of 2-aminoacetophenone in *Vitis labrusca*, as well as two furanones (4-hydroxy-2,5-dimethyl-3-furanone, commonly known as furaneol, and 4-methoxy-2,5-dimethyl-3-furanone) reminiscent of strawberries (Rapp *et al.*, 1980). Most of these compounds have also been identified in *Vitis vinifera* wines, but at lower concentrations (Guedes de Pinho, 1994; Moio and Etiévant 1995).

7.7 DEVELOPMENT OF GRAPE AROMAS DURING RIPENING AND THE IMPACT OF VINEYARD FACTORS

Free and bonded forms of terpenols accumulate in ripening grapes from the color change onwards. Some authors report a continuous accumulation of monoterpenes, even in overripe grapes (Wilson *et al.*, 1984; Park *et al.*, 1991). Others share the more widespread opinion that the free monoterpenes start to decrease before the maximum sugar level is reached (Marais, 1983; Günata, 1984). Park *et al.* (1991) suggested that vineyard conditions during ripening (especially temperature) may be partly responsible for the variations observed. The water supply to the vines may also be assumed to influence aroma development during ripening.

C_{13}-norisoprenoid derivatives develop according to a similar pattern. The carotenoid concentration decreases following color change. This correlates with increased concentrations of C_{13}-norisoprenoid derivatives (TDN, vitispirane, etc.), mainly in glycosylated form. These changes probably require the action of grape enzymes, initially in the oxidative degradation of carotenoids and later in glycosylation mechanisms (Razungles and Bayonove, 1996).

Exposure of the grapes to sunlight during ripening accelerates carotenoid breakdown and is accompanied by an increase in the glycosylated C_{13}-norisoprenoid derivative content. These phenomena have been observed in Riesling and Syrah grapes (Marais, 1993; Razungles and Bayonove, 1996). Thus, leaf thinning in the fruiting areas of Riesling vines leads to a greater concentration of glycosylated C_{13}-norisoprenoid derivatives, except in the case of C_{13}-norisoprenoid glucosides that produce β-damascenone (Marais *et al.*, 1992).

It has thus been demonstrated that the excessive hydrocarbon smells that sometimes develop as Riesling wines age are related to extremely high temperatures, especially during the grape ripening period. It has been demonstrated that, although hot climates are favorable for the accumulation of sugar, they are not necessarily best in terms of wine quality.

The highest concentrations of methoxypyrazines are found in unripe grapes, up to 100–200 ng/l in Sauvignon Blanc or Cabernet Sauvignon juice (Allen and Lacey, 1993; Roujou de Boubée, 2000). Concentrations gradually decrease during ripening. Australian Sauvignon Blanc and Cabernet Sauvignon from the coolest regions had the highest methoxypyrazine concentrations (Lacey *et al.*, 1991) (Figure 7.18). In Bordeaux, analyses of wines from the same estate, from the 1991 to 1995 vintages, also showed the effect of climate on methoxypyrazine concentration (Figure 7.19).

In the Bordeaux climate, soil has a decisive influence on the methoxypyrazine concentrations in Merlot, Cabernet Franc and Cabernet Sauvignon wines due to its effect on vegetative growth. Grapes grown on well-drained, gravel soils have the lowest concentrations. On limestone or clay–silt soils, Cabernet Sauvignon has a higher methoxypyrazine content, often expressed by a herbaceous character. In certain wine-producing regions, a 'green pepper' character, linked to the presence of pyrazines, is considered typical of Cabernet Sauvignon wines. In Bordeaux, a strong green pepper odor indicates lack of maturity in the grapes and is definitely considered a defect.

Under identical weather conditions, increasing the grapes' sun exposure during ripening reduced their methoxypyrazine content, probably due to the light sensitivity of these compounds (Heymann 1986; Maga, *et al.*, 1989).

Little observation has been made of the date of assays of the aroma precursors in this grape variety should make it possible to study its aromatic development in various soils and climates.

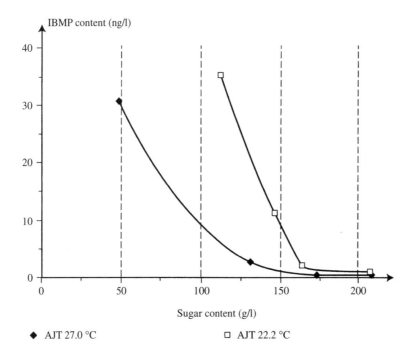

Fig. 7.18. Effect of temperature during ripening on the decrease in isobutylmethoxypyrazine (IBMP) concentrations in Sauvignon Blanc grapes in Australia (AJT = average January temperature) (Lacey *et al.*, 1991)

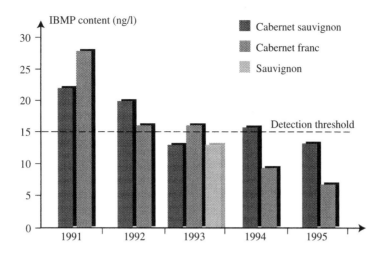

Fig. 7.19. Comparison of the IBMP concentrations of three wines made from different grape varieties grown on the same estate in different vintages (Roujou de Boubée, 2000)

In the Bordeaux vineyards, a combination of leaf thinning and removal of side-shoots in the fruiting zone is particularly effective in enhancing the aromatic maturity of Cabernet Sauvignon and Sauvignon Blanc grapes (Roujou de Boubée, 2000). It is important to trim the vines this way between fruit set and the time the bunches close, as, later, leaf thinning may boost the grapes' sugar content but may not always result in a sufficient decrease in IBMP (Table 7.10) content to avoid the formation of herbaceous aromas in the wine (Table 7.9).

Leaf thinning promotes the early photodegradation of IBMP by allowing more light to reach the grapes, as well as by reducing the main source of IBMP production: mature leaves. Roujou de Boubée (2000) clearly showed that the leaves contained IBMP at the time when the bunches closed up. In particular, mature leaves (the three or four leaves near the base of each shoot) had an IBMP content 8–10 times higher than that of young leaves and grapes. Transfer of IBMP from the leaves to the grape bunches has also been demonstrated experimentally. This compound is synthesized in the leaves prior to color change. It then migrates to the grapes, where it is stored, mainly in the skins.

The total IBMP content of the grapes can be considered to result from its synthesis in the leaves and transfer to the grapes, offset by its degradation due to exposure to light. To summarize, grapes will have a lower IBMP content if the

Table 7.9. Impact of leaf thinning dates (% difference compared to control) on certain compounds in Cabernet Sauvignon grapes at harvest time in Bordeaux in 1998

	Side-shoots removed and leaves thinned when the bunches closed	Side-shoots removed and leaves thinned after bunches closed and before color change	Side-shoots removed and leaves thinned after color change
Grape weight	−7.4	−4.4	+2
Total acidity	0	0	0
Reducing sugars	+8.5	+6.7	+3
IBMP	−68.4	−10.5	0

accumulation of this compound is restricted before color change, and its degradation from that point on is accelerated.

Until recently, it was not possible to study changes in the aromatic potential of non-Muscat grape varieties during ripening because of lack of knowledge of the chemistry of their aromas and aroma precursors. As a result, studies of the impact of soil, climate, and viticultural methods on the aromatic expression of these grape varieties were based solely on tasting wines, or the even less reliable method of tasting grapes. Winemakers knew, however, that it was difficult to make aromatic Sauvignon Blanc, or white wines in general, in excessively hot or dry climates and/or on soils with very low water reserves unless the vines were irrigated. Current knowledge of cysteinylated precursors of Sauvignon Blanc aromas has provided a scientific explanation for this empirical knowledge. Peyrot des Gachons (2000) showed that, as these grape varieties ripened in Bordeaux vineyards, their concentrations of 4-MMPOH precursors remained stable or increased while precursors of 4-MMP and 3-MH varied in a less regular way, with no marked tendency to accumulate or degrade. These variations in concentration depended on the S-conjugate, soil type, and climate. In the Bordeaux vineyards, on certain well-drained gravel soils with low water reserves, subject to early, severe water stress in dry summers, the cysteinylated precursor content of the grapes at harvest time was lower than that in grapes grown on limestone soils, which benefited from a better water supply via capillarity through the porous rock, thus ensuring that the vines were less severely stressed.

Choné (2001) analyzed the favorable impact of moderate water stress on the aromatic potential of Sauvignon Blanc grapes. One initial finding of this research was that stem potential (Ψ_T), measured in a "pressure chamber", provided an earlier indication of moderate water stress than basic leaf potential (Ψ_F) (Choné et al., 2001a). As described in Volume 1, Section 10.4.8, stem potential represents the sap pressure resulting from the difference between leaf transpiration and water absorption by the roots, whereas basic leaf potential indicates an equilibrium between vine and soil humidity at the end of the night.

Choné (2001) compared the aromatic potential of ripe grapes from Sauvignon Blanc vines subjected to two levels of water stress: unlimited water supply (deep soil) and a moderate water deficit obtained by covering the soil with a waterproof tarpaulin from early June until the harvest.

The grapes from Sauvignon Blanc vines subjected to moderate water stress had higher concentrations of cysteinylated precursors than those with an unlimited water supply (Table 7.10).

It is known that a moderate water deficit before color change leads to a significant increase in total phenolic content that has a negative impact on the aromatic stability of wine (Choné et al., 2001b). However, a moderate water deficit after color change leads to an increase in the concentrations of cysteinylated precursors without any significant variation in the must's total phenolic content (TPI). Results are not yet sufficient to define the precise period and intensity of the water deficit required to have a positive effect on the aromatic potential of Sauvignon Blanc. The most favorable moderate water deficits probably occur just after color change, provided that the weather is not excessively hot. It is conceivable that late ripening is more favorable in this respect than early ripening, when the grapes are picked in hot weather (late August or early September).

The vines' nitrogen supply also has a strong impact on the aromatic potential of Sauvignon Blanc, as demonstrated by Choné in the following experiment (2001).

The experiment was carried out on a plot of Sauvignon Blanc with relatively low vigor and a severe nitrogen deficiency that produced must with an available nitrogen content of 30 mg/l. Fertilizing part of the plot with mineral nitrogen (60 units), in the form of ammonium nitrate, at the beginning of fruit set led to a considerable increase in the nitrogen supply to the vines, resulting in a significantly higher nitrogen content being available in the must (Table 7.11). In this experiment, the water supply was not limited as the Ψ_T values, determined from July to August 2000, varied from -0.22 Mpa to -0.58 Mpa. An increase in the

Table 7.10. Impact of the vines' water supply on the composition of ripe Sauvignon Blanc grapes (SP: deep soil; SB: soil covered with a tarpaulin. The lines followed by a different letter had statistically different values.)

Water supply	Unrestricted (SP)	Restricted (SB)
Ψ_T (Mpa) on 29 July 2000 (color change)	−0.18 a	−0.70 b
Ψ_T on 10 August 2000	−0.22 a	−0.95 b
Ψ_T on 28 August 2000 (8 days before harvest)	−0.30 a	−1.10 b
Grape weight per vine (kg)	3.6 a	2.9 b
Primary leaf surface (m^2)	3.257 a	2.60 b
Secondary leaf surface (m^2)	3.804 a	1.63 b
Average weight per grape (g)	2.03 a	1.81 b
Reducing sugars (g/l)	178.4 a	210 b
Total acidity (g/l)	6.69 a	4.21 b
Malic acid (g/l)	4.95 a	2.44 b
Available nitrogen (g/l)	172 a	225 a
P-4-MMP (ng eq. 4-MMP/l)	1263 a	2548 b
P-4-MMPOH (ng eq. 4-MMPOH/l)	2226 a	2127 a
P-3-MH (ng eq. 3-MH/l)	7254 a	24 288 b
Phenolic compounds (TPI)	1.6 a	2.31 b

Table 7.11. Impact of the vines' nitrogen supply on vigor and grape composition. The lines followed by a different letter had statistically different values

	Nitrogen deficiency	Fertilized with nitrogen (60 units N)
Available nitrogen content in the must (mg/l)	29 a	174 b
Grape weight per vine (kg)	1.43 a	1.58 a
Primary leaf surface (m^2)	2.13 a	2.37 a
Secondary leaf surface (m^2)	0.40 a	1.44 b
Average weight per grape (g)	1.5 a	1.9 b
Reducing sugars (g/l)	202 a	199 a
Malic acid (g/l)	2.72 a	4.22 b
P-4-MMP (ng eq. 4-MMP/l)	405 a	715 b
P-4-MMPOH (ng eq. 4-MMPOH/l)	760 a	2059 b
P-3-MH (ng eq. 3-MH/l)	3358 a	14 812 b
Phenolic compounds (TPI)	0.28 a	0.21 b
Glutathion (mg/l)	17.9 a	120 b

nitrogen supply to the vine led to a significantly greater accumulation of cysteinylated precursors and glutathion in the grapes. Glutathion is a powerful reducing agent capable of protecting the varietal aroma of Sauvignon Blanc wines (Dubourdieu and Lavigne-Cruège, 2002). In Merlot and Cabernet Sauvignon (Choné et al., 2001b), nitrogen deficiencies also led to a reduction in grape size and an increase in the must's total polyphenol content. Consequently, an adequate nitrogen supply to the vine is indispensable for aromatic expression in Sauvignon Blanc, as it not only promotes synthesis of cysteinylated aroma precursors and glutathion but also restricts the production of phenolic compounds.

Finally, it has been reported that some vine sprays may have an unexpected impact on the varietal aromas of wines. In particular, owing to the reactivity of thiols with copper, the application of copper-based products on Sauvignon Blanc and Cabernet Sauvignon vines results in a significant decrease in the aroma of young wines made from these grape varieties (Hatzidimitriou et al., 1996; Darriet et al., 2001).

REFERENCES

Acree T.E., Lavin E.H., Nishida R. and Watanabe S. (1990) In *Flavour Science and Technology* (eds. Y. Bessière and A.F. Thomas). Wiley, New York, p. 49.

Allen M.S. and Lacey M.J. (1993) *Vit. Enol. Sci.*, 48, 211.

Allen M.S., Lacey M.S., Brow W.V. and Harris R.L.N. (1989) In *Actualités OEnologiques 89*, Compte-Rendus du 4th Symposium d'Œnologie de Bordeaux. Dunod Paris Ed.

Allen M.S., Lacey M.J. and Boyd S. (1994) *J. Agric. Food Chem.*, 42, 1734.

Allen M.S., Lacey M.J. and Boyd S.J. (1995a) *J. Agric Food Chem.*, 43, 769.

Allen M.S., Lacey M.J. and Boyd S.J. (1995b) In Australian Wine Industry Technical Conference, Adelaide, South Australia, 16–19 July 95.

Augustyn O.P.H., Rapp A. and Van Wyk C.J. (1982) *South Afric. J. Enol. Viticult.*, 3, (2), 53.

Austerweil G. (1946) *Ind. Parfum.*, 1, 195.

Ayran A.P., Wilson B., Strauss C.R. and Williams P.J. (1987) *Am. J. Enol. Viticult.*, 38, 182.

Aznar M., Lopez R., Cacho J.F. and Ferreira V. (2001) *J. Agric. Food Chem.*, 49, 2924–2929.

Bassols F. and Demole E. (1994) *J. Essent. Oil Res.*, 6, 481.

Baumes R., Cordonnier R., Nitz S. and Drawert F. (1986) *J. Sci. Food Agric.*, 37, 927.

Bayonove C. (1993) In *Les Acquisitions Récentes en Chromatographie du Vin*, (ed. B. Donèche). Lavoisier, Paris.

Bayonove C., Cordonnier R. and Ratier R. (1974) *C.R. Acad. Agric.*, 60, 1321.

Bayonove C., Cordonnier R. and Dubois P. (1975) *C.R. Acad. Sci. Paris*, 281, (D), 75.

Bayonove C., Günata Z. and Cordonnier R. (1984) *Bull. OIV*, 741.

Biron C., Cordonnier R., Glory O., Günata Z. and Sapis J.-C. (1988) *Conn. Vigne Vin*, 22, 125–134.

Bitteur S.-M., Baumes R., Bayonove C.-L., Versini G., Martin C.-A. and Dalla Serra A. (1990) *J. Agric. Food Chem.*, 38, 1210–1213.

Blanchard L. (1997) Mise au point et application d'une méthode de dosage du 3-mercapto-hexanol dans les vins rouges de Bordeaux. DEA OEnologie-Ampélologie. Université Victor Segalen Bordeaux II.

Blanchard L. (2000) *Recherche sur la contribution de certains thiols volatils à l'arôme des vins rouges étude. Etude de leur genèse et de leur stabilité*. Thèse de Doctorat, Université de Bordeaux 2.

Boidron J.-N. (1978) *Ann. Technol. Agric.*, 27 (1), 141.

Bouchilloux P., Darriet P. and Dubourdieu D. (1996) *J. Int. Sci. Vigne Vin*, 30, 23.

Bouchilloux P., Darriet Ph. and Dubourdieu D. (1998a) *Vitis*, 37(4), 177–180.

Bouchilloux P., Darriet Ph., Henry R., Lavigne Cruège V. and Dubourdieu D. (1998b) *J. Agric. Food Chem.*, 46, 3095–3099.

Buttery R.G., Seifert R.M., Guagagni D.G. and Ling L.C. (1969) *J. Agric. Food Chem.*, 17, 1322.

Calo A., Di Stefano R., Costacurta A. and Calo G. (1991) *Riv. Viticult. Enol.*, 3, 3.

Chatonnet P. and Dubourdieu D. (1997) Unpublished work.

Choné X. (2001) *Contribution à l'étude des terroirs de Bordeaux: Etude des déficits hydriques modérés, de l'alimentation en azote et de leurs effets sur le potentiel aromatique des raisins de Vitis vinifera L. cv. Sauvignon blanc*. Thèse de Doctorat, Université de Bordeaux II.

Choné X., Dubourdieu D. and Gaudillère J.P. (2001a) *Ann. Bot.*, 87, 477–483.

Choné X., Van Leeuwen C., Chery P. and Ribereau-Gayon P. (2001b) *S. Afr. J. Enol. Vitic.*, 22, 8–15.

Cordonnier R. (1956) *Ann. Techn. Agric.* 5, 75.

Cordonnier R. and Bayonove C. (1974) *C. R. Acad. Sci. Paris*, 278, 387.

Cordonnier R., Günata Z., Baumes R.L. and Bayonove C. (1989) *Conn. Vigne Vin.*, 3, 7.

Darriet Ph. (1993) L'arôme et les précurseurs d'arôme du Sauvignon. Thèse de Doctorat, Université de Bordeaux II.

Darriet Ph., Tominaga T., Demole E. and Dubourdieu D. (1993) *C.R. Acad. Sci. (Paris)*, série 3, 316 (11), 1332.

Darriet Ph., Tominaga T., Lavigne V., Boidron J.-N. and Dubourdieu D. (1995) *Flavour Fragrance J.*, 10, 395.

Darriet P., Bouchilloux P., Poupot C., Bugaret Y., Clerjeau M., Sauris P., Medina B. and Dubourdieu D. (2001) *Vitis*, 40, 93.

Demole E. and Berthet D. (1972) *Helv. Chim. Acta*, 54, 681.

Demole E., Enggist P., Säuberli U., Stoll M. and Kovats E. (1970) *Helv. Chim. Acta*, 53, 541.

Demole E., Enggist P. and Ohloff G. (1982) *Helv. Chim. Acta*, 65 (6), 1785.

Di Stefano R. and Maggiorotto G. (1993) In Comptes-Rendus du Symposium International on *Connaissance Aromatique des Cépages et Qualité des Vins*, Montpellier, 9–10 February 1993, Rev. Fr. Œnologie Ed.

Dubourdieu D. and Lavigne Cruège V. (2002) *13ème Symposium International d'œnologie*, International Association of Enologie Management and Wine Marketing, ENSA de Montpellier.

Du Plessis C.S. and Augustyn O.P.H. (1981) *S. Afr. J. Enol. Vitic.*, 2, 2.

Engel K.H. and Tressel R. (1991) *J. Agric. Food Chem.*, 39, 2249.

Enzel C.R. (1985) *Pure Appl. Chem.*, 57 (5), 693.

Etiévant P.-X., Issanchou S.N. and Bayonove C.-L. (1983) *J. Sci. Food Agric.*, 34, 497.

Falqué-Lopez E., Darriet Ph., Fernandez-Gomez E. and Dubourdieu D. (1994) In Comptes-Rendus 1st Congrès International de Vitiviniculture Atlantique, La Toja, Espagne, 16–19 May 1994.
Grossmann M. and Rapp A. (1988) *Deutsche Lebensmitt. Rund.*, 84, 35.
Grossmann M., Sukran I. and Rapp A. (1990) *Deutsche Lebensmitt. Rund.*, 83, 7.
Guedes de Pinho P. (1994) Caractérisation des vins de la région de Vinos Verdes au Portugal. Reconnaissance chimique et sensorielle des vins de cépages blancs non *Vitis vinifera*. Thèse de Doctorat, Université de Bordeaux II.
Günata Z. (1984) Recherches sur la fraction liée de nature glycosidique de l'arôme du raisin: importance des terpénylglycosides, action des glycosidases. Thèse Docteur-Ingénieur, Université. Sciences et Techniques du Languedoc, Montpellier.
Günata Z. (1993) *Progress in Flavour Precursor Studies* (eds. P. Schreier and P. Winterhalter), Würzburg, October 1992.
Günata Z., Bitteur S., Brillouet J.-M., Bayonove C. and Cordonnier R. (1988) *Carbohyd. Res.*, 184, 139.
Günata Z., Biron C., Sapis J.-C. and Bayonove C. (1989) *Vitis*, 28, 191.
Günata Z., Dugelay I., Sapis J.-C., Baumes R. and Bayonove C. (1990) *J. Int. Sci. Vigne Vin*, 24, 133.
Guth H. (1997) *J. Agric. Food Chem.*, 45, 3022.
Guth H. (1997) *J. Agric. Food Chem.*, 45, 3027–3032.
Harris R.L.N., Lacey M.J., Brown W.V. and Allen M.S. (1987) *Vitis*, 26, 201.
Hatzidimitriou E., Bouchilloux P., Darriet P., Clerjeau M., Bugaret Y., Poupot C. and Dubourdieu D. (1996) *J. Int. Sci. Vigne Vin*, 30 (3), 133.
Heymann H., Noble A.C. and Boulton R.B. (1986) *J. Agric. Food Chem.*, 34, 268.
Idstein H. and Schreier P. (1985) *J. Agric. Food Chem.* 33, 138.
Kolor M.G. (1983) *J. Agric. Food Chem.*, 31 (5), 1125.
Kotseridis Y and Baumes R. (2000) *J. Agric. Food Chem.*, 48, 400–406.
Kotseridis Y., Anocibar Beloqui A., Bayonove C., Baumes R.L. and Bertrand A. (1999) *J. Int. Sci. Vigne Vin.*, 1999, 33, 19–23.
Lacey M.J., Allen M.S., Harris R.L.N. and Brown W.V. (1991) *Am. J. Enol. Viticult.*, 42 (2), 103.
Maga J.A. (1989) In *Flavors and Off-flavors* (ed. Charalambous). Elsevier, Amsterdam, p. 61.
Marais J.S. (1983) *S. Afric. J. Enol. Viticult.*, 4 (2), 49.
Marais J.S. (1993) In Comptes-Rendus du Symposium International on *Connaissance aromatique des Cépages et Qualité des Vins*, Montpellier, 9–10 February 1993, *Rev. Fr. Œnologie Ed.*
Marais J.S., Van Wyck C.J. and Rapp A. (1992) *S. Afric. J. Enol. Viticult.*, 13, 33.
Masneuf I. (1996) Recherches sur l'identification génétiques des levures de vinifications. Applications oenologiques. Thèse Doctorat, Université Victor Ségalen Bordeaux II.
Moio L. and Etiévant P.-X. (1995) *Am. J. Enol. Viticult.*, 46 (3), 392.
Murat M.L., Tominaga T. and Dubourdieu D. (2001a) *J. Int. Sci. Vigne Vin.*, 35, 99–105.
Murat M.L., Tominaga T. and Dubourdieu D. (2001b) *J. Agric. Food Chem.*, 49, 5412–5417.
Murat M.L. Tominaga T., Saucier C., Glories Y. and Dubourdieu D. (2003) *Am. J. Enol. Vitic.*, 54, 135–138.
Murray K.E., Shipton J. and Whitfield F.B. (1970) *Chem. Ind.*, 897.
Park S.K., Morrison J.C., Adams D.O. and Noble A.C. (1991) *J. Agric. Food Chem.*, 39, 514.
Peynaud E. (1980) *Le Goût du Vin*. Dunod, Paris.
Peyrot des Gachons C. (2000) Recherches sur le potentiel aromatique des raisins de *Vitis vinifera* L.cv. sauvignon blanc. Thèse de Doctorat, Université de Bordeaux 2.
Peyrot des Gachons C., Tominaga T. and Dubourdieu D. (2000) *J. Agric. Food Chem.*, 48, 3387–3391.
Peyrot des Gachons C., Tominaga T. and Dubourdieu D. (2002a) *Am. J. Enol. Vitic.*, 53(2), 144–146.
Peyrot des Gachons C., Tominaga T. and Dubourdieu D. (2002b) *J. Agric. Food Chem.*, 50, 4076–4079.
Power F.B. and Chesnut M.K. (1921) *J. Am. Chem. Soc.*, 43, 1741.
Puglisi C., Elsey G., Prager R., Skouroumounis G. and Sefton A. (2001) *Tetrahedron Lett.*, 42, 6937.
Rapp A. (1987) In *Frontiers of Flavor* (ed. Charalambous). Elsevier, Amsterdam, p. 799.
Rapp A. and Pretorius P.J. (1990) In ≪ *Development in Food Science* ≫ 24, *Flavors and off flavors 89*. Elsevier Sciences Publishers B.V, Amsterdam, 1–21.
Rapp A., Knipser W., Engel L., Ullemeyer H. and Heimann W. (1980) *Vitis*, 19 (1), 13.
Razungles A. and Bayonove C. (1996) La Viticulture à l'aube du 3rd millénaire (eds. J. Bouard and G. Guimberteau). *J. Int. Sci. Vigne Vin*, hors série, 85.
Razungles A., Günata Z., Pinatel S., Baumes R. and Bayonove C. (1993) *Sciences des Aliments*, 13, 39.
Ribéreau-Gayon P., Boidron J.-N. and Terrier A. (1975) *J. Agric. Food Chem.*, 23 (6), 1042.
Rigaud J., Etiévant P., Henry R. and Latrasse A. (1986) *Sciences des. Aliments*, 6, 213.
Roujou de Boubée D. (1996) Optimisation du dosage de la 2-méthoxy-3-isobutylpyrazine dans les vins de Bordeaux. DEA Œnologie-Ampélologie, Université de Bordeaux II.
Roujou de Boubée D. (2000) Recherches sur le 2-méthoxy-3-isobutylpyrazine dans les raisins et dans les vins. Approches analytique, biologique et agronomique. Thèse de doctorat, Université Victor Ségalen Bordeaux II.
Roujou de Boubée D., Cumsille A.M., Pon M. and Dubourdieu D. (2002) *Am. J. Enol. Vitic.*, 53, 1–5.

Schneider R. (2001) *Contribution à la connaissance de l'arôme et du potentiel aromatique des vins du Melon B. (Vitis vinifera L.) et des vins de Muscadet*. Thèse de Doctorat, Université de Montpellier 2.

Schreier P., Drawert F. and Junker A. (1976) *J. Agric. Food Chem.*, 24 (2), 331.

Sefton M.A., Skouroumounis G.K., Massy-Westropp R.A. and Williams P.J. (1989) *Aust. J. Chem.*, 42, 2071.

Sefton M.A., Francis I.L. and Williams P.J. (1993) *Am. J. Enol. Viticult.*, 44, 359.

Sefton M.A., Francis I.L. and Williams P.J. (1994) *J. Food Sci.*, 59, 142.

Simpson R.F. (1978) *Chem. Ind.*, 1, 37.

Simpson R.F., Strauss C.R. and Williams P.J. (1977) *Chem. Ind.*, 15 663.

Skouroumounis G.K. and Winterhalter P. (1994) *J. Agric. Food Chem.*, 42, 1068.

Skouroumounis G.K., Massy-Westropp R.A., Sefton M.A. and Williams P.J. (1992) *Tetrahed. Lett.*, 33 (24), 3533.

Stahl-Biskup E. (1987) *Flavour and Fragrance J.*, 2, 75.

Strauss C.R., Wilson B., Gooley P.R. and Williams P.J. (1986) In *Biogeneration of aromas, Parliament* (ed. Croteau) American Chemical Society, Washington DC, p. 222.

Strauss C.R., Wilson B. and Williams P.J. (1988) *J. Agric. Food Chem.*, 36, 569.

Tominaga T. (1998) Recherches sur l'arôme variétal des vins de *Vitis vinifera* L.cv. sauvignon blanc et sa genèse à partir de précurseurs inodores du raisin. Thèse de Doctorat, Université de Bordeaux 2.

Tominaga T. and Dubourdieu D. (1997) *Flavour and Fragrance J.*, 12 (6), 373–376.

Tominaga T. and Dubourdieu D. (2000) *J. Agric. Food Chem.*, 48, 2874–2876.

Tominaga T., Masneuf I. and Dubourdieu D. (1995) *J. Int. Sci. Vigne Vin*, 29 (4), 227–232.

Tominaga T., Darriet P. and Dubourdieu D. (1996) *Vitis*, 35 (4), 207–210.

Tominaga T., Guimberteau G. and Dubourdieu D. (2003) *J. Agric. Food Chem.*, 51, 1373–1376.

Tominaga T., Furrer A., Henry R. and Dubourdieu D. (1998a) *Flavour and Fragrance J.* 13 (3), 159–162.

Tominaga T., Murat M.-L. and Dubourdieu D. (1998b) *J. Agric. Food Chem.*, 46, 1044–1048.

Tominaga T., Peyrot des Gachons C. and Dubourdieu D. (1998c), *J. Agric. Food Chem.*, 46, 5215–5219.

Tominaga T., Baltenweck-Guyot R., Peyrot des Gachons C. and Dubourdieu D. (2000) *Am. J. Enol. Vitic.*, 51, 178–181.

Versini G., Rapp A. and Dalla-Serra A. (1992) In Comptes-Rendus du Symposium International on *Flavour Precursors. Analysis–Generation–Biotechnology*, (eds. P. Schreier and P. Winterhalter) 30 September–2 October 1992. Würzburg, Allemagne, p. 243.

Voirin S., Baumes R., Bitteur S., Günata Z. and Bayonove C. (1990) *J. Agric. Food Chem.*, 38, 1373.

Williams P.J., Strauss C.R., Wilson B. and Massy-Westropp R.A. (1982) *Phytochemistry*, 21 (8), 2013.

Wilson B., Strauss C.R. and Williams P.J. (1984) *J. Agric. Food Chem.*, 32, 919.

Winter M., Velluz A., Furrer A. and Winterhagen W. (1990) *Lebensm. Wiss. u. Technol.*, 23, 94.

Winterhalter P. (1993) In *Flavour Precursors— Thermal and Enzymatic Conversion*, (eds. R. Teranishi, G. Takeoka and M. Güntert), *ACS series* 490, p. 98.

PART TWO

Stabilization and Treatments of Wine

8

Chemical Nature, Origins and Consequences of the Main Organoleptic Defects

8.1 Introduction	233
8.2 Oxidative defects	235
8.3 Effect of various forms of bacterial spoilage	238
8.4 Microbiological origin and properties of volatile phenols	242
8.5 Cork taint	256
8.6 Sulfur derivatives and reduction odors	261
8.7 Premature aging of white wine aroma	274
8.8 Organoleptic defects associated with grapes affected by various types of rot	277
8.9 Miscellaneous defects	279

8.1 INTRODUCTION

Progress in enology has led to considerable improvements in wine quality and made it possible to bring out the individual character of wine grapes, which is in turn related to the environmental conditions specific to each vineyard. Major defects that were once common have now practically disappeared. Winemakers the world over now know how to produce high-quality, healthy, clean wines, whose flavor fully expresses the quality of the grapes.

At the same time, tasting criteria have become increasingly demanding, which is also a good thing. Greater attention is now paid to minor departures from organoleptic perfection that diminish quality slightly without compromising it altogether.

These defects stand out even more in quality wines, as their highly-refined organoleptic characteristics are affected by the slightest problem.

Before they are marked enough to be perceived and identified in tasting, these defects detract from the finesse of normal fruity aromas. They then lead to a certain loss of character often described as heaviness. At higher concentrations, the defects themselves are clearly identifiable. It would be wrong to consider that these olfactory defects add complexity to a wine. One of the main objectives in enology over the next few years will certainly be to find out how to avoid these problems and their consequences.

This chapter deals exclusively with organoleptic defects that develop during aging. Some are of chemical origin (oxidation, reduction and contact with certain materials), but microbiological processes are often involved, even in the development of 'cork taint' and spoilage due to sulfur derivatives. The various problems caused by anaerobic lactic bacteria are described, as well as the role of acetic bacteria. The mycodermic yeasts responsible for *flor* are included in Section 8.3.

Fermentation yeasts are also likely to be responsible for problems of microbial origin. In view of its importance, the production of ethyl-phenol from cinnamic acid by *Brettanomyces* is described in some detail (Section 8.4.5). This same chapter also explores the formation of the sulfur derivatives responsible for reduction odors (Section 8.6.2), as well as the production of vinyl-phenols by *Saccharomyces cerevisiae* (Section 8.4.2). This yeast also represents a grave danger during the aging of sweet wines containing fermentable sugars that run the risk of refermenting. Other species, such as *Saccharomycodes ludwigii*, raise major problems due to their resistance to ethanol and sulfur dioxide. *Zygosaccharomyces* contamination has also been reported (Boulton *et al.*, 1995) in concentrated must stored under unsuitable temperature conditions with insufficient sulfuring. Contamination is transmitted when this must is used in the preparation of sweet wines. However, some fermentation yeast strains are capable of developing in dry wines, even in the bottle. These yeasts use up traces of residual carbohydrates, causing a slight turbidity that settles out as a deposit.

Microbial contamination is prevented by sulfur dioxide and certain other processes or adjuvants (Volume 1, Chapters 8 and 9), as well as sterilizing and 'low microbe' filtration (Section 11.3.4). Germs may also be destroyed by heating (Section 12.2.3).

It is also important to avoid, or at least to minimize, contamination of wine by microorganisms responsible for disease, mainly transmitted by contact with cellar equipment. These microorganisms may also be transmitted by wine brought in from outside the cellar that has not been properly tested. Poorly maintained barrels are another source of germs.

It is therefore essential to keep winery installations clean. This concept is well known in principle, but not always sufficiently applied in practice. There is a particular danger during fermentation, as microbes develop best in must, a sweet medium containing little alcohol. Lack of proper hygiene results in the appearance of contaminant populations that invade floors and walls, as well as the insides of pipes, hoses and containers. These populations multiply, especially in areas where cleaning access is difficult. In some, fortunately unusual, situations (underground pipes, vats heavily encrusted with tartrates, etc.) sterilizing products could not even reach the contaminated site. In some rare instances, wines have even been spoiled before the end of fermentation. This type of serious problem can only be solved by completely renovating the installation. A similar situation can occur when wines are contaminated by chloroanisoles produced by the microbial breakdown of chlorophenols used to treat roof timbers. In such instances, the entire roof may need to be replaced (Section 8.5.2).

Modern stainless-steel equipment is easier to maintain than old wooden or concrete vats. Furthermore, it is frequently equipped with efficient cleaning systems. Proper hygiene of barrels used to age wines certainly remains a constant concern, especially when they are stored empty (Section 13.6.2).

It is impossible to eliminate contamination completely by cleaning winery facilities, which can never be as absolutely sterile as a pharmaceutical laboratory. However, every effort should be made to keep the microbial populations present in cellars to a minimum. Regular cleaning of the premises is necessary, but it is vital for equipment that comes into contact with wine or must. It should always be borne in mind that the development of harmful microorganisms may be facilitated in a medium partially diluted by rinse water (Boulton *et al.*, 1995), which has a lower sugar content and higher pH than wine. Effective cleaning is absolutely necessary. During the fermentation period, thorough cleaning should be carried out every morning and evening. Cleaning should start by rinsing with water, followed by regular use of detergents (alkali, polyphosphates, etc.) and disinfecting agents (iodine or chlorine derivatives and quaternary ammonium). Cellar and vat design is extremely important to ensure the full effectiveness of cleaning procedures. Any areas inaccessible to disinfecting agents, such as porous surfaces and those covered by layers of tartrate, must be eliminated. Dangerous microorganisms may be very difficult to destroy, as they are often coated with a polysaccharide film that protects them from disinfecting agents.

8.2 OXIDATIVE DEFECTS

8.2.1 Role of Oxidation

The concept of oxidative defects is quite subjective, as there are many wines where an oxidative character (*rancio*) is considered desirable. Wines aged under a yeast bloom are a classic example, e.g. *Vin Jaune* from the Jura (Volume 1, Section 14.5). The yeast acts as an oxidation–reduction buffer and prevents excessive oxidation. *Vin Jaune* is well known for its highly oxidized character and contains large concentrations of free and combined ethanal (Figure 8.1) (Etievant, 1979). The same is true of Sherry.

This type of oxidized impression is not, however, considered acceptable in other types of wine, where freshness is an essential quality. This chapter does not deal with oxidasic casse, which causes a very fast enzymic oxidation of many components in must and wine (Volume 1,

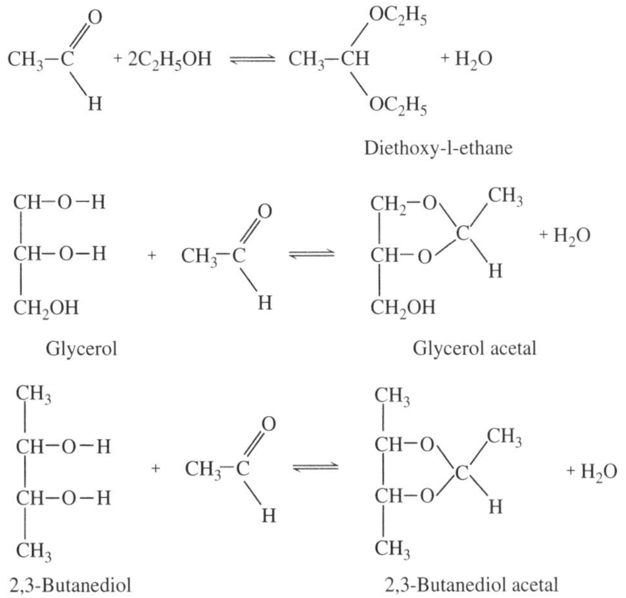

Fig. 8.1. Ethanal in the combined state with aliphatic and cyclic acetals

Section 11.6.2). In particular, color is affected and red wines take on a brownish hue. Modern winemaking techniques are capable of totally eradicating these enzymes. Since this defect should no longer occur in wine, the only remaining problem is protecting unfermented must.

All wines, however, may be affected by chemical oxidation due to dissolved oxygen. The role of oxygen (Section 13.7) has been interpreted in different ways at various times. For years, oxygen was regarded as the 'enemy of wine'. Pasteur was probably the first to consider that oxygen was indispensable for the development of red wine. Later, it was generally accepted that Pasteur's theories were excessive and that his conclusions should be tempered. In relatively recent times, it was even asserted that protection from any type of oxidation was an essential element of rational enology, but the prevailing attitude today is less extreme.

It is certainly necessary to protect must during fermentation, especially in the case of botrytized grapes. However, new techniques for total stabilization involve carefully controlled, high-level oxidation (hyperoxygenation).

An oxidative phase during aging is indispensable to ensure normal color development, particularly in red wines (Ribéreau-Gayon *et al.*, 1983). Excessive oxidation in any type of wine, however, results in an organoleptic defect known as 'flatness' (Section 8.2.3).

8.2.2 Oxidation 'Buffer' Capacity

The sensory impression of oxidation or reduction in wine indicates abnormal development. This is linked to the presence of an oxidizing (oxygen) or reducing agent, and is also related to the buffer capacity that protects wines to varying degrees from sharp variations in their oxidation–reduction potential.

The concept of buffer capacity is related to the oxidation–reduction potential (Section 13.2) that links oxidized and reduced forms in the medium:

$$E_H(V) = E_0 + m\frac{RT}{nF}\log\frac{[Ox]}{[Red]} - 0.06\frac{m}{n}\text{pH}$$

where E_0 is a constant known as the normal redox potential, R the perfect gas constant, T absolute temperature, F the Faraday constant (96 500 coulombs), n the number of charges exchanged and m the number of protons corresponding to the dissociation of the reduced form as an acid. The oxidation–reduction potential of an aerated wine is on the order of 400–450 mV. When it has been stored for a long time in the absence of air, the potential drops to a minimum of approximately 200–250 mV. The expression representing this potential involves the logarithm of the ratio between oxidized and reduced forms. It is clear that if there are high absolute concentrations of both the oxidized and reduced forms, variations in both of these forms will produce a limited variation in their ratio and, therefore, in the E_H potential. Wines with these properties have a high buffer capacity and are relatively well protected from oxidative defects. This is typically the situation in red wines with a high phenol content and good aging potential. As white wines are made under conditions that avoid releasing phenols, they have a low phenol content and, consequently, are considered to be much more vulnerable to oxidation. However, certain white wines with good aging potential have proved to be particularly resistant to oxidation. The types of reducing substances that protect these wines have not yet been identified, but phenols are unlikely to be the only molecules involved.

Independently of its disinfectant properties, sulfur dioxide is widely used to protect wines from oxidation (Volume 1, Section 8.7.2). It thus contributes to the oxidation–reduction buffer capacity and prevents an increase in potential that would otherwise occur when oxygen is dissolved. Due to their structure, white wines require a higher dose of SO_2 than red wines to ensure effective protection.

It is well established that reactions involving oxygen dissolved during handling operations carried out in contact with air are generally slow. These reactions are catalyzed by iron and copper ions. Sulfur dioxide acts as an irreversible antioxidant. It has been shown (Ribéreau-Gayon *et al.*, 1976) that a free SO_2 concentration of around 100 mg/l is necessary to provide full protection, i.e. all of the dissolved oxygen reacts with the SO_2. In practice, at doses on the order of 30 mg/l of SO_2 only half of the oxygen oxidizes the SO_2, while the rest oxidizes the wine's most oxidizable components. Reactions necessary for wine to develop may take place normally (oxidation of the phenols), without excessive oxidation producing any undesirable modifications.

In any event, SO_2 provides only limited protection. Although it is effective in protecting from controlled oxidation, SO_2 may be insufficient if there is a sudden influx of oxygen (e.g. during bottling) and may be unable to prevent flatness.

8.2.3 Flatness

While the development of bouquet in bottled wines is linked to reduction phenomena, flatness, on the contrary, corresponds to the appearance of oxidizing substances in aerated wines. Both of these transformations are reversible. Oxidized character may be desirable in some wines (Madeira, *rancio*, etc.) (Section 13.5.3), whereas in other wines it is considered a defect.

Flatness involves several transformations (Ribéreau-Gayon *et al.*, 1976). Initially, a few mg/l of oxygen combined with wine causes the disappearance or modification of certain odors. The wine also develops the freshly cut apple smell of ethanal. However, this combined oxygen is also responsible for the development of a bitter, acrid taste. The rate at which a wine becomes flat depends on temperature. The same reactions may take several days in winter but only a few hours in summer. When a flat wine is kept in an airtight container, this organoleptic defect disappears more rapidly at higher temperatures.

The presence of free ethanal and its derivatives is an essential aspect of flatness in wine (Figure 8.1). This aldehyde may be produced by direct oxidation of ethanol in the presence of a catalyst (Fe^{3+}, Cu^{2+}) or by mycodermic yeast (Section 8.3.4). It may also result from the dissociation of sulfur dioxide combined with ethanal, to restore the balance following the oxidation of free SO_2. The purpose of burning sulfur in barrels before they are filled with wine is precisely to combine the ethanal formed by oxidation of ethanol and that released by oxidation of the SO_2 in the wine. Ethanal is not always the only substance responsible for a wine's flat character. Indeed, it has been observed (Ribéreau-Gayon *et al.*, 1976) that a higher amount of SO_2 had to be added to a highly oxidized red wine than the dose theoretically required to combine with all the ethanal present before the flat odor was completely eliminated.

Even moderate contact between wine and air dissolves very small quantities of oxygen (on the order of mg/l), which are sufficient to produce the first signs of flatness within a few hours. The main operations likely to be responsible for this defect are racking, bottling and taking samples. In fact, these procedures are not the direct cause, but they almost inevitably involve aeration. Violent agitation of the wine or transferring it through pipes does not cause any organoleptic changes, provided the wine does not come into contact with air.

Oxidation caused by bottling is considered normal. This produces a flat character of varying intensity known as 'bottle sickness'. Although red wines generally have a high phenol concentration, they also have less added sulfur dioxide, so they are more sensitive to bottle sickness than white wines. This phenomenon is, however, widespread in Champagne shortly after the wine has been corked following disgorging. These wines must be stored in the cellar for a few months before shipment so that reduction phenomena can attenuate the flat character.

Bottle sickness may be minimized by increasing the free SO_2 level. However, in view of the sudden influx of oxygen during bottling, the doses required for complete protection may be unacceptably high. Ascorbic acid enhances the effect of SO_2 (Volume 1, Section 9.5.4) as it eliminates oxygen instantaneously.

The effects of aeration during bottling may be minimized by flushing oxygen out of the empty bottles with a low-pressure jet of inert gas and ensuring that the filler nozzle outlet is at the bottom of the bottle.

Other frequent examples of flatness occur in samples taken under poor conditions or from wines that are insufficiently sulfured. This has a strong negative impact on flavor and may compromise a wine's saleability.

Finally, another defect well known to winemakers is the premature aging of bottled white wines. Although this has been associated with oxidative development, it has not been fully explained by enologists. This defect can affect any type of white wine, dry or sweet, still or sparkling, and whatever the grape variety or origin. It leads to the early disappearance of varietal aromas, the absence of reduction bouquet and, above all, the appearance of a characteristic, heavy smell, reminiscent of rancid beeswax, stale honey and, in extreme cases, naphthalene. Neither the compounds responsible for these odors, nor the mechanisms that produce them, nor the means of preventing this phenomenon, have yet been elucidated.

8.3 EFFECT OF VARIOUS FORMS OF BACTERIAL SPOILAGE

8.3.1 Formation of Volatile Acidity by Bacteria

Excessive amounts of acetic acid in wine are due to the action of anaerobic lactic bacteria or aerobic acetic bacteria. Together with other molecules, this acid plays a major part in organoleptic defects of bacterial origin. On the other hand, mycodermic (Section 8.3.4) and *Brettanomyces* yeasts (Section 8.4.6) cause defects that do not involve accumulations of this acid.

Acetic acid is the main component of volatile acidity (Section 1.3.2), as defined in winemaking. Pasteur was the first to state that an analysis of volatile acidity could be used to assess the spoilage of a wine.

Yeast produces small amounts of acetic acid at the beginning of alcoholic fermentation. The level increases to a maximum and then starts decreasing. Malolactic fermentation is always accompanied by a slight increase, due to the breakdown of citric acid. Wine naturally has a volatile acidity of 0.3–0.4 g/l, expressed in H_2SO_4 (or 0.36–0.48 g/l, expressed in acetic acid). This value tends to increase slightly during aging. Higher values indicate bacterial activity. In view of the impact of total acidity on quality, all wine-producing countries have legislation setting an upper limit for volatile acidity. This value has been regularly reduced over the years, as progress in enology has made it possible to avoid bacterial problems. Current values in the EEC are rather high: 18 meq/l (0.88 g/l of H_2SO_4 or 1.07 g/l of acetic acid) for white and rosé wines and 20 meq/l (0.98 g/l of H_2SO_4 or 1.20 g/l of acetic acid) for red wines. There are also special exceptions for wines subject to a long period of barrel aging, as well as sweet wines made from grapes affected by noble rot (Volume 1, Section 14.2.5).

These exceptions may be linked to cases of difficult alcoholic fermentation, leading to abnormally high concentrations of acetic acid produced by yeast. It is, however, possible to detect by analysis whether the acetic acid in a wine was produced exclusively during fermentation, without the involvement of bacteria (Volume 1, Section 14.2.3). Lactic bacterial activity leads to concentrations higher than 200 mg/l of isomers of lactic acid and acetic bacteria produce ethyl acetate levels above 160 mg/l.

Volatile acidity is not easily detectable on the palate in normal wine if the concentration is below 0.60 g/l expressed in H_2SO_4 (0.72 g/l acetic acid), and probably has no effect on flavor. Above this value, the smell becomes acid and the flavor deteriorates, becoming harsh and bitter on the finish. It is incorrect to suggest that a certain level of volatile acidity may be useful in enhancing aromatic complexity. A wine may still seem good despite slightly high volatile acidity, but it would be even better without it. Volatile acidity never has a positive effect on a wine's organoleptic characteristics.

A physical method has been suggested for eliminating excess volatile acidity (Section 12.4.2). This involves treating the wine by reverse osmosis and then eliminating the acetic acid by passing it through an anion exchanger. However, current winemaking techniques should make it possible to avoid excessive volatile acidity, consequently obviating the use of this highly debatable process.

8.3.2 Spoilage Caused by Lactic Bacteria

Lactic bacteria from the genera *Leuconostoc* and *Lactobacillus* are likely to develop in wine, in spite of its high alcohol concentration and low pH. When there is no more sugar left in the wine, the most easily biodegradable molecule is malic acid. Consequently, malolactic fermentation is the first sign that lactic bacteria are developing. Malolactic fermentation improves the quality of red wines and certain white wines, producing small quantities of volatile acidity, mainly due to the breakdown of citric acid.

The same lactic bacteria, however, are likely to break down sugars. The consequences may be serious, especially if the wine has a high sugar concentration. The most common situation occurs when alcoholic fermentation stops, leaving the sweet medium open to lactic bacteria. For this reason, winemakers take great care to avoid stuck fermentations, although the bacteria may also take over just before the end of fermentation if it has slowed down. The initial result of this bacterial contamination is malolactic fermentation, but lactic spoilage may follow. This situation must be avoided, especially as the development of these bacteria may prevent the completion of alcoholic fermentation.

In the past, red wines were often affected by lactic spoilage, which had very serious consequences. In the Bordeaux area, it was particularly dangerous, as it usually occurred in hot years when the grapes were very ripe, i.e. good vintages. Volatile acidity of 1.0 g/l of H_2SO_4 (1.2 g/l of acetic acid) and even more was observed at the end of fermentation. Nowadays, it is extremely unusual to find as much as 0.60 g/l of H_2SO_4 (or 0.72 g/l of acetic acid), except in press wine. This value, however, already corresponds to unacceptable spoilage.

Besides acetic acid, lactic spoilage produces lactic acid and various secondary compounds that contribute to various olfactory defects. Some bacteria convert fructose into mannitol, explaining why this phenomenon used to be known as mannitic fermentation.

Even when alcoholic fermentation has been completed, wine always contains small quantities of residual sugar, on the order of 1–2 g/l. This consists of traces of glucose and fructose, as well as a few hundred mg/l of pentoses (xylose and arabinose) that cannot be fermented by the yeast, although they are broken down by lactic bacteria. The breakdown of pentoses is observed during malolactic fermentation. The consequence is a small but regular increase in volatile acidity in barrel-aged wines. The wine becomes more acidic, dryer and thinner. When wine is aged in the barrel under normal conditions for over a year, a regular increase in volatile acidity of 0.3 g/l of H_2SO_4 (0.36 g/l of acetic acid), up to a total of 0.50 or even 0.60 g/l of H_2SO_4 (0.60–0.72 g/l of acetic acid), is frequently observed. This is not really spoilage, as the bouquet is not affected, but the wine tends to dry out, losing its softness and fullness. Acetic bacteria (Section 8.3.3) are now considered to be involved in these slight increases in volatile acidity. This problem can be avoided by proper care: adequate maintenance and cleaning of all containers, clarifying the wine, sulfuring when appropriate and maintaining a sufficiently low temperature (15°C). A high pH increases the risk of spoilage. In any case, all necessary steps should be taken to ensure that, even after several months of barrel aging, wines have a volatile acidity on the order of 0.40–0.50 g/l of H_2SO_4 (0.48–0.60 g/l of acetic acid) at the time of bottling.

Lactic bacteria can break down other compounds in wine, causing very serious problems (Table 8.1). Fortunately, their development can be limited by taking appropriate care. Although these problems have practically disappeared, they are described below.

Table 8.1. Analysis report on simultaneous lactic fermentation of tartaric acid and glycerol (concentrations are in meq/l) (Ribéreau-Gayon *et al.*, 1976)

	Before spoilage	After spoilage	Variations
Glycerol (mmol/l)	70	46	−24
Reducing sugars (g/l)	1.5	0.5	−1.0
pH	3.93	3.96	
Total acidity	54	74	+20.0
Volatile acidity	11.2	33.8	+22.6
Tartaric acid	40.0	26.9	−13.1
Malic acid	0	0	
Citric acid	0.9	0.7	−0.2
Lactic acid	16.2	25.2	+9.0

'Tourne' is caused by rather rare bacteria that decompose tartaric acid, essential for a wine's acidity, flavor and aging capacity. Affected wines lose acidity, their pH rises, the color becomes dull and brown, and the carbon dioxide content increases. These wines have an unpleasant, lactic smell and are flat and flabby on the palate. Bacterial turbidity develops and wines sometimes seem to have silky, iridescent highlights when they are swirled around in the glass.

'Amertume' is due to the breakdown of glycerol. It is known to have caused considerable damage at some estates at the end of the 19th century, but is now extremely rare. It may possibly occur if grapes are rather unripe and spoiled. This defect more frequently affects wines with a low alcohol content, particularly press wine and lees wine. Glycerol may be broken down by bacteria in several different ways (Volume 1, Section 5.4.1), producing acetoin derivatives of lactic and acetic acid, or acrolein (Figure 8.2). The latter condenses with phenols and gives a characteristic bitter taste. This defect is detectable at concentrations as low as 10 mg/l. Acrolein passes into the distillates of wines with this problem, giving a pungent smell that may spoil the brandy.

'Graisse', or fatty degeneration, causes wines to become viscous and oily, and is particularly noticeable when wine is poured into a glass. This defect occurs in wines intended for brandy production as they are not sulfured. It may also occur in red or white wines during malolactic fermentation or even later, during storage in bottle. It is not really a type of spoilage. The wine's composition, in particular its volatile acidity, is unchanged, and no appreciable modifications in aroma or flavor are observed. 'Graisse' results from malolactic fermentation with specific bacterial strains (Volume 1, Section 5.4.4) that synthesize β-glucane polysaccharides, consisting of a repeated trisaccharide unit. This substance, often known as mucilage, surrounds the bacterial cells and holds them together, giving the wine an oily appearance. In wine, a few tens of mg/l of residual glucose are sufficient to produce this oiliness.

This problem generally develops during malolactic fermentation, at the end of the exponential growth phase, and may sometimes disappear spontaneously. It is treated by sulfuring the wine (6–8 g/hl) and whisking it vigorously. The glucanases, capable of breaking down the $\beta(1 \rightarrow 3 : 1 \rightarrow 6)$ glucane of *Botrytis cinerea* have no effect on the $\beta(1 \rightarrow 3 : 1 \rightarrow 2)$ glucane responsible for 'graisse'.

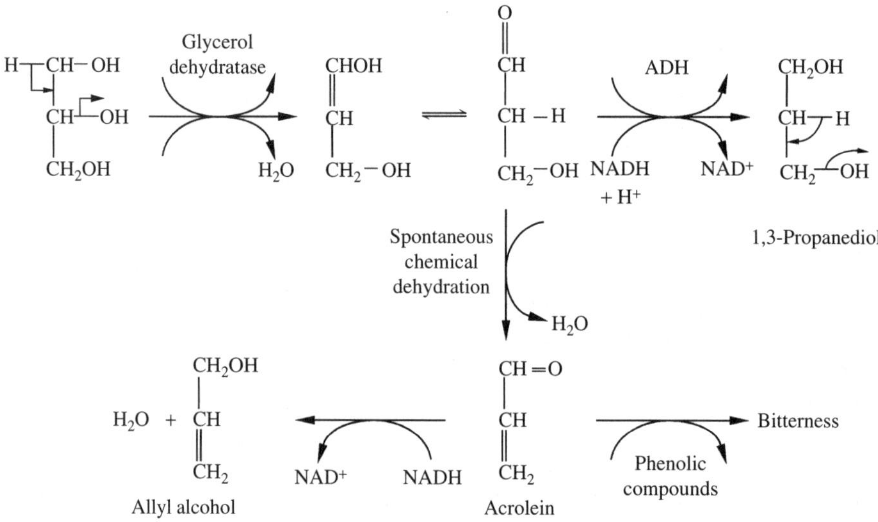

Fig. 8.2. Reaction sequences for the conversion of glycerol into acrolein by lactic bacteria

8.3.3 Spoilage Caused by Acetic Bacteria

There are several types of acetic bacteria (Volume 1, Chapter 7) with different metabolic properties. These are responsible for serious problems due to acescence, sometimes called 'acetic spoilage'. Wine is only affected by *Acetobacter*, or vinegar ferment. The main reaction consists of the oxidation of ethanol to produce acetic acid. In the presence of ethanol, this same bacterium may also esterify acetic acid to produce ethyl acetate. Acetic bacteria develop in the form of a white bloom that may take on various appearances. Prolonged development produces a viscous mass, known as 'vinegar mother'.

A great deal of air is required to oxidize ethanol. The development of acetic bacteria on a large contact surface between air and wine causes a major increase in volatile acidity. This surface must also be undisturbed, as agitation drowns the bacteria and inhibits their aerobic activity. It is also quite true that slight aeration, e.g. during racking, may be sufficient to reactivate the bacteria, sparking a growth in the population. The bacteria then become capable of producing a few tens of mg/l of acetic acid, even if there is no further contact with air, probably due to oxidation–reduction mechanisms. This phenomenon, combined with lactic bacterial activity (Section 8.3.2), is probably responsible for the slight increase in volatile acidity that is always observed when red wines are aged in the barrel.

At the same time, acetic bacteria are capable of esterifying the acetic acid that they form, producing ethyl acetate. The latter is responsible for the organoleptic characteristics of acescence, characterized by a very unpleasant, suffocating odor and an equally nasty impression of harshness and burning on the finish. The perception threshold of ethyl acetate (150 mg/l) is much lower than that of acetic acid (750 mg/l).

The sensation of acescence is not only governed by the ethyl acetate concentration. The wine's richness and aromatic complexity also play a role in the overall impression. All wines contain small quantities of ethyl acetate, formed during alcoholic fermentation. Apiculate and certain other yeasts produce larger quantities and should be avoided for this reason. All wines therefore contain a few tens of mg/l of ethyl acetate, while higher concentrations indicate acetic bacterial activity. It is estimated that this ester does not affect flavor at concentrations below 120 mg/l. Above this amount, it is not identifiable on the nose, but affects the aftertaste and accentuates an impression of harshness. At concentrations above 160–180 mg/l, ethyl acetate is identifiable on the nose and severely affects the wine's organoleptic characteristics, even if the volatile acidity is not excessively high. Maximum ethyl acetate content is not currently specified anywhere in the world, although many authors agree that legislation to that effect would be useful.

It is a well-established fact that acetic spoilage is primarily related to storage conditions. Containers must be kept full and perfectly sealed with airtight bungs. Barrels are positioned with their bungs on the side to maintain an airtight closure and prevent the development of acetic bacteria. The same result is obtained by using plastic bungs, but without positioning the bungs on the side. If wine is stored in containers that are not completely full, it must be protected with inert gas (Volume 1, Section 9.6). This is much more efficient than simply filling the empty space in the vat with sulfur dioxide. Only limited protection can be obtained by sulfuring wine, as the bacteria are active on the surface in contact with the air, whereas SO_2 disappears rapidly due to oxidation.

Other elements may be involved in acetic spoilage. Temperature is a major factor, as spoilage is twice as rapid at 23°C than at 18°C. The optimum storage temperature is 15°C. Acidity also plays a role. While spoilage is practically impossible at pH 3.0, it may easily occur at pH 3.4. Finally, a relatively high alcohol content also reduces this risk.

Cleanliness and proper maintenance of all containers is a vital factor. Infection is normally transmitted by containers, especially barrels that have been stored empty, as the oak may have become impregnated with acetic acid, ethyl acetate and contaminant bacteria. Used barrels must be disinfected with hot water (80°C) or steam at least once per year. When used barrels that have been stored empty are to be reused, it is advisable to disinfect them in the same way. They should also be filled with water for 24 hours before use, to eliminate any acetic acid and ethyl acetate from the wood.

Acescence is a serious defect, that ultimately makes wine unfit for consumption. Experimental methods for eliminating ethyl acetate, especially by using a vacuum (Ribéreau-Gayon et al., 1976), have not been developed into practical techniques. This problem, characteristic of wines stored without proper care, is easy to avoid by taking basic precautions.

Acetic bacteria are present everywhere: on grapes, in wineries, on walls and floors, as well as inside empty wooden containers. Even if steps are taken to minimize contamination, wine always contains small quantities of bacteria, especially if it is not sulfured. If a young wine is allowed to remain in contact with air, it starts to produce a bloom (Section 8.3.4) and then acetic spoilage occurs. In older wines, spoilage occurs immediately. Consequently, it is essential to keep wine under conditions where bacterial development is as limited as possible.

8.3.4 Mycodermic Yeast Contamination (*Flor*)

Pasteur was the first to include *flor* with bacterial problems, because, like acetic spoilage, it involves a mycodermic microorganism. However, *flor* is caused by a yeast (which Pasteur called *Mycoderma vini*) and not a bacterium (*Mycoderma aceti*). *Flor* should not be confused with the bloom formed by *Saccharomyces cerevisiae* under certain, specific, conditions, responsible for producing high-quality wines such as Sherry and *Vin Jaune* from the Jura (Volume 1, Section 14.5).

Low-alcohol wines are affected by *flor* and a bloom develops on the wine's surface that is in contact with air. This bloom consists of a strain of yeast (*Candida mycoderma*) with a high respiratory capacity, but practically no fermentation activity, so it does not affect sugars. The main transformation caused by *Candida mycoderma* is the oxidation of ethanol into ethanal. This reaction may continue until the ethanol has been completely oxidized into CO_2 and H_2O. This yeast also oxidizes some organic acids, producing a decrease in fixed acidity. Volatile acidity decreases slightly when the bloom starts to develop, and even more markedly when most of the alcohol has been broken down. When the bloom surface is sufficiently large, the wine seems flat and is dominated by the smell of ethanal. The flavor becomes flat and watery, and the wine is turbid. This problem occurs when a wine has not received proper care for a long time. It used to be common when the volume of wine corresponding to daily consumption was drawn from the same barrel over a period of several weeks.

Flor may also occur, to a lesser extent, during the aging of wines with a low alcohol content. The development of bloom rapidly becomes obvious whenever wine is left in contact with air. Although most of the wine is unaffected, it is preferable to avoid this type of contamination by using appropriate systems to prevent contact with air.

In the past (Peynaud, 1981), *flor* was observed in wines with a low alcohol content, when the corks left a relatively large volume of air in the necks of the bottles, which were stored upright. Spots of *flor* appeared rapidly in the neck of the bottle, especially if the storage temperature was rather high. This problem is prevented by fine filtration, combined with reducing the ullage in the bottles, and, above all, high-temperature bottling (Section 12.2.4). *Flor* has now become very rare, thanks to the care given to wines during aging and storage.

8.4 MICROBIOLOGICAL ORIGIN AND PROPERTIES OF VOLATILE PHENOLS

8.4.1 The Volatile Phenols Responsible for Olfactory Defects in Wines Known Collectively as the 'Phenol' Character

Although only trace amounts are present in must, wine contains volatile phenols at concentrations between a few tens and several hundreds of µg/l (Dubois, 1983; Chatonnet and Boidron, 1988). The most widely represented compounds are vinyl-4-phenol, vinyl-4-guaiacol, ethyl-4-phenol and ethyl-4-guaiacol (Figure 8.3). White wines contain variable quantities of vinyl-phenols but no ethyl-phenols. On the contrary, red wines only contain small quantities of vinyl-phenols and have variable concentrations of ethyl-phenols (Table 8.2). The volatile phenol composition of rosé wines is between those of red and white wines (Chatonnet et al., 1992b, 1993b).

Vinyl- and ethyl-phenols are responsible for certain olfactory defects in wine. The most unpleasant smelling are vinyl-4-phenol (reminiscent of pharmaceuticals, gouache paint and 'Band Aids') and ethyl-4-phenol (stables and sweaty saddles). Vinyl-4-guaiacol (carnations) and ethyl-4-guaiacol (smoky, spicy aromas) are much less unpleasant, but they are unfortunately always associated with vinyl-4-phenol and ethyl-4-phenol, respectively. The olfactory impact of the two vinyl-phenols or ethyl-phenols should therefore be considered together, in the proportions in which they are present in the wine. The majority of white wines (Figure 8.4) have a vinyl-4-phenol/vinyl-4-guaiacol ratio of 1:1. The ratio in some wines,

Table 8.2. Ethyl- and vinyl-phenol contents of different wines (µg/l) (Chatonnet et al., 1992b, 1993b)

Volatile phenols	White wines $n = 54$	Rosé wines $n = 12$	Red wines $n = 83$
Vinyl-4-phenol			
Minimum	73	3	0
Maximum	1150	215	111
Mean	301	71	35
Standard deviation (%)	79	99	75
Vinyl-4-guaiacol			
Minimum	15	4	0
Maximum	496	75	57
Mean	212	17.5	12
Standard deviation (%)	44	113	79
Ethyl-4-phenol			
Minimum	0	0	1
Maximum	28	75	6047
Mean	3	20	440
Standard deviation (%)	229	122	179
Ethyl-4-guaiacol			
Minimum	0	0	0
Maximum	7	15	1561
Mean	0.8	3	82
Standard deviation (%)	225	159	230

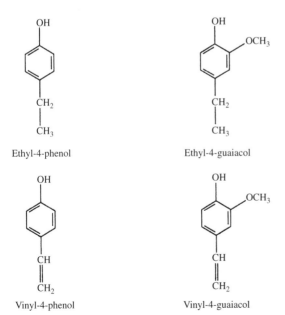

Fig. 8.3. Volatile phenols responsible for olfactory defects in wine known as 'phenol odors'

however, is 3:1, although the reason for this is not known. In red wines, the ethyl-4-phenol/ethyl-4-guaiacol ratio is more homogeneous (Figure 8.4), on the order of 8:1.

The perception threshold of an odoriferous compound is conventionally considered to be the minimum concentration at which its presence in a model dilute alcohol solution is detectable by 50% of trained tasters. The recognition threshold of an odoriferous compound corresponds to its perception threshold in wine. The preference threshold of a compound is the concentration above which the overall aroma of a wine is affected. In the case of vinyl- and ethyl-phenols, the preference thresholds have been estimated at 720 µg/l for a 1/1 mixture of vinyl-4-phenol and vinyl-4-guaiacol in white wines and 420 µg/l for a 10/1 mixture of ethyl-4-phenol and ethyl-4-guaiacol in red wines. These values are relatively close to the recognition thresholds (in wine) for the same mixtures of vinyl- and ethyl-phenols.

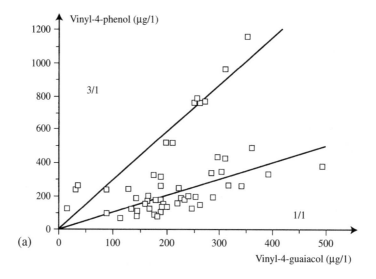

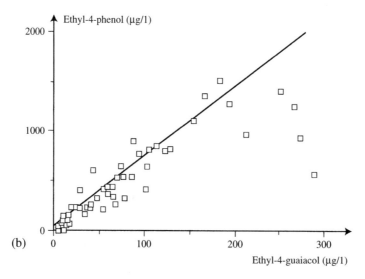

Fig. 8.4. Comparison of (a) vinyl-phenol concentrations in white wines and (b) the ethyl-phenol content in red wines (Chatonnet *et al.*, 1992b, 1993b)

This means that, as soon as they are detectable by tasters, vinyl- and ethyl-phenols have a negative impact on wine aroma.

Olfactory defects in wine attributable to volatile phenols are relatively common. In a recent study (Chatonnet *et al.*, 1993b), one hundred (mainly French) wines, from different appellations and vintages, were classified according to their volatile phenol content. Almost one-third of the red and white wines analyzed had volatile phenol concentrations above the perception threshold (Figure 8.5).

The detrimental effect of vinyl-4-guaiacol has also been demonstrated in South African white wines made from the Kerner grape (Van Wyk, 1993). On the other hand, Versini (1985) found that this compound could have a positive effect on the quality of certain Gewürztraminer wines.

Chemical Nature, Origins and Consequences of the Main Organoleptic Defects

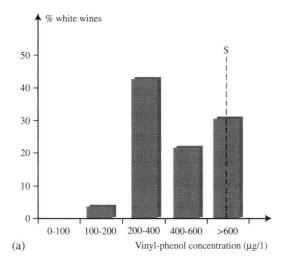

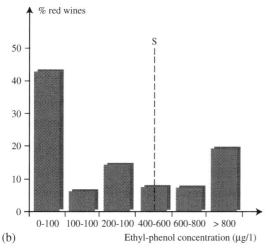

Fig. 8.5. Wines classified according to their volatile phenol concentrations. Organoleptic effect. Percentage of 'phenol' wines with a concentration S above the perception threshold (Chatonnet et al., 1993b)

8.4.2 Enzyme Mechanisms Responsible for the Production of Vinyl-Phenols by *Saccharomyces Cerevisiae*

The vinyl-phenols in white wines are formed due to enzymic decarboxylation by yeast of two cinnamic acids (*p*-coumaric acid and ferulic acid) in must, producing vinyl-4-phenol and vinyl-4-guaiacol, respectively (Figure 8.6).

The cinnamate decarboxylase (CD) of *Saccharomyces cerevisiae* is highly specific. These yeasts are incapable of converting benzoic acids into volatile phenols. Only certain acids in the cinnamic series (phenyl-propenoic acids) may be decarboxylated by this microorganism. Among the cinnamic acids in grapes, only ferulic and *p*-coumaric acids are affected by the CD activity. Caffeic (4,5-dihydroxycinnamic) and sinapic (4-hydroxy-3,5-dimethoxycinnamic) acids are not decarboxylated by *S. cerevisiae*. Cinnamic acid and 3,4-dimethoxycinnamic acid are affected by this reaction *in vitro*, but they are almost entirely absent from must. Finally, the CD of *S. cerevisiae* only catalyzes the decarboxylation of *trans* isomers in the cinnamic series. The CD of *S. cerevisiae* is endocellular, constitutive and active during alcoholic fermentation only.

For many years, it was assumed that the low vinyl-phenol concentrations in red wines were due to the fact that, after their formation by yeast, they were converted into the corresponding ethyl-phenols by *lactobacilli* during malolactic fermentation (Dubois, 1983). This interpretation is no longer accepted (Section 8.4.4). Indeed, it is now known that lactic bacteria are not involved in the production of ethyl-phenols or vinyl-phenols in red wines.

The low concentration of vinyl-phenols in red wines is mainly due to the inhibition of the CD of *S. cerevisiae* by certain grape phenols (Chatonnet et al., 1989, 1993b). This inhibition may be demonstrated by adding an extract of grape seeds and skins in ethanol to must containing phenol acids. Procyanidins, the most active compounds (Section 8.4.5, Table 8.5), may inhibit the enzyme completely, resulting in the total absence of synthesis of vinyl-phenols.

8.4.3 Influence of Winemaking Parameters on the Vinyl-Phenol Concentrations of White Wines

The vinyl-phenol content of a white wine depends on the concentration of phenol acid precursors in the must, on the one hand, and the CD activity of the yeast strain responsible for alcoholic fermentation, on the other hand.

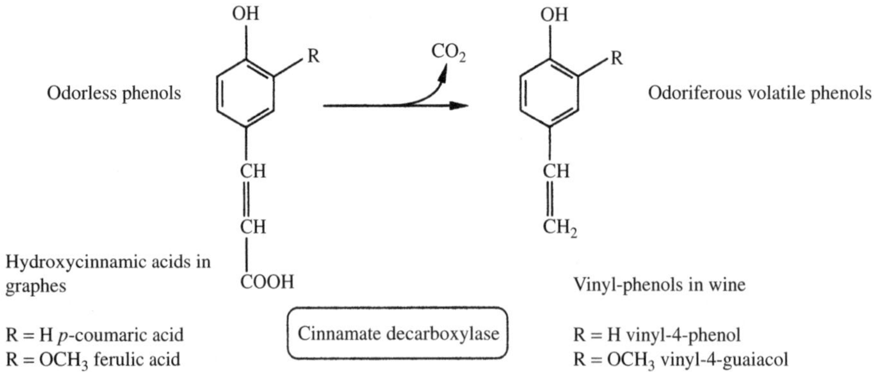

Fig. 8.6. Decarboxylation of phenol acids in must by *Saccharomyces cerevisiae* during alcoholic fermentation

The hydroxycinnamic acid concentration of must varies according to grape variety and ripening conditions. For example, Semillon or French Colombard often have a higher *p*-coumaric and ferulic acid content than Sauvignon Blanc. Concentrations are higher in ripe grapes and those grown in hot climates.

In a given batch of grapes, the phenol acid content in the must and, consequently, the vinyl-phenols in the wine depend on the extraction conditions and clarification of the must. Brutal mechanical handling of the grapes (dynamic juice separators, continuous presses, etc.), insufficient settling and, to a lesser extent, prolonged skin contact facilitate extraction of phenol acids from grape solids and thus the forming of vinyl-phenols during alcoholic fermentation.

The degree of oxidation of the must also affects its concentrations of phenol acids, substrates for the tyrosinase (polyphenoloxydase) activity in grapes. Thus, there is a marked decrease in the amount of *p*-coumaric acid in the wine if no sulfur dioxide is added to the must. This effect is exacerbated by hyperoxygenation (Volume 1, Chapter 13). Sometimes, wines made from hyperoxygenated must have a more distinctive aroma than a control wine, made from must protected from oxidation. This difference can be at least partially explained by the lower vinyl-phenol concentrations in wine made from hyperoxygenated must (Dubourdieu and Lavigne, 1990). However, this practice cannot be envisaged for certain grape varieties, such as Sauvignon Blanc, as the wine loses varietal aroma.

The use of certain pectolytic enzyme preparations to facilitate the extraction or clarification of white must may lead to an increase in the vinyl-phenol content of white wines and a deterioration of their aromatic qualities (Chatonnet *et al.*, 1992a; Dugelay *et al.*, 1993; Barbe, 1995). Indeed, certain industrial pectinases, made from *Aspergillus niger* cultures, have a cinnamyl esterase (CE) activity. This enzyme catalyzes the hydrolysis of tartrate esters of hydroxycinnamic acids in must during the pre-fermentation phase (Figure 8.7). Ferulic and *p*-coumaric acids are then converted into vinyl-phenols during alcoholic fermentation due to the cinnamate decarboxylase activity of *Saccharomyces cerevisiae*.

A study (Barbe, 1995) of the contaminant esterase activity of industrial pectinases made from *Aspergillus niger* identified three different enzymes (Figure 8.8):

1. Cinnamate esterase hydrolyzes cinnamic acid esters, such as chlorogenic acid, ethyl cinnamate, and tartaric acid esters.

2. Depsidase is specific to the ester bonds between two phenol cycles (e.g. digallic acid).

3. Phenyl esterase hydrolyzes phenol acid esters (benzoic series) and aliphatic alcohols, such as methyl gallate.

In the past, certain undesirable effects of must clarifying pectinases on wine aroma were

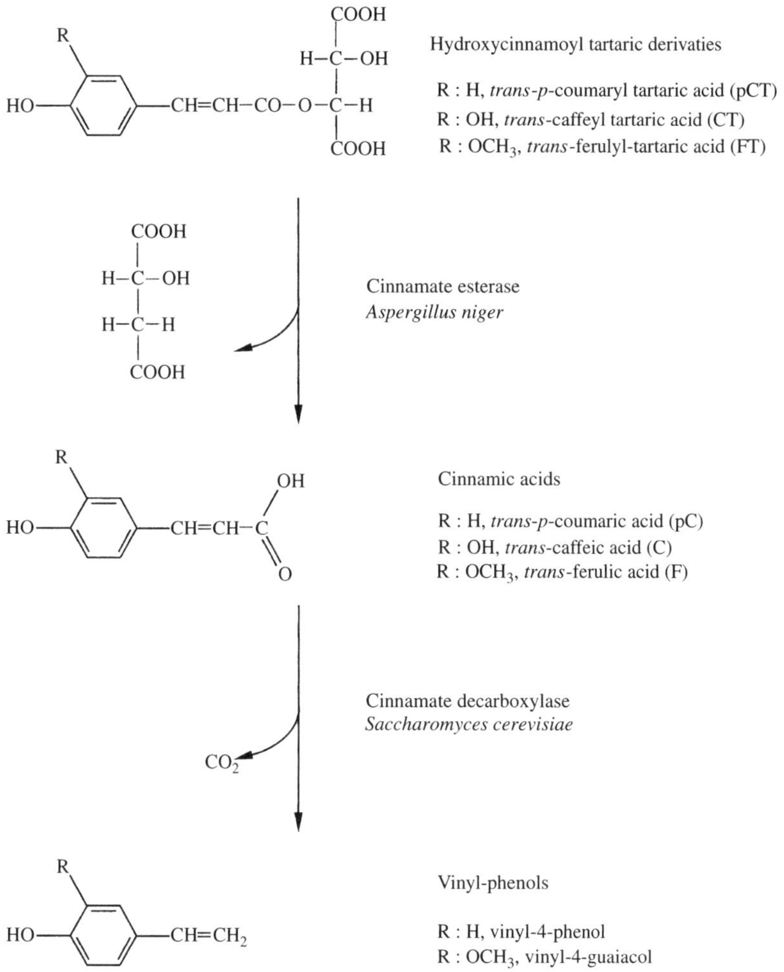

Chemical Nature, Origins and Consequences of the Main Organoleptic Defects

Fig. 8.7. Vinyl-phenol formation mechanism in must clarified using a pectinase preparation with cinnamate esterase activity (Barbe, 1995)

attributed to depsidase (Burckhardt, 1976) or chlorogenase activity (Maurer, 1987), which reduced the concentration of cinnamoyl-tartrate esters in the must. These compounds are assumed to contribute to the fresh flavor of white wines. This interpretation maintains a certain confusion between the various esterase activities of *Aspergillus niger*. It is now known that depsidase and phenyl esterase (collectively referred to as tannase) activities are not involved in the olfactory defects of white wines attributable to the use of pectinases. The only reaction responsible for these defects is due to cinnamate esterase hydrolyzing tartaric acid esters. There are no direct consequences, as these compounds do not have any organoleptic impact on the concentrations present in wine. However, the cinnamic acids released by CE increase the quantity of vinyl-phenols likely to be produced by cinnamate decarboxylase from the yeast. The vinyl-phenols, of course, are responsible for olfactory defects.

The cinnamate esterase in commercial pectinases has been purified (Barbe, 1995). It is a 240 000 Da glycoprotein, consisting of two subunits of 120 000 Da. It is strongly inhibited in the presence of ethanol, so it is only active during the

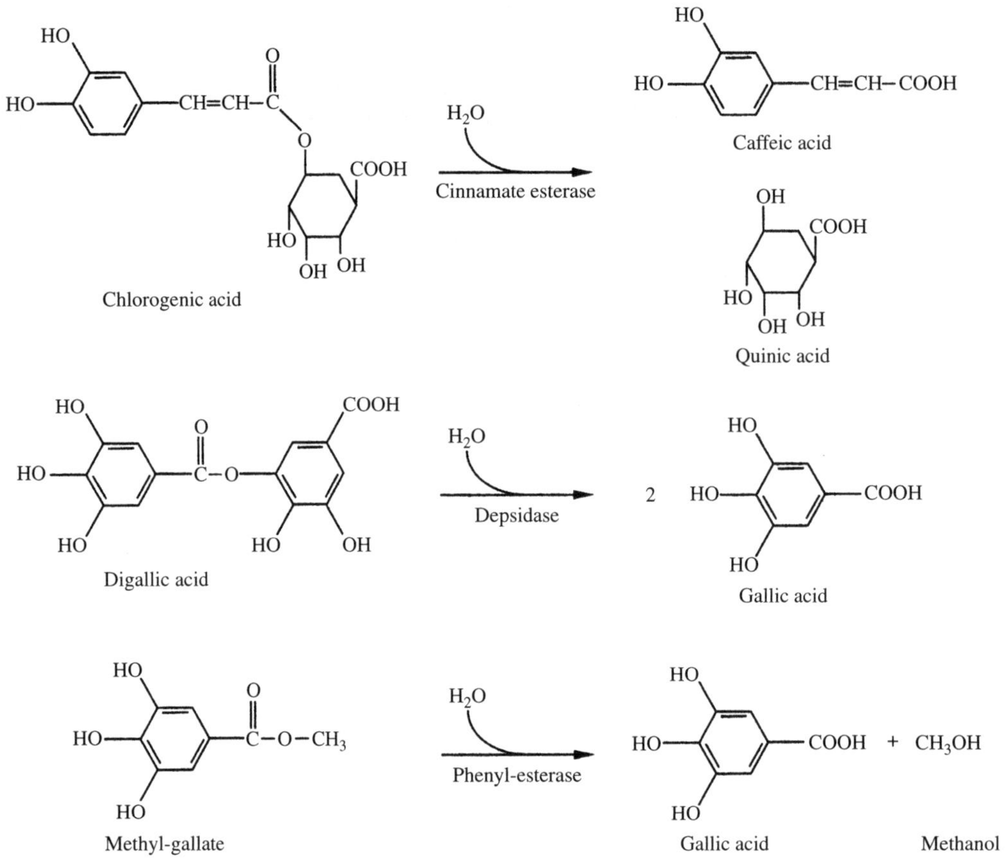

Fig. 8.8. Specific effects of various *Aspergillus niger* esterases in commercial pectinases (Barbe, 1995)

pre-fermentation phase and at the beginning of fermentation.

These findings have encouraged some pectinase producers to offer preparations without any contaminant cinnamate esterase, so that there is no longer a risk of increasing the volatile phenol content of white wines when pectinases are used as adjuvants in clarifying white wine must (Table 8.3).

The yeast strain also plays an essential role in determining the volatile phenol concentration in white wines. For many years now in the brewing industry (Goodey and Tubb, 1982), yeast strains have been selected for their low production of vinyl-phenols, as malt has a high phenol acid content. These are called *Pof-* (phenol off-flavor) strains. The selection of winemaking yeast has

Table 8.3. Impact of clarifying must with commercial pectinases on the vinyl-phenol content of Sauvignon Blanc wines (Chatonnet et al., 1992a)

Enzyme treatment of the must	CE activity (U g/l)	Vinyl-4-phenol (μg/l)	Vinyl-4-guaiacol (μg/l)
None	—	545	192
Pectinase-1 (not purified)	12	1900	218
Pectinase-2 (purified)	1.9	795	169

only recently included this character. One strain of *Saccharomyces cerevisiae* (CCI) has been selected by the Bordeaux Faculty of Enology for its low CD activity (Figure 8.9) and is marketed as 'Zymaflore VL1'. More recent findings have indicated that

Chemical Nature, Origins and Consequences of the Main Organoleptic Defects

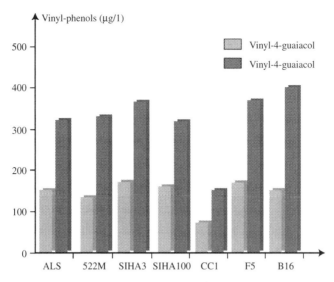

Fig. 8.9. Impact of the strain of winemaking yeast on the vinyl-phenol concentration of white wine (Chatonnet *et al.*, 1993b)

strains with a low CD activity are relatively rare, both in natural as well as commercial winemaking yeasts (Grando *et al.*, 1993). The use of *Pof-* yeast ensures that even white must with a high hydroxycinnamic acid content will produce a wine without any phenol off-odors.

A considerable decrease in vinyl-phenol concentration, and the resulting olfactory impact, occurs as white wine ages, especially during bottle aging. This is mainly due to radical polymerization of the vinyl-phenols into odorless polyvinyl-phenols (Klaren De Witt *et al.*, 1971). Vinyl-phenols may also be converted into ethoxy-ethyl-phenols with little odor by an addition reaction of ethanol in an acid medium (Dugelay *et al.*, 1995).

8.4.4 Conditions and Frequency of Ethyl-Phenol Defects in Red Wines

Until recently, the origin of the ethyl-phenols in wine was not very well known and was even rather controversial. These compounds are, however, responsible for a common aromatic defect in red wines. Ethyl-phenol concentrations may, in some cases, reach several mg/l, giving the wine a strong 'barnyard' smell. Even at lower concentrations, 600–700 mg/l, ethyl-phenols alter the aroma. The odor may be less unpleasant, but it still masks the fruit and bouquet, robbing wines of their personality.

Ethyl-phenols are only rarely formed during alcoholic fermentation, causing drastic spoilage accompanied by the rapid production of large quantities of acetic acid. The causes of this type of problem are not very well known. Fortunately, it is extremely rare. Nevertheless, insufficient sulfuring of the grapes as they are put into the vat and a lack of hygiene in the winery provide ideal conditions for this type of spoilage to occur.

The appearance of ethyl-phenols during aging is much more common, especially in used barrels, although phenol off-odors may also develop in red wines aged in new barrels, or even in vats. Figure 8.10 shows an example of the increase in ethyl-phenol concentrations in a red wine in the barrel during the summer months. This phenomenon is promoted by the rise in cellar temperature and the decrease in the wine's sulfur dioxide content during this period.

Ethyl-phenols may occasionally develop in the bottle. Some bottles may have abnormally high concentrations, whereas others in the same batch have hardly any.

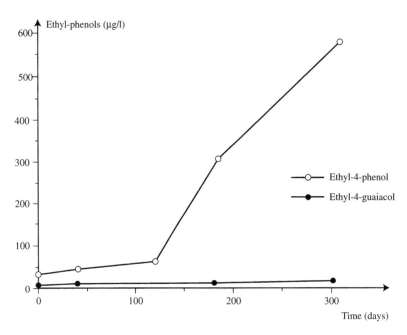

Fig. 8.10. Example of changes in the ethyl-phenol concentration of a red wine during barrel aging (Chatonnet *et al.*, 1992b)

Table 8.4. Ethyl-phenol content in ten vintages of various Bordeaux wines (Chatonnet *et al.*, 1992b)

Vintage	Σ Ethyl-phenols (µg/l)			
	Wine A	Wine B	Wine C	Wine D
1979	46	86	512[a]	512[a]
1980	6.5	—	266	666[a]
1981	30.5	253	395	995[a]
1982	276	106	629.5[a]	929.5[a]
1983	243.5	54	926[a]	726[a]
1984	—	5	401	1401[a]
1985	198	924[a]	46	779[a]
1986	15	975[a]	950[a]	1515[a]
1987	4	429[a]	714.5[a]	564.5[a]
1988	3	654[a]	274	987[a]
1989	5	147	655[a]	1695[a]
1990	14	3	—	2789[a]
Frequency of wines with concentrations above the perception threshold	0	1/3	1/2	1/1

[a] Wines with an ethyl-phenol content above the olfactory detection threshold.

Systematic analyses of bottled wines from different vineyards showed that this defect may occur several years running at some estates, but rarely, if ever, at others (Table 8.4). Similarly, in the same cellar, before blending, some batches of wine (in vats or barrels) may be contaminated and others unaffected. Sometimes, contamination from a few barrels with phenol off-odors may affect the entire vintage, as a result of blending prior to bottling.

8.4.5 Microbiological Origin and Synthesis Pathways of Ethyl-Phenols in Red Wines

For many years (Dubois, 1983), ethyl-phenols were thought to result from the metabolism of lactic bacteria. However, it was never possible to link their appearance to the completion of malolactic fermentation or to the storage of wines in the presence of lees containing bacteria (Di Stefano, 1985; Chatonnet et al., 1992b).

By isolating acetic bacteria, lactic bacteria and yeasts from red wines with phenol off-odors, it was demonstrated that *Brettanomyces/Dekkera* yeasts were the only microorganisms capable of producing several milligrams of ethyl-phenols per liter of wine. The species most prevalent in wine is *Brettanomyces bruxellensis* (Chatonnet et al., 1992b).

A study of the mechanisms by which *Brettanomyces* biosynthesizes ethyl-phenols demonstrated the sequential action of two enzymes (Figure 8.11). The first is a cinnamate decarboxylase that transforms cinnamic acids into vinyl-phenols. This enzyme, unlike that of *Saccharomyces cerevisiae*, is capable of decarboxylating sinapic acid (hydroxy-4-dimethoxy-3,5-cinnamic). Above all, it is not inhibited by the phenols that affect the *S. cerevisiae* enzyme (Table 8.5). The second enzyme is a vinyl-phenol reductase (VPR), which is totally absent from *S. cerevisiae*. These properties explain why *S. cerevisiae* is incapable of producing large quantities of volatile phenols in red wines. They also account for the aptitude of *Brettanomyces* to break down cinnamic acids into vinyl-phenols and then ethyl-phenols, under the same conditions.

Certain strains of *Pediococcus pentosaceus* and a few lactobacilli (*Lactobacillus plantarum* and *Lactobacillus brevis*) may be capable, to a very limited extent, of decarboxylating *p*-coumaric acid and ferulic acid into vinyl-phenols and then reducing them to the corresponding ethyl-phenols (Chatonnet et al., 1992b, 1995; Cavin et al., 1993). The decarboxylase activity of *Leuconostoc oenos* is only detectable in permeabilized cells, which explains the absence of any marked variation in volatile phenol concentrations during malolactic fermentation. Unlike the cytoplasmic, constitutive CD of yeast, the CD activity of lactic bacteria is membrane-based and induced by the substrate. The permeation of *p*-coumaric acid through the membrane is energy-dependent and requires a proton gradient.

The quantities of ethyl-phenols formed by lactic bacteria are always very small compared

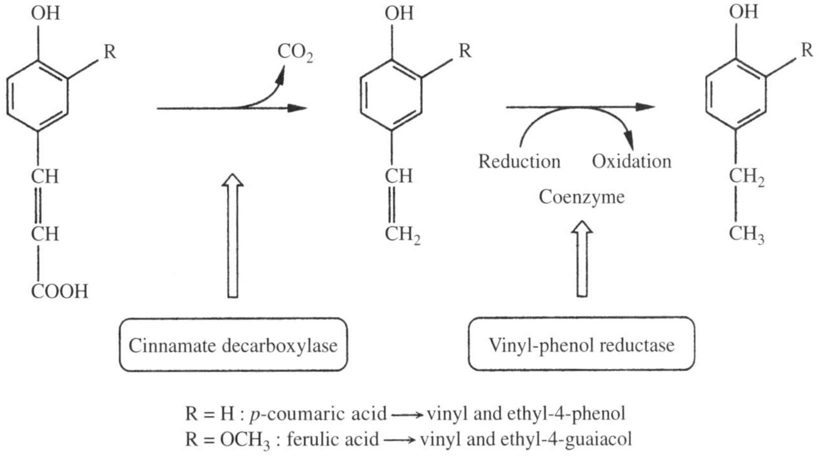

R = H : *p*-coumaric acid ⟶ vinyl and ethyl-4-phenol
R = OCH$_3$: ferulic acid ⟶ vinyl and ethyl-4-guaiacol

Fig. 8.11. Enzyme mechanism for the production of ethyl-phenols by *Brettanomyces* sp.

Table 8.5. Impact of polyphenols on the synthesis of volatile phenols by *Saccharomyces cerevisiae* and *Brettanomyces intermedius* (Chatonnet et al., 1992b)

Conditions	Vinyl-4-phenol (μg/l)	Ethyl-4-phenol (μg/l)	Fermentation time (days)	Inhibition (%)
Saccharomyces cerevisiae Control	770	0	15	0
Saccharomyces cerevisiae + procyanidins (2 g/l)	31	0	15	95
Brettanomyces intermedius Control	42	1100	45	0
Brettanomyces intermedius + procyanidins (2 g/l)	0	3080	18	0

Model fermentation medium, supplemented with 5 mg/l *p*-coumaric acid, anaerobic conditions, 25°C.

Table 8.6. Comparative study of the synthesis of volatile phenols by lactic bacteria and yeasts in a model medium supplemented with hydroxycinnamic acids (5 mg/l). Culture at 25°C for 2 weeks under anaerobic conditions (Chatonnet et al., 1995)

Microorganisms	Residual malic acid (g/l)	Volatile phenols (μg/l)			
		trans-Ferulic acid		*trans-p*-Coumaric acid	
		Vinyl-4-guaiacol	Vinyl-4-phenol	Ethyl-4-guaiacol	Ethyl-4-phenol
Control (not inoculated)	8.50	0	0	0	0
Lactobacillus hilgardii R771	0.01	57	44	2	10
Lactobacillus plantarum CHL	0.03	0	154	25	230
Lactobacillus brevis 8407	0.01	65	1909	0	3
Pediococcus pentosaceus 33 316	0.01	37	2063	10	2
Pediococcus damnosus 25 248	0.01	12	14	10	12
Leuconostoc œnos LALL1	1.20	3	9	11	0
Leuconostoc œnos LALL2	1.25	0	0	0	0
Leuconostoc œnos 8417	0.05	100	89	0	0
Dekkera intermedia MUCL 27 706	8.50	25	15	3947	2915
Saccharomyces cerevisiae EG8C	8.50	700	1185	0	0

to those produced by *Brettanomyces/Dekkera* yeasts (Table 8.6). Furthermore, when *Lactobacillus plantarum* is cultivated in a model medium enriched with certain phenols, or seeded directly into a red wine, its volatile phenol production is even more limited. Experimental results indicate that bacterial CD, like that of *S. cerevisiae*, is inhibited by phenols (Chatonnet et al., 1997).

Yeasts in the genus *Brettanomyces/Dekkera* are, therefore, the only microorganisms responsible for the phenol off-odor in certain red wines. These yeasts are capable of producing ethyl-4-phenol and ethyl-4-guaiacol from the hydroxycinnamic acids (*p*-coumaric and ferulic) in grapes. They may also form ethyl-4-syringol from the sinapic acid in oak. These results confirm those of Heresztyn (1986a), clearly demonstrating the capacity

Table 8.7. Laboratory experiment observing changes in the ethyl-phenol content of a red wine inoculated with *Brettanomyces* (10^5 cells/ml) and maintained at 20°C under anaerobic conditions (Chatonnet *et al.*, 1992b)

	Vinyl-4-phenol (μg/l)	Ethyl-4-phenol (μg/l)
$t = 0$ Control	100	7
$t = 30$ days Control (not seeded)	95	6
Seeded with *Brettanomyces intermedius*	301	1230

of *Brettanomyces* to form ethyl-phenols during the alcoholic fermentation of grape juice. These same yeasts may also multiply in red wine after alcoholic fermentation and produce large quantities of ethyl-phenols, independently of any fermentation process. In the experiment shown in Table 8.7, a red wine that had normally completed its alcoholic and malolactic fermentations, and was therefore dry and biologically stable, was aged in the barrel. It was then membrane-filtered and distributed between two sterile flasks. One flask was seeded with a culture of *Brettanomyces intermedius* (10^5 cells per milliliter). The other, unseeded, acted as control. After 30 days at 25°C under anaerobic conditions, the sample inoculated with *Brettanomyces* contained over a milligram of ethyl-phenol per liter. *Brettanomyces* is therefore capable of developing under totally anaerobic conditions in dry wines and producing large quantities of ethyl-phenols by breaking down very small quantities of residual sugars (glucose, fructose, arabinose and trehalose). The consumption of 300 mg/l of residual sugars by *Brettanomyces* is generally sufficient for the ethyl-phenol content (425 mg/l) to be clearly over the perception threshold. Although this phenomenon is very widespread during the aging of red wines, it went unnoticed until fairly recently.

Contamination by *Brettanomyces* has always been considered detrimental to wine quality. According to Ribéreau-Gayon *et al.* (1975), it may give the wine a butyric character, associated with a particular olfactory defect, reminiscent of acetamide and known as 'mouse odor'. Heresztyn (1986b) attributed this odor to the presence of acetyl-tetrahydropyridine, and reported that other microorganisms (*Lactobacillus brevis* and *L. hilgardii*) were also capable of producing these compounds. Curiously, there are no findings indicating a direct link between the 'barnyard' character of certain red wines, their ethyl-phenol concentration and the development of *Brettanomyces*. For many years, this microorganism was only held responsible for extremely rare olfactory defects, such as 'mousiness'. It was not realized until recently that contamination by this yeast caused the much more common problem of the phenolic 'barnyard' smell in wines.

8.4.6 Impact of Barrel-Aging Conditions on the Ethyl-Phenol Content of Red Wines

The sulfur dioxide concentration of wines is an essential parameter in controlling the risk of contamination by *Brettanomyces* during aging, especially in summer (Chatonnet *et al.*, 1993a). In practice, a free SO_2 concentration of 30 mg/l always results in the total elimination of all viable populations after 30 days.

Some observations seem to indicate that this is probably true for red wines with normal pH levels (3.4–3.5), containing 2.0–2.5% free SO_2 in active, molecular form (Volume 1, Section 8.3.1, Table 8.2). When pH reaches 3.8, there is only 1% active, molecular SO_2 left, raising the concern that 30 mg/l would be insufficient to eliminate contaminant populations.

In the example in Figure 8.12, the same batch of red wine (Graves Cabernet Sauvignon), stored in used barrels for 9 months, was analyzed in September, 3 months after racking and sulfuring. It was observed that the ethyl-phenol concentration in the wine varied considerably from one barrel to another and that it was higher when the sulfur dioxide concentration was low. Those barrels with free SO_2 concentrations above 18 mg/l when the analyses were carried out had ethyl-phenol

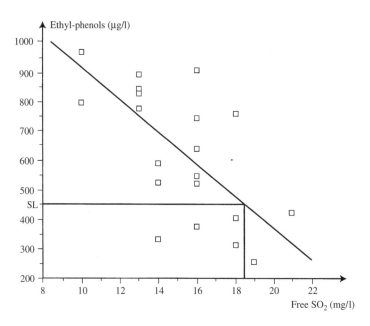

Fig. 8.12. Relationship between the free SO_2 and ethyl-phenol concentrations of several barrels of the same red wine after 9 months of aging. Analyses carried out in September, 3 months after the last racking and sulfuring (Chatonnet *et al.*, 1993a)

concentrations below the tolerance threshold and, consequently, none of the wines had a phenol off-odor.

When wines are stored in the barrel with the bung on the side, i.e. they are perfectly airtight, and there is no possibility of sulfuring between rackings, it is important to ensure that there is a relatively high quantity of free SO_2 (30–35 mg/l) when the barrels are filled. This is to ensure that, 3 months later, there will still be enough SO_2 left (approximately 20 mg/l) to inhibit the development of *Brettanomyces* (Table 8.8).

The rhythm of racking must also be taken into account. In a cellar where the barrels are stored with the bung on the side, racking and adjustment of the free SO_2 level, accompanied by disinfection of the barrels, must be carried out every 3 months during the first year of barrel aging. If racking is delayed, as shown in Table 8.9, the quantity of free SO_2 remaining at the end of the period is too small to protect the wine effectively. The wine is subject to rapid contamination by *Brettanomyces* yeasts and its ethyl-phenol content inevitably increases.

When wine is barrel-aged, sulfuring is not always sufficient to prevent the development of *Brettanomyces*. During racking, barrels must be disinfected with SO_2 gas or burning sulfur. This disinfects the top layers of wood that come into contact with the wine, and would otherwise provide a habitat for yeasts. Table 8.10 compares the ethyl-phenol concentrations of a red wine (Tannat, from Madiran) aged in new or used barrels, racked every 3 months. Some barrels were stored bung upwards with weekly topping-up and others with the bung on the side. Sulfuring was carried out by adding a sulfite solution directly into the wine or by burning sulfur in the barrels. After 6 months of barrel aging, it was observed that the volatile phenol concentration of the wine in barrels that had not been disinfected by burning sulfur was always much higher, whatever the age of the barrel or the position of the bung.

At least 5–7 g of sulfur must be burnt (in ring form) to disinfect a 225 l barrel. However, the effectiveness of disinfection varies according to the sulfuring conditions. The replacement of sulfur wicks that were formerly used by rings has greatly

Table 8.8. Impact of free SO_2 content when wine is put into the barrel on the development of *Brettanomyces/Dekkera* sp. and on the formation of ethyl-phenols in red wines (Chatonnet et al., 1993a)

Initial free SO_2 (mg/l)	Parameters measured	Aging time (months)		
		0 July	3 September	4.5 November
35	Free SO_2 (mg/l)	35	20	15
	Brettanomyces (cells/ml)	1	0	0
	Increase in ethyl-phenols (%)	0	0	0
30	Free SO_2 (mg/l)	30	19	10
	Brettanomyces (cells/ml)	1	0	1
	Increase in ethyl-phenols (%)	0	0	0
25	Free SO_2 (mg/l)	25	16	11
	Brettanomyces (cells/ml)	2	0	0
	Increase in ethyl-phenols (%)	0	0	0
15	Free SO_2 (mg/l)	15	10	6
	Brettanomyces (cells/ml)	6	510	1200
	Increase in ethyl-phenols (%)	0	4	120

Wine sulfured using a sulfite solution, aged in used barrels, bung on the side, pH 3.65.

Table 8.9. Influence of a delay in the third racking on the ethyl-phenol content of barrel-aged wines (Chatonnet, 1995)

Racking date	Ethyl-phenol content[a] (µg/l)	% of barrels with 'phenol' odor
September (after 3 months)	695	57
October (after 4.5 months)	1396	100

[a]Mean of 15 barrels.

improved the efficiency of sulfuring. Indeed, while the wick was burning, some of the sulfur dripped to the bottom of the barrel, where it was extinguished on contact with the damp wood. When sulfur rings are used, there may be considerable differences (50%) in the quantity of sulfur dioxide formed by burning the same weight of sulfur. This depends on the type of incombustible filler used in the rings and humidity levels during storage. It is quite common for water from the atmosphere to be fixed by the hygroscopic filler, forming sulfuric acid rather than sulfur dioxide. Sulfur rings must,

Table 8.10. Influence of aging conditions and sulfuring method (burning sulfur or adding SO_2 solution) on the ethyl-phenol content of a red wine (Chatonnet et al., 1993a)

Barrels used	Sulfuring method	Bung position	Free SO_2 at racking	Σ Ethyl-phenols (µg/l)	Phenol character (0–5)
New	Burning	Top	14	556	1
	Solution	Top	12	975	4
1 year old	Burning	Side	18	103	0
	Solution	Side	16	1432	5
2 years old	Burning	Top	28	89	0
	Burning	Side	24	331	0
	Solution	Side	18	1291	5

Analyzed in July after 6 months of aging, during the second racking, red wine, pH 3.5.

therefore, be stored in a dry place to ensure consistent sulfuring when they are burnt. Any other means of disinfecting the barrels (hot water or steam) could probably be used instead of, or in addition to, the sulfuring of barrels with sulfur dioxide gas.

The preceding discussions emphasize the role of sulfuring conditions in protecting wine from *Brettanomyces* infection and the consequent risk of a major increase in ethyl-phenol content. It is quite clear that several additional precautions must be taken. Firstly, care must be taken to ensure that alcoholic fermentation has consumed all the sugars as a few hundred mg/l residual sugars are sufficient for *Brettanomyces* to produce appreciable quantities of ethyl-phenols, even during bottle storage. Secondly, the wine must be aged at relatively low temperatures (below 15°C). Finally, it may be advisable to eliminate excessive populations of the contaminant *Brettanomyces* by clarification (filtration) or heating (flash-pasteurization).

8.5 CORK TAINT

8.5.1 Contamination in Wine due to Corks

It was not until the 17th century that glass bottles and corks came into use for wine storage. The considerable growth in high-quality (*appellation contrôlée*) wines over the past 25 years and the proliferation of estate bottling partly explain the constantly increasing volume of bottled wines distributed worldwide. Currently, 12–13 billion bottles of wine are corked every year. Although screw caps are used for some wines intended for rapid consumption, there has been little growth in this sector over the past 20 years. Not only are the technical characteristics of the corks preferred for high-quality wines, but they are also closely associated with the concept of great wines. Cork, however, is a natural material: it is difficult and expensive to produce, while quality can be heterogeneous. There have been a few attempts to replace cork, using synthetic materials with the same qualities but without its defects. Positive results have been obtained, but it is impossible to envisage giving up natural cork in the near future.

The advantages of corks are well known: they are easily inserted in bottles by a high-speed machine and they provide a perfect seal which makes long aging possible. Finally, uncorking the bottle presents no major difficulties.

The main risks involved in using natural cork are the leakage that can occur if the bottles are not perfectly sealed and contamination by malodorous substances released by corks that produce 'cork taint'. In France, approximately 3–4% of all bottled wine is estimated to be affected by cork taint. The organoleptic and economic harm caused by cork taint, the frequency of the problem and the constantly increasing use of cork in bottling no doubt explain the large number of extensive studies examining this issue.

Cork taint has long been a major concern (Riboulet and Alegoet, 1986). Since the first observations, dating back to the beginning of the 20th century, it has been associated with the development of mold (*Penicillium* and *Aspergillus*) occurring on cork oak trees, in sheets of cork during preparation or in corks themselves. The species of mold prevalent in cork are known for their capacity to draw energy for further growth from substrates that are not easily degradable. During the decomposition of the long carbon chains in the cork, they form many intermediate volatile molecules that are soluble in a dilute alcohol medium. Different compounds are produced according to the strain of mold and the conditions under which it develops. The problem of cork taint is obviously highly complex. It is therefore recommended to avoid excess humidity and high temperatures that facilitate the growth of fungi at all stages in cork manufacture. A number of stabilization processes (chemical treatments, γ radiation, etc.) have been proposed. However, these may only be efficient in protecting corks provided they have not previously been contaminated.

Initial observations led specialists to differentiate 'true cork taint' from 'moldy off-flavors'. Cork taint is fortunately very rare. It produces a very unpleasant, putrid smell that makes wine nauseating. Its origins are certainly related to cork, but the real cause is unknown. The cork defect known as 'yellow stain' has been blamed, but this hypothesis is not universally accepted.

'Moldy off-flavors' are attributed to the effects of many types of mold. Contamination may start on the tree and continue during storage and treatment of the cork sheets and cork manufacture, as well as in the bottles of wine as they age in the cellar. The microbial flora at each of these stages are not necessarily the same.

Over the past 15 years, much research has been done and over twenty articles of real scientific value have been published on this subject. These articles examine the composition of cork in terms of volatile compounds, especially those molecules likely to cause organoleptic defects in wine.

Recently, Mazzoleni et al. (1994) used gas-phase chromatography coupled with mass spectrometry (GCMS) to identify 107 volatile compounds in powdered new and used corks. These findings are not qualitative, but they certainly confirm the complexity of the cork taint issue. Cork is a biological medium, and therefore necessarily complex. Even after boiling and chloride treatment, it retains a nutritional value that mold, bacteria and even yeasts are capable of using, if the humidity becomes sufficiently high. In 1981, Dubois and Rigaud and Tanner et al. showed that chloroanisoles (Figure 8.13) are involved in the moldy off-flavors and cork taint found in wine. Lefebvre et al. (1983) and then Rigaud et al. (1984) attempted to specify the effect of microorganisms in cork on the olfactory spoilage of wine. Riboulet (1982) showed that a bacterium in the genus Streptomyces was capable of converting vanillin in cork into guaiacol, with its accompanying pharmaceutical smell (Figure 8.14).

Dubois and Rigaud (1981) confirmed that true cork taint was rare and attributed it to a tetra methyl tetra hydronaphthalene. These authors did not find that cork taint was related to yellow stain.

Amon and coworkers (1986, 1989) analyzed 37 wines affected by cork taint, using gas-phase chromatography, equipped with an olfactometry system and coupled with a mass spectrometer. They analyzed the compounds most frequently found, both in corked wine and their respective corks. Their findings are shown in Figure 8.13. These same authors ran triangular tests with a panel of twelve tasters and found that the perception thresholds for these compounds in a non-aromatic, dry white wine were particularly low (Table 8.11).

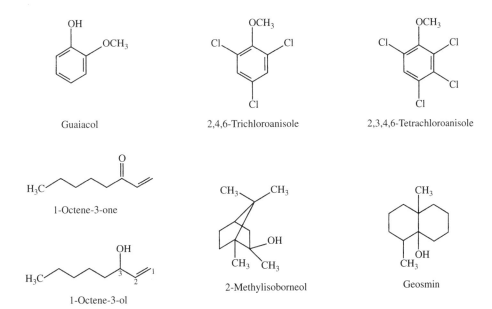

Fig. 8.13. The main compounds responsible for cork taint

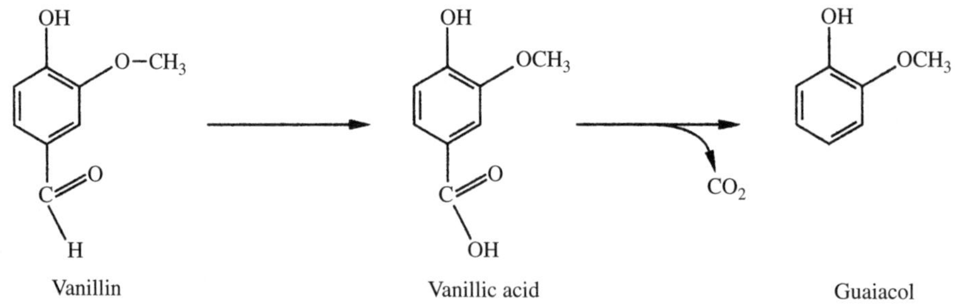

Fig. 8.14. Bacterial transformation of vanillin from cork into guaiacol

Indeed, four out of seven had perception thresholds of 30 ng/l (30×10^{-9} g/l) or less. The compound most often implicated in cork taint is 2,4,6-trichloroanisole, but that may be due to its extremely low perception threshold, on the order of 4 ng/l.

Recent findings confirm that chloroanisoles play a major role in moldy off-flavors, even if these molecules are not solely responsible for cork taint.

Chloroanisoles are produced when mold acts on chlorinated derivatives used to treat trees and prepare corks, as well as in cork manufacture. Chlorophenols are considered to be the main precursors of chloroanisoles (Figure 8.15). They are believed to be methoxylated by fungi without tetrahydrofolic acid (FH$_4$) deficiency. This acid is a growth factor known as a unit transporter and has a single carbon atom.

Furthermore, using simple culture media with a known composition, Maujean *et al.* (1985) showed

Table 8.11. Perception thresholds in dry white wine of the main compounds involved in cork taint and description of their odor

Compound	Perception threshold (ng/l)	Description of odor
1-Octen-3-one	20	Mushroom, metallic
1-Octen-3-ol	20 000	Mushroom, metallic
2-Methylisoborneol	30	Earthy, moldy, dirty
2,4,6-Trichloroanisole	4	Moldy, damp cardboard
Geosmine	25	Earthy, moldy, dirty
Guaiacol	20 000	Smoky, 'phenol', medicinal

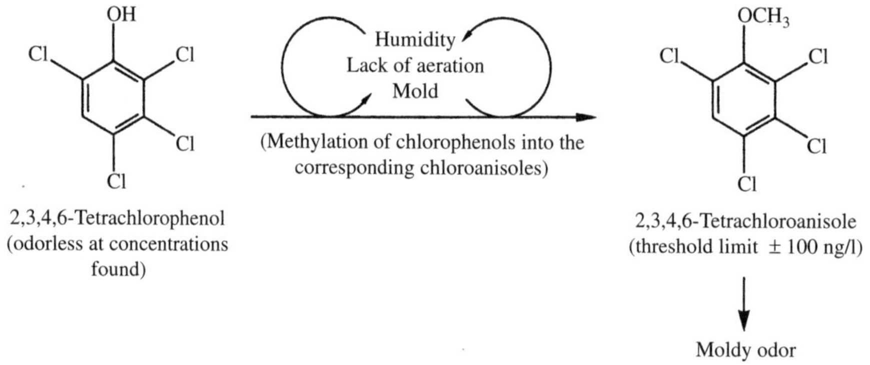

that *Penicillium*, isolated from corks, was capable of the total biosynthesis of chloroanisoles from glucose, via the pentose channel (Figure 8.16). The medium must contain chlorine (this may come from the bleach used in processing the cork) and methionine (this may come from the casein used

Fig. 8.15. Transformation of a chlorophenol into chloroanisole

Chemical Nature, Origins and Consequences of the Main Organoleptic Defects

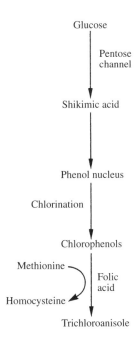

Fig. 8.16. Complete biosynthesis of 2,4,6-trichloroanisole (Maujean *et al.*, 1985)

to assemble rings of solid cork with agglomerated cork to produce sparkling wine corks).

Chemical treatment, using chlorine-based oxidizing, disinfecting and bleaching agents, such as hypochlorite, followed by reducing agents, such as oxalic acid, is gradually being phased out and replaced by high-temperature autoclaving. This should be an efficient way of reducing the frequency of 'corked' wines. In the same vein of preventive action, it could be possible to take steps even earlier, by avoiding chlorine-based insecticides in cork oak plantations.

Cork taint may also be caused by coatings used to lubricate the corks. As these have a certain nutrient content, they may act as a carbon source, facilitating the growth of mold. Fatty acids and even paraffins provide usable carbon sources for mold to develop on corks. These substances are converted into methylketones, which smell of cheese or even drains. Fortunately, these coatings are gradually being replaced. One group of substitute coatings consists of polymethylsiloxanes, in the form of non-reactive oils. Their viscosity increases with the number n, defining the length of the chain. The other alternatives are reactive oils that reticulate and polymerize, forming an elastomer with a three-dimensional network (Figure 8.17). Polymethylsiloxanes are hydrophobic and do not migrate into wine. Although they are well known for their foam-inhibiting properties, polymethylsiloxanes are still usable for treating corks for sparkling wines. Similarly, their nutrient potential in a liquid culture medium is very limited, as siloxanes are the only carbon source. Indeed, it took over 40 days before a strain of mold such as *Penicillium* started to develop.

Finally, research in progress is examining the composition of cork and its interaction with wine. While the migration of certain compounds from corks into wine is extremely prejudicial from an organoleptic and commercial standpoint, other volatile odoriferous compounds released by cork could be beneficial to wines with aging potential. Riboulet and Alegoet (1986) remarked that cork components are only perceptible for a short time after bottling. They subsequently blend in with the wine's aromas, and may contribute to the development of bouquet. Of course, if the cork character is too strong, it affects the wine and represents a serious defect. Although considerable numbers of cork components have been identified in wine, the olfactory impact of individual compounds is not yet very well known.

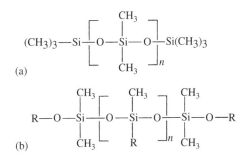

Fig. 8.17. Chemical structure of silicones used to coat corks. (a) Non-reactive silicone oil. (b) Reactive oil that reticulates and polymerizes to form an elastomer with a three-dimensional network. R = lateral radical or reactive terminal

8.5.2 Wine Contamination from Storage Premises

The preceding section clearly shows the major contribution of chloroanisoles to the olfactory defect known as cork taint. It is now widely accepted that cork is not solely responsible for chloroanisole contamination. Wurdig (1975) found a correlation between moldy smells in wine and the use of chlorophenols in fungicides and wood preservatives (pallets, roof timbers, etc.). Maujean *et al.* (1985) confirmed this result, by analyzing some Champagne base wines that had been refused at an official tasting to qualify for the right to the appellation on the grounds of cork taint, although they had never been in contact with cork!

The same types of mold (particularly *Penicillium*) may find a similar environment (humidity, temperature, presence of chlorinated derivatives, etc.) in cork as well as wooden containers and structures in the winery. This explains the possible confusion between authentic cork taint and 'moldy off-flavors', which have similar smells, but different origins. When cork is responsible, only a few bottles are affected, whereas all the bottles may be spoiled if the defect is of external origin.

Some very serious problems have been attributed to the use of chlorophenol insecticides to protect wood used in roof timbers and cellar insulation. The same contamination may occur due to pallet crates used for bottle storage, especially as these are often kept in the same area as wine in barrels or vats. These molecules are not highly odoriferous, but they may lead to the development of malodorous chloroanisoles which can contaminate the cellar atmosphere, and consequently the wine, during racking and other winery operations.

A detailed study of this question was carried out by Chatonnet *et al.* (1994). The following molecules were identified and assayed in contaminated wine:

TCP: 2,4,6-trichlorophenol
TCA: 2,4,6-trichloroanisole
TeCP: 2,3,4,6-tetrachlorophenol
TeCA: 2,3,4,6-tetrachloroanisole
PCP: 2,3,4,5,6-pentachlorophenol
PCA: 2,3,4,5,6 pentachloroanisole

Only TCA and TeCA have intense unpleasant odors. Their perception thresholds in water are 0.03 and 4 ng/l, respectively. The value for PCA is much higher: 4000 ng/l. The aroma of wine is considered to be significantly altered at 10 ng/l of TCA and 150 ng/l of TeCA. Chlorophenols are assayed as precursors of the chloroanisoles. These various molecules do not naturally occur in wine, so their presence, even in very small amounts, indicates contamination that may be due to cork, but may also be attributable to other causes. The figures in Table 8.12 show that even wines that have not been in contact with natural cork may be contaminated. TeCA is the main cause of this contamination. TCA bears a lesser responsibility, although its perception threshold is lower. In rare instances, PCA may also be involved.

The research into these various molecules has been carried out by analyzing solid samples from wooden articles likely to be contaminated, as well as air samples from the atmosphere in affected buildings. The findings show very clearly that wood treated with chlorophenols is the main source of pollution. Under certain conditions (high humidity levels and limited ventilation) the cellar and surrounding premises rapidly become

Table 8.12. Assay of chlorophenols and chloroanisoles in wines that have never been bottled, but have a moldy flavor (results in ng/l) (Chatonnet *et al.*, 1994)

	TCP	TCA	TeCP	TeCA	PCP	PCA
White wine in vat	0	0	2900	1 090	50	270
Rosé wine in vat	0.68	Traces	960	14 000	5900	6500
Red wine A in vat	0	0	1140	570	700	500
Red wine B in barrel	0	0	180	230	320	110

For TCP, PCA, etc., see the text and Figure 8.13.

contaminated, following the air circulation pattern. When the contaminant molecules (TeCA and PCA) are in gas phase, they may easily dissolve in wine during pumping, via the gas emulsion that inevitably occurs during this operation. Similarly, wine may be contaminated by contact with certain products and materials that have been stored in a polluted atmosphere. Cork is particularly sensitive to contamination by chlorophenols and chloroanisoles. Healthy corks stored in a polluted atmosphere may later contaminate bottled wine. It is possible, to a certain extent, to distinguish between 'true cork taint' caused by TCA and 'moldy cellar odor' resulting from TeCA. Table 8.13 shows the case of bottle 1, where the cork did not contain any chloroanisoles and the slight contamination (by TeCA and PCA) came from the winery atmosphere. The cork in bottle 2 had a high TCA content that was directly responsible for spoiling the wine (cork taint). Chatonnet *et al.* (2003) proposed a specific method for assaying chlorophenols and chloroanisoles at the same time, with the aim of determining the source of the pollution (corks or winery atmosphere).

It would be illusory to expect to avoid these serious problems by completely eliminating the microorganisms involved. Indeed, there are a number of very moldy cellars where wine ages perfectly well. Of course, it is nevertheless advisable to restrict mold populations by keeping premises clean and controlling humidity, as well as ensuring adequate ventilation and cool temperatures. The most important recommendation is to prohibit the use of chlorine derivatives as fungicides.

When a winery has become polluted, the source of contamination must obviously be eliminated. This may necessitate replacing the roof or destroying pallet crates, barrels, etc. These measures, coupled with improved air circulation, are generally sufficient to solve this serious problem.

8.6 SULFUR DERIVATIVES AND REDUCTION ODORS

8.6.1 Introduction

The presence of sulfur derivatives in must and wine is always a matter of concern for winemakers, who are fully aware of the risk that they may cause unpleasant smells. Much research is in progress on this topic as these phenomena are not clearly understood. Indeed, until certain modern analysis techniques were recently perfected, it was impossible to assay these substances as they are only present in trace amounts. Another reason for concern about these volatile sulfur compounds is the wide range of possible origins and the diversity of preventive actions.

Some sulfur-based compounds (especially those with thiol functions) make a positive contribution to the varietal aromas of certain grape varieties. The role of 4-methyl-4-mercaptopentanone in Sauvignon Blanc aroma is well known (Section 7.4.1). However, most sulfur derivatives have a very bad smell and have detection thresholds as low as 1 μg/l. These compounds are known as mercaptans, referring to their capacity to be precipitated by mercury salts.

Table 8.13. Demonstration of the dual contamination of a red wine by the atmosphere in the winery and the cork (wines bottled on the same day with a single batch of corks, analysis carried out after 12 months, results in mg per bottle or per cork) (Chatonnet *et al.*, 1994)

	TCP	TCA	TeCP	TeCA	PCP	PCA	odor
Bottle 1							
Wine	Traces	0	600	80	690	240	Not very clean
Cork	Traces	0	170	0	560	Traces	Clean
Bottle 2							
Wine	380	70	1180	140	1480	410	Moldy
Cork	3440	1240	690	150	1310	450	Moldy

For TCP, PCA, etc., see the text and Figure 8.13.

The production of sulfur derivatives by yeast during fermentation is described below. The various molecules and their formation mechanisms are discussed, as well as methods for eliminating them. Although red wines may be affected, this issue is particularly important for white wines. The effect of winemaking techniques and barrel aging on the presence of sulfur derivatives is described elsewhere (Volume 1, Section 13.9.1).

This section deals with those reduction defects attributable to vine sprays containing elemental sulfur and various fungicides or insecticides. The risk of sulfur compounds developing due to thermal or photochemical mechanisms ('sunlight flavor') is also discussed.

Organoleptic defects in wine due to the presence of thiols, or mercaptans, are often associated with a reduced character. This link between 'reduction flavors' and the presence of sulfur compounds is easily justifiable. Indeed, one characteristic of thiol-disulfide redox systems is its particularly low normal potential (E_0') values ($-270 \leq E_0' \leq -220$ mV), compared to the redox potential values of wines ($+220 \leq E_h \leq +450$ mV). It is, therefore, quite clear that the presence of thiols in a wine, and the corresponding hydrogen sulfide smells, require an abnormally low oxidation–reduction potential. This is totally consistent with the impression of reduction on the palate.

8.6.2 Volatile Sulfur Compounds Produced by Yeast Metabolism

Unpleasant odors may sometimes develop in wine during alcoholic fermentation, due to the formation of sulfur compounds by yeast. In view of the complexity of yeast's sulfur metabolism, there are many biochemical mechanisms capable of producing these malodorous molecules. For this reason, theories that attempt to explain the appearance of reduction defects in fermenting wines are often contradictory, and of practically no use to winemakers wishing to implement reliable preventive measures (Rankine, 1963; Eschenbruch, 1974).

The sulfides and thiols involved in this type of olfactory defect are divided into two categories: 'heavy' (boiling point above 90°C) and 'light' (boiling point below 90°C) (Figure 8.18, Table 8.14 and Table 8.15). The latter, particularly hydrogen sulfide (H_2S), have long been considered solely responsible for reduction defects described as 'rotten egg smell'.

It is true that 'light' sulfur compounds have particularly unpleasant smells (rotten eggs, garlic, etc.). Even at low concentrations (on the order of μg/l) these odors are likely to ruin a wine's aroma. Among the sulfur compounds identified and assayed in wine (Tables 8.14 and 8.15), mercaptans (H_2S, methanethiol and sometimes ethanethiol) play a decisive role in reduction defects. They are always present in 'reduced' wine at concentrations much higher than their perception thresholds.

Hydrogen sulfide and methanethiol are directly produced by yeast metabolism. The production of H_2S during alcoholic fermentation is controlled by the enzymes responsible for reducing sulfates and biosynthesizing certain sulfur amino acids (cysteine and methionine) (Figure 8.19). Methanethiol is synthesized by yeast from methionine (De Mora et al., 1986).

Abnormally high concentrations of H_2S may be produced during the fermentation of musts with nitrogen deficiencies. However, several authors suggest different mechanisms for the formation of H_2S in this type of must. According to Vos (1981), yeast protease activity is stimulated in must by nitrogen deficiency, causing sulfur amino acids to be released by proteins. Waters et al. (1992), however, demonstrated that fungal acid proteases have no effect on grape proteins. Ammonium sulfate is frequently added to must from hot regions to prevent H_2S from forming. This provides a source of easily assimilated nitrogen, so the yeast no longer breaks down sulfur amino acids, and H_2S is not released.

The amino acid composition of the must also affects the formation of H_2S by yeast. Besides cysteine and homocysteine, the following amino acids promote the production of H_2S: aspartic and glutamic acids, glycine, histidine, homoserine, lysine, ornithine, threonine and serine. Methionine prevents the formation of H_2S by retroinhibiting the activation channel and reducing sulfates.

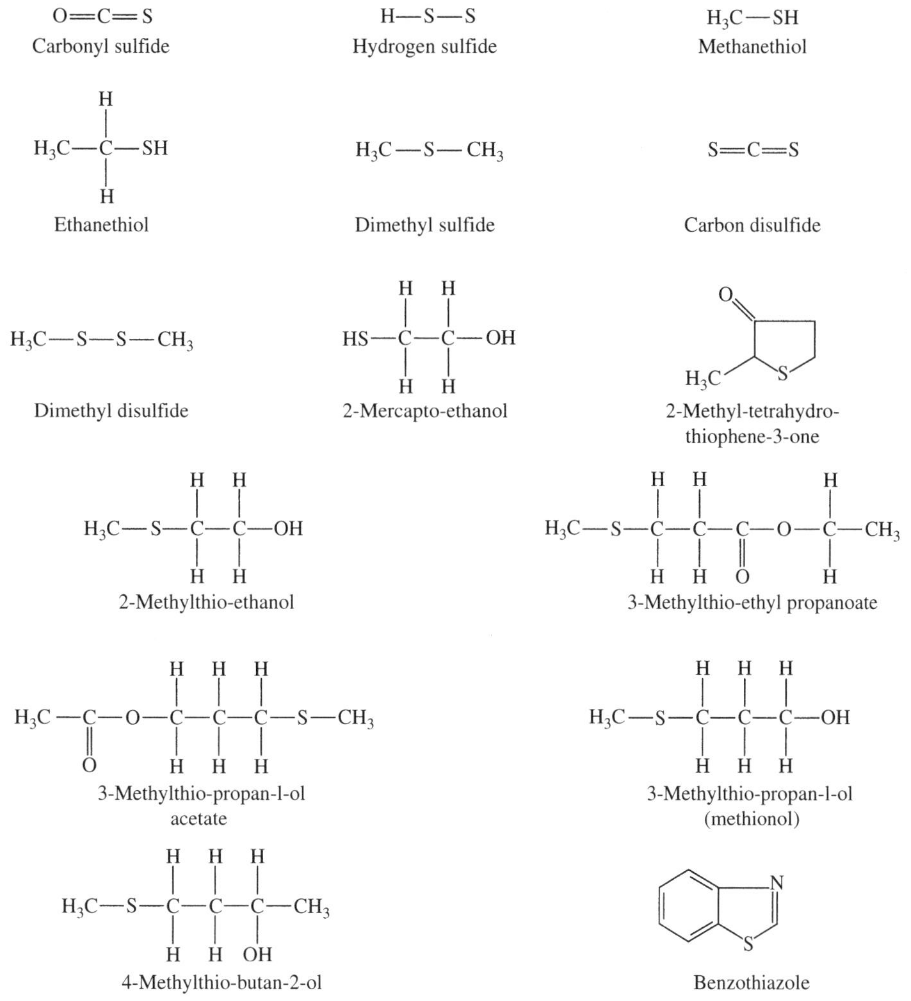

Chemical Nature, Origins and Consequences of the Main Organoleptic Defects

Fig. 8.18. Formulae for the main 'light' and 'heavy' sulfur derivatives (Tables 8.14 and 8.15) involved in reduction odors

Table 8.14. 'Light' sulfur compounds responsible for reduction odors

Substances	Perception thresholds (μg/l)	Description	'Clean' wine (concentrations in μg/l)	Wine with 'reduction' odors (concentrations in μg/l)	Boiling point (°C)
Carbonyl sulfide[a]		Ether	0.7	0.4	−50
Hydrogen sulfide	0.8	Rotten eggs	0.3	16.3	−61
Methanethiol	0.3	Stagnant water	0.7	5.1	6
Ethanethiol	0.1	Onion	0	10.8	35
Dimethyl sulfide	5	Quince, truffle	1.4	2	35
Carbon disulfide		Rubber	1.7	2.4	46

[a] measured by the ratio of the peak surface to that of the internal standard.

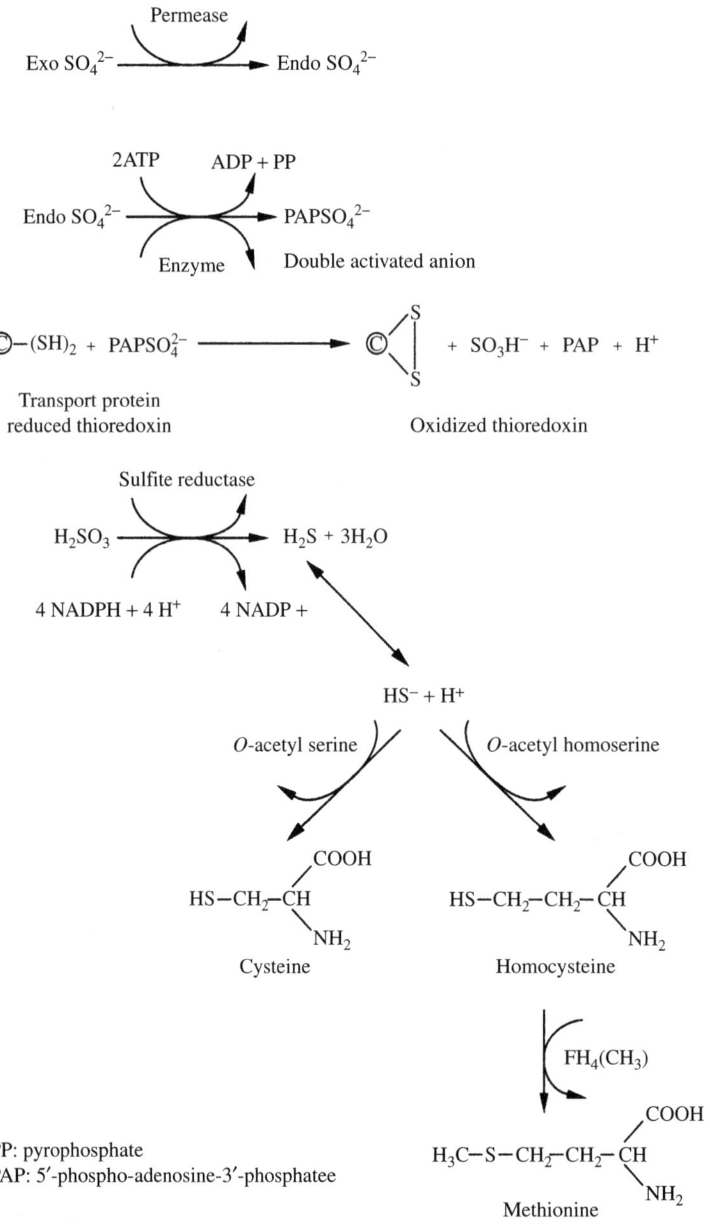

Fig. 8.19. Reduction of sulfates with production of sulfur dioxide, hydrogen sulfide and sulfur amino acids

The winemaking problems likely to cause reduction odors are described elsewhere (Volume 1, Section 13.9.1). When no such problems occur, the concentration of H_2S produced by alcoholic fermentation does not exceed 3–4 µg/l and the methanethiol content is below 1 µg/l. In the worst affected wines, the final concentration of either compound may be as much as several tens of mg/l. Ethanethiol concentrations higher than 2 µg/l may also be observed. This molecule is

Table 8.15. 'Heavy' sulfur compounds responsible for reduction odors

Substances	Perception thresholds in (µg/l)	Description	'Clean' wine (concentrations in µg/l)	Wine with 'reduction' odors (concentrations in µg/l)	Boiling point (°C)
Dimethyl disulfide	2.5	Quince, asparagus	0	2	109
2-Mercaptoethanol	130	Burnt rubber	72	124	157
Methyl-2-tetrahydrothiophenone	90	'Gas'	68	276	84
2-Methylthio-ethanol	250	Cauliflower	56	80	170
Ethyl methionate	300	'Metallic'	1	2	90
Methionyl acetate	50	Mushrooms	1.5	3	92
Methionol	1200	Cooked cabbage	838	1776	90
4-Methylthio-butanol	80	Earthy	36	35	96
Benzothiazole	50	Rubber	2	11	234

probably formed in wine by a direct chemical reaction between H_2S and ethyl alcohol (Maujean et al., 1993).

Other 'light' volatile sulfur compounds are less significant in reduction defects. Carbonyl sulfide is an odorless substance produced by a reaction between carbon dioxide and H_2S (Shaw and Nagy, 1981). Dimethyl sulfide, synthesized by yeast from cystine, cysteine or glutathion (Schreier et al., 1976), has no negative impact on wine aroma. Some authors even consider that it contributes to the bouquet (De Mora et al., 1987). Although carbon disulfide is not actually perceptible in the concentrations present in wine, at high concentrations it may modify a taster's impression of the aroma. It 'masks' pleasant aromas in wine by raising their perception thresholds and accentuates unpleasant odors. This compound's formation mechanism in wine is not very well known.

'Heavy' sulfur compounds in wine have rarely been studied. They are always produced by yeast metabolism during fermentation. Unlike light compounds that may increase after the end of alcoholic fermentation under certain conditions (Volume 1, Section 13.9.1), concentrations of these heavy compounds remain stable during aging, in the great majority of cases. It is, however, impossible to eliminate heavy sulfur compounds from wine, due to their high boiling point and the fact that they do not react with copper.

Among the many heavy sulfur compounds identified in wine, only a few play a significant role in reduction defects (Table 8.15). The most important of these is undoubtedly methionol. It is produced (Figure 8.20) by yeast from methionine in the must, via deamination, followed by decarboxylation (Ehrlich reaction) (Barwald and Kliem, 1971). The aldehyde thus formed (methional) is then reduced by an enzyme reaction into an alcohol (methionol). Settling has a decisive influence on methionol concentrations (Volume 1, Section 13.5.2). When a wine develops a reduction defect attributable to heavy sulfur compounds during alcoholic fermentation, its methionol content is always above the perception threshold. This compound therefore plays a major role in the reduction defects caused by yeast.

The 2-mercaptoethanol concentrations of some wines with reduction odors may also be in the vicinity of the perception threshold. This compound, produced by yeast from cysteine in must (Rapp et al., 1985), may also contribute to the unpleasant odors in certain wines. At concentrations over 90 µg/l, 2-methyl-tetrahydro-thiophenone tends to mask other flavors. The organoleptic impact of the other heavy sulfur compounds identified in wine is negligible. Indeed, although concentrations are higher in wines with reduction defects, they rarely reach the perception threshold.

Thus, the development of reduction defects in wine during alcoholic fermentation is mainly due to the yeast producing abnormally high concentrations of a small number of malodorous sulfur compounds. The most important of these

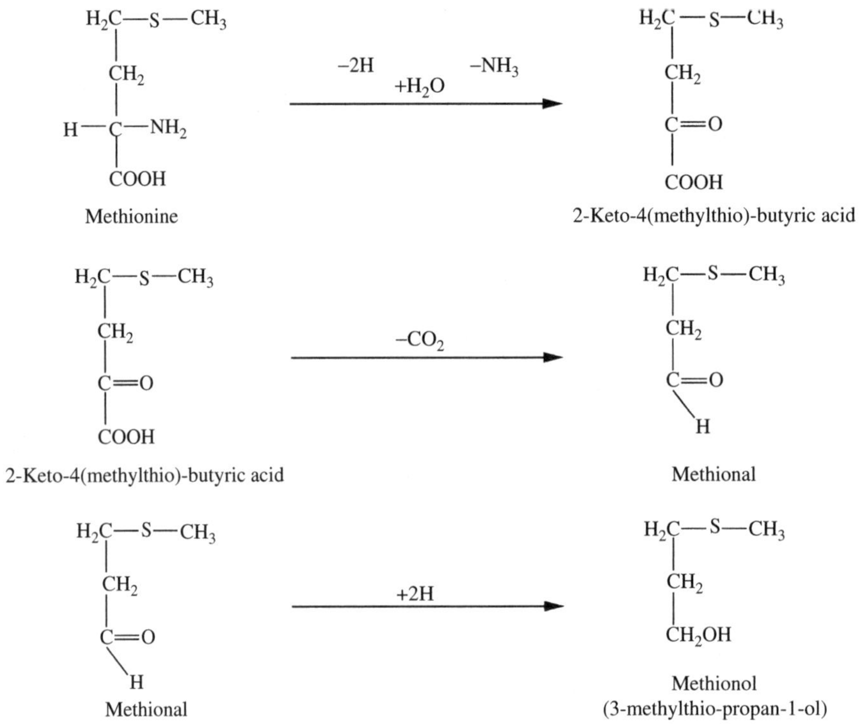

```
H₂C—S—CH₃           H₂C—S—CH₃
   |                   |
  CH₂                 CH₂
   |     −2H   −NH₃    |
H—C—NH₂  +H₂O   →    C=O
   |                   |
  COOH                COOH
Methionine      2-Keto-4(methylthio)-butyric acid


H₂C—S—CH₃           H₂C—S—CH₃
   |                   |
  CH₂                 CH₂
   |      −CO₂         |
  C=O      →         C=O
   |                  \
  COOH                 H
2-Keto-4(methylthio)-butyric acid    Methional


H₂C—S—CH₃           H₂C—S—CH₃
   |                   |
  CH₂                 CH₂
   |      +2H          |
  C=O      →         CH₂OH
  \
   H
Methional         Methionol
              (3-methylthio-propan-1-ol)
```

Fig. 8.20. Formation of methionol from methionine (Ehrlich reaction)

are H_2S, methanethiol and ethanethiol, as well as methionol. It is relatively easy to reduce the H_2S content of a wine by racking and aeration, thanks to its extreme volatility. This is by no means true of methanethiol, ethanethiol and, above all, methionol. The concentrations of these compounds remain stable or increase in wine during aging, contributing to a persistent reduction defect. It is therefore vital to prevent these malodorous compounds from forming in wine.

Research by Lavigne *et al.* (1992, 1993) showed that a few simple precautions during extraction of the juice (settling and sulfuring the must) and fermentation (selected yeast strains and sulfuring the wine) were effective in preventing this problem in dry white wines. It is indispensable to check for the appearance of hydrogen sulfide smells in young wines stored in large vats that have not been clarified. A sample must be taken from the lees at the bottom of the vat. If there is the slightest off-odor, the wine must be aerated, and it may be necessary to remove the lees. Early olfactory defects due to hydrogen sulfide disappear easily. The mercaptans that form at a later stage are more stable and resistant to aeration treatment. These problems rarely occur in barrel-aged wines, as the oxidation–reduction potential is maintained at a higher level.

It is quite true, however, that the lees gradually lose their capacity to reduce sulfur derivatives. This is probably due to the inactivation of the enzyme responsible for reducing sulfites to H_2S (sulfite reductase). It is then possible for methanethiol and ethanethiol to fix on fresh yeasts, with even more serious consequences (Lavigne and Dubourdieu, 1996). Disulfide cross-bonds are formed between cysteine from the yeast wall mannoproteins and the SH group of the sulfur derivatives (Figure 8.21). The copper adsorbed by the yeast lees is involved to a large extent in the formation of disulfide cross-bonds between the free thiols and the cysteine remains of the

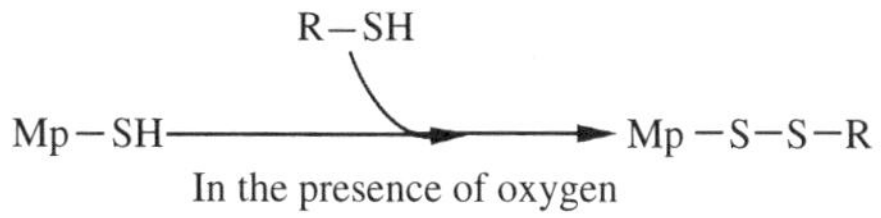

Chemical Nature, Origins and Consequences of the Main Organoleptic Defects

$$Mp-SH \xrightarrow[\text{In the presence of oxygen}]{R-SH} Mp-S-S-R$$

Fig. 8.21. Fixing sulfur derivatives (methanethiol, $R = CH_3$; ethanethiol, $R = CH_2-CH_3$) on the SH groups of mannoproteins (MP)

mannoproteins (Palacios *et al.*, 1997; Vasserot *et al.*, 2003; Maujean, 2001). Provided that certain precautions are taken, it is thus possible to barrel-age white wines on their yeast lees, without any risk of olfactory defects (Volume 1, Section 13.9.2).

Copper turnings are used to eliminate hydrogen sulfide odors produced by yeast metabolism, although their effectiveness is debatable. EEC legislation permits the use of 1 g/hl of copper sulfate, provided that the final copper concentration in the treated wine is no higher than 1 mg/l. Silver chloride and palladium chloride are also effective in eliminating hydrogen sulfide odors. Nitrogen scavenging is also recommended for this purpose, and its effectiveness is improved by repeated aeration. These methods eliminate the most volatile substances first. It should, however, be noted that copper has a negative effect on the varietal aromas of certain grape varieties such as Sauvignon Blanc. It is nevertheless always preferable to take sufficient care and monitor the wine closely enough, to prevent these olfactory defects from developing.

8.6.3 Volatile Sulfur Compounds from Vine Sprays

Elemental sulfur used to spray vines may cause the formation of H_2S during alcoholic fermentation. According to Wainwright (1971), this mechanism is purely chemical and depends on fermentation conditions. It has been demonstrated that the presence of 1 µg/l of sulfur in the must may produce a concentration of H_2S above the perception threshold (0.8 µg/l).

Many fungicides and insecticides also contain one or more sulfur atoms. This is certainly true of dithiocarbamates, reduced to form thiocarbamic acids due to the oxidation–reduction balance. These compounds are reputedly as unstable as their oxygenated counterparts. Thiocarbamic acids are precursors of isothiocyanates, which constitute the active ingredients in fungicides (Figure 8.22).

The mechanisms suggested in Figure 8.22 clearly demonstrate that hydrogen sulfide and carbon sulfide are present in commercial sprays (Maujean *et al.*, 1993). These findings are in agreement with previous observations. Products of this type are likely to contaminate wine.

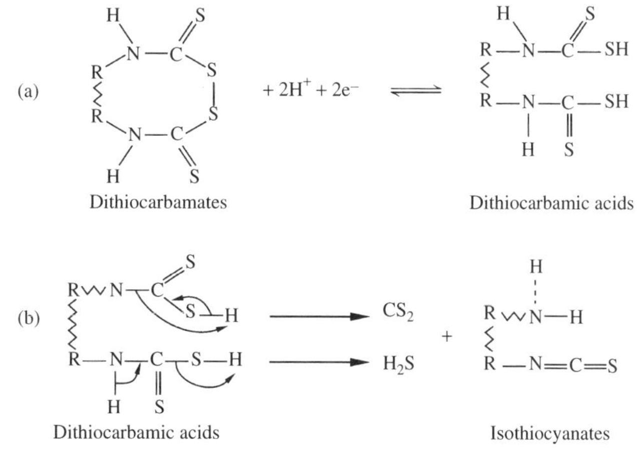

Fig. 8.22. (a) Formation of thiocarbamic acids from dithiocarbamates. (b) Their decomposition produces light sulfur compounds (CS_2 and H_2S)

The composition of other insecticides includes compounds with a methyl radical, linked to a sulfur atom, in turn linked to a carbon or phosphorus atom. Compounds (a) acephate and (b) lannate are examples of this type of product (Figure 8.23).

It has been shown by Chukwudebe (1984) and Rauhut *et al.* (1986) for *S*-methyl-*O*-methyl-*N*-acetyl phosphoramide and Dittrich (1987) for oxime carbamate that the hydrolysis of these two compounds (Figures 8.24 and 8.25) produces methanethiol. This is the principal substance responsible for olfactory defects. Methanethiol may be accompanied by its oxidation by-product, dimethyldisulfide, formed as a result of the oxidation–reduction balance:

$$2CH_3SH \rightleftharpoons CH_3-S-S-CH_3 + 2H^+ + 2e^-$$

These two volatile sulfur compounds appear gradually. Indeed, for kinetics reasons, due to the very low pesticide concentration, the hydrolysis reaction is extremely slow, especially in the case of *S*-methyl-*O*-methyl-*N*-acetyl phosphoramide (acephate). Problems have been observed with wines made from vines treated with orthene (a pesticide with acephate as the active ingredient). The young wine had no reduction defects. However, these developed slowly during aging, sometimes several months after bottling, reaching totally unacceptable levels in some instances.

The ethylated equivalents of these two volatile sulfur compounds may also be formed, as well as trisulfides such as dimethyltrisulfide (DMTS), generated by an oxidation–reduction balance (Nedjma, 1995) in two reactions coupled with copper. This is particularly likely to occur when wines are distilled in copper stills (Figure 8.26), presenting a risk of additional contamination in the brandy.

8.6.4 Heat-Generated Volatile Sulfur Compounds

It is important to take into account the possible presence of sulfur compounds generated by heat in must and wine. Indeed, certain conditions may promote Maillard reactions when wines are made using technologies such as thermovinification (heating the grapes) and high-temperature bottling. There is also a considerable risk when grape

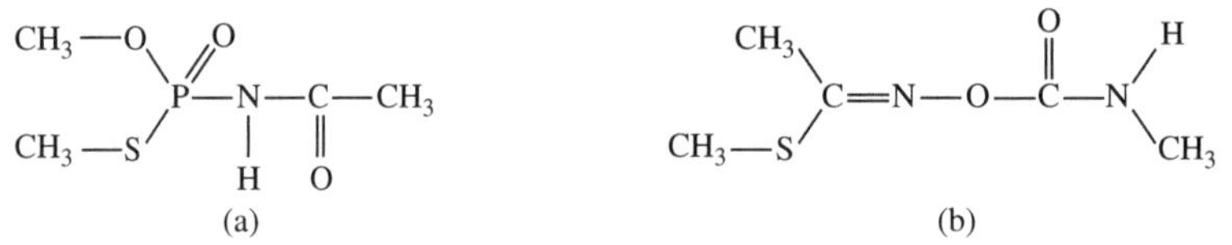

Fig. 8.23. Structure of: (a) acephate and (b) lannate, active ingredients in certain insecticides

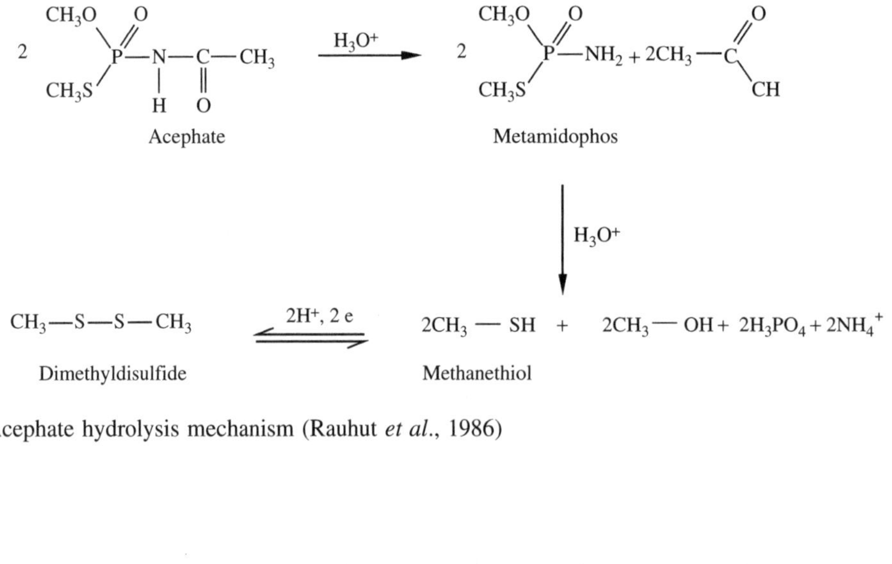

Fig. 8.24. Acephate hydrolysis mechanism (Rauhut *et al.*, 1986)

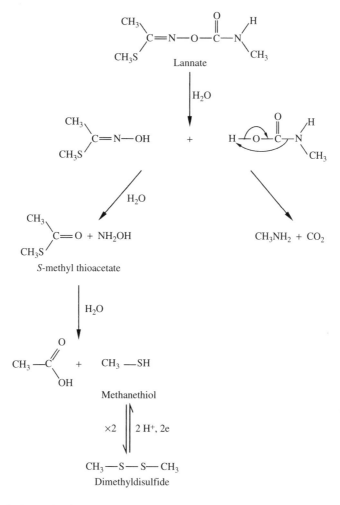

Fig. 8.25. Lannate hydrolysis mechanism

$$\begin{cases} CH_3-SH + H_2S \rightleftharpoons CH_3-S-S-H + 2H^+ + 2e^- \\ 2Cu^{2+} + 2e^- \rightleftharpoons 2Cu^+ \end{cases}$$

$$\begin{cases} CH_3-S-S-H + HSCH_3 \rightleftharpoons CH-S-S-S-CH_3 + 2H^+ + 2e^- \\ 2Cu^{2+} + 2e^- \rightleftharpoons 2Cu^+ \end{cases}$$

Fig. 8.26. Disulfide and trisulfide formation by oxidation–reduction, in the presence of copper

juice is shipped in tank trucks exposed to the sun, or when wine is stored at temperatures above 20 °C during distribution and at the point of sale. This reaction is generally carefully monitored in the agri-food business, as certain products of this Maillard reaction are toxic (e.g. hydroxymethylfurfural and nitrosamine).

The Maillard reaction is best known as the phenomenon responsible for the browning of untreated foodstuffs, by non-enzymic oxidation. It involves the condensation of amino acids on sugars, both aldoses (glucose) and ketoses (fructose) (Figure 8.27). When aldoses are involved, the primary condensation products are aldimines ($R_1 = H$) (or Schiff base), while ketoses produce ketimine ($R_1 \neq H$). Due to their enolizable character, these imines develop according to two tautomeric equilibriums into enaminol and

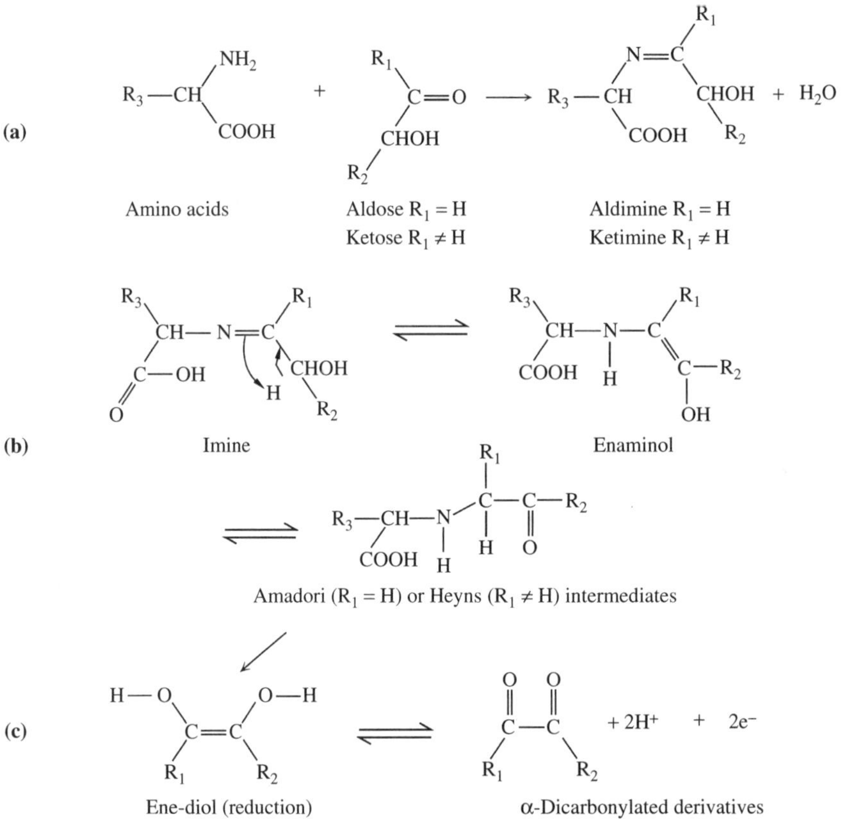

Fig. 8.27. Maillard reaction involved in the non-enzymic oxidative browning of plant tissues. (a) Formation of an imine by an amino acid reacting with an aldose ($R_1 = H$) or ketose ($R_1 \neq H$). (b) Enolization of the imine to enaminol, then to an Amadori ($R_1 = H$) or Heyns ($R_1 \neq H$) intermediate. (c) Breaking of the preceding intermediates, with the appearance of a reductone in redox equilibrium with an α-dicarbonylated compound, responsible for the non-enzymic oxidation phenomenon

then into Amadori ($R_1 = H$) or Heyns ($R_1 \neq H$) intermediates.

The decomposition of these intermediates leads to the formation of ene-diols, also known as reductones, in redox equilibrium with α-dicarbonylated compounds. These are responsible for the oxidation of plant tissues. The α-dicarbonylated compounds resulting from the rearrangement of the Amadori and Heyns intermediates may in turn add amino acids from must and wine (Figure 8.28). The corresponding addition products develop by intramolecular decarboxylation, according to the well-known Strecker breakdown reaction. Amino acids that become involved in this reaction ultimately become aldehydes.

Thus, if the amino acid (Figure 8.23) is alanine ($R_3 = -CH_3$), widely represented in must and wine, the corresponding aldehyde is ethanal. If the amino acid is methionine ($R_4 = CH_3-S-CH_2-CH_2-$), which is certainly only present in small quantities but is reputed to be highly reactive with carbonylated compounds, then methional, or S-methyl-3-propanal, is produced. This compound is thermally unstable and evolves rapidly, via a Retro-Michael reaction, into acrolein and methanethiol (Figure 8.28). These smell of cooked cauliflower, wet dog, etc. In wine, part of the methional returns to methionol via catalyzed reduction by alcohol dehydrogenase with NADH.

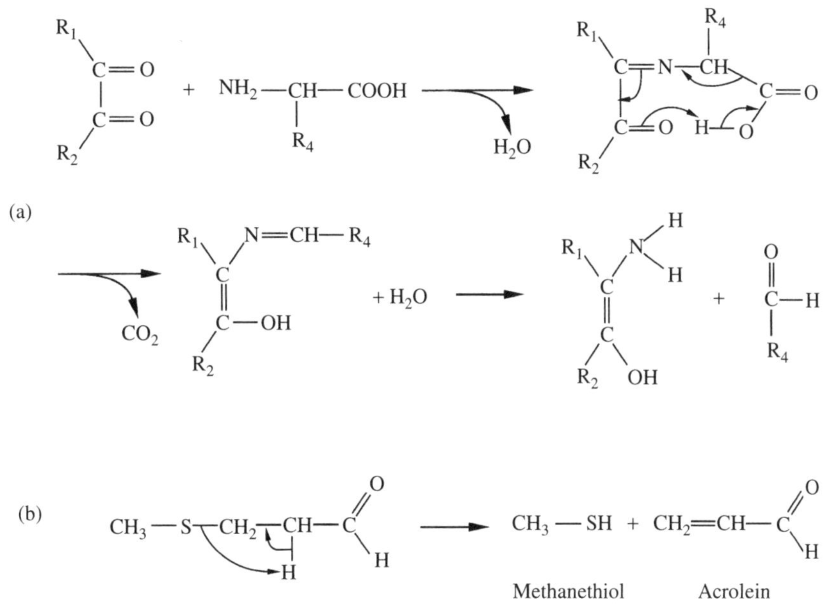

Fig. 8.28. Maillard reaction involved in the appearance of sulfur derivatives. (a) The α-dicarbonylated compound (Figure 8.27) adds an amino acid. The intermediate product is decomposed and an aldehyde is formed. (b) As the amino acid is tyrosine ($R_3 = -CH_2-CH_2-SCH_3$), the aldehyde breaks down and methanethiol is produced

The wines that most require protection from high temperatures (Marai, 1979) are those that contain residual sugar (*vins doux naturels* and sweet wines) and sparkling wines, due to the *dosage* added after the bottles are disgorged and before they are finally corked. It has been observed that Champagne stored for one year at room temperature in a dark place contained 70 times the quantity of total thiols as the same wine stored in a cellar at 10–12 °C.

Dry wines should also be stored at cool temperatures, as pentoses (not fermentable by yeast) are more reactive than hexoses.

8.6.5 Photochemical Origin of Volatile Sulfur Compounds

The oxidation–reduction potential of white wine decreases on exposure to natural light. This property is used to reduce copper and assess the risk of copper casse (Section 4.7.3). In the past, there was even a method of preventing copper casse based on this principle, but which is no longer acceptable.

The implicated wavelengths are centered around 370 and 450 nm (Maujean and Haye, 1978). These two wavelength fields, located on either side of the boundary between ultraviolet and visible light (400 nm), correspond precisely to the two absorption wavelengths of riboflavin, better known as vitamin B_2. Vitamin B_2 is a yellow coloring widely used in the agri-food industry under the code E 101. It is an oxidation–reduction coenzyme involved in the cytochromic bridge, and constitutes an electroactive, photosensitive, biological oxidation–reduction system in white wines in the absence of oxygen.

The decrease in redox potential produced by lamps with a solar emission spectrum, or fluorescent tubes, corresponds to the bleaching of white wines by reducing vitamin B_2 to its (colorless) form (Figure 8.29). The redox potential of a Champagne may drop by over 100 mV. The confusion between 'reduction flavor' and 'sunlight flavor' is, therefore, quite legitimate.

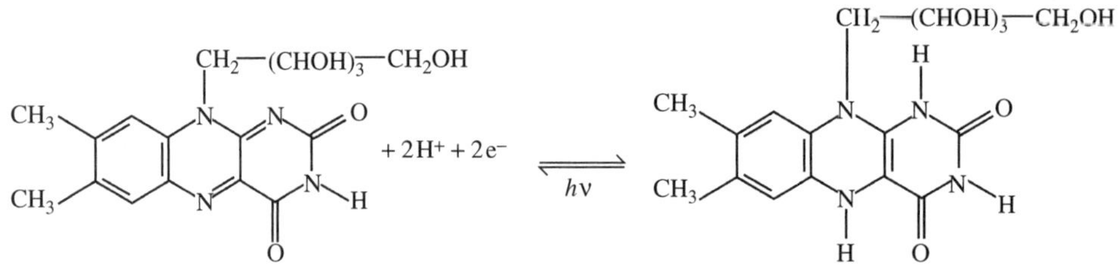

Fig. 8.29. Photochemical reduction of riboflavin involved in 'sunlight flavor'

Exposure to light, especially in the higher energy ultraviolet wavelengths that are not totally absorbed by bottle glass, puts vitamin B_2 in an excited, high-energy state. In this excited state, the riboflavin may dissipate its excess energy in several ways:

1. By emitting light through fluorescence or phosphorescence.
2. By releasing heat energy.
3. By transferring kinetic energy via collisions with molecules for which it has certain affinities.

These molecules include sulfur amino acids, which play an essential role in 'sunlight flavor'. This phenomenon is directly linked to the appearance of methanethiol and dimethyldisulfide in wines exposed to light, which give them cooked cauliflower or wet wool smells.

Investigation of the reaction mechanism (Maujean and Seguin, 1983) showed that 'sunlight flavor' was mainly due to the oxidative photodegradation of methionine (Figure 8.30). It was observed that methional was the primary product of oxidative photolysis of the amino acid.

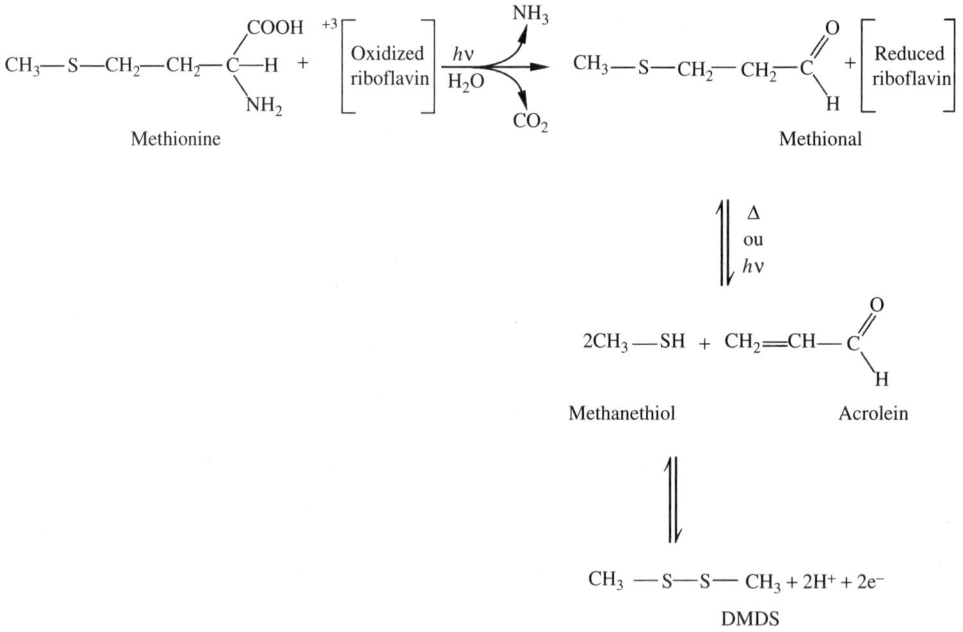

Fig. 8.30. Sequence of reactions involved in the development of 'sunlight flavor' (*3 indicates that the oxidized riboflavin reacts in triple state.)

This compound is photochemically and thermally unstable, evolving into acrolein and methane thiol as a result of these two factors. The methane thiol is then oxidized to a greater or lesser extent according to the wine's redox potential, producing dimethyldisulfide (DMDS). These sulfur derivatives are responsible for the 'sunlight flavor' found in white wines, particularly those bottled in clear glass with a low filter capacity for light, especially the most dangerous types of radiation, at wavelengths around 370 nm.

Among the adjuvants tested as preventive treatments for 'sunlight flavor', copper is the most useful. Copper complies with the three essential requirements: it is a legally permitted substance, it meets organoleptic criteria and, of course, it is effective, both chemically and photochemically. On this last point, there are theoretical grounds for predicting that copper will have a strong preventive effect, due to its affinity for riboflavin and its capacity for complexation with that compound.

The other additives likely to prevent 'sunlight flavor' in white wines include various dimeric and polymeric catechin oligomers isolated from grape seeds. These are effective at concentrations of 40 mg/l.

This action of catechic tannins is due to their high absorption capacity for ultraviolet light, especially that absorbed by riboflavin at 370 nm, which prevents it from reacting with methionine. This explains why red wines, with their high procyanidin content, are much less light-sensitive. Inhibition of the amino acid photolysis reaction may also be explained by the fact that phenols are scavengers of free radicals.

Another adjuvant for preventing, or at least delaying, the development of 'sunlight flavor', is ascorbic acid or vitamin C. (Volume 1, Section 9.5.4) Doses of up to 100 mg/l are legally authorized. Vitamin C is added to sparkling wines made by the traditional method in conjunction with sulfur dioxide in the *dosage* liquor, after the bottles have been disgorged.

Vitamin C is effective in preventing 'sunlight flavor' due to its photosensitive, reducing properties. Photochemical interaction between the two vitamins reduces riboflavin's interaction with methionine. The reversibility of the reaction shows, furthermore, that when wines are no longer exposed to light, the vitamin C is totally recovered. It may then revert to its antioxidant role.

Catechins and ascorbic acid are chemical means of preventing 'sunlight flavor'. The most satisfactory solution, however, is to use bottles made from glass that completely filters out wavelengths around 370 nm. Champagne bottle manufacturers are required to guarantee a protective light filtration capacity of $95 \pm 1\%$. Chemical protection may be envisaged for wines bottled in clear glass. Simple precautions should also be taken during storage and distribution, as well as at the point of sale. In modern wineries, it is recommended to replace neon tubes with sodium or incandescent lamps, as these light sources do not emit ultraviolet radiation. As far as retail sales premises are concerned, especially supermarkets, enologists must ensure that not only sales staff but also consumers are informed that wines are stored, sometimes for months, under neon lights, and often at relatively high temperatures. Such unsuitable conditions promote the production of methanethiol and dimethyldisulfide, both photochemically as well as thermally, via Maillard reactions.

Reduction reactions producing the olfactory defect known as 'sunlight flavor' occur in white wines, especially sparkling wines. As these reactions have not always been correctly identified, they are perhaps more common than was generally thought. They are certainly involved in certain olfactory defects (such as *rancio* odor) that develop in dry white wines during bottle aging, and were previously inaccurately attributed to winemaking errors.

Finally, some samples of very young wines still aging on the lees (mainly white but sometimes red) are extremely light-sensitive and develop very strong reduction odors. This defect gradually disappears as the wine ages and has not yet been fully explained.

8.7 PREMATURE AGING OF WHITE WINE AROMA

8.7.1 Type of Defect and Molecules Responsible for Defective Aging Aroma

When white wine aroma ages prematurely, they rapidly lose the fruity bouquet of young wines and develop heavier aroma reminiscent of resin, polish, camphor, or even honey and mead. This unusual odor, similar to that of oxidized white wines, may affect all types of dry and sweet white wines, irrespective of the vineyard region or grape variety. It develops to the detriment of the empyreumatic, mineral, and truffle nuances characteristic of the "reduction bouquet" of white wines.

The contribution of 2-aminoacetophenone (Figure 8.31) to the prematurely aged aroma of white wines, initially demonstrated in white German wines (Rapp *et al.*, 1993), has now been

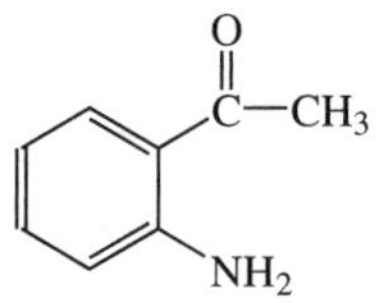

Fig. 8.31. 2-aminoacetophenone

established in many white wines from other vineyard regions. Indeed, the odor of this compound is reminiscent of prematurely aged white wines: naphthalene, acacia flowers, and "Mediterranean bouquet".

Two pathways have been envisaged for the biosynthesis of 2-aminoacetophenone from tryptophan, to explain how it is formed in wine. The first pathway is physicochemical, via indoleacetic acid, while the second is enzymatic, involving cynurease (Rapp *et al.*, 1998; Gebner *et al.*, 1998) (Figure 8.32).

Some prematurely aged wines contain up to 5000 ng/l of 2-aminoacetophenone (Rapp *et al.*,

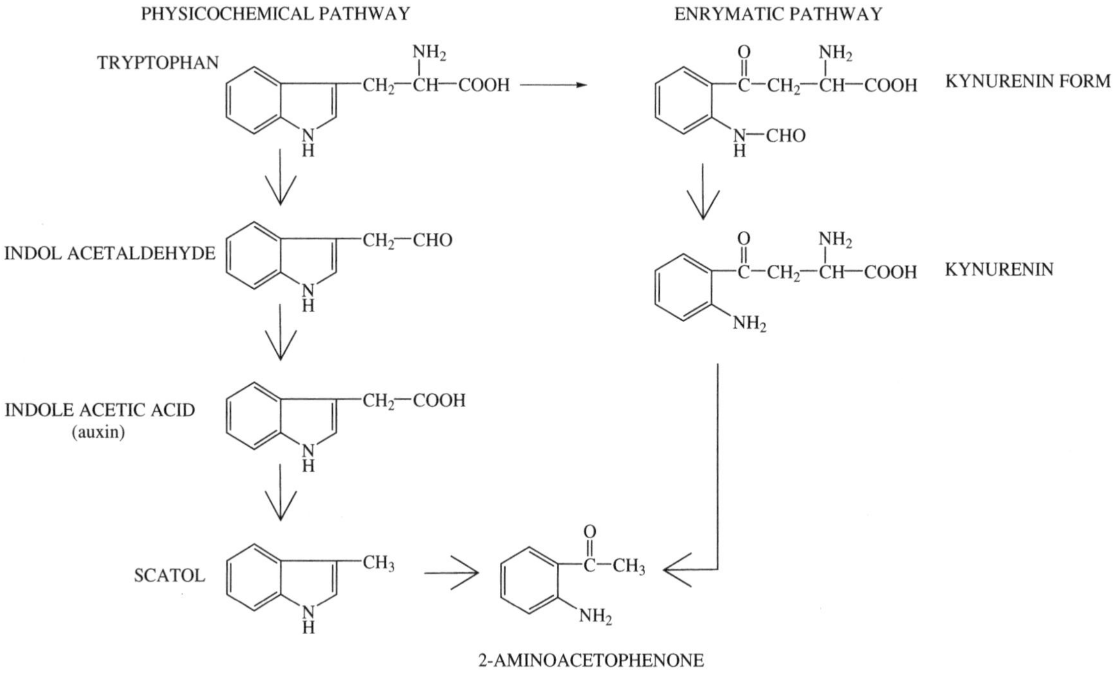

Fig. 8.32. 2-Aminoacetophenone formation pathways

1998), significantly above the perception threshold of 800 ng/l.

Wines with a 2-aminoacetophenone content close to the perception threshold are always described by tasters as "prematurely aged". However, some wines with this aroma defect do not contain any 2-aminoacetophenone, so this is not the only molecule responsible for white wines with prematurely aged aromas.

Sotolon (3-hydroxy-2(5H)-furanone) (Figure 8.33) has also been identified in prematurely aged white wines (Dubourdieu and Lavigne, 2002). Quantities are much lower than in sweet wines aged under oxidizing conditions (Volume 1, Sections 10.6.4, 14.2.3, 14.5.2) (Dubois et al., 1976; Guichard et al., 1993; Cutzach et al., 1998), but may still exceed the perception threshold (8 μg/l). A higher sotolon content that may be in the mg/l ranges contributes to the walnut, fig, and rancio aroma in Sherry- and Port-style wines. At the concentrations found in prematurely aged white wines, sotolon is more reminiscent of polish and this impression is reinforced by the presence of 2-aminoacetophone.

Several pathways leading to the formation of sotolon have been described, including two that are likely to occur in white wine. This compound may be formed from threonine in the presence of glucose and oxygen in an acid medium (Takahashi et al., 1976; Pham et al., 1995; Cutzach et al., 1998), or, as is the case in lemon juice, in the presence of ethanol, ascorbic acid, and oxygen (König et al., 1999).

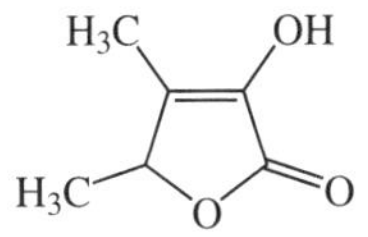

Unlike 2-aminoacetophenone, wines with prematurely aged aromas always contain sotolon.

A number of winemaking factors, such as maintaining them on their lees throughout barrel aging, preserve the fruity character of young dry white wines and minimize or delay their premature aging.

8.7.2 Impact of Aging Conditions on the Defective Aging of Dry White Wines

Dubourdieu and Lavigne (2002) demonstrated the protective effect of lees on the premature aging of Sauvignon Blanc aroma. When these wines are aged under reducing conditions, i.e. on total lees in used barrels, the loss of fruity aroma is limited and the formation of sotolon and 2-aminoacetophenone is attenuated. On the contrary, aging dry white wines in new oak without their lees promotes premature aging (Tables 8.16 and 8.17).

This phenomenon is only partially explained by the lees' capacity to combine oxygen (Salmon et al., 1999), as some wines develop prematurely aged aroma although they have been maintained on total lees in used barrels or in vat.

Fig. 8.33. 3-Hydroxy-2(5H)-furanone (Sotolon)

Table 8.16. Impact of barrel-aging conditions on the concentration of volatile thiols characteristic of Sauvignon Blanc varietal aroma

	Sampling date	4-MMP (ng/l)	3- MH (ng/l)
Used barrel	End of AF	11	1501
	April (following year)	13	1318
Used barrel + racking	End of AF	11	1501
	April (following year)	10.1	717
New barrel	End of AF	10	1406
	April (following year)	8.3	1235
New barrel + racking	End of AF	10	1406
	April (following year)	5.5	520

(4-MMP: 4-methyl-4-mercaptopentanone; 3-MH: 3-mercaptohexanol)
(AF: alcoholic fermentation)

Table 8.17. Impact of aging conditions (6 months) on sotolon and 2-aminoacetophenone formation in a Sauvignon Blanc wine

	Used barrel on lees	Used barrel: no lees	New barrel on lees	New barrel: no lees
Sotolon (µg/l)	1	2.6	4	8.2
2-aminoacetophenone (ng/l)	<20	75	80	128

8.7.3 The Role of Glutathion in the Premature Aging of White Wines

The presence of several mg/l glutathion, a sulfur-based peptide with antioxidant properties, in wines has already been reported (Adams and Liyanage, 1993; Lavigne *et al.*, 2003). This compound, present in grapes and must (Cheynier *et al.*, 1989; Dubourdieu and Lavigne, 2002), is released by the yeast at the end of alcoholic fermentation. The glutathion content of a wine at the beginning of aging depends on the initial concentration in the must as well as on the proper completion of alcoholic fermentation (Lavigne *et al.*, 2003). Little or no glutathion is released by the yeast if fermentation is sluggish.

The glutathion content of wine decreases inevitably during aging (Table 8.18), because of its strong propensity for reacting with oxygen (Adams and Cassol, 1995) and oxidized phenolic compounds (quinones) (Singleton *et al.*, 1984, 1985; Cheynier *et al.*, 1986). Naturally, the more reducing the conditions in the wine during aging (used barrels and aging on total lees), the better the glutathion is preserved.

Thus, the yeast lees release several milligrams per liter of glutathion into the wine at the beginning of the aging process. This reducing compound protects the white wine aroma from premature aging.

The important role played by glutathion in the development of white wine aroma during bottle-aging was demonstrated by the following experiment: 10 mg/l glutathion was added to a Sauvignon Blanc wine when it was bottled. Three years later, the volatile thiol and sotolon content was assayed and the yellow color (OD 420) measured (Table 8.19).

The wine bottled with the highest glutathion content clearly showed the least yellowing and oxidative aroma (sotolon content) and had best retained its fruitiness, assessed by assaying 3-MH.

Cheynier *et al.* (1989) and Liyanage *et al.* (1993) found that grapes contained large quantities of glutathion—up to 300 mg/l. Although the mechanisms for this compound's accumulation in grapes have not yet been fully elucidated, the vine's water and nitrogen supply are apparently decisive factors.

Insufficient nitrogen supply to the vine is known to result in musts with a deficiency in available nitrogen, required by the yeast metabolism (Volume 1, Sections 2.4.2, 3.4.2). When this is the case, the must also has a low glutathion content (Dubourdieu and Lavigne, 2002). It is possible to improve the grapes' glutathion content by using nitrogen-based fertilizer on the vines. Thus, fertilizing a vineyard suffering from severe nitrogen deficiency (low vigor and yellowing leaves) with ammonium nitrate (60 units) in June resulted in must with a nitrogen content comparable to that of

Table 8.18. Glutathion content (mg/l) of Sauvignon Blanc wines after eight months in barrel

	Used barrel on lees	Used barrel: no lees	New barrel on lees	New barrel: no lees
April	5.8	3.1	4.8	2

Table 8.19. Yellow color measurements as well as 3-mercapto-hexanol (3MH) and sotolon content of wines after three years in bottle

	Control	Wine supplemented with glutathion (10 mg/l)
OD 420	0.203	0.136
3-MH (ng/l)	320	445
Sotolon (µg/l)	9	3

a control must from a vineyard with naturally high nitrogen levels.

Water supply to the vine also seems to affect the accumulation of glutathion in the grapes. A moderate water deficit is more favorable to glutathion accumulation than severe water stress.

These findings indicate that the premature aging of white wines is frequently associated with particular vineyard conditions—vines suffering from a nitrogen deficiency or subjected to excessive water stress.

8.8 ORGANOLEPTIC DEFECTS ASSOCIATED WITH GRAPES AFFECTED BY VARIOUS TYPES OF ROT

8.8.1 Types of Defects Associated with Rot

Defects indicated by fungal, moldy, or earthy odors have long been reported in winemaking (Sémichon, 1905; Ribéreau-Gayon and Peynaud, 1964). These defects may result from contamination of the must or wine during the winemaking or aging process, via contact with contaminated materials (corks and treatment products) or containers (underground vats and barrels). More commonly, these defects are present in wines made from grapes affected by gray rot because of *Botrytis cinerea*, frequently associated with other types of rot (white, green, yellow, etc.). Heavy rains and hail damage are considered to be aggravating factors.

A significant increase in these types of problems has recently been observed in a number of wine growing regions, where grapes and wines have been contaminated with odors of damp earth, beetroot, and humus, sometimes with such intensity that quality drops off sharply. In other cases, smells of humus or camphor are not immediately perceptible to the nose, but develop retronasally. The common characteristic of these defects is that they concern ripe grapes, apparently affected by gray rot. In some cases, defective aromas detectable on grapes (fungal, moldy, earth, etc.) disappear during winemaking operations prior to fermentation, or during alcoholic fermentation, because of the metabolic activities of *Saccharomyces cerevisiae* yeast.

8.8.2 The Compound Responsible for the Main Earthy-Smelling Defect: (−)-Geosmin

Analysis of a number of wines with marked damp-earth odors, using gas-phase chromatography (GPC) coupled with olfactometric detection, revealed that the only compound with this smell that was present was geosmin, or *trans*-1,10-dimethyl-*trans*-decalol, a compound in the terpene family (Figure 8.34) (Darriet *et al.*, 2000). The name, invented by Gerber and Lechevallier (1965), is based on Greek roots (geo: earth + osem: odor).

Geosmin is a highly odoriferous compound (perception threshold: 10 ng/l in water, 40 ng/l in model solution with a similar composition to that of wine, and 50–80 ng/l in wine). It is present in wine in chiral form (−)-geosmin (Darriet *et al.*, 2001), at concentrations up to 400 ng/l (Table 8.20). Geosmin is relatively stable, degrading very slow at the pH of wine. A wine must be stored for 2 months at 20°C or for 8 months at 10°C to lose 50% of its geosmin content.

Geosmin is produced biologically, by *Actinomycetes* bacteria in the *Streptomyces sp.* genus and a number of fungi, particularly those in the *Penicillium sp.* genus (Mattheis and Roberts, 1992; Darriet *et al.*, 2001). Geosmin has been well known for many years as a pollutant in water (Gerber and Lechevallier, 1965) and a number of foods (Darriet *et al.*, 2001). Contaminated barrels and corks have also been reported as sources for the geosmin found in wine (8.5) (Amon *et al.*, 1987, 1989).

The geosmin in the wines analyzed was already present in the grapes, provided they had been at

Fig. 8.34. Formula for geosmin

Table 8.20. Geosmin content of red and white wines from several wine growing regions (Darriet *et al.*, 2000; La Guerche *et al.*, 2003a)

Origin	Grape variety	Geosmin concentration (ng/l)
Bordeaux		
Haut-Médoc Q1 (2000)[a]	Cabernet Sauvignon	63
Haut-Médoc C1 (1998)		300
Haut-Médoc G1 (1994)		60
Bordeaux P1 (1999)		120
Sauternes SA1 (2000)	Sémillon	216
Sauternes SA2 (1998)		82
Burgundy		
Hautes Côtes de Beaune (2002)	Pinot Noir	95
Pommard 1er cru (2001)		75
Beaujolais		
Beaujolais nouveau D85 (2002)	Gamay	130
Beaujolais P (2002)		400
Loire Valley		
Touraine T16 (2002)		230
Touraine T20 (2002)		95

[a]The appellation is given to indicate the origin and does not confirm that the wine concerned was approved for that AOC.

least partially affected by gray rot. This compound has never been isolated, however, from healthy grapes, even those from plots affected by geosmin. Analysis of the microflora on black and white grapes (Sémillon, Cabernet Sauvignon, Gamay, and Pinot Noir) from a number of wine growing regions (Bordeaux, Beaujolais, Burgundy, and Loire) containing geosmin showed that a fungus in the *Penicillium expansum* species was systematically present, together with *Botrytis cinerea* (La Guerche *et al.*, 2004; La Guerche *et al.*, 2003a) (Table 8.21).

The presence of geosmin in grapes is always associated with *Botrytis cinerea* infection. For this reason, all possible protective measures must be applied in the vines to prevent the grapes from being affected by *Botrytis cinerea*. Special care must be taken to ensure that the grapes remain

Table 8.21. Distribution of isolates of *Botrytis cinerea* and several species of *Penicillium sp.* on grapes from various grape varieties produced in a number of French wine growing areas (La Guerche *et al.*, 2005)

Microorganisms	Wine-producing region													
	Medoc (site 1)				Sauternes (site 2)				Beaujolais		Burgundy	Loire Valley		
									Sites analyzed in 2002					
	1999	2000	2001	2002	1999	2000	2001	2002	1	2	1	1	2	3
Botrytis cinerea	+++	+++	+++	++	+++	+++	+++	++	+++	+++	+++	+++	+++	+++
P-expansum*	[+++]	[+++]	[+++]	[++]	[+++]	[+++]	[+++]	[++]	[+++]	[+++]	[+++]	[+++]	[+++]	[+++]
P-thomii*	+	++	−	+	+	−	−	−	−	+	−	+	+	−
P-purpurogenum*	+	+	+	−	+	−	++	−	−	−	+	−	−	−
P-frequentans	−	−	−	−	+	+	+	−	−	−	++	−	−	++
P-stoloniferum	−	+	−	+	+	−	+	−	+	−	+	+	+	−
P-roquefori	[+]	−	−	−	−	−	−	−	−	−	−	−	−	−

− no microorganisms; + few isolates (<5); ++ some isolates (5–10); +++ many isolates (>10)
[+] geosmin production in model medium
*microorganism characterized by molecular biology

healthy, especially if the vineyards are subjected to heavy precipitation. Ongoing research is focused on studying the spatio-temporal dynamics of microbial populations and finding methods for controlling them in the vineyard.

8.8.3 Other Defects Associated with Rot on the Grapes

Other fungal, moldy, or earthy defects have been identified in juice made from black and white grapes affected by various kinds of rot. The microorganisms responsible have not always been identified, but a considerable number of odorous zones have been identified by GPC-olfactometry, and some of the odoriferous volatile compounds have also been identified (La Guerche et al., 2003b).

Some unsaturated alcohols and ketones with 8 carbon atoms, including 1-octen-3-ol, identified by Schreier et al. (1976), as well as 1-octen-3-one and 2-octen-1-ol, are systematically associated with the mushroom smell that is characteristic of grapes infected with gray rot (Table 8.22) (La Guerche et al., 2003b). These degradation products of unsaturated fatty acids are metabolites of many species of fungus (Badings, 1970; Tressl et al., 1982). Other complex compounds smelling of earth and camphor have been identified in black and white grapes picked following large-scale aerobic development of Botrytis cinerea in the vineyard. Among the earthy-smelling compounds, 2-methylisoborneol is apparently mainly responsible for the moldy odors of black grapes infected with Botrytis cinerea, often right to the center of the grape bunches (Table 8.22).

Table 8.22. Olfactory perception threshold of compounds identified in grapes infected with various types of rot (La Guerche et al., 2003b)

Compound	Odor	Olfactory perception threshold in water (μg/l)
1-octen-3-one	mushrooms	0.003
1-octen-3-ol	mushrooms	2
2-octen-1-ol	mushrooms	100[a]
2-methylisoborneol	earthy	0.012
(−)-geosmin	earthy	0.01[b]

[a]Kaminski et al., 1972;
[b]Polak and Provasi, 1992.

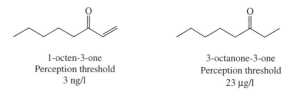

Fig. 8.35. Enzyme reduction of 1-octen-3-one to 3-octanone by Saccharomyces cerevisiae

Many of these compounds break down during alcoholic fermentation to form much less odoriferous substances. In this way, 1-octen-3-one is reduced by the yeast's enone-reductase activity to 3-octanone, with a perception threshold 1000 times lower (Wanner and Tressl, 1998) (Figure 8.35). 2-methylisoborneol is also rapidly degraded, so it cannot contaminate wine aroma. However, some compounds found in Pinot Noir, with a strong retronasal aroma of humus and camphor, remain stable over time. Defects such as these, as well as geosmin, indicate the emergence of new types of rot in the vineyards when they are affected by bad weather during ripening, leading to organoleptic defects of varying severity.

8.9 MISCELLANEOUS DEFECTS

8.9.1 The Breakdown of Sorbic Acid and 'Geranium Odor'

Issues relating to the properties and conditions for use of sorbic acid as a preservative in wine have been described elsewhere (Volume 1, Section 9.2.3). This adjuvant to SO_2 protects wine from fermentation in the presence of yeast, but has no effect on bacterial activity. Furthermore, it is broken down by lactic bacteria, developing a very unpleasant smell reminiscent of geranium leaves. All species of lactic bacteria occurring in wine seem capable of causing this reaction. The use of sorbic acid must therefore be restricted to sweet wines, in conjunction with a sufficiently high dose of SO_2 to prevent any bacterial activity.

A study of the breakdown of sorbic acid by bacteria (Crowel and Guymon, 1975) showed that the first molecule to be produced was the corresponding alcohol, 2,4-hexadiene-1-ol (Figure 8.36). Its

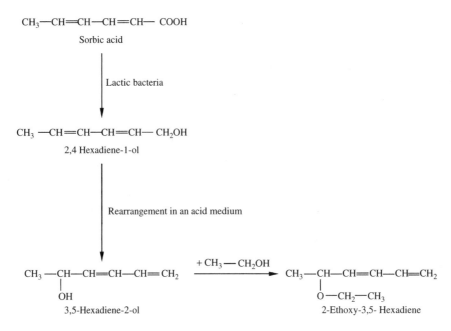

Fig. 8.36. Transformation of sorbic acid by lactic bacteria

aldehyde is probably an intermediate product, but this could not be demonstrated. Adding this alcohol, mixed with its lactate or acetate, to wine reproduces the characteristic unpleasant smell. Furthermore, in the hours following the addition of these compounds, the perception threshold drops to one-fiftieth, or even one-hundredth, of its initial value. This phenomenon was originally attributed to ethanol acting as a solvent for molecules that were not highly soluble in an aqueous medium. However, it is now known that ethanol causes a modification in the chemical structure of sorbic acid breakdown products (Figure 8.36). Initially, rearrangement of these molecules in an acid medium causes a shift in the double bonds and the alcohol function, producing 3,5-hexadiene-2-ol. An ether oxide is then produced in the presence of ethanol, leading to the formation of 2-ethoxy-3,5-hexadiene. This extremely odoriferous molecule plays a major role in 'geranium odor', perhaps in conjunction with other similar molecules.

8.9.2 'Mousiness'

This olfactory defect actually gives wine a smell reminiscent of mice. It occurs in wine stored under poor conditions, especially if it is insufficiently sulfured. It was no doubt for this reason that Schanderl (1964) linked 'mousiness' to an oxidation–reduction defect.

Mousiness is a strong stink perceptible on the aftertaste, which seems to be due to relatively non-volatile products. It is most readily identified by wetting a finger in the contaminated liquid and letting it dry. The odor appears when the liquid has been evaporating for some time (Ribéreau-Gayon *et al.*, 1975).

One hypothesis suggests that yeasts in the genus *Brettanomyces* may play a role in the appearance of this defect (Ribéreau-Gayon *et al.*, 1975; Heresztyn, 1986a). Although their participation in 'mousiness' has not been totally ruled out, it is now known that these contaminant yeasts are mainly responsible for phenol off-odors in red wines (Section 8.4.5). These off-odors result from the conversion of cinnamic acids into ethylphenols, which have an unpleasant odor reminiscent of horses and barnyards.

Apparently, lactic bacteria, and particularly lactobacilli (*Lactobacillus hilgardii*, *L. brevis* and *L. cellobiosus*), are the main microorganisms responsible for 'mousiness' (Heresztyn, 1986a, 1986b).

'Mousiness' has long been believed to be caused by the presence of acetamide (CH_3–CO–NH_2). Heresztyn (1986a, 1986b) associated this olfactory defect with two function isomers (I and II) of 2-acetyltetrahydropyridine (Figure 8.37), each a tautomer of the other. The perception threshold of these compounds in water is very low, around 1.6 ng/l.

The development of these substances would require highly specific environmental conditions (the presence of ethanol and lysine). Under certain conditions, *Brettanomyces* may be involved as well as *Lactobacillus*.

8.9.3 'Bitter Almond Flavor' Caused by a Material in Contact with Wine

A characteristic bitter almond odor is generally attributed to the presence of benzoic aldehyde, which has a perception threshold in water of 3 mg/l (Simpson, 1978). Gunata (1984) and Baumes *et al.* (1986) studied the aroma precursors in grapes and showed that benzoic aldehyde was always present in wine. Concentrations produced by alcoholic fermentation of must never exceed 0.5 mg/l. This value may be higher in wine made by carbonic maceration (Ducruet *et al.*, 1983). There is also a significant increase in concentrations of this aldehyde in Champagne as it ages (Loyaux, 1980).

These normal quantities, attributable to specific winemaking techniques, have no organoleptic effects. However, this substance may also be produced accidentally, causing a 'bitter almond flavor'. This defect corresponds to a benzoic aldehyde content that may be as high as 20 mg/l. This problem is related to the chemical composition of the walls of storage vats. Benzyl alcohol is used as a solvent in the resin base (epichlorhydrine and bisphenol A) and hardeners (aromatic amine) incorporated in epoxy resin linings (Blaise, 1986). If the vat lining is not correctly applied, residual benzyl alcohol from the resin may migrate after polymerization and penetrate into the wine, where it is oxidized to form the benzoic aldehyde responsible for this organoleptic defect (Figure 8.38).

Other cases of contamination have been due to the wine coming into contact with certain materials, especially those used in filtration (Section 11.10.1).

8.9.4 Eliminating Organoleptic Defects

Old enology textbooks describe many somewhat empirical processes likely to attenuate off-flavors and unpleasant smells (Ribéreau-Gayon *et al.*, 1977). Thanks to recent progress in winemaking and aging techniques that have made it possible to avoid these problems altogether, such corrective measures are of much less interest to winemakers and, indeed, legislation on this issue is becoming increasingly strict.

Absorbent charcoal has been used for many years. There are many different preparations (based on animal or plant charcoal), which have been subjected to various activation processes. These products are relatively suitable, either for eliminating unpleasant smells or for removing stains or maderized color from white wines. The use of doses up to 100 g/hl is permitted by European legislation for the treatment of white wines. Doses of 10–50 g/hl are generally sufficient to treat the color of white wines, while effective deodorization may require

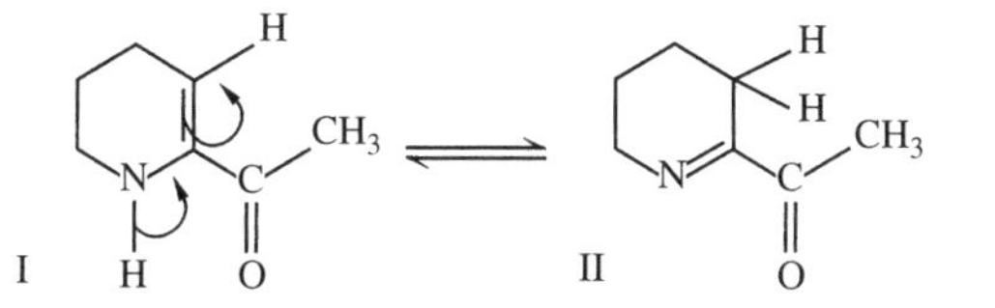

Fig. 8.37. Tautomeric forms of 2-acetyltetrahydropyridine, assumed to be responsible for 'mousiness'

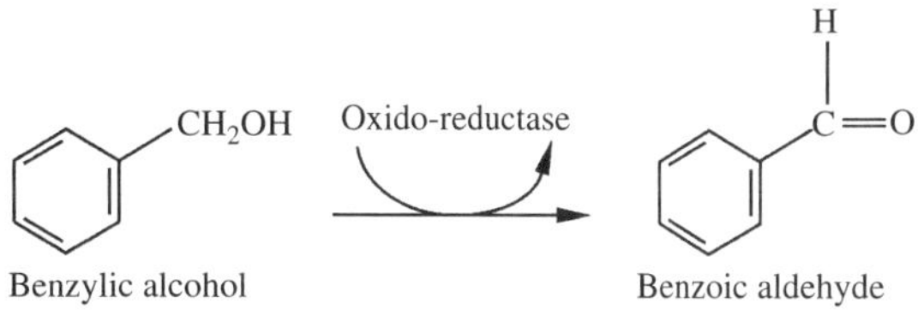

Fig. 8.38. Appearance of a bitter almond flavor due to the formation of benzyl aldehyde from residual benzyl alcohol in the epoxy resin lining of vat walls

up to 100 g/hl. The result depends on the type of defect and the quality of the charcoal. Preliminary laboratory tests are recommended before each treatment. Of course, wines treated in this way also lose their fruity aromas and freshness.

The effectiveness of treatment depends on maintaining a good mixture for several days. This is achieved by agitating the wine, but it is not easy to mix large quantities evenly. The charcoal is eliminated by fining and filtration. There is also an undeniable risk of oxidation.

Besides charcoal, ancient enology treatises mention other products likely to eliminate unpleasant smells and off-flavors: toasted barley or wheat, mustard flour, oil, milk, etc. All these have practically disappeared from use. Fresh yeast lees are permitted in treating wine and are effective in eliminating a number of olfactory defects. This has already been mentioned in connection with fixing certain thiols, such as methanethiol (Section 8.6.2). This treatment is also recommended for adsorbing chloroanisoles in moldy wines (Section 8.5.2).

Oil acts by extracting liposoluble substances. It is stirred vigorously to create an emulsion in the wine. This operation is repeated several times, until the oil is perfectly dispersed. When the mixture is allowed to rest, a layer of oil forms on the surface and is eliminated by careful racking. Liquid paraffin is mentioned in many books on this subject. It was not only recommended for eliminating unpleasant smells but also for isolating wine from the air in partly empty vats in order to protect it from acetic bacteria. In certain countries, solid paraffin discs, impregnated with allyl isothiocyanate, are used to create a sterile atmosphere.

Treatment with whole milk is sometimes assimilated to fining with casein, although the milk fat provides an additional deodorizing effect.

The prevention and treatment of reduction defects, due to sulfur derivatives produced by yeast during the fermentation of white wine, is described elsewhere (Volume 1, Section 13.6).

REFERENCES

Adams D.O. and Liyanage C. (1993) *Am. J. Enol. Vitic.*, 44 (3), 333.
Adams D.O. and Cassol T. (1995) *Am. J. Enol. Vitic.*, 46 (3), 410.
Amon J.M. and Simpson R.F. (1986) *Austral. Grape Grower Wine Maker.*, 268.
Amon J.M., Simpson R.F. and Vandepeer J.M. (1987) *Aust. N.Z. Wine Ind. J.*, 2, 35–37.
Amon J.M., Vandepeer J.M. and Simpson R.F. (1989) *Wine Industry J.*, 4.
Amon J.M., Vandepeer J.M. and Simpson R.F. (1989) *Aust. N.Z. Wine Ind. J.*, 4, 62–69.
Badings H.T. (1970) *Ned. Melk-Zuiveltijdschr*, 24, 147.
Barbe Ch. (1995) Thèse de Doctorat, Université de Bordeaux II.
Barwald G. and Kliem D. (1971) *Chemie Mikrobiologie Technologie der Lebensmittel*, 1, 27.
Baumes R., Cordonnier R., Nitz S. and Drawert S. (1986) *J. Sci. Food. Agric.*, 37, 9.
Blaise A. (1986) Thèse de Doctorat ès Sciences Pharmaceutiques, Université de Montpellier I.
Boulton R.B., Singleton V.L., Bisson L.F. and Kunkee R.E. (1995) *Principles and Practices of Wine Making*. Chapman and Hall Enology Library, New York.
Burckhardt V.R. (1976) *Deutsche Lebensmittel Rundshau*, 72 (12), 417.
Cavin J.F., Andioc V., Etievant P.X. and Divies C. (1993) *Am. J. Enol. Viticult.*, 44 (1), 76–80.
Chatonnet P. (1995) Influence des procédés de la tonnellerie et des conditions d'élevage sur la composition et la qualité des vins élevés en fûts de chêne, Thèse Doctorat, Université de Bordeaux II.
Chatonnet P. and Boidron J.-N. (1988) *Sci. Aliments*, 8, 479.
Chatonnet P., Boidron J.-N. and Dubourdieu D. (1989) *Conn. Vigne Vin*, 23, 59.
Chatonnet P., Labadie D. and Bouton S. (2003) *J. Int. Sci. Vigne Vin.*, 37 (3), 81.
Chatonnet P., Barbe Ch., Canal-Llauberes R.M., Dubourdieu D. and Boidron J.-N. (1992a) *J. Int. Sci. Vigne Vin*, 26 (4), 253.
Chatonnet P., Dubourdieu D., Boidron J.-N. and Pons M. (1992b) *J. Sci. Food Agric.*, 60, 165.
Chatonnet P., Boidron J.-N. and Dubourdieu D. (1993a) *J. Int. Sci. Vigne Vin*, 27 (4), 277.
Chatonnet P., Dubourdieu D., Boidron J.-N. and Lavigne V. (1993b) *J. Sci. Food Agric.*, 62, 191.
Chatonnet P., Guimberteau G., Dubourdieu D. and Boidron J.-N. (1994) *J. Int. Sci. Vigne Vin*, 28 (2), 131.
Chatonnet P., Dubourdieu D. and Boidron J.-N. (1995) *Am. J. Enol. Viticult.*, 46 (4), 463.
Chatonnet P., Viala C. and Dubourdieu D. (1997) *Am. J. Enol. Viticult.*, 48 (3), 443–448.
Cheynier V., Trousdale E., Singleton V.L., Salgues M. and Wylde R. (1986) *J. Agric. Food Chem.*, 34, 217.
Cheynier V., Souquet J.M. and Moutounet M. (1989) *Am. J. Enol. Vitic.*, 40 (4), 320.

Chukwudebe A.C. (1984) *J. Environ. Sci. Health B.*, 19 (6), 801.
Crowel E.A. and Guymon M.F. (1975) *Am. J. Enol. Viticult.*, 26 (2), 97.
Cutzach I., Chatonnet P. and et Dubourdieu D. (1998) *J. Int. Sci. Vigne Vin.*, 32 (4), 223. et 32 (4), 99.
Darriet Ph., Pons M., Lamy S. and Dubourdieu D. (2000) *J. Agric. Food Chem.*, 48, 4835.
Darriet Ph., Lamy S., La Guerche S., Pons M., Dubourdieu D., Blancard D., Steliopoulos P. and Mosandl A. (2001) *Eur. Food Res. Technol.*, 213 (2), 122.
De Mora S.J., Eschenbruch R., Knowles S.J. and Spedding D.J. (1986) *Food Microbiol.*, 3, 27.
De Mora S.J., Knowles S.J., Eschenbruch R. and Torrey W.J. (1987) *Vitis*, 26, 79.
Di Stefano R. (1985) *Vigne Vini*, 5, 35.
Dittrich H.H. (1987) *Mikrobiologie des Weines*, 2nd edn, Verlag Eugen Ulmer Ed., Stuttgart.
Dubois P.J. (1983) Volatile phenols in wine. In *Flavour of Distilled Beverages* (ed. J.R. Piggott). Soc. Chem. Ind., London.
Dubois P. and Rigaud J. (1981) *Vignes Vins*, 301.
Dubois P., Rigaud J. and Dekimpe J. (1976) *Lebensm.-Wiss. Technol.*, 9, 366.
Dubourdieu D. and Lavigne V. (1990) *Rev. Fr. Œnol.*, 124, 58.
Dubourdieu D. and Lavigne V. (2002) Rôle du glutathion sur l'évolution aromatique des vins blancs secs Actes du 13ème Symposium International D'œnologie, Montpellier, 9 au12 juin 2002.
Ducruet V., Flanzy C., Bouzeix M. and Chambrou Y. (1983) *Sciences des Aliments*, 3 (3), 413.
Dugelay I, Günata Z., Sapis J.C., Baumes R. and Bayonove C. (1993) *J. Agric. Food Chem.*, 41, 2092.
Dugelay I., Baumes R, Gunata Z., Razungles A. and Bayonove C. (1995) *Sciences des Aliments*, 15, 423.
Eschenbruch R. (1974) *Am. J. Enol. Viticult.*, 25, 157.
Etievant P. (1979) *Lebensm. Wiss. u. Technol.*, 12 (2), 115.
Gebner M., Christoph N. and Simat T. (1998) In *Intervitis Interfructa*, Innovations en Œnologie, 5 ème Symposium International, Messe Stuttgart, 11 et 12 Mai 1998.
Gerber N. and Lechevallier H.A. (1965) *Appl. Microbiol.*, 13, 935.
Goodey A.R. and Tubb R.S. (1982) *J. Gen. Microbiol.*, 128, 2615.
Grando M.S., Versini G., Nicolini G. and Mattivi F. (1993) *Vitis*, 32, 43.
Guichard E., Pham T.T. et Etievant P. (1993) Mise au point d'une méthode de dosage rapide du sotolon, molécule clé pour l'arôme des vins jaunes du Jura *Actes du Symposium International*, Montpellier, 360.
Gunata Z. (1984) Thèse de Docteur Ingénieur, Université des Sciences et Techniques du Languedoc, Montpellier.
Heresztyn T. (1986a) *Arch. Microbiol.*, 146, 96.

Heresztyn T. (1986b) *Am. J. Enol. Viticult.*, 37 (2), 127.
Kaminski E., Libbey L.M., Stawicki S. and Wasowicz E. (1972) *Appl. Microbiol.*, 24, 721.
Klaren De Wit M., Frost D.J. and Ward J.P. (1971) *Rec. Trav. Chim. Pays Bas, Belgique*, 90, 906.
König T., Gutsche B., Harlt M., Hübscher R., Schreier P. and Schawb W. (1999) *J. Agric. Food Chem.*, 47, 3288.
La Guerche S., Blancard D., Chamont S., Dubourdieu D. and Darriet Ph. (2003a) In 7ième *Symposium International d'œnologie*, (ed. A. Lonvaud). Tec-Doc Lavoisier, Paris.
La Guerche S., Pons M. and Darriet Ph. (2003b) 7ième *Symposium International d'œNologie*, (ed. A. Lonvaud). Tec-Doc Lavoisier, Paris.
La Guerche S., Garcia C., Darriet Ph., Dubourdieu D. and Labarère J. (2004) *Curr. Microbiol.*, 47, 1.
La Guerche S., Chamont S., Blancard D., Dubourdieu D. and Darriet Ph. (2005) *Antonie van Leeuwenhoe=ck*, 88, 131–139.
Lavigne V. and Dubourdieu D. (1996) *J. Int. Sci. Vigne Vin*, 30 (4), 201.
Lavigne V., Boidron J.-N. and Dubourdieu D. (1992) *J. Int. Sci. Vigne Vin*, 26 (2), 75.
Lavigne V., Boidron J.-N. and Dubourdieu D. (1993) *J. Int. Sci. Vigne Vin*, 27 (1), 12.
Lavigne V., Pons A., Choné X. et Dudourdieu D. (2003) 7ième *Symposium International d'œnologie*, (ed. A. Lonvaud). Tec-Doc Lavoisier, Paris.
Lefebvre A., Riboulet J.M. and Ribéreau-Gayon P. (1983) *Sciences des Aliments*, 3, 265.
Liyanage C., Luvisi D.A. and Adams D.O. (1993) *Am. J. Enol. Vitic.*, 44 (1), 8.
Loyaux D. (1980) Thèse de Docteur Ingénieur, Université de Dijon.
Marai J. (1979) *Vitis*, 18, 254.
Mattheis J.P. and Roberts R.G. (1992) *Appl. Environ. Microbiol.*, 58, 3170.
Maujean A. (2001) *J. Int. Sci. et Vigne Vin.*, 35 (4), 171.
Maujean A. and Haye B. (1978) *Conn. Vigne Vin*, 12, 277.
Maujean A. and Seguin N. (1983) *Sciences des Aliments*, 3, 589 and 603.
Maujean A., Millery P. and Lemaresquier H. (1985) *Cah. Scientif. Rev. Fr. Œno.*, 99.
Maujean A., Nedjma M., Vidal J.P. and Cantagrel R. (1993) *Elaboration et Connaissance des Spiritueux*. Tec. et Doc., Lavoisier, Paris.
Maurer R. (1987) *Rebe-Wein*, 2, 370.
Mazzoleni V., Caldentey P., Carem M., Mangia A. and Colagrande O. (1994) *Am. J. Enol. Viticult.*, 45 (4), 401.
Nedjma M. (1995) Thèse Doctorat, Université de Reims.
Palacios S., Vasserot Y. and Maujean A. (1997) *Am. J. Enol. Vitic.*, 48, 525.
Peynaud E. (1981) *Connaissance et Travail du Vin*. Dunod, Paris.

Pham T.T., Guichard E., Schlich P. et Charpentier C. (1995) *J. Agric. Food Chem.*, 45, 2616.
Polak E.H. and Provasi J. (1992) *Chem. Senses*, 17, 23.
Rankine B.C. (1963) *J. Food and Agric.*, 2, 79.
Rapp A., Guntert M. and Almy J. (1985) *Am. J. Enol. Viticult.*, 36 (3), 219.
Rapp A., Versini G., Engel L. and Ullemeyer H (1993) *Vitis*, 32, 62.
Rapp A., Versini G., Engel L. and Ullemeyer H. (1998) Intervitis Interfructa, *Innovations en Œnologie, 5 ème Symposium International*, Messe Stuttgart, 11 et 12 Mai 1998.
Rauhut D., Sponholz W. and Dittrich H.H. (1986) *Der deutsche Weinbau*, 16, 872.
Ribéreau-Gayon J. and Peynaud E. (1964) *Traité D'œnologie, Tome 1—Maturation du Raisin, Fermentation Alcoolique, Vinification*. Dunod, Paris.
Ribéreau-Gayon J., Peynaud E., Ribéreau-Gayon P. and Sudraud P. (1975) *Traité d'œnologie. Sciences et Techniques du Vin*, Vol. II: *Caractère des Vins, Maturation du Raisin. Levures et Bactéries*. Dunod, Paris.
Ribéreau-Gayon J., Peynaud E., Ribéreau-Gayon P. and Sudraud P. (1976) *Sciences et Techniques du Vin*, Vol. III: *Vinification—Transformation du Vin*. Dunod, Paris.
Ribéreau-Gayon J., Peynaud E., Ribéreau-Gayon P. and Sudraud P. (1977) *Sciences et Techniques du Vin*, Vol. IV: *Clarification et Stabilization. Matérials et Installations*, Dunod, Paris.
Ribéreau-Gayon P., Pontallier P. and Glories Y. (1983) *J. Sci. Food. Agric.*, 34, 505.
Riboulet J.M. (1982) Thèse Doctorat de Troisième Cycle, Université de Bordeaux II.
Riboulet M. and Alegoet C. (1986) *Aspects Pratiques du Bouchage Liège*. Bourgogne Publications, La Chapelle de Guinchay, France.
Rigaud J., Issanchon S., Sarris J. and Langlois O. (1984) *Sciences des Aliments*, 4, 81.
Salmon J.M., Fornairon C., Schreier P., Drawert F. and Junker A. (1999) *Z. Lebensm.-Unters.-Forsch.*, 161, 249.
Schanderl H. (1964) *Bull. Office Int. Vigne Vin*, 37, 399.
Schreier P., Drawert F. and Junker A. (1976) *Z. Lebensm.-Unters.-Forsch.*, 161, 249.
Schreier P., Drawert F., Junker A., Barton H. and Leupold G. (1976) *Z. Lebensm. Unters. Forsch.*, 162, 279.
Sémichon L. (1905) *Traité des Maladies des Vins*. Masson, Paris.
Shaw P.E. and Nagy S. (1981) In *The quality of Foods and Beverages. Chemistry and Technology* (eds. G. Charalambous and G. Infglett). Academic Press, New York.
Simpson R.F. (1978) *Vitis*, 17, 24.
Singleton V.L., Zaya J., Trousdale E. and Salgues M. (1984) *Vitis*, 23, 113.
Singleton V.L., Salgues J., Zaya J. and Trousdale E. (1985) *Am. J. Enol. Vitic.*, 36, 50.
Takahashi K., Tadenuma M. et Sato S. (1976) *Agric. Biol. Chem.*, 40 (2), 325.
Tanner H., Zanier C. and Buser H.R. (1981) *Schweiz. Z. Obst. Weinbam*, 117, 97.
Tressl R., Bahri D. and Engel, K.H. (1982) *J. Agric. Food Chem.*, 30, 89.
Van Wyk C.J. (1993). Association Internationale d'Oenologie et de Gestion d'Entreprise, 10th Symposium International d'Oenologie, Montreux, Suisse.
Vasserot Y., Steinmetz V. and Jeandet P. (2003) *Int. J. Gen. Mol., Microbiol.*, 83 (3), 201–207.
Versini G. (1985) *Vigne Vini*, 12 (1–2), 57 and 207.
Vos P.J.A. (1981) In *6th Internationales Onologisches Symposium* (eds. E. Lemperle and J. Franck), 28–30 April 1981, Mainz, Germany, pp. 163–180. Breisach: Eigenverlag der Internationalen Interessengemeinschaft für moderne Kellertechnik und Betriebsführung.
Wainwright T. (1971) *J. Appl. Bact.*, 34, 161.
Wanner P. and Tressl R. (1998) *Eur. J. Biochem.*, 255, 271.
Waters E.J., Wallace W. and Williams P.J. (1992) *J. Agric. Food Chem.*, 40, 1514.
Wurdig G. (1975) *Die Weinwirtschaft*, 111 (44), 1250.

9

The Concept of Clarity and Colloidal Phenomena

9.1 Clarity and stability	285
9.2 The colloidal state	287
9.3 Colloid reactivity	290
9.4 Protective colloids and gum arabic treatment	296

9.1 CLARITY AND STABILITY

9.1.1 Problems Related to Clarity

Clarity is an essential quality required by consumers, especially for white wines in clear glass bottles. Particles in suspension, either in forming a haze or dispersed through the liquid, not only spoil the presentation but usually also affect the flavor.

New wine has a very high particle content, consisting of yeast lees and other grape debris. Clarity is achieved by gradual settling, followed by racking to eliminate the solids. Other, more rapid, processes (filtration and centrifugation) may also be used.

Wine must not only be clear at the time of bottling but also retain its clarity during aging and storage for an indefinite period, whatever the temperature conditions. Besides the microbial problems and tartrate precipitations described elsewhere, turbidity detrimental to clear wine (precipitation of coloring matter and metallic casse) involves colloidal phenomena.

Traditionally, stable clarity was acquired during a long period of barrel aging. Transformations and precipitation took place spontaneously in the wine and any deposit was eliminated before bottling. Wine was usually bottled in the area where it was consumed. For a number of years now, thanks to progress in enology, winemakers have been able to assess the risk of turbidity and implement appropriate preventive measures before bottling. This has made it possible for bottling in the region of production to come into general use, providing an assurance of quality and authenticity to producers as well as consumers.

Nowadays, the only normally acceptable deposit is red coloring matter in old wines. Sediment should not appear until the wine is four or five years old, and then only in small quantities. It should be easy to eliminate by decanting. However, unreasonable consumer demands may sometimes necessitate treatments that enologists would prefer to avoid.

A distinction should be made in terms of cellar work between two separate issues. The aim, on the one hand, is to obtain total clarity by appropriate methods and, on the other hand, to achieve stability by means of efficient treatments. Wine treatments are differentiated by their purposes. For instance, filtration clarifies but does not stabilize, fining does both and treatment with gum arabic stabilizes wine but does not clarify it.

The mechanisms responsible for turbidity in red and white wines, as well as the processes for preventing it, are based on the properties of colloids: the conditions under which particles grow in size, resulting in flocculation and sedimentation. The main fields of practical winemaking involving colloidal phenomena are as follows:

- Clarification and limpidity
- Metallic precipitation (ferric casse and copper casse)
- Protein turbidity of white wines and bentonite treatment
- Precipitation of colloidal coloring matter in red wines
- Fining wine
- Involvement of protective colloids in clarification problems and the tartrate precipitation mechanism
- Treating wines with gum arabic

These mechanisms generally operate in two stages. Firstly, a purely chemical mechanism produces mainly colloidal particles (ferric phosphate, colloidal coloring matter, etc.) that remain in solution and leave the wine clear. Later, various factors cause them to combine, leading to flocculation. This produces turbidity that eventually settles out as sediment. This same mechanism is involved in certain treatment processes, e.g. protein flocculation during fining or the flocculation of ferric colloids as a result of ferrocyanide treatment. Flocculation has a stabilizing effect in these operations, as it eliminates invisible, but unstable, particles. It also has a clarifying effect, reacting with particles in suspension that are responsible for turbidity.

9.1.2 Observing Clarity

Turbidity in wine is due to the presence of particles in suspension that stop light rays and diffuse some of the light in other directions than that of the incident light beam. This makes the wine seem opaque to varying degrees.

Severe turbidity may be observed directly by looking through the wine. Slight turbidity is more difficult to identify and is assessed using diffused light. When the particles agglomerate, turbidity increases and light is more diffused. Indeed, the light diffused is proportional to nV^2 (n = number of particles, V = total particle volume). During agglomeration, nV remains constant (n decreases, V increases). Therefore, the light diffused is proportional to V. When the particles reach approximately 100 μm in size, the colloidal solution becomes a true suspension, with easily visible turbidity.

Turbidity due to the diffusion of light (Tyndall effect) exists in any colloidal solution through which a light beam is shone. When a solution is observed against a black background, perpendicular to the incident light rays, an opalescent bloom appears, even in an apparently clear solution. This is due to the diffusion of light by very fine particles that are invisible to the naked eye.

Relatively simple apparatus based on this principle has been used for many years to assess clarity (Ribéreau-Gayon et al., 1976). The observer does not see the light shining through the wine directly, as it is hidden by a mobile screen, but only light diffused by particles. A low-intensity (15–25 W) bulb must be used, as all wines seem to have a slight bloom if the light is too strong.

Nowadays, optical instruments, known as turbidimeters, provide objective measurements of light diffused in a given direction. If the measurements are made perpendicularly to the incident light, the apparatus is called a nephelometer. Results, expressed in NTU (nephelometric turbidity units), are correlated with the wine's appearance (Section 11.3.1). These instruments are very sensitive, which is especially useful in assessing the effectiveness of a treatment, e.g. filtration.

Another way of assessing turbidity in wine is by counting the particles electronically according to size. In fact, currently available systems are only capable of measuring objects larger or at least as large as colloidal particles, so they are not widely used in enology. However, they do make it possible to show that an apparently clear wine may contain several tens of thousands of particles per ml above a micrometer in size, and therefore larger than colloidal particles.

Several research techniques (ultrafiltration, gel chromatography, electrophoresis, etc.) may be used to separate colloidal particles and will help to add to knowledge on this subject. It is also possible to appreciate the quantity of particles by gravimetric analysis, once they have been separated from wine by ultracentrifugation.

Finally, colloidal particles may be observed directly by high-performance microscope systems (an ultramicroscope or optical microscope with differential interference contrast) (Saucier, 1993, 1997).

9.2 THE COLLOIDAL STATE

9.2.1 Classification of Dispersed Systems

'Ordinary solutions' are distinguished from 'colloidal solutions' and 'standard suspensions' according to particle size (Table 9.1). Of course, the limits between these different classes are not perfectly defined. In particular, the upper size limit for colloidal particles is between 0.1 and 10 μm, according to the criteria retained.

A wide range of unrelated substances with very different origins and chemical compositions are capable of forming colloidal dispersions. They all share certain properties, although there

Table 9.1. Classification of dispersed systems

	Particle size (nm; 10^{-6} mm)	Approximate number of atoms per particle	Particle properties
Ordinary solutions (or molecular dispersions)	<2	10^3	Pass through filters and ultrafilters, are not visible under a microscope or ultramicroscope, are dispersed in the solution and dialyze, do not settle
Colloidal solutions (or dispersions)	2–1000	10^3–10^9	Pass through filters but not ultrafilters, visible under an ultramicroscope but not a microscope, disperse in the solution with some difficulty and dialyze very slowly, settle very slowly
Standard suspensions	>1000	>10^9	Do not pass through filters, visible under a microscope, disperse in the solution with great difficulty, do not dialyze, settle very rapidly

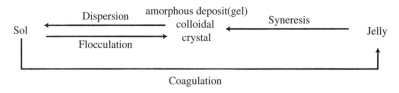

Fig. 9.1. Diagram of colloidal transformations

is no clear dividing line between colloids and non-colloids (formerly known as 'crystalloids'). Colloidal chemistry is more a matter of a set of shared properties than a group of compounds with similar structures. Furthermore, some compounds may exist in both states, e.g. sodium chloride forms a true molecular solution in water and a colloidal solution in alcohol.

The vocabulary used to describe colloid science and colloidal phenomena varies from one author to another. However, it is generally accepted (Ribéreau-Gayon *et al.*, 1976) that they may form solutions (known as 'sols' or 'colloidal solutions') or 'gels' also called 'colloidal crystals' (Figure 9.1). A 'sol' is fluid, with particles that move freely in relation to each other. The particles in a gel are not mobile, but gathered together in a mass that prevents Brownian motion. However, colloidal particles may form a deposit, arranged in a regular pattern like molecules in crystals; the expression 'colloidal crystal' is used. An amorphous substance that swells in an appropriate liquid is known as a 'jelly'.

When a sol flocculates it becomes a gel. This process is reminiscent of the precipitation of a salt and leads to the appearance of colloidal turbidity. The reverse phenomenon, called dispersion, is similar to the dissolving of a salt. Coagulation and syneresis correspond to the formation and disappearance of jellies.

9.2.2 Different Types of Colloids

A colloidal solution therefore consists of small solid particles, maintained dispersed in a liquid by a set of forces that prevent their aggregation and flocculation. It includes two phases (liquid and solid), with a mutual boundary that constitutes an interface. Exchanges between the two phases take place at the interface. It is obvious that certain properties (adsorption) of a two-phase system are more marked if the interface is larger (with a constant volume of liquid and mass of solid). The interface may be as large as several square meters in 1 ml of liquid. The total interface is one of the factors governing the physicochemical properties of colloidal solutions.

Two groups of colloids may be distinguished according to their properties, but they are clearly differentiated by their composition:

1. *Association colloids* (formerly known as '*micellar colloids*') are formed by aggregates or particles consisting of a large number of simple molecules, held together not by covalent chemical bonds but by low-energy physical bonds (Van der Waals, hydrogen, hydrophobic, etc.). The stability of colloidal dispersions may be ensured by the fact that the particles are electrically charged (Section 9.2.4) and repel each other. However, these particles are not pure, as they may adsorb other substances that are in solution in the liquid at the solid–liquid interface. Association colloids may be formed in wine either naturally, during aging (condensed phenols and colloidal coloring matter), accidentally (ferric phosphate and copper sulfide) or as a result of certain treatments (ferric ferrocyanide and copper sulfide). When the forces holding them together (electrolytes with opposite charges) are suppressed they flocculate and then precipitate. This phenomenon is involved in most spontaneously occurring turbidity in wine. It is also part of the mechanism of the various treatments. The instability of these association colloids is partly due to their hydrophobic character.

2. *Macromolecular colloids* consist of macromolecules such as polysaccharides (Section 3.6) or proteins (Section 5.5), in which only covalent chemical bonds are involved. They generally have an electrical charge that may be due to the dissociation of acid or basic functions. These colloids may be hydrophilic and, as a result, dissolve easily in water. This property leads to hydration, giving them a second stabilizing factor in addition to the repellent effect of the electrical charges. Some of these substances (polysaccharides) may even communicate this stability to association colloids, by coating them and protecting them from the precipitating effect of the electrolytes. In this instance, they are known as 'protective colloids' (Section 9.4.1). The flocculation of proteins, on the contrary, is widely used in fining wine.

9.2.3 Properties of Colloids

Association and macromolecular colloidal solutions have a number of common properties (Ribéreau-Gayon *et al.*, 1976):

1. All of the molecules in a solution are subjected to agitation forces, known as Brownian motion, that tend to make them occupy the maximum amount of available space. A solid that dissolves in a liquid is dispersed throughout the entire volume and is thus uniformly distributed. The Brownian motion of colloidal particles is slower. If they are put into the bottom of a container, they diffuse very slowly through the mass of the liquid.

2. In view of their size, colloidal particles have difficulty passing through dialysis membranes. The largest colloid particles are stopped by the finest filter membranes and some of them (protective colloids) have a high fouling capacity (Section 9.4.1).

3. When salts are extracted from a solution, they produce crystallized residues. Colloids, however, generally produce amorphous residues or precipitates, with no recognizable structure. However, structural analyses using X-rays or microscopes have, in some cases, detected a regular arrangement of atoms or colloidal particles, at least in certain directions. A few macromolecules (proteins) have also been obtained in crystallized form. Crystallized colloids have been found in wine, together with more usual crystals, such as tartaric acid.

4. The freezing and boiling points of even concentrated aqueous colloidal solutions are close to those of pure water (0°C and 100°C). Raoult's well-known law of molecular solutions is not applicable. It is as though the substance in a colloidal solution was not really dissolved. There are two distinct phases, a liquid phase and a dispersed phase.

5. Unlike that of normal molecules, the composition of the particles in an association colloid is not perfectly defined. Composition is variable from one solution to another and depends on the preparation method. In water at least, there is a single sodium chloride, whereas there may be a whole series of ferric phosphates with dimensions varying from 1 to 10. Furthermore, the ions present in the solution are fixed to variable degrees by adsorption at the interfaces. The components of macromolecular colloids, however, are less variable.

6. The flocculation of colloids in a solution is due to a different mechanism from that governing precipitation of salt: (a) flocculation may take place in dilute solutions, (b) no specific agent is required and (c) there is no set relationship between the proportions of the colloid and the precipitating reagent, so flocculation may occur at very low concentrations.

7. Colloidal solutions diffuse light, but the particles must reach a sufficient size in relation to the total quantity of colloids present for turbidity to appear.

8. The reactions involved in the appearance of colloidal turbidity are not only governed by the mass action law. Precipitation does not occur systematically when values exceed the solubility product.

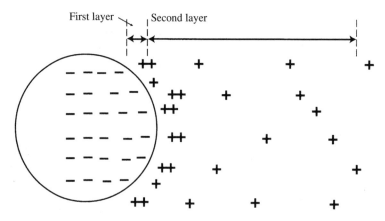

Fig. 9.2. Distribution of charges in a 'double layer' around a charged colloidal particle (Saucier, 1993)

9.2.4 Electrical Charges on Colloidal Particles

It is easy to observe the existence of electrical charges on colloidal particles by running a continuous electrical current through the solution. If the liquid is turbid, the movement of the particles toward one of the electrodes is visible to the naked eye. A chemical assay is necessary to characterize the particles that leave the solution clear. Of course, the particles that migrate toward the anode (+) are negatively charged while those that migrate toward the cathode (−) are positively charged.

In the case of particles consisting of neutral molecules, the charge results from the fixation or adsorption of ions from the solution on the particle surface. These ions give the particle a positive or negative charge, depending on whether they are anions or cations. Two electrical layers develop in the vicinity of the particle. The first consists of counter-ions adsorbed on the particle, while the second is more diffuse, also consisting of counter-ions, but in solution around the particle (Figure 9.2). At a certain distance from the particle, the resulting charge is zero.

In the case of charged polymers, the charge comes from the dissociation of acid or basic functions. According to the pH, some molecules may be acid (−) or neutral (pectins), or both acid (−) and alkaline (+) (proteins). Proteins have both functions, with an isoelectric pH (or isoelectric point, i.p.), where they are neutral. In a solution with a pH < i.p. (normally the case for proteins in wine), most of the alkaline functions are neutralized and dissociated, giving an excess (+) charge, so the proteins are positively charged. Inversely, protein molecules are negatively charged in solutions with a pH > i.p. *Botrytis cinerea* laccase is an example of this phenomenon in grapes and wine (Volume 1, Section 11.6.2). Its i.p., in the vicinity of 2.5, is responsible for its stability, particularly with regard to bentonite.

Among the colloids found in wine, proteins and cellulose fibers are positively charged, while yeast cells and bacteria, colloidal coloring matter, ferric phosphate, copper sulfide, ferric ferrocyanide and bentonite are negatively charged.

9.3 COLLOID REACTIVITY

9.3.1 Colloid Stability and Flocculation

The agglomeration of particles in a colloidal solution is due to instability and is responsible for most turbidity and sediment occurring in wine. This phenomenon, also known as flocculation, corresponds to the separation of the colloid into a colloidal crystal (gel) and a liquid. The end result is the formation of various types of flakes. In order to understand the particle agglomeration mechanism, which causes the solution, i.e. wine, to go from a clear state to a turbid state that is resolved by

The Concept of Clarity and Colloidal Phenomena

forming a deposit, it is first necessary to understand the opposite mechanism, which keeps these same particles in suspension although their density is generally above that of the liquid. The same problem is raised for particles in suspension in natural turbidity (e.g. yeast) that may be relatively stable and sometimes remain in suspension for long periods of time without settling out.

Even if the phenomenon is less marked than it is in molecular dispersions (Section 9.2.1), colloidal particles are subject to heat energy (Brownian motion). This may be a stabilizing factor as it prevents the particles from gathering together, promotes their dispersion throughout all the available space and inhibits sedimentation to the bottom of the container. It may also be a destabilizing factor, as it makes it easier for particles that naturally attract each other to come together.

Colloidal particles are also subjected to other forces, some of which are repulsive forces that add their effects to those of heat energy. Other forces attract and contribute toward instability. The system is stable if the resultant of these forces has a higher energy than that of the Brownian motion, as explained below:

1. The first forces to be taken into account are known as the 'Van der Waals attraction'. These attractive forces contribute toward the buildup of aggregates in association colloids. They originate in dipolar interactions between atoms. These attractive forces contribute toward the buildup of aggregates in association colloids. It has been shown that these forces are proportional to the diameter of the particles and inversely proportional to the distance between them. The Van der Waals attraction may balance the forces due to the thermal effect, or not, depending on the distance. When the distance between the particles is smaller than their radius, the energy due to the Van der Waals forces is greater than the thermal energy, which is not then capable of separating the particles. The Van der Waals forces therefore tend to promote attraction between colloidal particles, causing them to increase in size until aggregates are formed and precipitated.

2. The stability of a colloidal solution such as wine therefore requires the presence of repulsive forces to counterbalance these attractive forces. These repulsive forces are mainly electrostatic interactions due to surface charges on the particles (Section 9.2.4). These charges create an electrostatic potential around the particle, which decreases as the distance from the particle increases. Figure 9.2 shows an example of a negatively charged particle. It is surrounded by a cluster of ions with the opposite charge, in the form of a 'double layer'. Counter-ions (+) density is high in the vicinity of the charged particle surface. Thermal agitation tends to decrease the density of these (+) charges as the distance from the particle increases. Unlike the Van der Waals forces, these electrostatic forces that keep colloidal particles apart are highly dependent on conditions in the medium (type of solvent) and the type of particle surface.

It is possible to calculate the forces involved in these electrostatic interactions, especially those that vary according to the concentration of salts. It has thus been demonstrated that the scope of electrostatic interactions decreases with the salt concentration. When the medium is saturated with salts, the electrostatic forces become negligible as compared to the Van der Waals forces, so the particles tend to agglomerate and precipitate. This explains why proteins precipitate in an aqueous solution saturated with ammonium sulfate. It is also clear that colloids, occurring either naturally or following a treatment, are flocculated by salts in the wine.

Attempts (Hunter, 1993) have been made to interpret colloidal stability according to Van der Waals and electrostatic interactions, on the basis of the DLVO theory (named for its authors: Deragyuin, Landau, Verwey and Overbeek). Their calculations take the total of the two forces into account. When two particles with the same radius (100 nm) are brought very close together by thermal agitation (less than 5 nm), the repulsive forces are weak and precipitation is easy. However, before they reach this position, the particles must

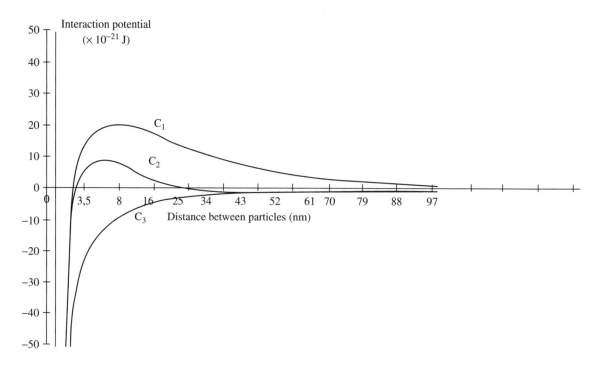

Fig. 9.3. Simulation of the impact of salt concentration ($C_1 < C_2 < C_3$) on the interaction potential of two particles. At high salt concentrations (C_3), electrostatic repulsion becomes negligible in relation to the Van der Waals attraction forces and the colloidal solution is unstable. (According to Saucier, 1997)

pass through an energy barrier when they are around 5–20 nm apart. If the salt concentration is low, this energy barrier is strong, due to the relative strength of the electrostatic interactions as compared to the Van der Waals forces. Under these conditions, thermal energy (Brownian motion) is not sufficient for the particle to pass this barrier, and the medium is stable (Figure 9.3). At higher salt concentrations, the electrostatic interactions are much weaker and no longer compensate for the Van der Waals forces, so there is no energy barrier. Irreversible aggregation is observed in this situation. When the particles are very large (250 nm instead of 100 nm) and a certain distance apart, the total energy in the medium reaches a secondary minimum. This may cause flocculation, which is, however, reversible. The aggregates may be broken up by agitation or changes in the physicochemical conditions.

The presence of macromolecular colloids (carbohydrate polymers) may also affect the stability of association colloids. Carbohydrate polymers may either act as protective colloids, preventing flocculation, or possibly destabilize the colloids and cause them to precipitate (Section 9.4.1).

9.3.2 Stability and Flocculation of Macromolecular Colloids

According to standard enological theory, macromolecular colloids owe their stability to their charge as well as hydration. Flocculation is considered to require the elimination of both of these stabilizing factors (Figure 9.4). Thus, when gelatin, positively charged in wine, comes into contact with tannins, it is said to form a negatively charged tannin-protein complex corresponding to "denaturation", attributed to dehydration of the protein

The Concept of Clarity and Colloidal Phenomena

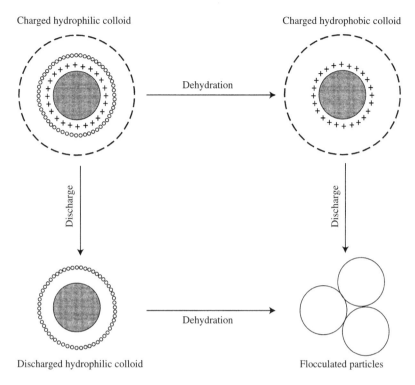

Fig. 9.4. Diagram of the flocculation of a hydrophilic colloid by elimination of the two stability factors: electrical charge and hydration (Ribéreau-Gayon *et al.*, 1976)

by adsorption of tannins. Flocculation was considered to be due to the loss of electric charges in the contact with of the cations.

It is now more readily accepted that this denaturation results from the adsorption of tannins without dehydration (Kawamoto and Nakatsubo, 1997) (Figure 9.5). The new complex is an electronegative hydrophobic colloid. It remains stable in a clear solution if no salts are present, otherwise it flocculates. In the same way, natural wine proteins are denatured by heating and flocculate as the liquid cools.

In general, the precipitation of proteins requires the presence of alcohol, tannins or heating. Furthermore, it only takes place in the presence of electrolytes. The role of alcohol, tannin or heating is to 'denature' the protein. The protein, a hydrophilic colloid, becomes a hydrophobic colloid that can be flocculated by salts. A large mass of electrolyte (ammonium sulfate) may be sufficient to transform the protein directly from a stable hydrophilic colloid into a discharged hydrophobic colloid, capable, therefore, of precipitating.

In a more recent theory on the behavior of colloidal tannins, hydration of hydrophilic colloids is not considered. When tannin molecules combine to form colloidal particles, the Van der Waals forces between tannins and proteins increase considerably, producing a non-specific adsorption phenomenon (Figure 9.6) (Saucier, 1997). The mechanisms involved are as follows:

1. Tannins form colloidal particles by hydrophobic interactions.

2. The tannin particles are likely to be destabilized by proteins due to the Van der Waals

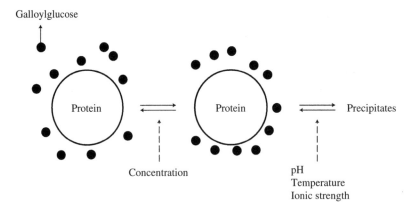

Fig. 9.5. Two-stage mechanism by which tannins precipitate proteins (galloylglucose). Influence of physicochemical conditions (Kawamoto and Nakatsubo, 1997)

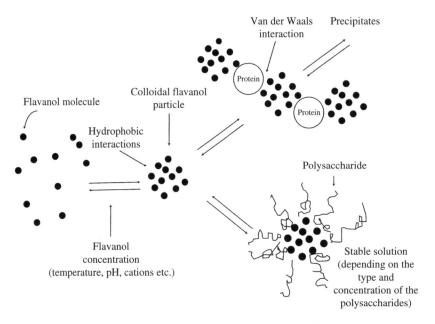

Fig. 9.6. Model of the colloidal properties of flavanols (tannins) (Saucier, 1997)

attraction, forming aggregates that precipitate (fining mechanism in wine).

3. Cations, especially iron, promote agglomeration of tannins to form colloidal particles.

4. The formation of aggregates of tannin particles, or tannins and proteins, may be inhibited by the presence of polysaccharides (macromolecular colloids). This observation has been confirmed by several authors (Riou *et al.*, 2002; de Freitas *et al.*, 2003).

9.3.3 Mutual Flocculation of Colloids

When two colloids with the same electrical charge are in the same solution, they are held apart by electrostatic forces so they do not precipitate. If, however, they have opposite charges, precipitation of both colloids may be produced by 'reciprocal' or 'mutual' flocculation. Even if precipitation does not occur spontaneously, the system becomes highly sensitive to the electrolyte's precipitating effect.

Mutual flocculation is very important in winemaking, as it is the most significant mechanism in fining. When the protein fining agent flocculates, particles in suspension and colloidal particles are eliminated as a result of mutual flocculation. Thanks to this mechanism, fining achieves clarification and stabilization at the same time. This explains the role of fining with a positively charged protein in the flocculation and precipitation of negatively charged ferrous phosphate, ferric ferrocyanide and copper sulfide colloids. This also applies to bentonite, which is a negatively charged suspension.

Ferrocyanide treatment is a particularly significant case (Section 4.6.5). This product reacts with ferric iron to produce ferric ferrocyanide (Prussian blue), a negatively charged colloid that remains in solution and passes through filters. It can only be eliminated by mutual flocculation with a positively charged protein. It reacts slowly with the ferrocyanide due to complexation of the iron, so it is necessary to ensure that all the ferrocyanide has reacted before the fining agent is added. This is one of the aims of the preliminary test that also defines the correct dose to use.

9.3.4 Adsorption Phenomena

Adsorption phenomena are another aspect of colloid activity. Adsorption is the reversible fixation on a solid surface of a body in solution (liquid or gas). This fixation does not involve any chemical reactions and is governed by the thermodynamic equilibrium. Adsorption phenomena occur during winemaking (Ribéreau-Gayon *et al.*, 1976) and have an effect on colloidal chemistry. These phenomena are more extensive when the adsorbent body is divided into smaller units, as the interface is proportionally larger.

Colloids have a relatively large surface area, so they may act as adsorbents. Colloidal sediment formed in wine due to natural settling or treatment generally contains various substances that were not involved in the colloidal flocculation mechanisms that caused the deposit. Thus, for example, ferric phosphate deposits frequently contain calcium. At one time, it was even supposed that ferric-calcium casse had occurred. In fact, the calcium is not involved in flocculation as an electrolyte, but is rather fixed by adsorption.

Secondly, colloids may be adsorbed. For example, enological charcoal removes most colloids from solutions. In wine, it acts on tannins, coloring matter and proteins. Bentonite fixes proteins by the same mechanism.

Adsorption is due to surface phenomena that do not necessarily involve electrical charges. Adsorption is limited and reaches a balance. It is proportionally more efficient if there is a low concentration of the substance adsorbed in the solution. High adsorption may be observed when only traces of dissolved matter are present. One example of this effect is the use of charcoal to remove discoloration from white wines.

Adsorption mechanisms are very complex. A standard example from winemaking is the action of tannins on gelatin. No clearly defined gelatin tannate is formed, but rather an adsorption compound. This compound's tannin content is higher when there is a larger proportion of this compound in the solution (in relation to the amount of gelatin); e.g. the quantity of tannins removed by adding 25 mg/l of gelatin are as follows:

(a) 5 mg/l if the initial tannin concentration was 0.1 g/l,
(b) 15 mg/l if the initial tannin concentration was 0.5 g/l,
(c) 50 mg/l if the initial tannin concentration was 3.0 g/l.

Tannins are not fixed according to a specific ratio, and neither is the amount proportional to its concentration in the solution. This is not a stoichiometric reaction.

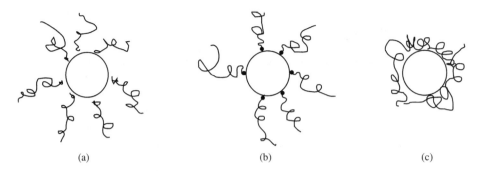

Fig. 9.7. Various mechanisms by which polysaccharides protect colloidal particles from flocculation: (a) block polymers, (b) grafted polymers (covalent bonds), (c) linear polymers

9.4 PROTECTIVE COLLOIDS AND GUM ARABIC TREATMENT

9.4.1 Composition and Properties of Protective Colloids

In some instances, an entire solution may be stabilized when a macromolecular colloid (polysaccharide) and an unstable colloid are put together. Macromolecular colloids with this property are known as 'protective colloids'.

This protective effect is attributed to a coating of the colloid particles that prevents them from agglomerating. Several mechanisms may come into play (Figure 9.7). The protective polymer must meet two, apparently contradictory, conditions. On the one hand, it must be adsorbed on the particle while, on the other hand, it must spread as much as possible in the solution to maintain a separation between the various colloidal particles.

Electrical charges play a secondary role. In all cases, stability is assured when there is a sufficiently high concentration of polymer to cover the entire surface of all the unstable colloid particles. However, if the carbohydrate polymer (protective colloid) concentration is insufficient, it may bond the particles together in pairs, by a cross-bonding phenomenon without preventing them from precipitating (Figure 9.8).

There is another situation in which the carbohydrate polymer may cause colloidal precipitation, instead of its usual protective effect. When the carbohydrate polymer content is much greater than the quantity necessary to coat the unstable particles,

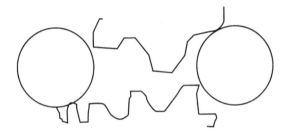

Fig. 9.8. Flocculation by cross-bonding of two colloidal particles in the presence of an excess of polysaccharides

it may cause a flocculation phenomenon known as 'depletion' (Figure 9.9). The excess polymers exert an osmotic pressure that tends to bring the particles closer together until they agglomerate and flocculate (Asakura and Osawa, 1954). This phenomenon may be responsible for the precipitation of colloidal coloring matter when red wines have a naturally high polysaccharide content.

Most turbidity occurring in wine is due to the flocculation of colloidal particles caused by chemical reactions that leave the solution clear. It is certain that the presence of natural polysaccharides, with their protective colloid properties, prevent the formation of turbidity and deposits. It is also clear that, in some cases, it may be useful to enhance this protective effect by adding a colloid such as gum arabic.

A typical example is that of ferric precipitation in white wines (Section 4.6.2). Aeration of wine leads to the oxidation of ferrous iron to ferric iron. Relatively insoluble ferric phosphate is formed. The molecules agglomerate, forming colloidal

The Concept of Clarity and Colloidal Phenomena

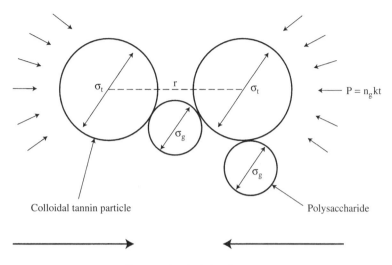

Fig. 9.9. Depletion phenomenon accounting for the precipitation of colloidal particles in the presence of an excess of polysaccharides (Saucier, 1993)

particles that are initially sufficiently small for the wine to remain clear. Electronegative particles may then be flocculated by cations or protons in the wine. The presence of protective colloids inhibits flocculation, but ferric phosphate is still formed, as it may be separated out by ultrafiltration.

This process accounts for most of the turbidity likely to form in wine. It shows quite clearly that there are two stages in the overall mechanism. The first stage consists of a chemical reaction, forming a colloidal substance that remains in clear solution. During the second stage, the colloids agglomerate into particles that flocculate and cause turbidity. Each of these stages is governed by various factors that may be modified to avoid the appearance of turbidity. Protective colloids take effect in the second stage.

Crystalline tartrate precipitation may be assimilated to colloidal phenomena (Section 1.5.1). Indeed, the natural colloids in wine, particularly mannoproteins (1.7.7), have a protective effect that inhibits tartrate precipitation, even when concentrations are higher than the solubility product. This phenomenon is particularly marked in red wines.

Furthermore, while protective colloids (mannoproteins) prevent the appearance of colloidal turbidity in clear wine, they also inhibit the clarification of wines where turbidity is already visible. Particle sedimentation is considerably slowed. Fining is difficult, as the fining agent does not flocculate very well and filter surfaces are rapidly fouled.

9.4.2 Natural Protective Colloids in Wine

Most wines certainly contain mucilaginous substances that act as protective colloids mannoproteins. Their existence is demonstrated by the elimination of the protective effect after fine ultrafiltration or dialysis. This phenomenon is well known in red wines, where colloidal coloring matter and tannins inhibit tartrate precipitation. It also exists in white wines and may be attributed to neutral polysaccharides (gum). According to the desired result, these substances may either be eliminated by fine filtration (e.g. to facilitate tartrate stabilization) or, on the contrary, protective colloids such as gum arabic may be added to a clear wine just before bottling to compensate for insufficient natural protection.

Many natural polymers, especially carbohydrates, probably have protective colloid properties, but they are not yet very well known. Furthermore, these same substances are likely to have a positive effect on a wine's organoleptic qualities, as demonstrated by the differences observed after very fine filtration (Escot *et al.*, 2003).

One fact that has been clearly demonstrated, even if it has not yet been fully explained, is the increase in the protective capacity of natural wine colloids following heating (Section 12.2.2). There is a remarkable parallel between the effects of heating and those of adding gum arabic. After heating, clear wines have improved stability as regards aeration (ferric casse) and cooling (precipitation of coloring matter in red wines). They are also better protected from copper casse. However, in a turbid wine, particle sedimentation is slower, filtration is more difficult and the flocculation of gelatin and albumin for fining purposes becomes practically impossible (isinglass and casein are less sensitive to heating). The effects of heating become noticeable at relatively low temperatures (40–50°C) and are accentuated at higher temperatures and after longer exposure. The maximum effect is obtained after heating to approximately 75°C for 30 mn. However, heating alone rarely results in total stability.

All of these important phenomena deserve a theoretical interpretation. It may be assumed that heating causes the enlargement of colloidal particles, which, however, remain sufficiently small to leave the wine clear. Indeed, when fine ultrafiltration or dialysis has eliminated all the colloids, heating no longer produces any protective effect.

Another classic example of natural protective colloids is that of wines made from grapes affected by rot. *Botrytis cinerea* secretes a polysaccharide (β-glucane) that is largely responsible for the difficulty of clarifying these wines, especially by filtration (Section 11.5.2).

This polysaccharide (glucane) is synthesized inside the grapes, forming a viscous jelly between the flesh and the skin. Mechanical systems for handling the harvested grapes (crusher-stemmer, pump, etc.), which treat the grapes rather roughly, disperse the glucane through the mass of must, resulting in wines that are difficult to clarify. If grapes affected by rot are pressed gently without crushing, the must has a low glucane content and the resulting wine is easily clarified. The same wineries always have problems clarifying their wines made from botrytized grapes. This is obviously a consequence of poorly designed equipment that treats the grapes too roughly, not only those affected by rot, but probably other grapes as well.

The fouling capacity of the *Botrytis cinerea* glucane depends on the alcohol concentration. The presence of glucane does not hinder filtration of the must, but it does affect filtration of the resulting wine to a greater extent if it has a high alcohol content. It may be supposed that alcohol acts on this polysaccharide by increasing the size of the colloidal aggregates. When they reach a certain size, the polysaccharide precipitates. This precipitation starts at 17% vol EtOH and is total at approximately 23% vol EtOH. Precipitation sometimes occurs spontaneously in vats of sweet wines. This phenomenon is widely used to isolate and purify this polysaccharide in the laboratory.

Various processes have been suggested for eliminating this protective colloid and facilitating clarification of the wines (Section 11.5.2).

9.4.3 Using Gum Arabic to Stabilize Clarity

Gum arabic has long been known as a particularly efficient protective colloid for stabilizing clarity (Ribéreau-Gayon *et al.*, 1977). Treatment with gum arabic is permitted in many countries and is authorized by EEC legislation. Gum arabic is a natural product with a perfectly neutral flavor, commonly used in the food industry. It does not affect the organoleptic quality of wine, even at doses much higher than those normally used. The only possible reservation concerns its use in wines intended for long aging.

Gum arabic is made from the natural exudation of branches of certain trees in the acacia family. It may also be produced by manual bark removal. The most common variety is Verek acacia. There are several qualities of gum. That used in winemaking must be as pure as possible. It is available in the form of hard white or reddish fragments of various sizes. One of its characteristics is its aptitude to break cleanly. Finely powdered industrial gum arabic is easy to dissolve into a solution.

Gum arabic is a macromolecular colloid, consisting of a polysaccharide with a molecular weight on the order of 10^6 Da. Acid hydrolysis causes it to release D-galactose (40–45%), L-arabinose (25–30%), L-rhamnose (10–15%) and D-glucuronic acid. The main chain consists of D-galactose links. The polysaccharide is associated with a protein fraction (approximately 2%), in which hydroxyproline and serine are the main amino acids.

Gum arabic is easily dissolved, even in cold water, although warm water is preferable. However, the natural product contains an insoluble fraction and the properties of the solution depend on the preparation conditions. For this reason, solutions (150–300 g/l) are prepared by specialized laboratories, stabilized by sulfuring and supplied ready for use. These preparations are checked to ensure that their purity complies with the Enological Codex standard (optical rotation) and that they have the expected protective effect in wine. Preparations should not affect turbidity, nor should they increase a wine's capacity to foul filter surfaces to any great extent.

Gum arabic is added to stabilize a clear wine that is ready for bottling. Indeed, if turbidity were to develop, for one reason or another, in a wine treated with gum arabic, clarification would be rendered much more difficult by the presence of this protective colloid. Particle sedimentation would be considerably slower and large quantities of protein fining agent would be required to obtain satisfactory clarification. However, even relatively fine filtration is not impossible when normal doses (10–20 g/hl) of a good-quality product are used. For this reason, gum arabic is generally mixed into the wine just before final filtration prior to bottling. Even membrane filtration is possible. Filtration is more difficult, but the protective effect is not significantly affected.

Gum arabic is a preventive treatment for many problems involving colloidal precipitation. It is effective in treating copper casse and was widely used when wines often contained excessive amounts of copper due to contact with bronze cellar equipment (Section 4.7.3). Doses of 10–15 g/hl were effective in preventing this problem, provided that wines did not contain more than 1.0 mg/l of copper. If the copper content was higher, it was preferable to eliminate the excess copper by an appropriate treatment. Even in this case, gum arabic was recommended as a back-up treatment. Gum arabic is more efficient at a higher pH.

If varying doses of gum arabic are added to samples of a wine with a high copper content and copper casse is caused by exposing the wine to light, the turbidity is significantly less opaque in samples with a higher gum arabic content. Indeed, turbidity is inversely proportional to the amount of gum added. If the colloidal copper sulfide is then eliminated from these same samples by fining, the deposits from all of the samples are found to contain the same quantity of copper. This experiment shows that gum arabic does not affect the formation of colloidal copper sulfide, but rather prevents it from flocculating.

Gum arabic is less effective in preventing ferric casse in white wines. Indeed, the unstable colloidal ferric phosphate that is precipitated has a much greater mass than the copper sulfide involved in copper casse. A much larger quantity of gum arabic would therefore be required to provide proper treatment, and this is likely to affect the wine's turbidity. Gum arabic is effective to a certain extent, but the effect is variable from one wine to another and is, in any case, insufficient to provide total protection. Recommended doses range from 20 to 25 g/hl as a supplementary treatment (Section 4.6.3).

Gum arabic is also at least partially effective as a treatment for ferric casse in red wines. It does not prevent the appearance of a dark, bluish color, due to the formation of colloidal ferric tannate, but it does stop the colloid flocculating. It acts differently from citric acid, which prevents color from changing, as it produces a soluble complex with iron that is no longer capable of reacting with tannins (Section 4.6.2). These two treatments are often complementary (Section 4.6.3).

The most important application of gum arabic in winemaking is in preventing the precipitation of phenols and coloring matter in red wines. It is well known that the coloring matter in red wines is partially colloidal and thus liable to precipitate at cold temperature. These problems were traditionally avoided, at least in young wines, by fining (egg albumin or gelatin) to eliminate this unstable coloring matter by mutual flocculation. Gum arabic is just as effective, although it acts by preventing the unstable coloring matter from flocculating, rather than by eliminating it.

By comparison with normal fining techniques, gum arabic treatment has the following characteristics:

1. It is instantaneous and therefore suitable for wines that must be bottled rapidly.

2. It does not attenuate color, as it does not reduce the total quantity of coloring matter.

3. It has a permanent effect. As colloidal coloring matter is known to form regularly during aging, a wine may again be unstable at cold temperatures only a few months after fining.

4. It inhibits the normal transformations that occur in certain wines with good aging potential. The deposit normally found in old wines is formed due to colloidal phenomena. These deposits cannot form in the presence of gum arabic, but the wine may take on a milky appearance, losing its normal clarity.

The previous assertion implies that the use of gum arabic in red wines should be restricted to those intended for rapid consumption. They should also be treated just before bottling, to avoid hindering the normal transformations that occur during aging in the barrel and in vat. Normal doses are between 10 and 20 g/hl. If the dose is too low, it does not stop the coloring matter from precipitating, but it prevents the particles that appear from forming unsightly sheets that stick to the glass inside of the bottle. This serious presentation defect may be observed in wines that have been bottled without any treatment to prevent the precipitation of colloidal coloring matter, i.e. neither fining nor gum arabic treatment. When too high a dose is used, 100 g/hl or more, the effect may be the reverse of the desired protection, as an excess of gum arabic actually promotes precipitation. These high concentrations may be to soften wines with tannins that are still too aggressive at the time of bottling.

Another application for gum arabic is in the production of *vins de liqueurs, rancio* wines, *apéritifs*, vermouth, port, *Pineau des Charentes*, etc. As these products are frequently stored in contact with air and their aging process includes deliberate oxidation, the formation and precipitation of colloidal coloring matter is the main cause of turbidity. Gum arabic, at doses of 20–25 g/hl, prevents flocculation of the coloring matter. This treatment is not recommended for wines of this type intended for long bottle aging.

REFERENCES

Asakura S. and Osawa F. (1954) *J. Chem. Phys.*, 22, 1255.

De Freitas V., Carvalho E., Mateus N. (2003) *Food Chem.*, 8 (4), p. 503.

Escot S., Gonzalez E., Feuillat M., Charpentier C. (2003) Actualités Œnologiques 2003, *VII ème Symposium International d'Œnologie*. Lavoisier, Tec et Doc, Ed. Paris.

Hunter R.J. (1993) *Foundations of Colloid Science.*, Oxford University Press, Oxford.

Kawamoto H. and Nakatsubo F. (1997) *Phytochemistry*, 46 (3), 479.

Ribéreau-Gayon J., Peynaud E., Ribéreau-Gayon P. and Sudraud P. (1976) *Sciences et Techniques du Vin*, Vol. III. Dunod, Paris.

Ribéreau-Gayon J., Peynaud E., Ribéreau-Gayon P. and Sudraud P. (1977) *Sciences et Techniques du Vin*, Vol. IV. Dunod, Paris.

Riou V., Vernhet A., Doco T., Moutounet M., 2003, *Food Hydrocolloids*, 16 (1), p. 17.

Saucier C. (1993) Approche colloïdale de l'interaction tanins-polysaccharides dans les vins. Mémoire pour le Diplôme d'Études Approfondies Œnologie-Ampélologie, Université de Bordeaux II.

Saucier C. 1997. Les tanins du vin: étude de leur stabilité colloïdale. Thèse Doctorat, Université de Bordeaux II.

10

Clarification and Stabilization Treatments: Fining Wine

10.1	Treating wine	301
10.2	Sedimentation of particles in suspension	303
10.3	Racking: role and techniques	304
10.4	Theory of protein fining	307
10.5	Tannin-protein interactions	312
10.6	Effect of fining on the organoleptic quality of wine: concept of overfining	315
10.7	Products used in fining	316
10.8	Fining techniques	322
10.9	Bentonite treatment	324
10.10	Miscellaneous clarification treatments	328

10.1 TREATING WINE

Clarity is one of the leading consumer quality requirements. It is an important aspect of a consumer's first contact with a wine and a key element in visual satisfaction. It also enhances the impression of quality on the palate, unaffected by particles in suspension or precipitates. Turbidity is undeniably a major negative factor in assessing a wine.

Turbidity in a liquid results from an optical phenomenon known as the Tyndall effect, caused by the presence of particles in suspension that deflect light from its normal path. The measurement of clarity is, therefore, related to estimations of turbidity (Section 9.1.2), depending on the number and size of particles in suspension. Wine may be clarified in the short term by eliminating these particles. However, the effect is not necessarily permanent, due to the many naturally occurring phenomena in wine that are often accompanied by the formation of turbidity or deposits.

The objective of stabilization is to ensure long-term clarity and prevent deposits, whatever the temperature, oxidation or lighting conditions where the wine is stored. The chemical and biological mechanisms likely to cause turbidity or deposits are now well known and may be predicted by laboratory tests. Efficient treatments are available for stabilizing wines, when necessary, before bottling.

Table 10.1 summarizes the treatments that promote clarification and stabilization, although some treatments are not permitted in certain countries. In view of the complexity and diversity of these phenomena, the various aspects are all covered in these two volumes. Table 10.1 gives paragraph references for the description of each type of treatment.

Several treatments are described elsewhere, in conjunction with the specific problems they treat or as applications of more general processes (physical treatments). This chapter presents clarification by sedimentation and racking, as well as protein fining and a few other treatments not discussed in other sections.

Table 10.1. The main treatments available for clarifying and stabilizing wines. They are not all recognized by legislation in every country and several are not permitted in the European Union

Clarification	Sedimentation and racking (Sections 10.2, 10.3), fining with gelatin, isinglass, casein, albumin from eggs or blood (not permitted in the EU), plant proteins, alkaline alginates, (Sections 10.4 to 10.8), siliceous earths (Section 10.10), filtration (Sections 11.2 to 11.10), centrifugation (Section 11.11)
Biological stabilization	Heating (Section 12.2.3 and Volume 1, Section 9.4), sulfur dioxide (Volume 1, Sections 8.6, 8.8), sorbic acid (Volume 1, Section 9.2), fatty acids (Volume 1, Section 9.3), dimethyldicarbonate (Volume 1, Section 9.4), and lysozyme (Volume 1, Section 9.5).
Preventing oxidation	Sulfur dioxide (Volume 1, Section 8.7.2), ascorbic acid (Volume 1, Section 9.5), PVPP (Section 10.10.3), blanketing with inert gas (Volume 1, Section 9.6.1)
Preventing tartrate precipitation	Cold stabilization (Sections 1.7.2 to 1.7.4, 12.3.2), electrodialysis (Section 12.5), ion exchange (Section 12.4), metatartaric acid (Section 1.7.6), mannoproteins (Section 1.7.7), carboxymethylcellulose (Section 1.7.8).
Preventing turbidity due to proteins in white wine	Bentonite (Sections 5.6.2, 5.6.3, 10.9.3), tannin (Section 10.7.8), cold stabilization (Section 12.3.3), heating (Section 12.2.1)
Preventing turbidity due to coloring matter in red wine	Cold stabilization (Section 13.3.3), fining (Sections 10.4 to 10.8), bentonite (Section 10.9.4), gum arabic (Section 9.4.3)
Preventing metallic casse	
Ferric casse:	Citric acid (Section 4.6.3), gum arabic (Sections 4.6.3, 9.4.3), ascorbic acid (Section 4.6.4 and Volume 1, Section 9.5.3), potassium ferrocyanide (Section 4.6.5), calcium phytate (Section 4.6.6)
Copper casse:	Bentonite (Sections 4.7.3, 10.9.3), gum arabic (Sections 4.7.3, 9.4.3), potassium ferrocyanide (Section 4.6.5), heating (Section 12.2.1)
Improving color and aroma	Charcoal (Section 8.9.4), casein and milk (Sections 8.9.4, 10.7.6), fresh yeast lees (Section 8.9.4)

Spontaneous clarification, i.e. through settling, is due to the sedimentation, by gravity, of the particles in suspension and their adsorption on container walls. After malolactic fermentation, young red wines contain particles from grape must, yeast, bacteria, salts, colloids and amorphous substances. External factors, such as temperature, oxygen and ellagic tannins from oak wood either promote or inhibit precipitation. Clarification may be achieved simply by racking, especially if the wine is stored in small containers. Natural settling is relatively fast in red and dry white wines, but occurs much less readily in sweet white wines and certain red wines made from grapes affected by rot.

Fining consists of adding a substance that induces flocculation and settling in turbid wines or wines with colloidal instability (coloring matter in red wines). This substance captures the particles responsible for turbidity or instability in the wine, thus clarifying and stabilizing it. Fining products are often a mixture of denatured proteins that precipitate on contact with tannins, cations or acidity. They may also be of mineral origin and flocculate on contact with cations in wine. From an organoleptic standpoint, fining leads to either positive or negative changes. According to the type and quantity of fining agent used, it may make a wine softer and more elegant or, on the contrary, thinner and less attractive.

Bentonite treatment is used to prevent protein problems in white wines (Sections 5.6.2 and 5.6.3), but it is also very effective for clarifying red wines and stabilizing colloidal coloring matter. Finally, siliceous earths and polyvinylpolypyrrolidone (PVPP) may also be useful for clarifying certain wines.

10.2 SEDIMENTATION OF PARTICLES IN SUSPENSION

10.2.1 Conditions for Sedimentation

Particle sedimentation in a clear, still liquid is subject to various factors:

1. Gravity $F = V(dp - dl)g$, which depends on the difference in density between the particle (dp) and the liquid (dl), as well as particle volume (V).

2. The resistance of the liquid to the particle's descent, depending on viscosity (μ), particle surface area (S), the particle's downward speed (v) and the distance to be covered (y):

$$R = \mu S \frac{v}{y}$$

Stokes' law gives the following expression for the terminal settling velocity of the particle:

$$v = \frac{2r^2}{9\mu}(dp - dl)g$$

Terminal settling velocity depends on the squared radius of the particle (r) and the density of the liquid (dl). It is inversely proportional to viscosity (μ).

Given that yeasts have a diameter of between 1 and 10 μm, while that of bacteria is between 10^{-2} and 10^{-1} μm, a yeast cell's settling velocity in a simple medium is 10^6 times higher than that of a tiny bacterium. In wine, the difference in behavior between these two microorganisms is significant, but much less marked (25–30 times).

Variations in a wine's viscosity, caused by increased ethanol content, and its density, produced by adding sugar, only very slightly reduce the particle sedimentation velocity. However, when these particles are negatively charged kaolin fragments, variations in pH have a significant effect on the settling velocity, which decreases as pH increases. Particle charge is, therefore, an important factor in these phenomena.

Nevertheless, this does not provide a sufficient explanation for the differences observed in wines. When kaolin gel is added, certain wines remain very turbid, whereas others are properly clarified. The presence of protective colloids in the medium (Section 9.4.2) is the decisive factor in hindering clarification, causing these differences in behavior. Colloids in wine consist mainly of long-chain polysaccharides that form networks, preventing sedimentation and clogging filters.

The origin of these polysaccharides is twofold:

1. The development of *Botrytis cinerea* may cause the release of glucane, which is responsible for colloidal problems in wines made from rotten or spoiled grapes. Precautions must be taken to avoid the release of glucane into the must. Adding a glucanase to the medium (Section 11.5.2) breaks up the chain and facilitates clarification by reducing clogging in the filters.

2. Excessive extraction from red grape skins by mechanical means during winemaking leads to the solubilization of components in the cell walls that act as protective colloids and precipitate to a variable extent during aging. The techniques responsible for this effect include: rotation, pushing down the cap, violent pumping-over, high-temperature vinification pressing, etc. The behavior of the molecules released from the cell walls depends on the ripeness of the grapes and the resulting cell breakdown level. Adding enzymes (cellulases and pectinases) may improve clarification.

10.2.2 Factors Affecting the Formation of Deposits

Particles must have a higher density than wine in order to settle and form a deposit. They must also be sufficiently large, although small particles may settle by entrainment, due to mechanical or electrical effects.

The rate at which deposits form depends on any movement of the liquid inside the container, as well as the temperature gradient, the release of CO_2, floor vibrations and the type of container. It is therefore important to avoid drafts in aging cellars, especially when metal vats are used, and to ensure that there are no major variations in pressure during racking.

Sedimentation and clarification are generally more efficient in oak barrels than in vats. However, their smaller size as compared to vats is not the only factor, as the composition of the surface in contact with the wine also plays an important role. The oak releases ellagitannins that modify the structure of the particles (by oxidation and combination) and also has adsorption sites to which some components become attached. One traditional method for clarifying turbid wines in the vat consisted of adding poplar wood shavings. Although stainless steel is chemically inert, it has a charge due to modifications in the crystalline structure around the welds. The wall may then act as an 'electron gun' and inhibit clarification of the medium.

The presence or absence of protective colloids (Section 9.4) is a vital factor in the settling of particles in suspension as they prevent precipitation and maintain persistent turbidity that is difficult to eliminate. Turbidity in wine may be the cause of microbiological (fermentation, bacterial development) and organoleptic problems (herbaceous flavors and loss of character). It is therefore absolutely necessary to clarify wine, although care must be taken that the processes used do not strip it of flavor and character.

10.3 RACKING: ROLE AND TECHNIQUES

10.3.1 Role of Racking

Repeated racking produces the clarity required in wine, especially if it is aged in the barrel. Of course, the most important aspect of racking is the decanting process, which eliminates 'waste' from the wine (yeasts and bacteria, grape fragments, potassium bitartrate, ferric phosphate and cuprous sulfide).

Besides clarification, racking also provides suitable conditions for oxygen to dissolve in the wine, at a rate varying from 2.5 to 5 mg/l (Vivas and Glories, 1993) according to the technique. Oxygen eliminates certain unpleasant reduction smells (H_2S), as well as iron (ferric casse). It also facilitates the fermentation of trace amounts of residual sugar. The presence of dissolved oxygen is also responsible for intensifying color, due to its effect on the colorless anthocyanin complexes formed during alcoholic fermentation. Furthermore, the ethanal formed from ethanol stabilizes this color. Slight oxygenation can lead to a significant organoleptic improvement, especially in young red wines.

Other useful effects can be attributed to racking: degassing by eliminating CO_2 and homogenization of the wine, especially when large vats or barrels of various ages and origins are used. Racking provides an opportunity for monitoring the hygiene of aging containers and disinfecting them, as well as adjusting free SO_2 levels, which must be maintained to prevent microbiological problems.

10.3.2 Frequency of Racking

It is obvious that the rhythm of racking must be adapted to each wine (Peynaud, 1975). However, by applying the preceding principles, it is possible to define general guidelines according to region, cellar temperature and type of wine.

During their first year, red wines with good aging potential should be racked at the following times: when the alcoholic and malolactic fermentations are completed (to clarify and degas the wine), at the end of winter (to eliminate sediment) and before summer (to adjust the free SO_2 level). In Bordeaux, barrels are positioned with the bung on one side after the summer racking, so that the wine keeps the bung damp. This maintains a perfectly airtight seal, and the barrels no longer require regular topping up. Silicone bungs are now available, providing the same perfectly airtight closure, without having to turn the barrels. Thus, barrel-aged red wines require a minimum of three rackings in the first year.

During the second year of aging, a racking after summer eliminates any sediment and provides an opportunity for organoleptic monitoring and chemical analysis of the wine. One racking is required before fining to facilitate this operation and one or two more to remove the fining agent and obtain a brilliant appearance.

In any event, it is recommended to rack wines regularly every three months and burn sulfur in the empty barrels to adjust the free SO_2 content. This prevents the development of *Brettanomyces* and the formation of volatile phenols (Section 8.4.6).

Aeration should be fairly intensive at the beginning of barrel-aging, minimized towards summer and moderate before fining. When a wine is aged in the barrel, the oxygen that penetrates through the bung hole and barrel staves is complementary to that provided by racking. When wines are aged in airtight vats, racking is the only source of oxygenation, so it should be more frequent. It is especially necessary at the beginning of aging, to promote stabilization reactions.

Red wines intended for early consumption should be racked according to the same schedule as 'second-year' wines. Further clarification of 'nouveau-style' wines is obtained by filtering shortly after fermentation.

Racking is not advisable for light, fresh, aromatic, dry white wines with high CO_2 levels, generally aged in the vat for a few months after fermentation. If this type of wine is racked at all, care must be taken to keep aeration to a minimum.

Dry white wines fermented in the barrel and aged on the lees may be racked at the end of fermentation, in order to eliminate the gross lees while retaining the fine lees.

Barrel-aged sweet white wines are racked in the same way as red wines. The first racking takes place when fermentation stops, to eliminate most of the yeast (Volume 1, Section 14.2.5). The second racking is carried out a few weeks later.

10.3.3 Racking Techniques

The objective of this operation is to separate clear wine from the sediment at the bottom of the container, and also from deposits on the sides, especially in wooden barrels. The wine is transferred to another, clean container and the free SO_2 concentration is adjusted. Oxygenation occurs naturally during this operation.

Racking from one vat to another is simple, as the wine is run off by means of a tap located above the layer of sediment. It is pumped or gravity-fed through an intermediate container, where it is oxygenated.

Racking from one barrel to another is more complex. Traditionally, wine in barrels turned on their sides was run off by gravity from the lower bung hole on one end (*trou d'esquive*). The other end of the barrel was gradually raised to ensure a constant flow. The wine was collected in a tub and transferred to another barrel. When inspection by candlelight showed signs of turbidity appearing in the wine, racking was stopped. Wine is then transferred by applying excess pressure (supplied by a compressor) to the upper bung hole so that the wine runs out through the lower bung hole (see Figure 10.2). Usually, a plunger is lowered from the top of the barrel, displacing the wine by excess pressure. A lower level is used to adjust the racking height according to the estimated volume of the lees (Figure 10.1). If the barrel is on its side, it must be turned so that the bung is at the top one week before racking, so that the plunger can be inserted from the top. Racking must be sufficiently slow (5–6 m for a 225 l barrel) to avoid stirring up the lees. The wine cannot be pumped out of the top bung hole as the deposit would be stirred up into suspension. In racking from barrel to barrel, the wine is properly clarified, but there is a relatively large volume of lees, as particles are loosened from the sides when the liquid level drops. The dissolved oxygen level may be adjusted by introducing the wine directly into the bottom of the container (minimum oxygenation), pouring it in from a higher position or spraying it through a funnel (maximum oxygenation) (Figure 10.2).

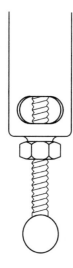

Fig. 10.1. End of a plunger for adjusting the volume of lees left at the bottom of the barrel after racking (Peynaud, 1975)

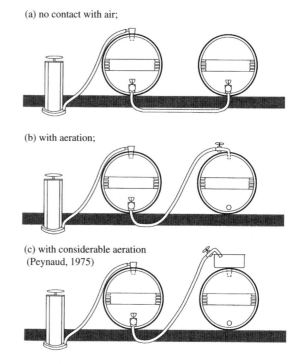

Fig. 10.2. Influence of the racking method on aeration of the wine: (a) no contact with air; (b) with aeration; (c) with considerable aeration (Peynaud, 1975)

Each barrel has its own special character, and a set of barrels is never homogeneous. Racking must be carefully planned so that there is always a clean, perfectly drained barrel ready when it is time to start racking the next full one. Free SO_2 should be adjusted by burning sulfur or injecting gas, thus sterilizing the wine and the barrel. This is vital to avoid contamination by *Brettanomyces* (Section 8.4.6).

To simplify operations, the wine may be transferred from barrels into vats, using the process described above. Once the barrels have been cleaned and drained, they are refilled. Clarification is generally less effective due to the size of the hoses used. However, this technique has the advantage of homogenizing the wine and ensuring that sulfuring is evenly distributed, taking into account the additional SO_2 from the gas in the empty barrel. The double transfer method may also be used, if necessary, to maximize oxygenation.

Once the clear wine has been run off, the barrels and vats must be cleaned to remove the deposit and disinfected, if necessary. A high-pressure water jet is generally sufficient to eliminate most of the precipitates in metal, polyester or lined concrete (epoxy resin) vats. In concrete vats coated with tartaric acid, the problem is more complex. These vats are difficult to clean and disinfecting them with SO_2 is likely to damage the coating.

Cleaning barrels before filling is even more difficult (Section 13.6.2). They must be rinsed with high-pressure water jets, drained for at least 20 minutes or, preferably, dried in a draft to improve the effectiveness of sulfuring (Volume 1, Section 8.8.5). Barrels are sulfured by burning 5–10 g of sulfur. Higher doses must be used if the barrels are to be stored empty for a long period. It is sometimes recommended to close the empty barrel for about 10 minutes after sulfuring and before filling to disinfect the wood more thoroughly. The quantity of sulfur burnt must be adjusted to obtain a free SO_2 concentration of around 25 mg/l in red wines, and this value should never drop below 15 mg/l between two rackings (Section 8.4.6). The cleanliness of a barrel and the condition of the inside surface may be inspected using a lamp inserted through the upper or lower bung holes.

10.4 THEORY OF PROTEIN FINING

Fining involves introducing a protein (fining agent) into a wine. This flocculates, gathering the particles that cause turbidity in the wine, as well as others that are likely to do so. Fining therefore has a clarifying and stabilizing effect. In view of the complex behavior of proteins in wine, many theories have been advanced to provide a chemical interpretation of the fining mechanism.

10.4.1 Background Research

The first theoretical approach to fining wine (Rüdiger and Mayre, 1928a, 1928b, 1929) presented fining as a series of charges and discharges of colloidal particles. These authors showed, by electrophoresis, that gelatin particles were positively charged at the pH of wine and that the particles responsible for turbidity were negatively charged. The result of fining depended on the reciprocal discharge of the particles present. Flocculation and clarification were more efficient if there was a full discharge. Tannins played a secondary role in this mechanism, and it was considered that an ideal dose of proteins would neutralize the turbidity of each wine and ensure optimum clarification.

Research by Ribéreau-Gayon, starting in 1934 (summarized by Ribéreau-Gayon et al., 1977), showed that fining mechanisms were, in fact, much more complex. The process can be divided into two stages:

(a) flocculation, produced by interactions between tannins and proteins,

(b) clarification, by eliminating matter in suspension from the wine.

In the first stage, flocculation was held to result from the reaction between proteins in the fining agent (e.g. gelatin) and tannins in red wine. This converted proteins, positively charged hydrophilic colloids, into negatively charged hydrophobic colloids. Complexes were formed between proteins and tannins, depending on many factors (pH, temperature, tannin and protein concentrations, etc.). These complexes were stable in a clear solution but precipitated in the presence of metal cations that caused discharges. This reaction produced or increased visible turbidity. Tannin-protein reactions produced flocculation, by associating particles and forming flakes that grew, clumped together and precipitated. The phenomenon depended on two parameters: electrical neutralization and dehydration (Section 9.3.2).

Clarification corresponds to the elimination of matter in suspension. This process consists of complex phenomena involving interaction between the fining agent and the components responsible for turbidity (Figure 10.3). Proteins that have not yet reacted with tannins may combine with particles in suspension or in colloidal solution, most of which are negatively charged. This mutual

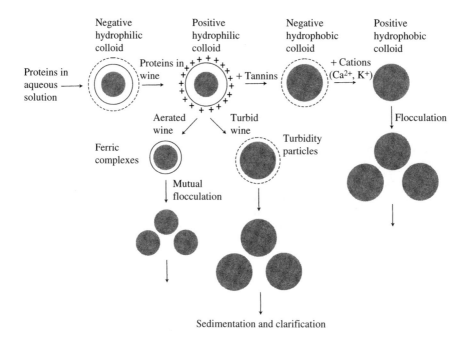

Fig. 10.3. Diagram of the flocculation mechanism of proteins in wine during fining (Ribéreau-Gayon *et al.*, 1977)

flocculation occurs during clarification in the absence of tannins.

The theory put forward by Salgues and Razungles (1983) covered the various points mentioned in the preceding research, extending the role of tannins to include that of 'wine particles'. They also took into account concepts such as the strength of chemical bonds and the reversibility of certain stages. According to these authors, fining involves reactions between colloids in wine and the fining agent: attraction, repulsion, hydration and dehydration of particles smaller than 0.1 μm to which the laws of chemistry governing true solutions are not applicable.

10.4.2 Measuring the Charges of Particles Involved in Fining

According to recent theories, the mechanisms occurring during fining depend on streaming potential and surface charge density (Lagune-Ammirati and Glories, 1996a, 1996b).

Particles in an aqueous solution are surrounded by ions. The particle/ion configuration is described by the double-layer model (Section 10.3.1). The charged surface of particles in contact with an aqueous phase are surrounded by a first layer of ions with the opposite charge. This layer, strongly bonded to the particle surface, is known as the fixed layer. This particle-fixed layer system is surrounded by a second layer of counterions, whose mobility increases in direct proportion to the distance from the particle (Figure 10.4). However, the fixed layer of counterions strongly bonded to the particle only partially compensates for the particle's initial charge. Residual charges are therefore responsible for the difference in potential at the solid/liquid interface (Sp), which decreases as the distance from the solid increases (Figure 10.5) (Hunter, 1981). The zeta potential, ζ, is defined as the potential at the plane that separates the fixed layer from the diffuse layer of counterions, known as the cut plane of the system.

The zeta potential is involved in the interaction and adsorption mechanisms between particles and ions, as well as their coagulation, flocculation and sedimentation behavior. This potential may be calculated from measurements of the streaming

Clarification and Stabilization Treatments: Fining Wine

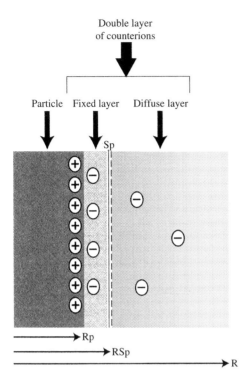

Fig. 10.4. Diagram of the double layer of a charged particle and the electrostatic phenomena. R, distance from the particle; Rp, radius of the charged particle; RSp, hydrodynamic radius (including the layer of strongly bonded counterions, i.e. the fixed layer); Sp, edge of the cut plane between the fixed and diffuse layers of counterions (Lagune, 1994)

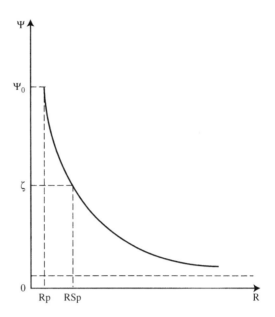

Fig. 10.5. Changes in electrostatic potential in the vicinity of a double layer. Ψ, potential; Ψ_0, surface potential; ζ, zeta potential = potential at the cut plane (Lagune, 1994)

potential, PE. PE is the potential created between the particle-fixed layer system and the diffuse layer when it moves away from the particle due to an external force. It may be measured using a particle charge detector. The zeta potential depends on the streaming potential PE, but is independent of the conditions in the medium.

PE is measured using a particle charge detector (PCD-O2, Müteck). This consists of a cylindrical polytetrafluoroethylene (PTFE) bath, equipped with two silver electrodes, located at the top and bottom and linked to an amplifier. A PTFE piston mounted in the bath oscillates vertically at a constant frequency, making the liquid flow along the sides of the bath (Figure 10.6). This apparatus is connected to an automatic titrator used to add polyelectrolyte.

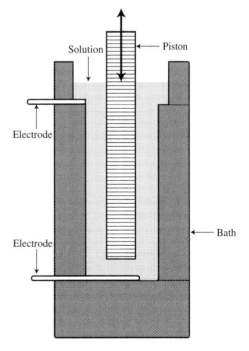

Fig. 10.6. Particle charge detector measuring system (Lagune, 1994)

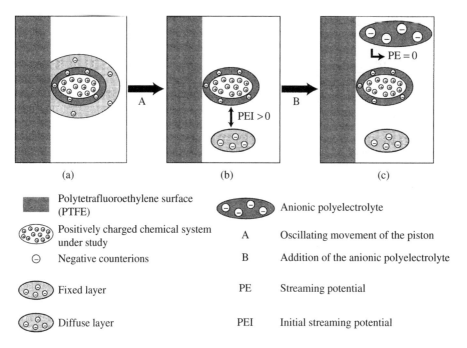

▓	Polytetrafluoroethylene surface (PTFE)
(positively charged symbol)	Positively charged chemical system under study
⊖	Negative counterions
(fixed layer symbol)	Fixed layer
(diffuse layer symbol)	Diffuse layer
(anionic symbol)	Anionic polyelectrolyte
A	Oscillating movement of the piston
B	Addition of the anionic polyelectrolyte
PE	Streaming potential
PEI	Initial streaming potential

Fig. 10.7. Mechanisms operating in the solution of a positive species during titration (Lagune, 1994)

When a solution of ionic particles is placed in the detector, the particles are surrounded by a double layer of counterions (Figure 10.7a). The Van der Waals force is then responsible for adsorption phenomena on the bath and piston surfaces.

The oscillation of the piston (A) streams the liquid phase along the walls, gathering counterions in the diffuse layer into a cloud that moves away from the particle-fixed layer system (Figure 10.7b). A difference in potential is thus created between the diffuse layer cloud and the particle-fixed layer system. This is known as the 'initial streaming potential' (PEI). It is measured by the two electrodes built into the vat and expressed in mV. It indicates the charge of the particles under investigation.

The addition of ions with the opposite charge (polyelectrolyte) (B) neutralizes the charge and cancels out the PEI potential (Figure 10.7c). The quantity of polyelectrolyte necessary to neutralize this charge is used to calculate the 'surface charge density', d, expressed in meq of polyelectrolyte g^{-1} or ml^{-1}. This is a characteristic of the system under defined conditions. If the system has a positive charge, the polyelectrolyte is anionic (sodium polyethensulfonate, or PES-Na). If the charge is negative, the polyelectrolyte is cationic (polyallyldimethylammonium chloride, or polyDADMAC).

The titration is represented by a curve (Figure 10.8). In the case of a negative system, V_0 (ml) is the volume of polyelectrolyte necessary to obtain PE = 0. This volume is used to calculate the surface charge density of the system, expressed in meq/l or meq/ml. according to the type of system.

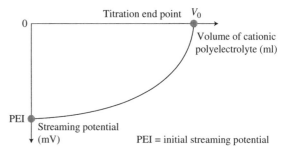

Fig. 10.8. Determining the surface charge density of a negative species by titration with a cationic polyelectrolyte (Lagune, 1994)

Clarification and Stabilization Treatments: Fining Wine

10.4.3 Applications in Fining Wine

This technique for determining surface charge density has been applied to various fining products, phenolic fractions extracted from grapes, and wines. The results have made it possible to characterize these various systems and envisage modeling the mechanisms that occur during fining.

The initial streaming potential of red wine is negative. This indicates that the compounds in wine have an excess of negative surface charges. The titration curve of a red wine with a cationic polyelectrolyte (Figure 10.9) is used to calculate the surface charge density of the wine, expressed in meq/l (Lagune, 1994).

The results in Table 10.2 show an excess of negative charges attributable to tannins and other compounds, such as polysaccharides (Tobiason and Hoff, 1989; Tobiason, 1992; Ferrarini *et al.*, 1995; Vernhet *et al.*, 1995). The surface charge density of red wines is neither proportional to the total phenols nor to the tannin concentration. The differences observed are probably due to grape varieties, the richness of the grapes (polysaccharides) and the wine's state of development.

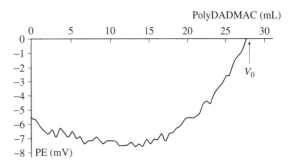

Fig. 10.9. Titration of a red wine with polyDADMAC polyelectrolyte (Lagune, 1994)

Models of fining assume that the negative surface charge of red wines is neutralized by adding proteins (fining agent), considered to act as positive electrolytes. To reproduce the conditions during fining, a positively charged gelatin solution was added regularly to the wine. The absolute value of the potential decreased, tending towards 0. It was neutralized at a specific volume, V_0, of gelatin solution and then became positive (Figure 10.10). During neutralization of the charges, the initially

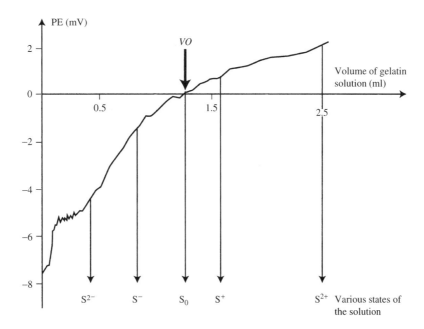

Fig. 10.10. Changes in the streaming potential of a wine according to the volume of gelatin solution added. S^{2-}, S^-, etc., represent the different states of the solution (Lagune, 1994)

Table 10.2. Surface charge densities (d) of particles from several samples of red wine according to their total phenol and tannin contents (Lagune-Ammirati and Glories, 1996)

Wines	Appellations	Total phenols (index)	Tannins (g/l)	d(meq/l)
A	Cahors	63	3.71	−2.79
C	Bordeaux Supérieur	40	2.18	−2.74
D	Puisseguin Saint-Emilion	54	3.19	−2.40
B	Saint-Emilion	52	3.03	−2.09
E	Côtes de Saint Mont	63	3.46	−1.89
F	Madiran	93	5.20	−1.60

negative solution (S^{2-}) gradually became neutral (S_0) and then positive (S^{2+}). This phenomenon is related to a series of interactions between polyphenols and gelatin (Figure 10.11).

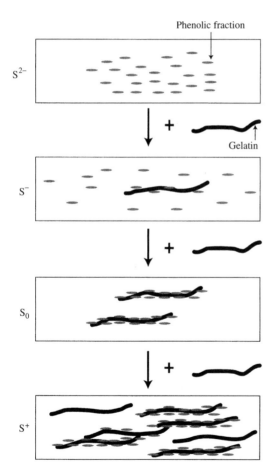

Fig. 10.11. Modeling the fining of a red wine with gelatin, using a surface charge detector (Lagune, 1994)

When the gelatin is first added, the compounds in the wine are bonded to the surface proteins by one or more sites and then gradually form a monolayer. During this time, equilibrium is established and aggregation and precipitation phenomena occur. The formation of cross links between various proteins is superposed on the other reactions (McManus *et al.*, 1985. Ozawa *et al.*, 1987. Haslam and Lilley, 1988. Haslam, 1995).

However, it has been observed that a very large quantity of gelatin, 100 times greater than that used normally in fining, is necessary to neutralize the charge (600 g/hl). This corresponds to the start of titration, with the formation of negative S^{2-} complexes only.

10.5 TANNIN-PROTEIN INTERACTIONS

A great deal of research has shown that tannins combine with proteins by hydrogen bonds and hydrophobic interactions (Section 6.3.4), depending on the characteristics of the tannins, those of the proteins and conditions in the medium.

10.5.1 Description of Tannin-Protein Interactions

Hydrophobic effects (Van der Waals attractions) occur between tannins and the non-polar regions of the proteins (Martin *et al.*, 1990. Haslam, 1993). Certain authors (Oh *et al.*, 1980) even consider that this is the predominant interaction mode, due to the hydrophobic nature of the tannins, as demonstrated by their adsorption on uncharged polystyrene resins. These reactions seem to be the origin of the complexation reinforced by hydrogen bonds, for example, between the carbonyl group of the secondary amine function of the proline and the phenol OHs (Martin *et al.*, 1990. Haslam, 1996). These surface phenomena depend not only on the number of phenol groups on the periphery of the molecule (Haslam and Lilley, 1985) but also on the relative proportions of each of the two families (Section 6.3.4).

Furthermore, at low protein concentrations, polyphenols bond to the surface of the protein at one or more sites, forming a monolayer that is less hydrophilic than the protein alone. This is followed by aggregation and precipitation. When the protein concentration is high, an identical phenomenon occurs, with the superposed formation of cross-bonds between various protein molecules. This explains the non-stoichiometry of the tannin-protein reaction observed by many authors (Ribéreau-Gayon *et al.*, 1977).

Tannin-protein complexation is reversible, provided that covalent bonds are not involved and that both condensation and aggregation are limited. If this is not the case, quinoid intermediaries are formed. These are highly reactive with proteins and the combinations formed are insoluble and irreversible (Beart *et al.*, 1985; Gal and Carbonell, 1992; Metche, 1993).

Tannin-protein interactions depend on the characteristics of the tannins: size, structure, charge, etc. These interactions increase with the degree of polymerization of the procyanidins (Asano *et al.*, 1984; Ricardo da Silva *et al.*, 1991; Cheynier *et al.*, 1992) and also according to their galloylation rate (Charlton *et al.*, 1996). Tannin-protein interactions also vary according to the composition of the tannins: condensed tannins formed from procyanidins linked by ethyl cross-bonds, tannins combined with anthocyanins or tannin–polysaccharide complexes. At pH 3.5, these molecules have different charges depending on their origin (Table 10.3). Furthermore, it has been observed that this surface charge density is affected by the pH of the solution. The higher the pH, the more charged the flavanols, with a variation on the order of 100% between pH 3 and 4. This phenomenon is obviously useful in fining.

The characteristics of the proteins (amino acid composition, structure, size, charge, etc.), like those of the tannins, obviously play a major role in these reactions. Indeed, proteins with a high proline content have a great affinity for tannins (Hagerman and Butler, 1980a; Mehansho *et al.*, 1987; Asquith *et al.*, 1987; Butler and Mole, 1988; Austin *et al.*, 1989; Butler, 1989; Charlton *et al.*, 1996). The importance of proline is probably due to its incapacity to form helixes, leaving the protein open and accessible to tannins (Hagerman and Butler, 1981). On the other hand, small, compact proteins have a low affinity for tannins (Hagerman and Butler, 1980b).

Proteins in fining agents have a positive charge at pH 3.5. Its strength depends on the isoelectrical pH and the degree to which the molecules are broken down. In gelatin, molecules may have a range of surface charge densities between $+100 \times 10^{-3}$ meq/g and $+1000 \times 10^{-3}$ meq/g, depending on the hydrolysis of the collagen.

Table 10.3. Surface charge densities (d) at pH 3.5 of phenolic fractions extracted from grape seeds and red wine (Section 6.4.6) (Lagune, 1994)

Procyanidins from grape seeds	(d) (10^{-3} meq/l)	Flavanols from red wine	(d) (10^{-3} meq/g)
Monomers	−22	(+) Catechin	−4.2
Oligomers	−347	Oligomeric procyanidins + condensed tannins	−30 to −120
Polymers	−676	Polymerized procyanidins + tannin–polysaccharides	−350 to −900

10.5.2 Influence of the Medium on Tannin-Protein Interactions

1. When a standard quantity of proteins is added, the quantity of tannins taken up generally increases with the tannin concentration of the wine, with certain exceptions. For example, Lagune (1994) showed that 5 g/hl of gelatin eliminated 120 mg/l of tannin from a red wine that initially contained 1.72 g/l. Only 40 mg/l was eliminated from another Bordeaux red wine with a much higher tannin content (3.54 g/l).

2. In general, the larger the quantity of proteins added, the more tannins are eliminated. However, the reaction depends on the type of proteins, and no direct correlation has been observed between the quantity of protein added and the quantity of tannin eliminated (Table 10.4). Turbidity (Siebert *et al.*, 1996), as well as the type and quantity of tannin-protein precipitates, depend on the relative concentrations of the various components (Calderon *et al.*, 1968).

3. At pHs ranging from 2 to 4, tannin-protein flocculation is faster and the particles precipitate better at lower acidity (Ribéreau-Gayon *et al.*, 1977). When the same dose of fining agent is added, the quantity of tannins eliminated increases according to the wine's pH. In red wine, this amount almost doubles between pH 3.4 and 3.9 (Glories and Augustin, 1992).

4. The presence of Na^+, K^+, Ca^{2+}, Mg^{2+} and especially Fe^{3+} cations is indispensable for flocculation and the precipitation of tannins and proteins (Ribéreau-Gayon, 1934). Negatively charged tannin–iron complexes react with positively charged proteins (Ribéreau-Gayon *et al.*, 1977). Dissolved oxygen promotes flocculation, as it facilitates the formation of trivalent iron. Thus, the aeration resulting from racking improves the effectiveness of fining.

5. Different types of polysaccharides have highly variable effects. These polymers may have a 'protective' action that prevents flocculation and precipitation, and, therefore, clarification. This is true of glucane and gum arabic, which may even make fining impossible (Section 11.4.3). Polysaccharides may also have an 'activating' effect. The presence of pectins, arabinogalactans and polygalacturonic acids increases the intensity of turbidity and is favorable to fining, while neutral polysaccharides have no effect.

6. Calderon *et al.* (1968) reported a decrease in the affinity of tannins for gelatin in media with a high alcohol content, and stated that the complexes they formed were soluble. Personally, we did not find any significant differences at alcoholic strengths between 11 and 13% (by volume).

7. A low temperature (15°C) enhances precipitation and clarification, due to the decrease in Brownian movement that facilitates flocculation of the colloids. It is generally recommended to carry out fining in winter.

Table 10.4. Influence of the quantity and type of gelatin on the elimination of tannins from wine during fining. Tannin content of the wine: 1.62 g/l (Lagune, 1994)

Gelatin added (mg/l)	Tannins eliminated (mg/l) by the following gelatin:	
	Heat-soluble	Cold-soluble
50	50	50
100	160	90
200	310	230

10.6 EFFECT OF FINING ON THE ORGANOLEPTIC QUALITY OF WINE: CONCEPT OF OVERFINING

10.6.1 Effect of Fining on the Phenolic Compounds in Wine

Fining a red wine with gelatin clarifies and stabilizes it by eliminating unstable colloidal coloring matter. The decrease in color intensity and phenol content is not very large. It mainly affects combined anthocyanins (PVP index) and most highly polymerized (HCl index) and bulky (dialysis index and EtOH) tannin molecules. The consequence is a decrease in gelatin index that corresponds to a softening of the wine's flavor.

Analysis of the tannins by capillary electrophoresis before and after fining shows preferential elimination by the fining agent of tannin–polysaccharide combinations and condensed tannins, i.e. those molecules with the highest charge density. A relatively large accumulation of soluble tannin-protein complexes is also observed, depending on the fining agent. Monomeric flavanols as well as dimeric and trimeric procyanidins are not affected by fining.

Fining eliminates those tannin molecules that react most readily with proteins, and are the most aggressive from an organoleptic point of view. Fining also removes molecules that contribute to the impression of body and volume on the palate. It is, therefore, unsurprising that wines should be softened by fining, and may also seem thinner. The presence of soluble complexes after fining corresponds to a deactivation of the tannins and is beneficial to quality.

It is easy to understand the advisability of carrying out preliminary trials with different doses of various fining agents before full-scale fining (Section 10.8.1).

10.6.2 Effect on Aroma

Losses of volatile compounds during fining are relatively limited and almost imperceptible. They depend on the wine, the type of fining agent and the dose used. Each fining agent has a particular affinity for certain aromatic compounds. It may also have an indirect effect, by fixing substances that act as supports for aromatic compounds: β-ionone, ethyl octanoate (Lubbers et al., 1993).

Fining may cause a perceptible decrease in aromatic intensity, but this is compensated by greater finesse. According to Siegrist (1996), variations in volatile compounds due to fining are on the order of 8% for gelatin and egg albumin and 11% for blood albumin.

10.6.3 Overfining

A wine is said to be 'overfined' if, after fining, some of the added proteins have not flocculated. Overfined wines are initially clear, but generally become turbid following the addition of tannins. Overfining should not be confused with poor efficiency of the fining agent, responsible for persistent turbidity and generally caused by the presence of protective colloids. Overfining is most frequent in white wines, when the fining agents used require a great deal of tannin to flocculate, e.g. gelatins and, possibly, isinglass.

There is always a risk of turbidity in overfined wines, especially if tannins are added during later winemaking procedures:

(a) blending with another wine that has a higher tannin content,

(b) addition of enological tannins,

(c) barrel aging, where the wine absorbs ellagic tannins from the oak,

(d) using natural cork stoppers that release ellagitannins.

To avoid overfining, it is recommended not to use gelatin, or to use only doses of 1–3 g/hl in those relatively rare white wines with high tannin contents (100–200 mg/l). Adding tannins (100 mg/l) to white wine before fining could be a solution, but it has the disadvantage (Section 10.7.8) of making the wines harder and reducing quality. Better results are obtained with siliceous earths (Section 10.10.1). Egg white is also unsuitable for fining white wines, as it requires a great deal of tannin to flocculate. On the other hand, casein, blood albumin and isinglass at low doses almost never cause overfining.

In fining, the tannin-protein reaction is not stoichiometric and flocculation of the proteins is incomplete. Both reagents (tannins and proteins) may be present at the same time in a clear solution. In the case of overfining, potential turbidity appears after the addition of either tannins or proteins. This emphasizes the importance (Section 10.6.1) of preliminary trials (Section 10.8.1) before fining, followed by analysis, to check not only the efficiency of clarification but also the extent to which the wine has been stabilized.

Overfining of red wines is rare and is generally due to poor flocculation, fining too rapidly or at too high a temperature, or the presence of colloidal turbidity. The deposit corresponding to overfining of red wines coats the inside of the bottle. In the case of white wines, overfining produces a precipitate.

There are various ways of treating overfined white wines:

1. The addition of tannin used to be advised, to trigger the flocculation of excess proteins. In reality, it is difficult to achieve total elimination, even with doses of tannins as high as 100–150 mg/l that noticeably harden the wine.

2. Bentonite (Section 5.6.2) is capable of eliminating proteins almost completely at high doses (up to 100 g/hl). In practice, only partial elimination is obtained, using doses of 30–50 g/hl, in order to avoid turbidity.

3. Cooling wine to 0°C, followed by low-temperature earth filtration, reduces the risk of turbidity by partially eliminating proteins.

4. Siliceous earths (Section 10.10.1) may be used at the same time as gelatin, or at a later stage, to avoid overfining.

10.7 PRODUCTS USED IN FINING

10.7.1 Protein Fining Agents

Traditionally, products used for fining are proteins of animal origin: egg albumin, blood albumin, casein (milk), isinglass (fish) and gelatins (collagen). Several inorganic products (bentonite and siliceous earth) are also used in clarification and stabilization. Although the expression 'bentonite fining' is used, 'bentonite treatment' would probably be preferable, to show that it is not a protein fining process (Section 10.9).

Every product used in protein fining has a specific action, according to its origin, and therefore its composition.

The issues involved with bovine spongiform encephalopathy (BSE) in animals and its possible transmission to humans have led to a restriction in the use of products of animal origin for fining wine. Legislation in several countries, particularly the European Union, has been updated, banning the use of dried blood powder and blood albumin. Egg and milk albumin are now the only animal albumins permitted. The use of gelatin has also been challenged, even though it is mainly a pork by-product. It is, however, still widely used for its excellent clarification and stabilization capacities, particularly in red wines. Winemakers would like to have substitute products with similar qualities, so there are incentives for developing alternative fining agents and at least two possibilities are currently being explored: plant proteins and egg albumin derivatives.

Irrespective of their origin, commercial fining agents are available in liquid, as well as several solid forms. Solid fining agents must be dispersed in water prior to use, at a concentration and temperature specific to each product.

In liquid form, pure products are available in 'colloidal solutions', at varying concentrations. Average concentrations of gelatin solutions are between 10 and 50%. There is no link between the concentration of a solution, its enological properties and its effectiveness, as these factors depend on the conditions under which the solution is prepared. These products may be used in a pure state or diluted in water, as required.

The same substance may be presented in various forms: gelatin in sheet, granule, powder or liquid form, isinglass in chip, powder or gel form, and egg albumin in fresh, frozen or lyophilized form.

Combinations of products are also used. These are commercial products specific to each manufacturer, consisting of blends of different fining agents adapted to specific purposes, e.g. bentonite + casein, which is known to enhance the freshness and finesse of white and rosé wines, as well as eliminating proteins.

10.7.2 Gelatins

Gelatins are produced by the almost complete hydrolysis of collagen from pig skins and animal bones. Their main components are: glycine, proline, hydroxyproline and glutamic acid. Industrial production dates from the early 18th century and several different types are now available, produced by acid, alkaline and enzymic hydrolysis. These industrial gelatins are classified according to their jellifying power (between 50 and 300 Bloom units) and solubility. Three categories have been defined, as follows: heat-soluble gelatins consist mainly of proteins with high molecular weights, while cold-soluble and liquid gelatins contain no proteins with high molecular weights. There is a special classification for gelatins hydrolyzed by enzymes (ASF, DSF and SPG) (Sanofi Bioindustrie).

Gelatins have a wide range of applications in the pharmaceutical, photographic, paper, cosmetics and especially food industries, which currently use the largest volumes. Enological gelatins represent only a very small share (1–5%) of the food gelatin market and they are not always suited to winemaking needs. Manufacturers are therefore obliged to prepare special gelatins for fining wine.

The enological codex includes this three-category classification. Compositions and charges were defined by Lagune (1994), as follows:

1. Heat-soluble gelatins (SC) have 30–50% proteins with a molecular weight above 10^5 and a strong charge, 0.5–1.2 meq/g.

2. Liquid gelatins (L), produced by intense chemical hydrolysis, have medium-weight molecules ($M < 10^5$), a weak charge and many highly charged peptides.

3. Cold-soluble gelatins (SF), produced by enzymic hydrolysis, have a very weak charge, a low peptide content and lightweight proteins: $M < 10^5$.

As compared to normal standards, the recommendations in the enological codex specify gelatin's nitrogen content (above 14% by dry weight) and 'precipitation by tannin' number. This number represents 'the quantity of tannins necessary to precipitate all of the gelatin'. It is not very significant, in view of the diverse structures of tannins and the non-stoichiometric nature of tannin-protein reactions (Section 10.5.1).

The mass distribution of proteins in gelatins (Hrazdina *et al.*, 1969; Cerf, 1973; Ricardo da Silva *et al.*, 1991), as well as their charge (Ribéreau-Gayon *et al.*, 1977; Lagune, 1994) are much more accurate, useful characteristics for interpreting the effect of these fining agents on wine. The more highly charged the gelatin, the more active it is in relation to the various groups of tannins found in red wine. Gelatins are therefore capable of eliminating all the negative tannins (TP and CT). If the gelatin proteins have a sufficiently high molecular weight, they will also precipitate. If wines have a high phenol content, fining softens them and makes them more elegant. However, if a wine is initially lacking in body, the same fining agent may make it hard and thin. In less robust wines, gelatin with a low or medium charge is best, as it only reacts with the most highly charged, reactive tannin molecules, without disturbing the tannic structure.

Solid gelatin (SC) is dissolved by stirring into hot water (40–50°C). The other preparations are cold-soluble (SF) or used as supplied (L). Doses vary from 3 to 10 g/hl in red wine.

Gelatin is used in conjunction with silica gel (Kieselsol, Bayer) or siliceous earth (Klebosol 30Vn, Hoechst) (Section 10.10.1) to treat white wines. This avoids overfining and takes advantage of the positive properties of gelatins in wines with a high polysaccharide colloid content, especially those made from grapes affected by noble rot (Wucherpfennig and Passmann, 1972; Wucherpfennig *et al.*, 1973). Kieselsol is a 30% colloidal solution of silica in water ($d = 1.20$, pH = 9). It is added to the wine either before or just after gelatin. The appropriate Kieselsol/gelatin ratio for good clarification is between 5 and 10. The fining agent is removed two weeks after the treatment and produces heavy, bulky lees.

10.7.3 Isinglass

This fining agent has been in use since the 18th century. Together with milk and, above all, egg white, it has replaced the powdered, fired clay that had been used to treat wines since ancient times. Isinglass is a raw, unprocessed product from the swim bladder of certain fish, such as sturgeon. It consists mainly of collagen fibers and is available in sheets, strips, whitish chips or coarse vermiculated powders. Preparation is long and laborious: the dry isinglass must be soaked in acidulated, sulfured water (0.5 ml HCl/l + 200 mg/l SO_2) for about ten days at a cool temperature and then sieved to obtain a homogeneous jelly. The vermiculated form of isinglass swells easily, without lumps. It must, however, be used immediately after preparation or hydrolysis converts it into gelatin.

As sturgeon are not readily available and there are a number of problems involved in using this product, enological product manufacturers currently offer this fining agent in the form of ready-to-use jelly, prepared from fish cannery waste (skin, cartilage, etc.).

The normal dose is from 1.25 to 2.5 g/hl for white wines. This concentration enhances their brilliance and reinforces the yellow color. However, the light, bulky lees are a disadvantage as they make racking more difficult and clog filter surfaus. Isinglass does not tend to overfine as it requires very little tannin to flocculate.

10.7.4 Albumin and Egg White

Egg albumin consists of several proteins and represents 12.5% of the weight of a fresh egg white. Ovalbumin is the main component. Besides fresh or frozen egg white, egg albumin may be used in the form of flakes. These vary in color from white to golden yellow.

Egg albumin is the oldest protein fining agent. It has always been presented as the only fining agent for great red wines. However, it may make some wines thinner. From a colloidal point of view, egg albumin is a fining agent that does not flocculate a great deal, but precipitates a compact deposit. It is recommended for softening wines with a high tannin content and excess astringency. Albumin must be used with care on light wines and is not recommended for white wines.

When fresh egg white is used, 3–8 egg whites are required per 225 liter barrel. One egg white corresponds on average to 4 g of dry matter. The whites must be mixed and dissolved in a quarter of a liter of water, producing as little foam as possible. Dissolving may be facilitated by adding a little sodium chloride, as this maintains the globulins in solution.

Egg albumin is also available in solid form (flakes or powder), obtained by desiccating fresh egg white. Sodium carbonate is added to facilitate dissolving in water. A paste is made with the egg powder and sodium carbonate, then gradually diluted. Dried albumin has a slightly different composition from that of fresh egg white, as certain proteins with high molecular weights are eliminated by the drying process (Ikonomou-Potiri, 1985). The results are often different from those obtained with fresh egg white. Doses required are between 5 and 15 g/hl. Egg albumin may be spoiled by heat, so it is not advisable to warm the preparation to facilitate dissolving the powder. It is also available commercially in ready-to-use, sterilized, liquid form.

Frozen egg white produces similar results to those obtained using fresh egg white. The frozen product is left to defrost at room temperature and used immediately, at average doses of 75–200 ml/hl. They exist also as commercial product, sterilised and in a liquid stab, ready for use.

Egg white contains non-negligible quantities (9 g/l) of lysozyme. The amount added during fining, on the order of 5 mg/l, is theoretically sufficient to destroy some of the lactic bacteria (Amati *et al.*, 1989). However, all of the work on this subject (Ribéreau-Gayon *et al.*, 1977) shows that, in fact, fining with egg white has no effect on lactic bacteria, as the lysozyme probably precipitates with the albumin, due to the effect of the tannins (Volume 1, Section 9.5).

Egg white, or egg albumin, is universally recognized for its qualities as a fining agent for red wines. However, it produces the best results in full-bodied wines that have already aged for some time, where there has already been partial spontaneous clarification and stabilization. This fining agent performs less well in young wines or those with a lighter tannic structure, which are likely to lose body in the process. Lagune-Amirati and Glories (2001, 2002) subjected a commercial liquid egg albumin solution (Albucoll, Laffort, 126 Quai de la Souys, 33100 Bordeaux) to a variety of treatments to modify its mass and surface charge density. This resulted in a range of fining products with different characteristics, all with good stabilizing and clarifying properties. Some are likely to be suitable for replacing gelatin in fining fragile, young red wines.

10.7.5 Blood By-products

Blood by-products are currently prohibited for health reasons in many countries, including those in the European Union.

Their effectiveness had made them popular for many years. The fresh, liquid blood that was initially used was the first product to be banned. It was replaced by dried blood or blood albumin in powder form, both more recently prohibited.

From an enological standpoint, these products give good results in fining young red and white wines. It is highly effective and attenuates any herbaceous character. It is not very sensitive to protective colloids and does not require much tannin to flocculate, so the risk of overfining is minimal. Bitter, stalky, young red wines, with a robust tannic structure, are nicely softened. The dose must be adjusted from 10 to 20 g/hl according to the wine's tannin content. Herbaceous white wines, with an intense, heavy aroma, lose some of their coarseness after fining with doses of 5–10 g/hl.

10.7.6 Milk and Casein

Casein, a heteroprotein containing phosphorus, is obtained by coagulating skimmed milk. It is an excellent fining agent for white wines and has a 'refreshing' effect on their color and flavor. It not only has a curative effect on yellowing and maderization, but may also be used preventively. One characteristic of this fining agent is that flocculation occurs exclusively due to the acidity of the medium, but the presence of tannins is necessary for precipitation and clarification. This property is both positive, as this type of treatment never produces overfining, and negative, as it makes this fining agent rather difficult to use. It must be rapidly distributed through the entire mass of wine before it flocculates, which occurs in a very short time. An injection pump is the best solution, making it possible to avoid losing any of the fining agent through partial flocculation before it is completely dispersed in the wine.

Casein powder is not very soluble in pure water, but dissolves better in an alkaline medium, produced by adding potassium or sodium bicarbonate or carbonate, or possibly potash. The normal dose is from 10 to 20 g/hl, although in curative treatment 50 g/hl or more may be used. Casein powder's preventive action is not fully understood, but it affects phenols, either by eliminating them or, more probably, by protecting them from oxidation.

Fining with whole milk is not permitted in the EU, but it may be effective in certain cases. It improves the color of white wines and eliminates reduced, moldy odors. The treatment's effectiveness is due to the milk's fat content. Skimming reduces the adsorption capacity, but increases the clarifying effect. One liter of cow's milk contains approximately 30 g of casein and 10–15 g of other proteins that are likely to increase the risk of overfining if too high a dose is used (above 0.2/0.4 l/hl).

10.7.7 Plant proteins

In response to winemakers' interest in replacing fining agents of animal origin with plant-based products, the Martin Vialatte research company (BP 1031, 51319 Epernay, France) started studying the properties of plant proteins and assessing the possibilities of using them as fining agents for wine (Lefebvre et al., 2000). Initial results have been found to be promising with several powdered products.

In 2003, Maury et al. carried out a study using a protein extracted from white lupine, two wheat gluten-based preparations, and two chemical hydrolysates of gluten. Experiments were carried out using two unfiltered wines and a model solution prepared with phenolic compounds extracted from Syrah wine. All the fining agents tested precipitated relatively low levels of phenolic compounds. As is the case with gelatin, selective precipitation affected only condensed tannins. Molecular weight is a major factor in the effectiveness of these proteins. Gelatin generally fines the wine more efficiently, although some plant proteins precipitated galloylated tannins under the same conditions.

To conclude, it should be possible to use plant proteins as fining agents in wine, but each preparation behaves in a specific way. It will, therefore, be necessary to test a large number of products to determine which ones give the best results with different types of wine and define the most effective doses, likely to be around 10–20 g/hl.

An application has been submitted to the appropriate authorities (Office International de la vigne et du Vin, OIV) for approval of these products and authorization to use them in fining wine (Lefebvre et al., 2003).

In view of the fact that wheat gluten is capable of producing allergic phenomena, it was important to ensure that no residues were left in wine fined with plant proteins and, that there was no risk of triggering allergic reactions. Lefebvre et al. (2003) showed that, even red wines treated with 50 g/hl of the fining agent did not contain any wheat gluten residues. The treated wine was also tested for immunoreactivity and presented no risk of triggering allergic phenomena. A study in progress is examining the treatment of white wine with pea and lupine extracts. The results indicate that, from a health standpoint, there are no objections to using plant proteins for fining wine.

10.7.8 Alkaline Alginates

Sodium alginate is an alginic acid salt. It is extracted from various phaeophyceae algae, especially kelp, by alkaline digestion and purification. It may be effective in clarifying wine, although it is not a protein fining agent.

It is available as a practically odorless, flavorless, white or yellowish powder, consisting of fiber fragments that are visible under a microscope. When sodium alginate is mixed with water, it produces a viscous solution with a pH between 6 and 8. It is insoluble in alcohol and in most organic solvents.

When a 20% calcium chloride solution (10^{-1}) is added to a 1% aqueous sodium alginate solution, a gelatinous calcium alginate precipitate is formed. If the calcium chloride is replaced by 10% dilute sulfuric acid, gelatinous matter also precipitates, due to the formation of alginic acid. Sodium alginate is a polymer of mannuronic acid, consisting of chains with a basic motif consisting of two mannuronic cycles.

Alginic acid has a pK of 3.7. It is displaced from its salts by relatively strong acids and then precipitates at acid pH values ≤3.5, as it is insoluble in water. Flocculation is generally good, but is nonexistent or incomplete at pH >3.5. Tannins are not involved in the process.

The alginates used in enology have molecular weights between 80 000 and 190 000. Doses used range from 4 to 8 g/hl. Flocculation is very fast if the wine has a sufficiently high acidity, but the deposit settles slowly as the particles are very light. Clarification is irregular and inorganic substances are not fixed very well.

Adding 5–10% gelatin or blood meal accelerates settling and improves clarity. However, alginates are much less effective in clarification than normal fining agents. Their main advantage is that they make it possible to filter wines just a few hours after fining.

The solution is prepared by adding small amounts of cold water to the powder until it forms a paste and then adding more water to produce a solution at 10–15 g/l. This solution is poured into the wine while stirring energetically, left to settle and then filtered 5 hours later.

10.7.9 Enological Tannins and Their Role in Fining

The official definition of tannin (*acidum tannicum*) given by the Enological Codex is as follows:

> Enological tannin is whitish-yellow or buff-colored, with an astringent taste. It is soluble in water, and partially soluble in ethanol, glycerol, and ethyl acetate. The commercial product is made from gall nut, wood with a high tannin content such as oak or chestnut, or grape pomace. It produces stable combinations with proteins. At a pH between 3 and 5, tannin solutions produce a blue-black precipitate in the presence of ferric salts. Tannin solution also precipitates alkaloids (cinchonine sulfate).

Commercial tannins are mixtures (Table 10.5), classified into two groups: procyanidin-based condensed tannins from grapes and ellagitannin- and gallotannin-based hydrolyzable tannins from oak and chestnut wood, or gall nuts. Tannin made from the latter is the most widely available commercially, although it is quite different from wine tannins. From an organoleptic standpoint, they have a bitter, green, astringent character. They do not give wine the same structure and body as natural condensed tannins.

Gallotannins may be used to prevent oxidation in must made from botrytized grapes. Seed tannins stabilize anthocyanins and wine color during fermentation, deepen the color of new wine by co-pigmentation, and facilitate ageing. Tannins also cause partial precipitation of excess protein matter and may be used to facilitate clarification in new wine and fining in white wines. However, adding tannin to white wines is controversial. It is common practice in certain regions (Champagne), while in other areas it is found to toughen the wine. The use of bentonite or fining agents that do not cause overfining is recommended for eliminating excess proteins in the above instances. Doses

Table 10.5. Phenolic composition of some commercial enological tannins

Origin	Extraction method	OD 280	Proanthocyanidins (mg/g)	Ellagitannins (mg/g)	Gallotannins (mg/g)	Scopoletin (µg/g)	Acetic acid (mg/g)
Oak	Water	24	1	680	2	8	2
Chestnut	Water	20	2	230	2	2	2
Gall	Water	24	Traces	0	780	0	0.7
Gall	EtOH	24	1	0	670	0	5
Gall	Et-O-Et	31	0	0	240	1	7
Grape pomace	Water	27	260	0	0	Traces	2
Grape seeds	Water	92	630	0	0	0	4
Quebracho	Water	26	45	14	0	0.7	1
Myrobolans	Water	14	3	85	148	0	3

used vary from 5 to 10 g/hl for red wines and approximately 5 g/hl for white wines.

The quality of commercial tannins depends on the conditions under which they are extracted from the plant matter and the way the powder is dried. Oxidative phenomena cause a rapid breakdown of these products.

10.8 FINING TECHNIQUES

10.8.1 Preliminary Trials

The specific behavior of various fining agents in different wines has been repeatedly emphasized (Sections 10.6.1 and 10.6.3). These variations affect clarification and colloidal stabilization (coloring matter and possible overfining), as well as tasting characteristics.

In view of these remarks, it is advisable to carry out a laboratory test before fining a particular wine to assess the behavior of various products, possibly at several different doses. It should, however, be taken into account that fining in small volumes does not always reproduce the conditions found in the full-scale process. The complexity of this operation explains why preliminary fining trials are relatively little used in normal cellar practice.

Peynaud (1975) advised using 750 ml clear glass bottles or, preferably, test tubes 80 cm long and 4 cm in diameter. The following parameters are observed:

(a) the time flocculation occurred,

(b) settling speed,

(c) clarity obtained after resting,

(d) the thickness of the layer of lees.

If there is a risk of overfining, it is advisable to test for excess, non-flocculated proteins (Section 5.5.4). Heating to 80°C is not always sufficient to show up the excess proteins corresponding to overfining, so it is advisable to add tannins as well. Red wines may also be chilled to assess the stabilizing effect of fining on the precipitation of colloidal coloring matter.

Finally, this preliminary testing is recommended in order to assess the organoleptic consequences of different fining procedures. It is advisable to find the minimum amount of fining that clarifies and stabilizes the wine, while softening its tannic structure, but without making it taste thin. Good clarification prior to fining makes it possible to reduce the dose of fining agent and may have a beneficial effect on the results.

A large number of clarifying agents are available, and each one reacts differently (Table 10.6), according to the type of protein in the fining agent, the wine's phenol composition, its colloidal structure and the type of particles in suspension. Fining agents derived from plant proteins will be added to this table when more detailed information becomes available on the effectiveness of the various products in red and white wines, as well as on the doses required. A single fining agent is not always sufficient to obtain good results. A combination of several protein and inorganic fining agents may be more effective (e.g. gelatin or egg albumin with bentonite). Furthermore, certain wines with a high concentration of protective colloids 'do not react well to the fining agent', i.e. added proteins flocculate poorly and leave the wines turbid. In this case, prior clarification of the wine is recommended to facilitate fining. The addition of pectolytic enzymes, filtration, or possibly a combination of both, promote the effectiveness of fining agents.

10.8.2 Fining Procedures

Successful fining depends on the rapid mixing of the fining agent with the wine. The difficulty of this operation varies according to the volume to be treated. The fining agent must be dispersed throughout the entire mass of wine immediately, otherwise it is likely to finish coagulating before it is completely mixed with the wine, thus reducing its effectiveness. It is recommended to use fining agents diluted in water (0.25 l/hl). Of course, fining agents must not be diluted in wine, as they would coagulate and lose their clarifying effect.

When small volumes (225 l barrels) are to be fined, the wine is first agitated with a whisk.

Table 10.6. Summary of products used to clarify wine

Type of product	Doses used	Characteristics
White wine		
Isinglass	1–2.5 g/hl	Good clarity. Intensifies yellow color. Light flakes, bulky, settles slowly
Bloodmeal	5–10 g/hl	Good clarity. Attenuates herbaceous character. Compact flakes, settles quickly
Casein	10–50 g/hl	Good clarification. Treats and prevents yellowing (maderization). No overfining
Bentonite	20–100 g/hl	Average clarification. Treats and prevents protein and copper casse. Facilitates racking with proteins. Avoids overfining
Siliceous earths	20–50 ml/hl 50–100 ml/hl	Act on protective colloids in wines that are difficult to clarify. Used with protein fining agents, prevents overfining and facilitates settling of the lees
Tannins	3–10 g/hl	Prevents and treats overfining
Red wine		
Gelatins	3–10 g/hl	Very good fining agent for tannic wines. Affects only the most aggressive tannins. May make wine softer or thinner
Bloodmeal	10–20 g/hl	Good results on bitter, young wines with a high phenol content. Bloodmeal (Not authorized under EU legislation)
Egg white	5–15 g/hl (powder) 3–8 fresh egg whites per barrel (225 l)	Very good fining agent for tannic wines with some age. Sensitive to protective colloids
Bentonite	20–50 g/hl	Clarification of young wines. Eliminates colloidal coloring matter. Facilitates sedimentation of protein fining agents

The fining agent is then injected into the mass with a syringe, and mixed thoroughly by energetic stirring.

It is even more difficult to mix the fining agent properly when large volumes of wine are to be treated (vats containing several hundred hectoliters). There are various ways of achieving a homogeneous mix (Figure 10.12) that are suitable for all types of products. Only metering pumps provide good distribution throughout the mass of wine, by pumping the fining agent into a hose as the wine flows through it. It is, of course, necessary to ensure that the metering pump is synchronized with the pump circulating the wine.

Racking is carried out after fining to separate the clear wine from the lees. If the wine is fined in a barrel during aging, the fining agent may be removed by as little as 1–5 weeks after it is added, depending on the type of product used. During this first racking, deposit from the sides of the barrels may drop into the wine, so a second racking may be required one month later to clarify the wine completely (turbidity <5 NTU).

When fining is carried out in large-capacity vats, the deposit does not settle as well. Not only must the particles settle from a greater height, but also small differences in temperature from one point to another create convection currents

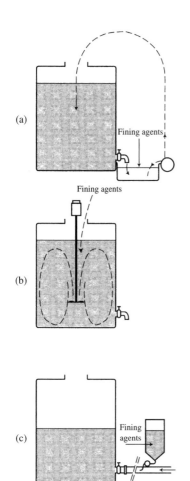

Fig. 10.12. Various methods for mixing fining agents into wine: (a) in a vat during pumping-over, (b) in a vat by stirring with a mobile propeller, (c) in a vat using a metering pump (Ribéreau-Gayon *et al.*, 1977)

that maintain the particles in suspension. However, commercial considerations often dictate that wines must be prepared rapidly for bottling. Siliceous earths (Section 10.10.2) (Siligel and Klebosol) speed up the clarification process and settling of the lees produced by the fining agent, so that the wine is ready for racking after 24–48 hours. These products facilitate fining, but must be added to the wine before the protein fining agents. Bentonite also promotes sedimentation. In all cases where rapid clarification is required, fining must be followed by filtration to achieve perfect clarity. In view of the number of treatments, there is cause for concern that the organoleptic effects may not always be positive, at least temporarily.

10.9 BENTONITE TREATMENT

10.9.1 Structure of Bentonite

Bentonites (Ribéreau-Gayon *et al.*, 1977) are hydrated aluminum silicates, mainly consisting of montmorillonites with simplified formulae, e.g. Al_2O_3, $4SiO_2$, nH_2O. Furthermore, bentonites contain exchangeable cations (Mg^{2+}, Ca^{2+}, Na^+) that play a major role in their physicochemical properties. These vary according to geographical origin. Bentonites from Germany or North Africa contain mainly calcium, while those from the United States (Wyoming) contain sodium. The latter are the most widely used as they are considered to be the most effective in treating wines.

Montmorillonite is structured in separate flakes (Maujean, 1993), thus distinguishing it from the more compact kaolinite, and also giving it remarkable colloidal properties. Montmorillonite, which swells considerably in an aqueous medium, has a large adsorption surface and a strong negative charge.

The flakes are organized in a fairly regular pattern. Each flake consists (Figure 10.13) of two rows of tetrahedra chained together. They have oxygen atoms at the nodes and a silicon atom in the center. Between these two rows, there is a series of octahedral structures linked together by oxygen atoms or hydroxyl radicals. In the center, three octahedra out of four contain Al^{3+} or Mg^{2+}. The difference in charges with the rows of tetrahedra containing Si^{4+} creates a negative charge on the surfaces between the flakes (Figure 10.13). This keeps the flakes apart and creates a gap that varies according to the origin of the bentonite and can be measured by X-ray diffraction. The exchangeable cations are taken up into this space by adsorption, as are the water molecules responsible for the swelling, jellifying and flocculation properties of bentonites. They form gelatinous pastes with water and, at high dilutions, stable colloidal

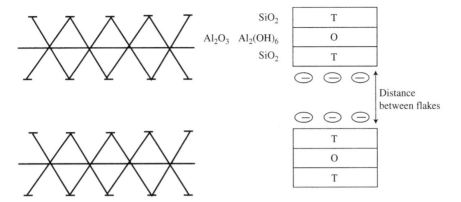

Fig. 10.13. Flake structure of montmorillonite (bentonite). (T, tetrahedron; O, octahedron)

suspensions. Furthermore, thanks to the large surface area of their flake configuration, bentonites have significant adsorption properties.

Not all bentonites are suitable for treating wines. Some have a coarse structure and are likely to give wines off-flavors. Others do not have sufficient adsorbent and clarifying capacity.

Bentonites recommended for treating wine have variable chemical compositions, unrelated to their enological properties. Rarely used in their natural state, they are usually activated with sulfuric acid or alkaline salts. In view of their strong ion exchange capacity, it is possible to load bentonites with H^+, Na^+ or Ca^{2+} ions, to form acid, calcium or sodium bentonites.

Acid and calcium bentonites are easy to disperse without forming lumps. Suspensions settle rapidly, leaving the liquid turbid but with a relatively light deposit. Protein adsorption is limited.

Sodium bentonite is most frequently used to treat wines. The flakes are more widely spaced (100 Å) than those of calcium bentonite (10 Å), so they swell more in wine and have a higher protein adsorption capacity. Sodium bentonite flakes are relatively difficult to mix into suspension in water, but the suspensions have a very stable colloidal character. When added to wine, they produce flocculation and settle out as a flaky deposit, leaving a clear liquid. The natural proteins are completely eliminated and the wine is protected from protein (Section 5.6.2) and copper casse (Section 4.7.3).

10.9.2 Physicochemical Characteristics of Bentonites

1. *The swelling number* represents the ratio between the volume of 5 g of bentonite powder and the volume occupied by 5 g of bentonite left in 100 ml water for 24 hours. Various forms of bentonite behave differently (Table 10.7) and swelling is also affected by the type of water, especially in the case of sodium bentonite (Maujean, 1993). The Ca^{2+} and Mg^{2+} cations in tap water take the place of the smaller Na^+ and K^+ cations in the bentonites, thus increasing the swelling number. This phenomenon is not observed in calcium bentonite, where, on the contrary, the swelling volume decreases.

2. *The specific adsorption surface* depends on the way the flakes are arranged. It may be measured using methylene blue, according to the formula:

$$S(m^2/g) = \frac{20.93 V(\text{ml})}{P(g)}$$

where V is the volume of methylene blue solution required to obtain a light blue ring centered around the deposit from a drop of suspension (bentonite-methylene blue) on an ashless filter paper containing a dry weight P of the bentonite to be tested. The results are highly variable, with surface areas ranging from 20 to 650 m^2/g (Maujean, 1993).

Table 10.7. Influence of the type of fixed ions on the specific swelling volume of various bentonites (Maujean, 1993)

Various bentonites	Na^+/Ca^{2+} ratio	Specific swelling volume (mg/g)		Proteins adsorbed by the bentonite (mg BSA/g)
		Distilled water	Tap water	
Sodium				
B1	1.8	13.8	87.5	571
B6	1.5	13.0	92.0	571
B2	1.4	13.0	45.5	408
B3	1.4	12.0	26.5	417
B7	1.4	4.5	9.0	379
Calcium				
B8	0.5	10.0	7.0	271
B5	0.3	4.0	4.0	300
B10	0.3	5.0	5.0	242
B11	0.2	5.0	4.0	258
B4	0.1	3.6	2.5	129
B12	0.04	2.5	3.5	333

The quantity of proteins adsorbed (egg or BSA blood albumin) may also be measured. Sodium bentonite has a higher adsorption capacity (Table 10.7) than calcium bentonite, which explains why it is preferred for treating wine.

10.9.3 Using Bentonite to Treat Wine

Clay has been used to clarify wine for many years. In the 1930s, kaolin was used in the first rational treatments for protein casse in white wines (Ribéreau-Gayon *et al.*, 1977). It later became apparent that bentonite was more effective, and it came into widespread use as early as the 1950s. This treatment made it possible to bottle white wines early, preserving their fruitiness while protecting them from protein turbidity in bottle. The effect of bentonite is negligible up to doses of 40 g/hl. The risk of losing aromatic character cannot be ruled out at higher doses, especially if the treatment is repeated.

Sodium bentonite has a high protein adsorption capacity and is also relatively chemically inert. It may release a few tens of mg/l of Na^+, but this does not have any organoleptic impact, at least in moderate doses.

All white wines naturally contain grape proteins, which may cause turbidity and deposit if they flocculate (Section 5.5.2). There are tests for forecasting instability (Section 5.5.4) and defining whether treatment is needed. Bentonite is currently the most widely used treatment for eliminating excess proteins (Section 5.6.2).

In view of the involvement of a protein support in the colloid flocculation occurring in copper casse in white wine, bentonite may be used to treat this problem (Section 4.7.3), provided that the copper concentration does not exceed 1 mg/l. The same is not true, however, of ferric casse (Section 4.6.2) as proteins are not involved, so bentonite is ineffective.

Bentonite was later found to be capable of fixing the coloring matter in red wines and *vins de liqueur*. It is as effective as standard fining techniques.

Bentonite is particularly useful in treating red wines with a high concentration of colloidal coloring matter due to heating or rough mechanical treatment of the grapes. These practices may be responsible for precipitation during aging. The colloids involved consist of anthocyanins in the form of flavylium with a positive (+) charge, as well as tannins, polysaccharides and possibly proteins (also positively charged). The addition of negatively charged (−) bentonite (20–50 g/hl) eliminates these unstable complexes and stabilizes the wine. This is nevertheless accompanied by a non-negligible loss of color.

In general, the effect of bentonite on red wines is comparable to that of protein fining. After flocculation and sedimentation, wines treated with doses of 25–40 g/hl of bentonite remain stable at low temperatures, staying brilliant and free of turbidity, even after several months. This treatment may be even more effective than gelatin (12 g/hl) or blood albumin (18 g/hl). It should, however, be taken into account that bentonite fixes anthocyanins, not only in discolored white wines but also in rosés and young red wines.

As bentonite flocculates in wine, with a behavior analogous to that of protein fining agents, experiments have been made in using it to clarify wine. In fact, its effectiveness in clarification depends on the type of bentonite and the composition of the wine. The only wines that may be properly clarified are reds or whites with low concentrations of polysaccharides and other protective colloids (Section 9.4.1), as these inhibit the flocculation and settling of the bentonite particles. In some wine-growing areas, bentonite is well suited to clarifying dry white wines, but elsewhere white wines may be more turbid after bentonite treatment than before.

In general, bentonite treatment is recommended for stabilizing red and white wines. Clarification is obtained by fining or filtration at a later stage. Bentonite treatment makes filtration difficult, while protein fining doubles the filtration yield of wines previously treated with bentonite. Fining red wines with a combination of gelatin and bentonite often produces good results.

When bentonite treatment is exclusively intended for stabilization, the bentonite suspension may be prepared directly in the wine. As bentonite flocculates as soon as it is dispersed in the wine, its clarifying capacity is diminished. In order to take advantage of its clarifying potential, a dilute suspension must be prepared in water and incorporated rapidly into the wine.

10.9.4 Bentonite Treatment Techniques

It may occasionally be advisable to use bentonite on white grape must before fermentation (Section 5.6.2). This process reduces the number of treatments required later in the process, and is justified if permanent stability is obtained. It is not compatible with barrel-aging white wines on yeast lees.

White wines are generally treated before bottling, following evaluation of their protein instability (Section 5.5.4). In red wines, treatment generally takes place at the same time as fining.

Commercial bentonites are available in either powder or granule form. The former may be used directly by sprinkling over the wine in a vat equipped with a thorough agitation system. However, preliminary swelling, indispensable for bentonite granules, is advisable in all cases.

A suspension is prepared in water (5–15%) and left to swell. The bentonite must be poured on the surface of the liquid as it is agitated to avoid lumps. Swelling is faster in hot water (50–60°C). Of course, if the bentonite suspension is prepared in wine, it coagulates immediately and loses some of its clarifying properties.

The bentonite suspension is put into the wine during racking. Pumping breaks up any lumps and homogenizes the mixture. Care must be taken to minimize the amount of oxygen dissolved. Unlike fining, clarification is more effective at 20°C than 10°C. It is also better at low pH.

Treatment in short (as opposed to tall) vats is recommended to facilitate sedimentation. It is advisable to use a protein fining agent, such as casein (5–10 g/hl), a few days after treatment to improve clarity and take full advantage of bentonite's clarifying capacity. The flocculation of the casein settles the fine bentonite particles and also improves the color of white wine. The wine is clarified by racking two or three weeks later. It may then be cold-stabilized, as the effectiveness of this process is improved by bentonite treatment. Filtration through coarse plates or diatomaceous earth also gives good results.

Another way of improving the clarification of wines treated with bentonite consists of fining after a few days with a combination of siliceous earth (30 ml/hl) and gelatin (5 g/hl) (Section 10.10.1). The mutual flocculation of the negative (siliceous earth) and positive (gelatin) particles eliminates the finest bentonite particles from the wine. Siliceous earth does not affect the aromatic qualities of the wine. Sometimes, this fining even attenuates the dampening effect of high doses of bentonite on Sauvignon Blanc aroma.

10.10 MISCELLANEOUS CLARIFICATION TREATMENTS

10.10.1 Properties of Siliceous Earths (Siligel, Klebosol)

A 'sol' (Section 9.2.1) is a fluid colloidal dispersion with free particles, subject to Brownian motion. Siliceous earths are stable, concentrated aqueous suspensions of non-aggregated particles of silica produced by the *in situ* growth of silica microcrystals. Particles of different sizes are obtained by controlling their growth. They are by-products of the glass industry.

Siliceous earths are stabilized by small quantities of bases that hydroxylate the surface of the particles and create a negative surface charge. This is necessary to balance the positive charge of the stabilizer. The fact that the particles repel each other enhances the stability of the sol. However, as the strongly alkaline pH of Klebosol, stabilized by Na^+, makes them excellent culture media for bacteria, they must always be kept in the dark, in airtight containers.

The three criteria for selecting a Klebosol are:

(a) specific surface in m^2/g or particle size (7–50 nm),

(b) SiO_2 content (15–50%) and

(c) the type of stabilizer (Na^+, NH_4^+).

As the pH decreases, part of the stabilizer is neutralized. The repulsion between the particles is then weaker and they may react to form a gel (pH 4–7). The increase in the silica concentration reduces the average distance between particles and decreases stability. To a certain extent, dilution with water produces the opposite effect. At temperatures $<0°C$, the water in siliceous earths crystallizes and the medium gels.

10.10.2 Use of Siliceous Earths in Winemaking

Silica particles are negatively charged. When they are neutralized by the positively charged proteins used in fining, they flocculate and settle, pulling down the particles in suspension. This facilitates the clarification, by fining, of grape must, fruit juice and wines with low tannin contents (white and rosé wines). The effect that siliceous earths have on the flocculation of protein fining agents is similar to that of tannin.

Siliceous earths associated with any protein fining agent produce the following effects:

(a) acceleration of the clarification process,

(b) optimum clarification and compacting of the fining agent lees, which minimizes wine loss and facilitates racking,

(c) removal of all the fining agent, therefore eliminating the risk of overfining,

(d) rapid elimination of ferric ferrocyanide after the treatment of white wines,

(e) improved filterability of fined wine which facilitates preparation for bottling.

When samples are tasted, it is apparent that, unlike tannins, siliceous earths do not harden the wine and its organoleptic qualities remain intact.

Siliceous earths may be used with all types of fining agents. The best results, however, are obtained in combination with gelatins (liquid and cold soluble) and isinglass. These products are particularly suitable for clarifying wines treated with bentonite. Siliceous earths should always be used before protein fining agents but after other treatments (ferrocyanide, bentonite).

Suggested doses are listed in Table 10.8, but it is always advisable to run preliminary tests to determine the appropriate quantities for a particular wine.

Clarification and Stabilization Treatments: Fining Wine 329

Table 10.8. Quantities of siliceous earth, gelatin and isinglass appropriate for different types of wine

Characteristics of the wine	Siliceous earth (ml/hl)[a]	Gelatin (g/hl)	Isinglass (g/hl)
Normal wine	20–30	2.5	1.5
Young wine, difficult to clarify	30–50	5	2
Wine that clogs the system, very difficult to clarify	50–100	10	3

[a]Siligel or Klebosol 30 V

10.10.3 Polyvinylpolypyrrolidone (PVPP) Treatment

The polymerization of vinylpyrrolidone produces water-soluble polyvinylpyrrolidone (PVP). However, if polymerization occurs in the presence of an alkali, the pyrrolidone cycle is broken, producing insoluble polyvinylpolypyrrolidone (PVPP) (Figure 10.14).

These products have a strong affinity for polyphenols. Like gelatin, PVP precipitates and flocculates when it comes into contact with tannins. Depending on the degree of polymerization, flocculation may be incomplete and cause overfining. PVP is not very useful for treating wine—gelatin is undoubtedly better. However, the insolubility of PVPP in dilute alcohol solutions makes this high polymer particularly adapted to eliminating phenols.

PVPP, sold as 'Polyclar AT', has been used since 1961 to stabilize beer and reduce concentrations of tannoid substances. Other applications include treating wine to reduce the phenol content, and fractionating and analyzing the phenols in red wine (Hrazdina, 1970; Glories, 1976). In France, it is approved for use in wine, at a maximum dose of 80 g/hl.

PVPP is useful for minimizing a tendency to browning in white wines. Alone, or combined with casein, it inhibits maderization by eliminating tannins, oxidizable cinnamic acids and the quinones formed when they oxidize. It acts differently from casein, as it eliminates the oxidizable phenols, whereas casein inhibits the oxidative phenomena (Table 10.9). The doses of 20–30 g/hl required to prevent browning do not produce any negative organoleptic changes. On the contrary, PVPP attenuates the bitterness of certain wines.

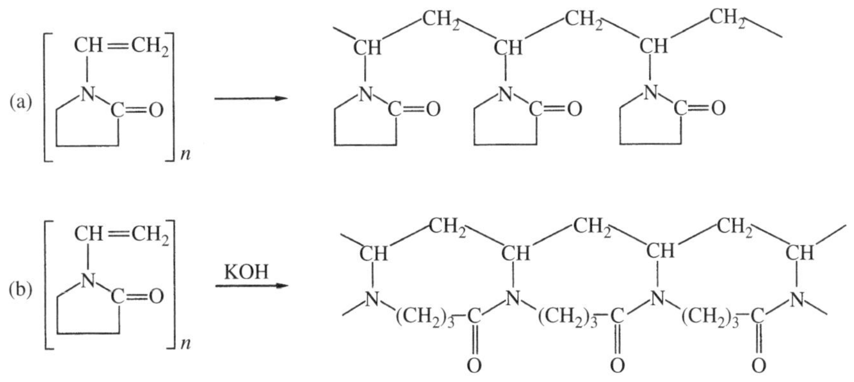

Fig. 10.14. Polymerization of vinylpyrrolidone into: (a) polyvinylpyrrolidone (PVP) and (b) polyvinylpolypyrrolidone (PVPP)

Table 10.9. Effect of various treatments on the phenol content of a white wine (Sapis and Ribéreau-Gayon, 1969)

	Total phenols[a]	Tannins[a]	OD after browning
Control	0.325	0.27	0.35
Casein treatment 10 g/hl	0.310	0.22	0.28
PVPP treatment 10 g/hl	0.255	0.16	0.31

[a] The values give the optical density (OD) obtained in the presence of an appropriate reagent.

PVPP may also be used to correct discoloration in white wines, either those made from red grape varieties or those made with white grape varieties that have been stained with red wine. It is used to eliminate unwanted pigments either on its own or, more effectively, in combination with decolorizing vegetable carbon.

Finally, PVPP reduces astringency and softens excessively tannic red wines. It fixes the most reactive tannins (200–300 mg/l of tannin for 250 mg/l of PVPP), although it has less effect on anthocyanins.

REFERENCES

Amati A., Boschelle O., Manzano M., Pitotti A. and Zironi R. (1989) *Actualités Œnologiques*, Compte-Rendu du 4th Symposium International d'OEnologie de Bordeaux. Dunod, Paris.

Asano K., Ohtsu K., Shinagawa K. and Hashimoto N. (1984) *Agric. Biol. Chem.*, 48 (5), 1139.

Asquith T.N., Uhlig J., Mehansho H., Putman L., Carlson D.M. and Butler L. (1987) *J. Agric. Food Chem.*, 35, 331.

Austin P.J., Suchar L.A., Robbins C.T. and Hagerman A.E. (1989) *J. Chem. Ecol.*, 15 (4), 1335.

Bayonove C., Cabaroglu T., Dufour C., Razungles A., Sapis J.-C., Baumes R. and Günata Z. (1994) Compte-Rendu 21st Congrès de *la Vigne et du Vin*. OIV, Paris.

Beart J.E., Lilley T.H. and Haslam E. (1985) *J. Chem. Soc. Perkin, Trans II*, 1439.

Butler L.G. (1989) Effects of condensed tannin on animal nutrition. In *Chemistry and Significance of Condensed Tannins*. (eds. R.W. Hemingway and J.J. Karchesy). Plenum Publishing Corporation, New York.

Butler L.G. and Mole S. (1988) *Bull. du Groupe Polyphénols*, 111.

Calderon P., Van Buren J. and Robinson W.B. (1968) *J. Agric. Food Chem.*, 16 (3), 479.

Cerf P. (1973) Contribution à l'étude de l'action des produits oenologiques à base de gélatine et de tanins. Thèse de Docteur Ingénieur, Université de Dijon.

Charlton A.J., Baxter N.J., Lilley T.H., Haslam E., McDonald C.J. and Williamson M.P. (1996) *FEBS Lett.*, 382, 289.

Cheynier V., Rigaud J. and Ricardo da Silva J. (1992) In *Plant Polyphenols* (eds. R.W. Hemingway and P.E. Laks). Plenum Press, New York.

Ferrarini R., Celotti E. and Zironi R. (1995) *Actualités Œnologiques*, Compte-Rendu du 5th Symposium International d'OEnologie de Bordeaux, Lavoisier, Paris, p. 433.

Gal J.Y. and Carbonell F. (1992) *Bull. du Groupe Polyphénols*, 358.

Glories Y. (1976) *Conn. Vigne Vin*, 10, 51.

Glories Y. and Augustin M. (1992) *INRA Viti*, 37.

Hagerman A.E. and Butler L.G. (1980a) *J. Agric. Food Chem.*, 28, 944.

Hagerman A.E. and Butler L.G. (1980b) *J. Agric. Food Chem.*, 28, 947.

Hagerman A.E. and Butler L.G. (1981) *J. Biol. Chem.*, 256, 4494.

Haslam E. (1993) In *Polyphenolic Phenomena* (ed. A. Scalbert). INRA Editions, Paris, p. 25.

Haslam E. (1995) *Actualités Œnologiques*, Compte-Rendu du 5th Symposium International d'Œnologie de Bordeaux, Lavoisier, Paris.

Haslam E. (1996) *J. Nat. Prod.*, 59, 205.

Haslam E. and Lilley T.H. (1985) *Ann. Proc. Phytochemical Soc. Europ.*, (eds. C.F. Van Sumere and P.J. Lea) Vol. 25, p. 237.

Haslam E. and Lilley T.H. (1988) *Critical Reviews in Food Science and Nutrition*, 27 (1), 1.

Hrazdina G. (1970) *J. Agric. Food Chem.*, 18, 243.

Hrazdina G., Van Buren J.P. and Robenson W.B. (1969) *Am. J. Enol. and Viticult.*, 20, 66.

Hunter R.J. (1981) *Zeta Potential in Colloid Science, Principles and Applications*. Academic Press, New York.

Ikonomou-Potiri M. (1985) DEA OEnologie-Ampélogie, Université de Bordeaux II.

Lagune L. (1994) Etude des gélatines oenologiques et des mécanismes du collage dans les vins rouges. Thèse de Doctorat, Université de Bordeaux II.

Lagune-Ammirati L. and Glories Y. (1996a) *Rev. Fr. Œnol.*, 161, 17.

Lagune-Ammirati L. and Glories Y. (1996b) *Rev. Œnologues*, 157, 158.

Lagune-Amirati L. and Glories Y. (2001) *Rev. Fr. Oenol.*, 191, 25.

Lagune-Amirati L. and Glories Y. (2002) *Rev. Fr. Oenol.*, 194, 18.

Lefebvre S., Gerland C., Maury C. and Gazzola M. (2000) *Rev. Fr. Oenol.*, 184, 28.

Lefebvre S., Restani P. and Scotto B. (2003) *Rev. Fr. Œnol.*, 202, 28.

Lubbers S., Voilley A., Charpentier C. and Feuillat M. (1993) *Rev. Fr. Œnol.*, 144, 12.

Martin R., Cai Y., Spencer C.M., Lilley T.H. and Haslam E. (1990) *Bull. du Groupe Polyphenols.*

Maujean A. (1993) *Rev. Fr. Œnol.*, 143, 43.

Maury C., Sarni-Manchado P., Lefebvre S., Cheynier V. and Moutonnet H. (2003) *Am. J. Enol. Vitic.*, 542, 205.

McManus J.P., Davis K.G., Beart J.E., Gaffney S.H., Lilley T.H. and Haslam E. (1985) *J. Chem. Soc. Perkin*, 1429.

Mehansho H., Butler L.G. and Carlson D.M. (1987) *Ann. Rev. Nutr.*, 7, 423.

Metche M. (1993) In *Polyphenolic Phenomena* (ed. A. Scalbert). INRA Editions, Paris.

Oh H.I., Hoff J.E., Amstrong G.S. and Haff L.A. (1980) *J. Agric. Food Chem.*, 28, 394.

Ozawa T., Lilley T.H. and Haslam E. (1987) *Phytochemistry*, 26 (11), 2937.

Peynaud E. (1975) *Connaissance et Travail du Vin*. Bordas, Paris.

Ribéreau-Gayon J. (1934) *Bull. Soc. Chim.*, 1, 483.

Ribéreau-Gayon J., Peynaud E., Ribéreau-Gayon P. and Sudraud P. (1977) *Traité d'Œnologie, Sciences et Techniques du Vin*, Vol IV. Dunod, Paris.

Ricardo da Silva J., Cheynier V., Souquet J.M. and Moutounet M. (1991) *J. Sci. Food. Agric.*, 57, 111.

Rüdiger M. and Mayr E. (1928a) *Zeits. ange. Chemie*, 29, 809.

Rüdiger M. and Mayr E. (1928b) *Kolloid-Zeitschrift*, 46, 81.

Rüdiger M. and Mayr E. (1929) *Kolloid-Zeitschrift*, 47, 141.

Salgues M. and Razungles A. (1983) *Vititechnique*, 64, 32.

Sapis J.-C. and Ribéreau-Gayon P. (1969) *Conn. Vigne Vin*, 3, 215.

Siebert K.J., Troukhanova N.V. and Lynn P.Y. (1996) *J. agric. Food Chem.*, 44, 80.

Siegrist J. (1989) *Viti*, 137, 126.

Tobiason F.L. (1992) In *Plant Polyphenols* (eds. R.W. Hemingway and P.E. Laks). Plenum Press, New York.

Tobiason F.L. and Hoff L.A. (1989) In *Chemistry and Significance of Condensed Tannins* (eds. R.W. Hemingway and J.J. Karchesy). Plenum Publishing Corporation, New York.

Vernhet A., Pellerin P. and Moutounet M. (1995) *Actualités Œnologiques*, Compte-Rendu du 5th Symposium International d'OEnologie de Bordeaux, Lavoisier, Paris.

Vivas N. and Glories Y. (1993) *Rev. Fr. Œnol.*, 142, 33.

Wucherpfennig K. and Passmann P. (1972) *Flussigas Obst.*, 39.

Wucherpfennig K., Passmann P. and Bassa K. (1973) *Flussigas Obst.*, 40, 488.

Yokotsuka K. and Singleton V.L. (1995) *Am. J. Enol. Viticult.*, 46 (3), 329.

11

Clarifying Wine by Filtration and Centrifugation

11.1 Principles of filtration	333
11.2 Laws of filtration	334
11.3 Methods for assessing clarification quality	336
11.4 Filtration equipment and adjuvants	338
11.5 How filter layers function	342
11.6 Filtration through diatomaceous earth (or kieselguhr) precoats	346
11.7 Filtration through cellulose-based filter sheets	351
11.8 Membrane filtration	356
11.9 Tangential filtration	358
11.10 Effect of filtration on the composition and organoleptic character of wine	361
11.11 Centrifugation	364

11.1 PRINCIPLES OF FILTRATION

Filtration is a separation technique used to eliminate a solid in suspension from a liquid by passing it through a filter medium consisting of a porous layer that traps the solid particles. 'Filtering' generally refers to the clarification of a liquid, while 'filtration' is more often used to describe the technical process. However, both words are often used to mean the same thing.

The first problem in filtering wine is that of ensuring clarification quality (Ribéreau-Gayon et al., 1977; Gautier, 1984; Molina, 1992; Guimberteau, 1993). All the particles must be retained, without causing any modifications in chemical structure likely to affect flavor. Other major issues are filtration throughput and the clogging of filter surfaces. These criteria control the efficiency of the operation, its cost and, consequently, its practicability.

There are several types of filtration, using different filter media mounted on appropriate equipment. The following are used in winemaking:

1. Filtration through a diatomaceous earth precoat (kieselguhr) formed by continuous accretion.

2. Filtration through cellulose sheets or lenticular modules. These are permeable boards consisting of cellulose fibers with incorporated granular components (diatomaceous earth, perlite, cation resins, polyethylene fibers, etc.)

3. Filtration through synthetic polymer membranes, with calibrated pores.

4. Tangential filtration through inorganic or organic membranes. Unlike the standard clarification technique with frontal flow, the liquid flows parallel to the filter surface in tangential filtration, thus minimizing clogging.

An untreated wine is not usually perfectly clarified in a single operation—only tangential filtration (Section 11.9) is capable of achieving this result. Filtration through fine filter media leads to rapid clogging, whereas, if the medium is too coarse, all the particles are not removed. Each filtering operation fits into an overall clarification strategy, including the other techniques that contribute towards ensuring total clarity (spontaneous sedimentation, fining, centrifugation, etc.).

Wines that are barrel-aged for several months, or even years, have fairly low turbidity by the time they are bottled, but are still often capable of causing significant clogging. A single sheet filtration is generally sufficient. In the case of great red wines, some winemakers take the risk of not filtering at all. Their reservations about this technique, alleged to make wine taste thinner, are probably excessive.

Wines that are bottled relatively young are subjected to a greater number of clarification operations. Wine may be filtered through a diatomaceous earth precoat one or more times to prepare it for bottling. Sheet, lenticular module or possibly membrane filtration are used, resulting in low microbe levels, or even totally sterile wines. All these operations are not always necessary. Clarification techniques should be adapted to each wine and kept to a minimum.

11.2 LAWS OF FILTRATION

11.2.1 Introduction

The flow rate of a non-clogging liquid circulating through the pores of a filter medium is governed by Poiseuille's law:

$$q = \frac{dV}{dt} = K\frac{SP}{E}$$

K is a constant, proportional to the pore diameter multiplied by a power of 4 and the number of pores per unit area, but inversely proportional to the viscosity of the liquid, S is the surface of the filter layer, E is the thickness of the filter layer and P is the filtration pressure. This law simply expresses the proportionality between the flow rate and surface area, on the one hand, and pressure, on the other hand. It also shows that the flow rate is inversely proportional to the thickness of the filter layer.

It has been observed experimentally, and explained theoretically (Serrano, 1981), that the filtration behavior of a not very concentrated suspension, such as wine (particle content less than 1%), obeys different physical laws according to the type of porous material used to remove the solids. A mathematical model expresses the variations in volume filtered over time, at constant pressure, for each of these laws. The behavior of a given product in industrial filtration may be predicted on the basis of laboratory tests, by applying the corresponding equation.

These laws of filtration take the following parameters into account:

V = instantaneous volume filtered at time t

V_{max} = maximum volume that can be filtered before total clogging

t = filtration time

$q = \dfrac{dV}{dt}$ = instantaneous flow rate at time t

$q_0 = \left(\dfrac{dV}{dt}\right)_0$ = initial flow rate at time t_0

The flow rate remains constant provided that none of the filter pores are blocked.

11.2.2 Filtration with Sudden Clogging of the Pores

This is the simplest case. The filter behaves like a series of capillary tubes that are gradually blocked by individual particles. Filtration under these conditions is governed by the equation:

$$q = -K_1 V + q_0 \quad (11.1)$$

The results may be plotted in a straight line, indicating the variation in flow rate according to the volume filtered.

In the case of filtration at constant pressure, flow volume recordings over time are used to calculate the flow rates. The origin of the vertical axis represents q_0 and, as V_{max} corresponds to a zero flow rate:

$$V_{max} = \dfrac{q_0}{K_1}$$

This law of filtration does not apply to filtering wine.

11.2.3 Filtration with Gradual Clogging of the Pores

Particles deposited inside the pores during filtration cause a gradual decrease in their diameter. A filtration process governed by the law of gradual clogging of the pores is governed by the equation (Figure 11.1):

$$\dfrac{t}{V} = K_2 t + \dfrac{1}{q_0} \quad (11.2)$$

Filtering wine with sheet or lenticular module filters, as well as standard membrane systems, is governed by this law under clearly defined operating conditions. The volume that can be

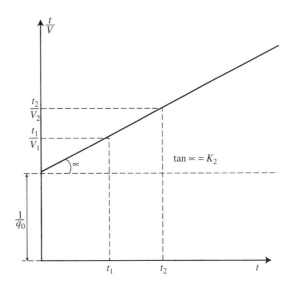

Fig. 11.1. Graph $t/V = k_2 t + 1/q_0$

filtered in a given length of time through a given filter medium may be predicted from the results of a laboratory test carried out at constant pressure.

The maximum volume that can be filtered before clogging is calculated using Eqn (11.2):

$$\dfrac{1}{V} = K_2 + \dfrac{1}{q_0 t}$$

When time t tends towards infinity, $1/q_0 t$ tends towards zero, and V tends towards $1/K_2$, which is the cotangent of angle α:

$$V_{max} = \dfrac{t_2 - t_1}{t_2/V_2 - t_1/V_1}$$

The V_{max} value is used, together with the fouling index (Section 11.8.3), to predict the behavior of an industrial filtration process.

11.2.4 Deep-bed Filtration

In this type of filtration, the particles are trapped in the mass of a filter cake, which constantly increases in thickness due to the continuous addition of filter medium. It has been demonstrated (Serrano, 1981) and experimentally verified that the variation

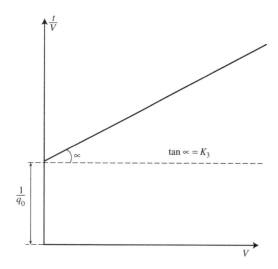

Fig. 11.2. Graph $t/V = k_3 V + 1/q_0$

t/V is proportional to V, at constant pressure (Figure 11.2):

$$\frac{t}{V} = K_3 V + \frac{1}{q_0} \quad (11.3)$$

Of course, filtering wine through a diatomaceous earth precoat follows this law. Although the principle is different, the same law also governs tangential filtration (Mietton-Peuchot, 1984), where the particles, concentrated along the membrane, behave like a cake.

In the special case of filtration through a precoat with continuous accretion, the metal frame of the filter layer has a negligible resistance to flow, so q_0 is very large and $1/q_0$ tends towards 0. Equation (3) becomes

$$\frac{t}{V} = K_3 V$$

or

$$\log V = \tfrac{1}{2} \log t + \text{Cste}$$

Volume throughput over time may be recorded in a laboratory test carried out at constant pressure. Plotting the straight line ($\log V = \tfrac{1}{2} \log t + \text{Cste}$) on logarithmic paper gives a direct readout of the volume filtered over a given time.

It is thus possible to optimize an industrial filtration process in the laboratory and compare the effectiveness of diatomaceous earths with different permeabilities used to filter the same wine.

11.2.5 Filtration with Intermediate Clogging of the Pores

It is generally accepted that there is an intermediate empirical law, between deep-bed filtration and filtration with gradual clogging of the pores. This situation is characterized by a linear graph, at constant pressure, of $1/q$ over time. The equation is as follows:

$$\frac{1}{q} = K_4 t + \frac{1}{q_0} \quad (11.4)$$

This law is apparently not applicable to the filtration of wine.

11.3 METHODS FOR ASSESSING CLARIFICATION QUALITY

11.3.1 Measuring Turbidity

The effectiveness of filtration processes may be assessed by measuring various parameters indicative of clarity.

Turbidity is measured by appreciating the disturbance in the diffusion of light caused by contact with particles in a liquid. A turbidimeter measures the intensity of the diffused light. A turbidimeter that makes measurements at a 90° angle is also known as a nephelometer. These apparatus are calibrated in NTU (nephelometric turbidity units).

These simple measurements, requiring only inexpensive equipment, are being increasingly used. Table 11.1 shows a scale of correspondences between turbidity measurements and visual inspections. Between the two extreme values, wines may be considered clear, cloudy or dull.

Table 11.1. Correspondence between turbidity measurements (NTU) and appearance

	Brilliant	Turbid
White wine	<1.1	>4.4
Rosé wine	<1.4	>5.8
Red wine	<2.0	>8.0

All the various clarification treatments (fining, centrifugation and filtration through precoats) leave wines brilliant, with a turbidity rating of 1 or lower. Sheet or membrane filtration just before bottling results in values between 0.10 and 0.65 NTU. Independently of the possible aim of totally eliminating microorganisms from the wine, clarification to these levels may be necessary in view of the regular increase in turbidity during bottle aging, even if there is no later contamination.

11.3.2 Determining the Solid Content

This defines the quantity of particles, by weight or volume, as a percentage of the total volume. The solids may be collected by: (a) centrifugation, (b) filtration through a glass fiber membrane, (c) filtration through a 0.45 μm membrane that retains colloids or (d) by evaporating dry in order to determine the total quantity of dry matter.

The process normally used in winemaking is centrifugation at 3000 rpm for 5 min, in a special test tube, directly graduated in percentage of total volume. This process is mainly used for very turbid liquids such as white must, deposits left when must has settled, yeast lees, fining lees (particle concentrations above 3% by volume) and new wines (particle concentrations between 0.5 and 2% by volume).

11.3.3 Particle Counts

This technique is used to assess the respective quantities of particles of various sizes (above 0.5 μm). Different types of special measuring apparatus are based on various principles: electrical conductivity, X-ray absorption, laser light diffusion and laser diffraction. Currently available systems are expensive and, consequently, restricted to research laboratories. Furthermore, most of them are incapable of counting particles smaller than 0.5 μm, although these also affect clarity.

11.3.4 Microbiological Analyses

These analyses are essential, not only as they provide a good assessment of the effectiveness of clarification but also due to the fact that residual yeast and bacteria are likely to affect biological stability.

The total number of microorganisms was formerly counted under a microscope using a blood-counting chamber (Malassez cell) either directly, if the population is sufficiently large or after concentration by centrifugation. If centrifugation is used, the technique is rather long and inaccurate.

A viable microorganism count used to be the most useful analysis.

Techniques for counting yeasts and bacteria (Lafon-Lafourcade and Joyeux, 1979) are based on the microorganisms' capacity to develop on specific agar nutrient media. Colonies that develop in Petri dishes are counted visually, while individual cells are counted using an epifluorescence microscope.

Agar gel is added to the nutrient medium to culture yeast prior to counting. It is supplemented with biphenyl (0.015% in ethanol) to prevent mold development and 0.01% chloramphenicol to inhibit bacterial growth. Cycloheximide (0.1%) may be added to select "non-*Saccharomyces*" (*Brettanomyces*, etc.) yeasts, as they alone survive this treatment. The incubation period is around 2–3 days for the *Saccharomyces* genus and 7–10 days for "non-*Saccharomyces*" yeasts, at a constant temperature of 25°C.

Dubois medium is used for bacteria counts. Pimaricine (0.01%) is added to eliminate any yeast cells.

For selective counts of acetic bacteria, the lactic bacteria are inhibited by adding 0.001% penicillin. Incubation time: 5–7 days.

For selective counts of lactic bacteria, the acetic bacteria are inhibited by incubating the sample in Petri dishes under anaerobic conditions (with CO_2 under pressure). Incubation time varies from 7 to 12 days, depending on the species under investigation. Incubation temperature: 25°C.

Each sample was diluted in decimal stages to ensure the reliability of the end result. The dilute solutions were seeded evenly on the surface of the Petri dishes with sterile beads. This technique is necessary if the cells are to be identified by DNA/DNA hybridization (Volume 1, Section 4.3.5) on colonies or PCR (Volume 1, Section 4.3.6). If the wine contains very few microorganisms, it is filtered onto a 0.45-μm membrane, which is then deposited on the specific culture medium. Samples with 30–300 colonies are ready for counting. Counts are expressed in Colony Forming Units (CFU)/ml, which does not correspond exactly to the number of living cells.

Cell counts by epifluorescence microscopy are carried out on a filter membrane, using markers. This protocol is based on the Chemeunex system. A 10/ml sample of the medium to be analyzed is filtered on a 0.4-μm membrane. The organisms collected are then incubated at 30°C for 15–30 minutes on a non-fluorescent substrate that penetrates the cells. The substrate is then split by an intracellular enzyme system, releasing a fluorochrome, which then accumulates. This molecule emits green fluorescence when it is excited by light of the appropriate wavelength. The intensity of the fluorescence depends on membrane integrity and metabolic activity in the cells. No fluorescence is emitted by dead cells. The marked cells are then analyzed using an epifluorescence microscope. The results are obtained by counting the microorganisms in 20 fields, each containing 30 to 100 cells. Counts are expressed in viable cells per ml (cells/ml). Epifluorescence is being increasingly widely used, as it provides a very rapid evaluation of cultivable and even non-cultivable viable populations (Millet *et al.*, 2000). This distinction is based on evidence (Millet *et al.*, 2000; Millet and Lonvaud-Funel, 2000) that some bacterial cells are viable but non-cultivable (VNC). They cannot be cultured in Petri dishes under experimental conditions, but they are capable of developing in wine and causing turbidity problems. This distinction is essential when microbiological analyses are carried out during the wine-aging process (Volume 1, Section 6.3.2).

Immediately after the end of fermentation, there is a viable population on the order of 10^6/ml of microorganisms. This value decreases rapidly to 10^3 or 10^4/ml due to the effects of settling and racking. After filtration through a diatomaceous earth precoat, wines generally have viable populations on the order of 10^2/ml. Populations may remain quite large in wines that receive no preliminary clarification treatment.

The final sheet, lenticular module or membrane filtration just before bottling is conventionally described as 'low microbe' if the residual population is below 1 cell per 100 ml and 'sterile' if there are fewer than 1 viable microorganisms per bottle. Of course, filtration must be carried out under extremely clean, germ-free conditions to achieve this level of purity.

11.4 FILTRATION EQUIPMENT AND ADJUVANTS

11.4.1 Properties

Two parameters define the performance of a filter medium: porosity and permeability. Porosity expresses the percentage of empty space in a porous structure, in relation to total volume. Porosity is an indication of the total volume likely to trap impurities. The more porous the filter, the greater its capacity to retain contaminants.

Furthermore, porosity is directly related to the pressure drop in the filter and, therefore, to the energy required to force the liquid through the filtration medium. High porosity results in double savings by lengthening the operational life of the filter and reducing operating costs. The porosity of flat-sheet filters, membranes and filtration adjuvants such as kieselguhr may be as high as 80% or here.

Permeability describes the property of a filter medium to let liquid through at higher or lower speeds. It is expressed in Darcy units. One Darcy corresponds to the permeability of a filter material 1 cm thick with a surface area of 1 cm^2 that lets through 1 ml/s of a liquid with a viscosity of 1 centipoise under a differential pressure of 1 bar. Filter materials have varying permeability and the following values are given as an indication:

Sterilizing flat-sheet filter	0.017 Darcy
Fine clarifying flat-sheet filter	0.15 Darcy
Coarse clarifying flat-sheet filter	1–2 Darcy
Kieselguhr	0.5–5 Darcy
Rapid filter medium	2–7 Darcy

This property is mainly used to categorize kieselguhr.

Another characteristic of filter media is their cutoff, indicating the size of particles that their pores are capable of retaining. In the case of membranes, there is an absolute cutoff, which corresponds to the largest size particle that can pass through the filter. The expression 'nominal cutoff' is also used, taking into account the heterogeneous distribution of different-sized pores. This indicates the size of particles normally trapped, although a few larger particles may come through.

In the case of membranes, the reduction ratio (RR) is also calculated from measurements of the proportion of microorganisms retained under perfectly defined operating conditions:

$$RR = \frac{\text{number of microorganisms before filter}}{\text{number of microorganisms after filter}}$$

Pall membranes used in winemaking, with a cutoff of 0.65 μm, have an RR of 10^5 for *Leuconostoc oenos* (Oenococcus oeni).

11.4.2 Cellulose

Cellulose is a macromolecule resulting from the polymerization of a large number of glucose molecules. It consists of long chains of elementary molecules with a periodic structure, mostly aligned in one direction to form small fibers.

The cellulose mixtures used in filtration are made from wood (pine, birch and beech) subjected to special shredding and chemical breakdown treatments to dissolve the lignin and release the fibers. The raw wood pulp is washed with water and then undergoes several additional stages of purification. The purified pulp is formed into sheets and dried. The fibers are isolated by mechanical treatment and may be broken up into powder. A range of different particle sizes and filtration efficiencies are produced by varying the intensity of the mechanical processing.

The cellulose used in filtering wines is in fiber form, and is commercially available as filter sheets or powder. The latter may be used alone or mixed with other filter media to prepare precoats.

This cellulose is relatively pure, but may contain traces of cations. Although cellulose is theoretically neutral, it is advisable to wash the filter with water to avoid any paper flavor that may be communicated to the wine.

Until 1980, cellulose with a negative electrokinetic charge was mixed with asbestos to filter liquids. This decreased the porosity of the cellulose, which has rather large channels, and increased the filter surface. These two factors improved retention of microorganisms and colloidal particles in suspension. Since the banning of asbestos for hygiene reasons in 1980, flat-sheet filters have been made with pure cellulose. Adjuvants such as diatomaceous earth, perlite and polyethylene may be used, in which case the cellulose must have a positive electrokinetic charge.

11.4.3 Kieselguhr or Diatomaceous Earth

Diatomite is a siliceous sedimentary rock, resulting from the accumulation of microscopic fossil algae shells, or diatomaceous earth, with dimensions ranging from a few μm to several hundred μm. Each diatom consists of a single cell covered with a siliceous shell that becomes impregnated with the silica dissolved in water. When the cells die, the hydrated silica shells are left behind and accumulate to form a soft rock known as diatomite. These rocks have different microscopic compositions depending on their marine or lacustrine origins, and are thought to be from 60 to 100 million years old. There are many deposits in the United States, especially California, as well as in Europe and North Africa. There are widespread deposits in France, located in ancient lake beds in the Massif Central. These fossil earths are ground up to produce a siliceous powder, known as diatomaceous earth, infusorial earth or kieselguhr ('small silica particle' in German).

Diatomaceous earth has been used as a filtration adjuvant since the late 19th century, due to the extreme porosity of the powder obtained by processing the rock. The filter layer represents 80% of the total mass, with a surface of $20-25$ m^2/g. These characteristics are highly favorable for filtration. Around 1920, a new treatment process was developed for making high-permeability diatomaceous earths.

Three types of diatomaceous earth are currently used:

1. Natural diatomaceous earth, gray in color, is crushed and dried to form fine particles.

Filtration is very fine with good clarification, but throughput is very low and this medium is hardly ever used today. It may also contain residues of organic matter.

2. Diatomaceous earth calcined at 1000°C, pink or red in color, is crushed and sorted to produce powders free of organic matter, with coarse particles that are capable of fine filtration at satisfactory flow rates.

3. Fritted diatomaceous earth, i.e. activated by calcination at 1100/1200°C in the presence of a flux (calcium chloride or carbonate) is sorted to produce a white powder with even larger particles and looser structure. Filtration is less fine but faster.

There are different qualities of kieselguhr, differentiated by particle size, which controls permeability (Section 11.4.1), i.e. the rate at which a liquid passes through the material. In wine filtration, a practical distinction is made between 'coarse kieselguhr', above 2 Darcy, and 'fine kieselguhr', below 1 Darcy.

It is important to store kieselguhr in a dry place. It must also be kept away from odoriferous products, as it may easily fix volatile substances that could later be released into wine.

11.4.4 Perlite

This consists of spherical, pearl-shaped, aluminum silicate particles, made by processing volcanic rock. This rock contains 2–5% interstitial water and occluded gases, giving it the property of expanding 10–20 times on heating to 1000°C. This treatment reduces the density of the powder and increases its porosity. After grinding and sorting, a range of light, white powders of varying particle sizes is obtained by adjusting the processing conditions.

Perlite makes it possible to run longer filtration cycles as it is much more porous than diatomaceous earth and its low density (20–30% lower) reduces the weight of adjuvant required. However, perlite has a lower adsorbent capacity and is most efficient in a fine precoat.

Perlite is used to filter must and liquids with a high solid content. It is abrasive and may cause rapid wear to injection pumps.

11.4.5 Flat-sheet Filters and Lenticular Modules

Deep flat-sheet filters consist of permeable boards made of plant cellulose fibers combined with granular compounds, such as diatomaceous earth or perlite, and possibly cation resins to increase the electrical charge.

Asbestos was widely used in flat-sheet filters for many years, as it was highly effective. It reduced the cutoff and increased the separation capacity. Current technology is capable of producing sheets with the same level of performance without asbestos, which has been banned for hygiene reasons. The inhalation of asbestos fibers, naturally very widespread in air, is thought to be carcinogenic.

Flat-sheet filters may be mounted on tray filters or built into closed filters that prevent leaks. These are known as 'lenticular modules'.

Depending on the cutoff required, cellulose fibers are ground coarsely or finely, the granular components are added and the preparation is suspended in water. The manufacturing system consists of a belt filter for vacuum-filtering the suspension, which is constantly agitated by vibration. The layer of filter material is dried and cut to the required dimensions. Variations in the composition of the initial mixture and the machine settings produce sheets with different cutoffs, permeability levels and physical strengths.

The filter pores are distributed asymmetrically, with the largest on the input side. This structure is comparable to a three-dimensional sieve, with a large number of very fine channels. The pore volume represents 70–85% of total filter volume. This means that the liquid moves relatively slowly through the many channels where the particles, microorganisms and colloids are retained not only by screening, but also by adsorption due to the difference in electrokinetic potential between the positively charged pore walls and negatively charged particles. This phenomenon enhances the retention achieved by mechanical screening. This specific retention, attributed to the existence of an electrokinetic potential, is known as the zeta potential. It depends on pH, temperature, filtration rate and electrokinetic charge.

These filters have large internal surfaces capable of retaining considerable volumes of turbid liquid (up to 3 l/m^2), achieving performance levels unequaled by membranes.

11.4.6 Membranes

Synthetic membranes with calibrated pores are used for various operations in the wine industry: ultrafiltration, front-end microfiltration, tangential microfiltration and reverse osmosis. Electrodialysis and pervaporation, special separation techniques described elsewhere in this book (Section 12.5.1), also make use of membranes.

Reverse osmosis is used to separate solutes with molecular dimensions comparable to those of the solvent (approximate pore diameter 0.001–0.01 µm). Solutes with molecules ten times larger than those of the solvent are separated by membrane ultrafiltration (approximate pore diameter 0.002–0.1 µm). Microfiltration (approximate pore diameter 0.1–10 µm) is used to eliminate even larger particles. In practice, it is not easy to distinguish between ultrafiltration and microfiltration membranes. On the one hand, the pores of ultrafiltration membranes may be distorted under strong pressure and allow particles larger than the nominal size to pass through. On the other hand, impurities may form a polarization layer on the surface of microfiltration membranes, gradually clogging the pores and stopping finer and finer particles under ever-increasing pressure. Membranes may also be defined by their absolute and nominal cutoffs (Section 11.4.1).

Pore sizes are expressed in µm for microfiltration membranes (1.2, 0.65 and 0.45 µm are standard for wine filters). Pore diameter is less well-defined and less consistent in ultrafiltration membranes, which are more frequently identified by their cutoff: the size of the smallest molecules they trap (expressed as molecular weight in Dalton).

Membrane characteristics include:

(a) separation efficiency, i.e. a well-defined cutoff and a known, homogeneous pore diameter,

(b) a high permeate flux,

(c) good physical, chemical and heat resistance.

Microfiltration membranes consist of a thin filter layer deposited on a base of the same (asymmetrical membranes) or a different type (asymmetrical and composite membranes).

The first membranes used, based on cellulose acetate, were not very resistant to microorganisms, shocks, temperature or pH. Second-generation membranes, made from polysulfone or polyacrylonitrile polymers, were much tougher. Current third-generation inorganic membranes have good chemical, physical and heat (temperature >100°C) resistance characteristics. They have almost unlimited lifetimes and are easily cleaned and disinfected. The operating parameters are as follows: (a) transmembrane differential pressure, (b) temperature, which affects viscosity, (c) flow rate and (d) retentate outlet rate.

These membranes are manufactured by evaporating a solvent that creates pores through the surface of the material. Their porosity depends on the number and size of these holes. In reality, these membranes are more like sponges than sieves. Membranes are pleated to increase their surface and assembled into modules, which may have a relatively large surface area (0.82 m^2). Several modules (1–4) may be assembled to form a strong, perfectly airtight cartridge. After sterilization, these filtration systems are neutral in terms of their effect on flavor and do not require any special preparation.

There are various types of synthetic membranes:

1. Cellulose ester membranes (diacetate or triacetate): these membranes are highly permeable, so they have a good filtration capacity. They are inexpensive and easy to implement. However, they have a few drawbacks: sensitivity to temperature and pH, and risk of degradation by microorganisms. Cellulose acetate and nitrate

blends are biologically inert, autoclavable and chemical resistant.

2. Polyamide/polyimide membranes: these have greater stability to heat and chemicals, as well as better physical strength than the preceding type. Membranes made of nylon 66 are well known in winemaking.

3. Polyvinylidene fluoride membranes: these membranes, consisting of dihalogen fluoroalcane, have good temperature, chemical and physical stability.

4. Polytetrafluoroethylene membranes: these microporous membranes, used in microfiltration, are obtained by drawing or extruding partially crystallized, polymerized films. They have good temperature, chemical and physical stability and may be heat-sterilized.

5. Polypropylene membranes: the structural depth of this material provides a number of filtration levels within the thickness of the membrane. It is also used for prefilters.

6. Glass fiber membranes: these may be used for prefiltration and final filtration. Cutoffs range from 1 to 40 µm. They are physically strong (4 bar pressure differential at 80°C). The cutoff may be lowered by coating the fibers with food-grade resins.

7. Inorganic ceramic membranes: the advantage of these membranes is that they are inert and imperishable. The filter unit consists of a macroporous base on which superimposed layers of ceramics of varying particle sizes and thickness are deposited, providing great physical strength and low resistance to the flow of liquid. The outside layer is the most active in terms of particle retention and has the smallest diameter pores. The smaller the pores, the thinner this layer will be (a few µm). These membranes may be used for tangential filtration. These inorganic membranes are made from a wide variety of materials (aluminum, zirconium and titanium oxides, sintered metal, etc.). Inorganic membranes are used for microfiltration and ultrafiltration.

11.5 HOW FILTER LAYERS FUNCTION

11.5.1 Filtration Mechanisms

The retention of particles by a filter layer depends on two mechanisms, screening and adsorption. It is quite obvious that, in general, both of these mechanisms operate concurrently.

When a yeast suspension is filtered through a layer of cellulose at low pressure, the fractions collected become decreasingly clear. This is a good example of adsorption. The yeast cells have a smaller diameter than the pores so they are adsorbed inside the filter. When the adsorption capacity is saturated, the yeast is no longer retained and the liquid is still turbid at the filter outlet. If the same filtration is carried out at higher pressure, compression of the cellulose reduces the size of the pores, so a screening phenomenon is involved in retention of the yeast. The fractions collected are much less turbid over time.

Asbestos is an ideal material for filtration by screening. When the same yeast solution is filtered through asbestos, the liquid remains clear until the filter becomes clogged. The flow rate is much lower than it would be with a cellulose filter. In this case, the yeast cells are larger than the filter pores, so they cannot penetrate inside. When all the pores are blocked by yeast the filter is clogged. Following the banning of asbestos, the same results have been obtained using mixtures of cellulose and positively charged kieselguhr.

Filtration through diatomaceous earth involves both adsorption and screening.

The various mechanisms for retaining particles in wine filtration are summarized in Figure 11.3. They may operate simultaneously.

11.5.2 Effect of the Type of Turbidity

The type of particles responsible for turbidity affect both clarification quality and filtration flow rates, especially clogging. It has been observed that wines behave in different ways. Some cause very little clogging, and it is possible to filter several hundred hectoliters on a 5 m^2 filter using a kieselguhr precoat. Other wines, not necessarily

Clarifying Wine by Filtration and Centrifugation

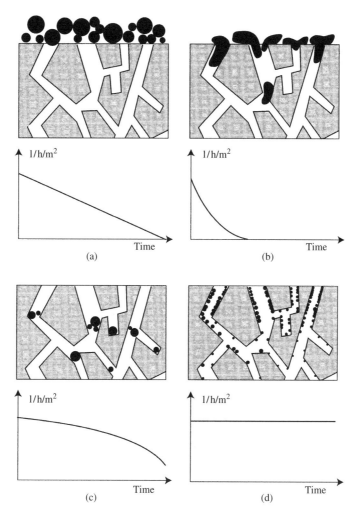

Fig. 11.3. The filtration mechanism (after Ribéreau-Gayon *et al.*, 1977). (a) Screening: the solid particles are rigid and larger than the channel diameter. A porous cake is formed on the surface and gradually clogs the filter. The volume filtered decreases gradually until it reaches zero. (b) Screening: the particles are the same size as those in the previous example, but deformable (under high pressure). They penetrate inside the channels and block them. The flow rate decreases rapidly and the system is soon completely clogged. (c) Adsorption and screening: the particles penetrate inside the pores and are then trapped, either by adsorption on the inside surface or mechanically by building up at certain sites. The empty spaces slowly fill up until the filter is blocked, but filtration continues for a relatively long time. (d) Adsorption: small particles penetrate the filter layers quite easily. They are adsorbed on the insides of the channels. When all of the adsorption sites are saturated, the wine can still flow through the filter, but it is almost as turbid at the outlet as it was initially

the most turbid, clog the filter after only a few hectoliters of wine have been processed.

Each wine has a specific clogging behavior, even if the same filter surface is used under the same conditions. No correlation has been observed between a wine's turbidity and its 'fouling capacity'. Wines with low turbidity are not necessarily more easily filtered. It is possible to measure a conventional 'fouling index' for different types of filtration and, thus, to predict the behavior of certain industrial filtration operations (Section 11.8.3).

Clogging depends more on particle size than the intensity of the turbidity. Coarse particles form a porous layer on the filter surface and cause little clogging. Finer particles penetrate the filter layer and block it rapidly.

When yeast is added to an ultrafiltered wine, little clogging occurs. It may be concluded from this observation that the difficulty in filtering new wines is due to the presence of mucilaginous colloids rather than yeasts. Bacteria have a variable fouling capacity, but it may be rather high. Some acetic and lactic bacteria (ropy wines) produce polysaccharides and mucilaginous matter with a high fouling capacity.

A study of various chemical problems in wine showed severe clogging due to turbidity produced by ferric casse in white wines, protein flocculation caused by heating and precipitation of coloring matter in red wines. Some substances responsible for turbidity cause less fouling if they have been flocculated by preliminary fining. However, the lees of certain fining agents, especially bentonite, clog filters quite rapidly.

In view of the involvement of polysaccharide colloids in these fouling phenomena, pectolytic enzymes have been used in an attempt to improve filtration throughput. It may be assumed that they do not decompose the clogging colloids, but rather destroy the pectin layer that coats them and which acts as a protective colloid. Good results have been obtained in clarifying certain young red wines, press wines and wines made from heated grapes (the natural enzymes have been destroyed by heat). Traditional fining is ineffective. Filtration is hindered by low flow rates and rapid clogging. Treatment with pectolytic enzymes (approximately 4 g/hl) increases the volume filtered through cellulose sheets, per unit time, by approximately a factor of 4.

Wines made from grapes affected by rot are also difficult to clarify due to highly clogging colloids. It has been known for many years that *Botrytis cinerea* secretes a colloid of this type into grapes and that the resulting wines are particularly difficult to clarify by filtration. The colloid in question is a polysaccharide that has been known in winemaking for many years. It belongs to the family of dextranes

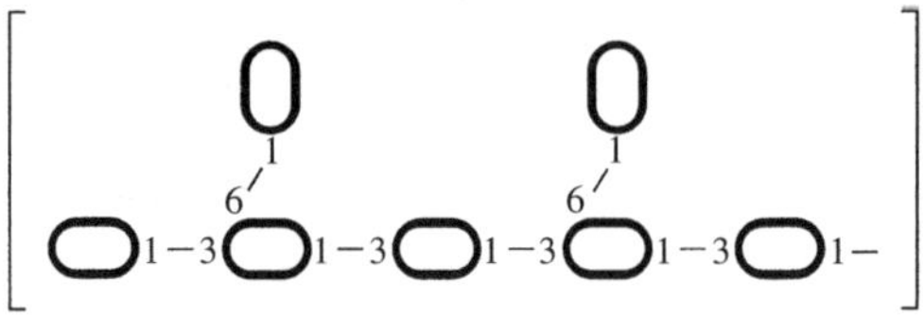

$$\left[\underset{1-3}{\bigcirc} \underset{1-3}{\overset{\overset{\displaystyle O}{\underset{6'}{|}}}{\bigcirc}} \underset{1-3}{\overset{\overset{\displaystyle O}{\underset{6'}{|}}}{\bigcirc}} \underset{1-3}{\bigcirc} \underset{1-}{\bigcirc} \right]$$

Fig. 11.4. Structural unit of the glucane molecule in *Botrytis cinerea* (or cinereane) showing the concatenation of glucose molecules (Dubourdieu, 1982)

that consist of a chain of glucose molecules linked by $\alpha(1 \rightarrow 6)$ bonds. Dubourdieu (1982) demonstrated that the polysaccharide produced by *Botrytis cinerea*, responsible for problems in clarifying wines, is a glucane consisting of a principal chain with glucose molecules linked by $\beta(1 \rightarrow 3)$ bonds. Branches consisting of a single glucose molecule are fixed at $\beta(1 \rightarrow 6)$, leaving one or two non-branched glucose molecules that alternate along the principal chain. This polysaccharide consists of repeats of the basic unit shown in Figure 11.4. Its molecular weight is on the order of 9×10^5.

The fouling capacity of the *Botrytis cinerea* glucane is shown by the graph in Figure 11.5. It depends on the alcohol concentration and the conditions under which the grapes are processed (Section 9.4.2). It also depends on temperature. At 4°C and below, the macromolecules grow larger as flocculation starts, so they are more easily trapped. Filtration cycles can continue longer and clarity is improved. At normal and especially at high temperatures (30–40°C), the colloidal particles are smaller, less likely to flocculate and clog the filter more rapidly.

Research was carried out to find a solution for removing excess glucane from wine. Prior filtration, even on a coarse filter, decreases the wine's fouling capacity, especially at low temperatures, but it may be rather time consuming.

Ultradispersion, a rough physical treatment, improves filtrability by breaking down the colloidal aggregates, but it is not sufficiently effective. The best solution is to use glucanase, produced from a *Trichoderma* culture and marketed by Novo (Switzerland) as Novozyme 116. It is authorized by European Community legislation.

Figure 11.6 shows the effect of adding glucanase on the filtrability of a white wine made from grapes

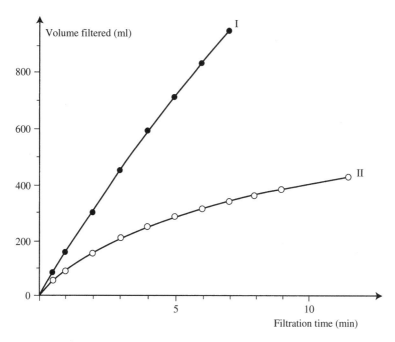

Fig. 11.5. Effect of glucane from *Botrytis cinerea* on flat-sheet filtration (Dubourdieu, 1982) I. Dry white wine made from healthy grapes. II. The same wine +200 mg/l of glucane

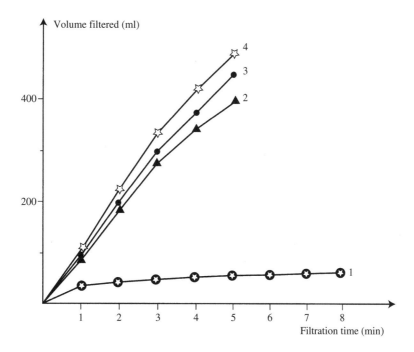

Fig. 11.6. Effect of adding various doses of glucanase SP 116 during fermentation on the filtrability through a flat-sheet filter of a white wine made from grapes affected by rot (Dubourdieu, 1982). 1, Control; 2, addition of 2 g/hl of SP 116; 3, addition of 4 g/hl of SP 116; 4, addition of 6 g/hl of SP 116

affected by rot. Even repeated fining with organic fining agents or bentonite cannot eliminate the protective colloids, so clogging is not alleviated.

11.6 FILTRATION THROUGH DIATOMACEOUS EARTH (OR KIESELGUHR) PRECOATS

11.6.1 Introduction

'Earth filtration' has been widely used to clarify wines for many years. Initially, this involved precoating a filter cloth. The diatomaceous earth, in suspension in wine or water, was deposited on the surface of the cloth, thus constituting the filter layer. Filtration really started when this stage was completed. This process has now been replaced by considerably more advantageous continuous accretion techniques, where diatomaceous earth is continuously added to the turbid wine before it enters the filter. The filter layer grows thicker throughout the process, the impurities are distributed through the mass and the outside layer is never blocked.

Diatomaceous earth of varying permeability, as well as mixtures of diatomaceous earth and cellulose, make filtration through precoats suitable for a wide range of applications. Table 11.2 shows the clarification of a turbid white wine filtered through three different types of earth. Filtration behavior may be predicted by laboratory tests (Section 11.6.2). This type of filtration is generally restricted to untreated wines, as one of the first stages in clarification. However, currently available fine earths may also be used to prepare wines for bottling. Table 11.3 indicates the quality of diatomaceous earth used in various situations and the quantities required at various stages in filtration.

One disadvantage of this type of filtration is that it involves discharging large volumes of diatomaceous earth that represent a source of environmental pollution. Furthermore, staff handling these filters work in an atmosphere contaminated with dust. Tangential flow microfiltration (Section 11.9.1) may be a suitable replacement technique.

11.6.2 Laboratory Filtration Tests

The equipment in Figure 11.7 is used to measure the volume filtered over time, at a constant pressure. According to the theoretical equation for deep-bed filtration (Section 11.2.4),

$$\log V = \tfrac{1}{2} \log t + \text{Cste}$$

If the filter's surface area is known, the straight line giving the volume filtered during a normal industrial filtration cycle may be plotted on logarithmic paper from two or three experimental points.

A method for measuring the fouling index will be described elsewhere (Section 11.8.3). It is not, however, usable at this stage in clarification, as the wine generally has an excessively high fouling capacity.

11.6.3 Filtration Equipment

Precoat filtration equipment consists of vertical or, more frequently, horizontal trays, which are

Table 11.2. Characteristics of a white wine after filtration on three diatomaceous earths with different permeabilities (Serrano, 1993)

	Diatomaceous earth filtration		
	Coarse (1.5 Darcy)	Average (0.35 Darcy)	Fine (0.06 Darcy)
Average throughput (hl/h/m^2)	20	15	7
Fouling index	250	50	22
Turbidity (NTU)	1.33	1.04	0.36
Viable yeasts (per 100 ml)	5000	4500	500
Viable bacteria (per 100 ml)	7700	3000	1500

Control wine: turbidity (NTU) = 21, viable yeasts (per 100 ml) 270 000, viable bacteria (per 100 ml) 180 000.

Clarifying Wine by Filtration and Centrifugation

Table 11.3. Precoat filtration: quantity and quality of adjuvants required to treat different products (Paetzold, 1993)

Products to be filtered	First precoat (time: 10–20 min)		Second precoat (time: 10–20 min)		Continuous accretion		Flow rate (hl/h/m²)
	Quality (Darcy)	Quantity (kg/m²)	Quality (Darcy)	Quantity (kg/m²)	Quality (Darcy)	Quantity (g/hl)	
New wine, first filtration (December)	2–3	0.5–1	2–3	0.5	2–3	200–300	5
Press wine	2–3	0.5–1	2–3	0.5	2–3	200–400	5
Wine aged for at least one winter	1–2	0.5	1–2	0.5	1–2	50–200	10
Wine sheet or lenticular module filtered before bottling	1	0.5	0.4–1	0.5	0.4–1	20–50	15
Wine membrane filtered before bottling	1	0.5	0.06–0.4	0.5	0.06–0.4	20–50	15

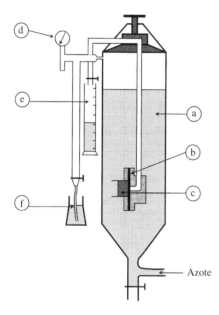

Fig. 11.7. Diagram of a 4 l test chamber capable of withstanding up to 7 bar pressure, used for filtration tests through diatomaceous earth. The useful surface area of the filter medium varies from 4 to 20 cm². It is used to compare the behaviors of different samples of diatomaceous earth (Ribéreau-Gayon et al., 1977): (a) mixture of wine and diatomaceous earth, (b) stainless-steel cloth, (c) layer of diatomaceous earth and filtration cake, (d) pressure gauge, (e) recovering and measuring filtrate volume. A nitrogen stream maintains pressure and keeps the diatomaceous earth in suspension. Bubbling is adjusted and pressure maintained using a bubbler (f)

easier to clean. Filter trays are usually made of stainless-steel mesh, but sometimes of synthetic fabric, metal cartridges or cellulose sheets. The filter is also equipped with a feed pump and a metering pump for injecting the diatomaceous earth suspension into the wine before it enters the filter.

Modern filters are equipped with a residual filtration unit, used to filter and recover any wine remaining in the filter bell at the end of the cycle. They are also equipped with systems for dry extraction of the filtration residues, recommended to avoid pollution. Most systems use centrifugal force. The horizontal trays are spun to eject the earth cake, which is then removed through a hatch at the bottom of the bell.

Modern filters are all made of stainless steel, which facilitates cleaning and maintenance, especially when it is kept polished.

11.6.4 Preparing Filter Layers and Operating Filters (Figure 11.8)

A two-layer precoat must be prepared on the filter rack prior to starting filtration. The second layer activates the filtration cycle. The first, mechanical, layer is made using a coarse adjuvant (permeability above 1 Darcy), with the possible addition of 10% of a cellulose-based product. The quantities

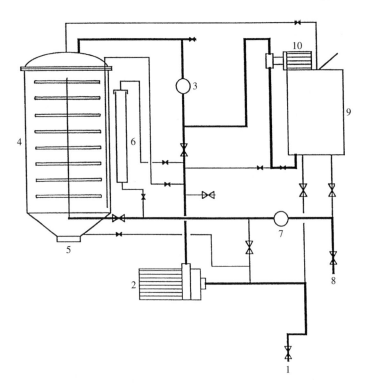

Fig. 11.8. Diagram of the circuits in a diatomaceous earth filter with continuous accretion: 1, inlet of wine to be clarified; 2, main feed pump; 3, inspection glass for wine to be clarified; 4, filtration vat with horizontal filter units; 5, filter cake removal; 6, external residual filtration unit; 7, inspection glass for clarified wine; 8, clarified wine outlet; 9, tank containing the filtration adjuvant in suspension; 10, filtration adjuvant metering pump

required are shown in Table 11.3. This mechanical layer acts as a base for the filter layer. Filtration efficiency depends on the proper preparation of this layer. Sudden changes in pressure, produced by quickly opening and closing the valves, are recommended during the preparation phase, as the precoat will be more stable, with a less compressible structure. However, such pressure changes should be avoided during the filtration process.

It is recommended that both precoats should be prepared with water or filtered wine. Filtration of the wine may start as soon as the filter precoat has been prepared. The outer surface of the filter layer is constantly renewed by continuous accretion, generally with the same adjuvant or mixture of adjuvants. This prevents rapid clogging of the precoat and increases the length of the filtration cycle.

When wine is filtered just before bottling, it may be advisable to use a finer earth for accretion than the grade used in the precoat (Table 11.3). The quantity of earth added ranges from 20 to 200 g/hl, and may exceptionally be higher when clarifying very turbid wines.

The differential pressure is initially low and increases gradually, depending on adjustments to the accretion process. Optimum filtration conditions require a pressure increase of 0.1–1 bar per hour throughout the cycle.

When accretion is insufficient, filter efficiency is low: clogging occurs, pressure increases sharply and the filtration cycle is shortened. If accretion is excessive, pressure rises slowly, but the cycle is also shortened by unnecessary filling of the filtration chamber. Filtration flow rates according to the type of diatomaceous earth are shown in Table 11.3.

Clarification quality monitoring is recommended throughout the entire filtration process. This

Clarifying Wine by Filtration and Centrifugation

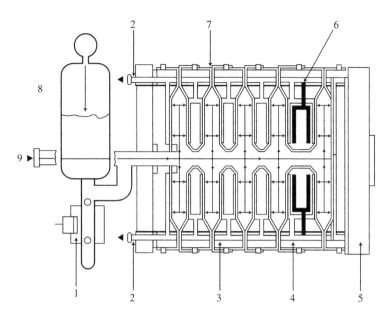

Fig. 11.9. Schematic diagram of a filter press: 1, piston pump; 2, filtrate collector; 3, standard tray; 4, membrane tray; 5, solid stainless-steel frame; 6, compressed air circuit; 7, polypropylene cloth; 8, tank with pneumatic buffer; 9, central input of product to be clarified

operation may be automatic. The major cause of insufficient clarification is that the adjuvant is too coarse and does not trap all the finer particles. Sudden changes in pressure, combined with errors in handling the filter, may damage the filter layer, releasing particles that increase the turbidity of the wine leaving the filter. Insufficient clarification may also be caused by clogging of the filter trays, as the liquid no longer circulates and the precoat does not form in the clogged areas. When the pressure rises, the obstruction is forced out, and wine goes through these areas of the filter without being clarified, as the filter layer is nonexistent.

At the end of each cycle, the filter is cleaned and dried, after filtration of the residual wine from the bell. Regular chemical cleaning and tartrate removal is essential.

11.6.5 Operating a Filter Press

This system is used to clarify liquids containing large quantities of solid particles, such as the deposits resulting from static settling of white must, fermentation lees or even the fining agents recovered after wines have been fined and racked.

A filter press (Figure 11.9) consists of a set of trays, generally made of polypropylene, set in a steel (or, preferably, stainless steel) frame, held tightly together by a hydraulic jack. These trays are covered with cloth and designed to form filtration chambers between the trays that receive the turbid liquid, thus making it possible to collect the filtered liquid. The filter is fed by a high-pressure piston pump. At the end of the operation, a compressed air circuit dries the filtration residues.

The wine to be clarified requires no particular preparation prior to processing in the filter press. An adjuvant (Table 11.4) is added and the mixture is fed into the filter. The impurities, mixed with the adjuvant, are directly retained by the cloth. This is a self-regulating filtration system, as the impurities retained by the cloth act as a filter layer. The filtrate is drained off through internal collectors. Extremely turbid liquids are clarified reasonably well, but it is not possible to achieve very low turbidity. Table 11.4 gives an indication of flow rates observed. In order to filter 240 hl in 8 hours, with a flow rate of 1 hl/h/m^2, a 30 m^2 filter is required, or approximately 45 (80 cm × 80 cm) trays.

Table 11.4. Filtration adjuvants in a filter press (Paetzold, 1993)

Products to be filtered	Adjuvant used	Permeability (Darcy)	Quantity (kg/hl)	Average throughput (hl/h/m^2)
Sediment from white must	Perlite	2–5	1–2	0.5–2
Sediment from protein fining	Kieselguhr	1–3	0.5–2	1.5–3
Bentonite lees	Kieselguhr	1.2	0.5–2	1.5–3
Lees from racking after fermentation	Kieselguhr	1–3	0.5–2	0–1

This type of very robust filter is easily adaptable, by adding or removing trays, and is capable of providing a large filtration surface. It is easy to operate and gives good results in clarifying turbid liquids.

Cleaning used to be considered a difficult operation, but has been greatly improved and it now takes only 20–80 min to clean a 30–100 m^2 filter. This operation may also be fully automated.

11.6.6 Operating a Rotary Vacuum Filter

This equipment has the same applications as a filter press in filtering turbid liquids. It is more complex and difficult to use, requiring a certain level of technical expertise.

The flow rate is constant and relatively high throughout the filtration cycle, thanks to the constant renewal of the filter layer. There are, however, grounds for concern that the vacuum may cause changes in the composition of the product, especially the loss of volatile compounds. In particular, decreases in concentrations of free SO$_2$ and carbon dioxide have been observed.

A rotary filter (Figure 11.10) consists of a cylindrical drum covered with a perforated sieve which supports a filter cloth. The drum rotates around its horizontal axis at adjustable speed, in a tank equipped with an agitating device to homogenize the liquid and keep the filtration adjuvant in suspension during preparation of the filter layer. Diatomaceous earth may be used, but perlite also gives good results at lower cost. A vacuum, created inside the drum by a vacuum pump, draws the liquid in. A layer of filtration adjuvant is deposited on the drum during each

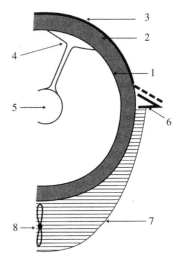

Fig. 11.10. Cross-section diagram of a rotary filter, used for must or wine lees: 1, metal filter cloth; 2, filter layer of diatomaceous earth or perlite; 3, film of trapped impurities; 4, vacuum cups distributed across the entire surface; 5, axis and filtered liquid outlet; 6, adjustable scraper blade; 7, tank containing the liquid to be filtered; 8, agitator maintaining the diatomaceous earth in suspension

rotation, building up a prefilter medium that may be 5–10 cm thick. Preparation of the precoat takes 60 min.

During filtration, any impurities are retained on the surface of the filter layer. These impurities are regularly eliminated by a scraper blade that constantly removes a fine layer of adjustable thickness (a few tenths of a millimeter).

The filtration surface is continually renewed and the flow rate is approximately constant throughout the filtration period. The cycle time is limited by the thickness of the filter layer and the forward

Table 11.5. Flow rates and adjuvant consumption for various rotary vacuum filter applications (Paetzold, 1993)

Liquid filtered	Flow rates (hl/h/m^2)	Adjuvant consumption (kg/hl)
Wine	4–6	0.20–0.60
Lees	0.5–2	1–2
Sediment	2–3	1–2.5
Must	3–5	0.75–1.5

speed of the scraper knife. Table 11.5 indicates average flow rates and adjuvant consumption in different types of filtration.

11.7 FILTRATION THROUGH CELLULOSE-BASED FILTER SHEETS

11.7.1 Introduction

Flat-sheet filtration is widely used just before bottling, to ensure that wines are perfectly clear and microbiologically stable. Flat-sheet filters (Section 11.4.5) are supplied as cardboard cartons, 40, 60 or 100 cm square. They retain particles by screening and adsorption. A distinction is generally made between 'clarifying' and 'sterilizing' filter sheets. The latter have a higher specific retention and some are even capable of eliminating all microorganisms, thus achieving absolute sterility. Several manufacturers offer a range of products in each category, with a variety of characteristics.

The properties of flat-sheet filters may be defined (Section 11.4.1) by a nominal cutoff expressed in μm. It is also possible to determine the maximum quantity of microorganisms in suspension likely to be retained per cm^2 of filter surface under specified operating conditions. Bacteria are much less efficiently trapped than yeast cells. The flow rate of the finest 'sterilizing' filter sheets is naturally lower than that of 'clarifying' sheets and they are also more susceptible to clogging.

The sheets are mounted on standard filters (Figure 11.11), making it possible to vary the total filtration surface by modifying the number of sheets. This equipment is made of stainless steel, with stainless-steel or plastic trays to hold the flat-sheet filters. Filters with reversing chambers (Figure 11.11) make it possible to use two sets of sheets with different performances on the same system.

11.7.2 Preparing Wines for Flat-sheet Filtration

The wine should be properly clarified prior to flat-sheet filtration at the time of bottling to ensure a satisfactory flow rate. This preliminary clarification may involve spontaneous settling, fining, centrifugation (Section 11.11) or filtration through a diatomaceous earth precoat (Section 11.6).

Flat-sheet filtration is subject to the law of gradual clogging of pores under well-defined conditions. A test (Serrano, 1981) to check a wine's aptitude for clarification by flat-sheet filtration may be carried out using the apparatus shown in Figure 11.12.

The maximum volume that can be filtered before total clogging (Section 11.2.3) is calculated as follows:

$$V_{\max}(\text{ml}) = \frac{t_2 - t_1}{t_2/V_2 - t_1/V_1}$$

where $t_1 = 1$ hour and $t_2 = 2$ hours, $V_1 =$ volume filtered in 1 hour, $V_2 =$ volume filtered in 2 hours and pressure $= 0.5$ bar. In most instances, a normal, one-day filtration cycle will be completed without the filter becoming totally blocked, so $V_{\max}$ is never reached.

On the other hand, it is interesting to find out the volume that can be filtered in 8 h in order to assess the operation's cost-effectiveness. The straight line of the variation t/V over time (Section 11.2.3, Equation 2) is plotted on the basis of three points, obtained after 1 h, 1 h 30 min and 2 h. The volume filtered in 8 hours is then obtained by extrapolation.

The values recommended by manufacturers for an 8-hour cycle are as follows:

- 5600–7200 l/m^2 for clarifying filter sheets
- 2800–4000 l/m^2 for sterilizing filter sheets

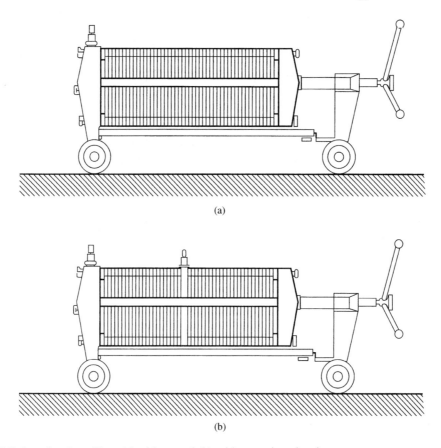

Fig. 11.11. Cellulose flat-sheet filter, (a) without and (b) with reversing chamber

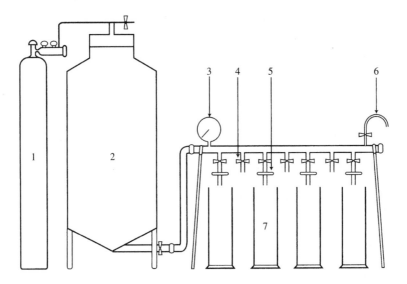

Fig. 11.12. Diagram of a system used to determine filtration characteristics: 1, compressed air source; 2, feed vat; 3, pressure gauge; 4, valve; 5, single disk filter (surface area of 22 cm^2); 6, purge valve; 7, graduated container

Table 11.6. Contamination of a sweet white wine during storage, after diatomaceous earth filtration (Serrano, 1981)

Sample description	Turbidity (NTU)	Viable yeasts (10^3/100 ml)
Immediately after kieselguhr filtration	0.9	2
After 15 days	5.5	320
After 1 month	6.9	480

If these criteria are not satisfied, this means that the wine has not been sufficiently clarified in advance to ensure that flat-sheet filtration will be efficient and cost-effective.

After filtration through a diatomaceous earth precoat, even relatively coarse, it is generally possible to carry out flat-sheet filtration, and possibly sterilizing filtration, with good flow rate and clarification quality. However, during aging, wines undergo modifications, especially in their colloidal structure, that lead to an increase in turbidity and the number of viable germs (Table 11.6). It is advisable to filter the wine through a diatomaceous earth precoat less than one week before flat-sheet filtration.

The heterogeneity of the filter sheets makes it impossible to obtain a direct result for sheet filtration, so the fouling index measurement on membranes (Section 11.8.3) may be used. Wines should have the following characteristics to be ready for flat-sheet filtration:

- Turbidity: <1.0 NTU (Section 11.3.1)
- Fouling index: IC < 200
- Number of viable microorganisms: <100 per 1 ml

These criteria are necessary to ensure that flat-sheet filtration will provide proper clarification and a satisfactory elimination of microorganisms, combined with an adequate flow rate (Table 11.7).

11.7.3 Selecting Filtration Parameters

Table 11.8 shows an example of industrial filtration, assessing the quality of clarification and, consequently, making it possible to select the filter sheets best adapted to this process. The results show that it is indispensable to use sufficiently fine filters to achieve perfect clarification just before bottling. The resulting loss of polysaccharides, a negative effect of filtration on quality, is negligible.

In order to ensure a satisfactory flow rate with fine filter sheets, the wine must be properly prepared, as described earlier in this chapter. If this has not been done, the wine may be filtered twice in a single operation, using a filter equipped

Table 11.7. Successive stages in the clarification of a sweet white wine until almost total sterility is obtained, using sterilizing filter sheets (Serrano, unpublished data)

	Untreated wine, before filtration	Filtration through diatomaceous earth precoat	Filtration through sterilizing filter sheets
Fouling index	Not measurable	250	24
Average filtration throughput (l/h/m^2)		2000	420
Turbidity(NTU)	21	1.33	0.33
Viable yeasts (10^3/100 ml)	270	5	<1
Viable bacteria (10^3/100 ml)	180	8	<1

Table 11.8. Filtration trial to find the best type of filter sheet for clarifying a specific red wine (Serrano, unpublished data)

	Before filtration	Clarifying filter sheet no. 3	Clarifying filter sheet no. 5	Clarifying filter sheet no. 7	Clarifying filter sheet no. 10	Sterilizing filter sheet
Turbidity (NTU)	1.0	0.78	0.69	0.44	0.34	0.34
Viable yeasts (cells per 100 ml)	800	50	15	5	<1	<1
Viable bacteria (cells per 100 ml)	9500	2100	900	130	<1	<1
Polysaccharides (% reduction)		0	0	0	5	5

with a reversing chamber (Figure 11.11) (Serrano and Ribéreau-Gayon, 1991). The first filtration eliminates the larger particles and makes it possible to achieve a sufficiently high flow rate during the second filtration through fine filter sheets to produce the required clarity.

It is advisable to eliminate holding vats between the filter and the bottle filler to avoid microbial contamination. The filter must, therefore, operate at a constant flow rate, governed by the throughput of the bottle filler. The values generally recommended by manufacturers are as follows:

- Clarifying filter sheets: 700 l/h/m^2 or 100 l/h per 40 × 40 cm sheet
- Sterilizing filter sheets: 350 l/h/m^2 or 50 l/h per 40 × 40 cm sheet

Newer designs operate effectively at higher flow rates, i.e. 900 l/h/m^2 for clarifying plates and 500 l/h/m^2 for sterilizing plates.

If wines are properly prepared, these flow rates may be maintained for 8 hours, without the differential pressure in the filter exceeding 0.5–0.7 bar (Serrano, 1981). If this is not the case, filtration will have to stop after only 4 or 5 hours. An excessive increase in pressure may even be required to continue filtering for that length of time. The effectiveness of the filter sheets is guaranteed up to 3 bar for clarifying sheets and 1.5 bar for sterilizing sheets. Clarification quality may be good at these high pressures, but they should be avoided as they tend to cause liquid to leak from the filter.

The number of sheets required in the filter depends on the throughput of the bottle filler. Calculations show that 23 clarifying sheets or 45 sterilizing sheets (40 cm square) are required for a bottling line with a capacity of 3000 bottles/h.

If wine has been properly clarified prior to filtering, the filter sheets will not be clogged after an 8-hour day and the flow rates will remain satisfactory. There is an obvious economic advantage in using the same filter sheets for several days. This is only possible if no contamination occurs during the night when the system is shut down. Industrial trials showed that it was possible to use the same sheets for several days, provided that the filter was emptied at the end of the day, cleaned and fully sterilized by running hot water at 85°C through the entire system for 20 min, either in the same direction as the filtration flow or as a backwash. Furthermore, this operation unclogs the flat-sheet filters. This is possible when the filter is operating as a prefilter prior to membrane filtration. If this is not the case, and especially if the wine must be absolutely sterile after filtering, the filter sheets really must be changed every day.

11.7.4 Sterilizing Equipment

It is vital to sterilize all equipment, and especially the filter and filter sheets, every morning prior to starting filtration and bottling. Table 11.9 shows the importance of sterilization. In particular, it is necessary in order to achieve perfect yeast retention, which is indispensable for sweet wines. If the system has not been sterilized, the first

Table 11.9. Impact of filter sterilization on filtration quality (Serrano, 1984)

	Unsterilized filter	Sterilized filter
Turbidity		
$t = 5$ min	0.97	0.56
$t = 4$ h	0.87	0.62
$t = 8$ h	0.84	0.66
Viable yeasts (cells/100 ml)		
$t = 5$ min	70	<1
$t = 4$ h	20	<1
$t = 8$ h	10	<1

Control wine: turbidity-1.25 NTU, viable yeasts = 500 cells/100 ml.

few liters of wine filtered will pick up any contamination from the equipment.

The system is sterilized with steam or hot water at 90°C, circulating at low pressure (0.2 bar) in the normal direction of filtration. This operation must continue for 20 min, starting from the time the filter reaches sterilization temperature. The filter is then cooled with cold water ($\Delta p < 0.2$ bar). The use of prefiltered fluids is recommended, both for sterilization and cooling. This minimizes the risk of blockage (which might occur if the water contained any particles) and the danger of microbial contamination during cooling. The volume of water used in this operation is generally sufficient to 'prime' the filter sheets as well, thus avoiding any organoleptic deterioration of the wine. However, the elimination of any unpleasant smells or off-flavors must be checked by tasting during cooling.

At the end of the operation, the water is drained from the filter and, at the same time, it is filled with wine. However, after flushing with the quantity of water necessary for sterilization, a 40 cm × 40 cm filter sheet retains approximately 0.85 l of liquid. It is, therefore, essential to eliminate the first wine that is filtered (at least one liter per sheet), as it is highly diluted and may have slight organoleptic defects.

At the end of the day, the filter must be drained, disassembled (unless the same filter sheets are used for several days) and rinsed with hot water. Chemical cleaning should be carried out weekly, using detergent.

All these operations are vital to avoid microbial contamination of the wine, which may easily become a major problem.

11.7.5 Lenticular Module Filtration

In a lenticular module, cellulose-based filter media identical to those in flat-sheet filters are mounted in sealed units, ready for use (Figure 11.13). Implementation is simpler and there is no risk of leaks at high pressure (a common problem in tray filters).

Modules are available in two sizes: 284 mm (12 inches) in diameter with a filter area of 1.8 m² and 410 mm (16 inches) in diameter with a filter area of 3.7 m². It is possible to install one to four modules in the same case, to adapt the filter area to the required flow rate.

In order to maintain reasonable operating costs, the same modules must be able to be used for several days. To obtain satisfactory results, it is necessary to regenerate the filter every evening by running hot water at 45°C through the system in the same direction as filtration. This is followed by sterilization with water at 90°C.

These lenticular filters provide satisfactory clarification in a single operation, even after several days in operation, provided that the wines have already been partially clarified, e.g. barrel-aged red wines. However, sterile filtration of sweet wines is by no means a guaranteed success (Serrano and Ribéreau-Gayon, 1991).

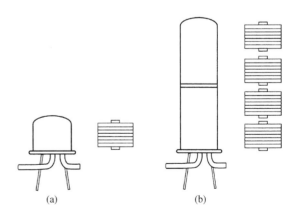

Fig. 11.13. Lenticular filter. Casings fitted with (a) one module and (b) four modules

11.8 MEMBRANE FILTRATION

11.8.1 Introduction

Membrane filtration is used at the time of bottling, mainly in cases where sterile, or at least low-microbe bottling is required (Section 11.3.4). The wine must have been properly prepared so that this operation can be run at a satisfactory flow rate, without excessive clogging.

The composition of the membranes used to filter wine has been presented above (Section 11.4.6). Membrane filters are supplied as ready-to-use cartridges. The flow rate depends on the number of cartridges installed in parallel in each unit.

A membrane's efficiency in trapping particles, or its retention value, depends on pore diameter, i.e. a membrane with a retention value of 0.5 μm will retain all particles with a diameter above 0.5 μm. Wine filters are in the microfiltration range, with pore diameters ranging from 0.45 to 1.2 μm.

A prefilter is normally used to protect the membranes and avoid excessively rapid fouling. An industrial filtration system includes a 'prefilter unit' and a 'final filtration unit', assembled in series on the same base.

11.8.2 Prefilter Cartridges

There are two categories of prefilters. 'In-depth prefilters' are coarse filters consisting of glass fiber or polypropylene, either alone or mixed with diatomaceous earth or cellulose. They trap the particles inside the filter layer by adsorption and screening, and have a high retention capacity.

'Surface prefilters' retain the particles on their surface. They are made from cellulose esters or layers of polypropylene. They also have good specific retention. Prefilters are not always defined by their retention value. The value given, e.g. 3 μm, frequently corresponds to a nominal retention value. In this case, a variable proportion of particles larger than 3 μm in diameter may pass through the filter.

Specific retention is measured using the same procedure and expresses the total quantity of particles that the filter is capable of retaining before it becomes blocked. Specific retention depends on the compactness of the filter: the more tightly packed the fibers, the faster the filter becomes clogged. Capacity and efficiency are two opposing, yet complementary, characteristics.

Prefilters are designed to improve the throughput of final filtration. They cannot guarantee perfect clarification quality or total retention of microorganisms.

11.8.3 Preparing Wines for Filtration: Filtration Tests

In order to achieve good results with membrane filters, the larger impurities must first be removed from the wine to reduce its fouling index, so that the flow rate will be satisfactory. This may involve filtration through a diatomaceous earth precoat. However, the 'coarse earth' (1.5 Darcy) systems, suitable for preparing wines for flat-sheet filtration, are not effective in this instance. Membrane filter flow rates (on the order of 150 l/h/m^2) are too low, even at high pressures (3 bar), and the filter clogs rapidly. Relatively fine earths (0.06 Darcy) must be used for prefiltration to ensure satisfactory flow rates (400 l/h/m^2) during membrane filtration.

A system similar to that in Figure 11.12 (Section 11.7.2) is used for filtrability tests that predict the wine's behavior during membrane filtration, i.e. the fouling index and V_{max} (maximum volume filtered before clogging).

The fouling index (IC) is obtained by measuring the difference in the time taken to filter 200 and 400 ml wine through a membrane with a pore diameter of 0.65 μm and a surface area of 3.9 cm^2, at a pressure of 2 bar. The formula is as follows:

$$IC = T_{400} - 2T_{200}$$

It is not always possible to collect 400 ml of filtrate if the wine clogs the system very rapidly. If this is

Clarifying Wine by Filtration and Centrifugation

the case, the volume throughput in 5 mn is noted. Fouling index measurements made on membranes are also used to predict flat-sheet filtration behavior (Section 11.7.2).

V_{max} (Gaillard, 1984) is calculated using the same formula as that used for flat-sheet filtration (Section 11.7.2), although the experimental method is different. The volume throughput of the membrane at a pressure of 1 bar is noted after 2 and 5 mn. The formula (Section 11.2.3) is the following:

$$V_{max} = \frac{5-2}{5/V_5 - 2/V_2}$$

or

$$V_{max} = \frac{3(V_5 \times V_2)}{5V_2 - 2V_5}$$

Experience has shown that, in order to obtain good clarification with a satisfactory flow rate, a wine must have a fouling index (IC) lower than 20, or possibly 30, with a V_{max} higher than 4000 ml or at least 2500 ml.

11.8.4 Selecting Filtration Parameters

Membranes with a pore diameter of 1.2 μm are used when only yeast has to be eliminated, while 0.65 μm or even 0.45 μm membranes are required when both yeast and bacteria must be removed. These membranes are very thin (150 μm). Adsorption may be considered negligible due to their very high porosity. They operate by screening and stop all particles larger than the pore diameter at the membrane surface.

The theoretical flow rate specified by the filter manufacturer for properly prepared wines is 800 l/h/m² or 1440 l/h for a 1.8 m² cartridge (30-inch diameter). However, to increase the life expectancy of the filter medium before total clogging, it is advisable to oversize the system so that it operates at half capacity, i.e. 400 l/h/m² or 720 l/h for a 1.8 m² cartridge. These flow rates may be maintained with a differential pressure below 1 bar. It is advisable to operate at low differential pressures to minimize clogging, although the membranes are designed to withstand 7 bar.

As in flat-sheet filtration, the constant flow rate of the bottle filler dictates the number of cartridges to use. It has been calculated that three 30-inch cartridges (filter surface of 1.8 m² each, flow rate: 720 l/h) are required to supply a bottling line operating at 3000 bottles/h or 2250 l/h. Filter membranes must be used for several weeks, or even months, before they become completely blocked in order to make this system cost-effective.

The system must be sterilized before filtration starts every morning, as described in the Section on flat-sheet filters (Section 11.7.4). The sterilizing fluid, either steam or hot water at 90°C, is circulated at low pressure in the same direction as filtration. Water must be prefiltered to avoid damaging the membranes. Once the equipment has been sterilized, it is cooled with filtered cold water.

Every day, before the filtration system starts operating, two tests (diffusion test and bubble point test) should be carried out to check the integrity of the damp membranes and to inspect the watertight seals, once the filter has cooled down. Details are given in the membrane manufacturer's instructions. These inspections are indispensable to ensure that the filter media operate at optimum efficiency (Table 11.10).

Membranes must be regenerated after each daily filtration cycle to ensure an optimum lifetime. Filtered hot water at 40°C is circulated through the system for about 15 mins, generally in the same direction as filtration. Water temperature is then increased to 90°C and the filter is sterilized for

Table 11.10. Characteristics of Pall membranes (Gautier, 1984)

Membrane pore diameter (μm)	Bubble point (bar)	Integrity test (bar)
0.45	1.3	1
0.65	1.1	0.9
1.2	0.7	0.6

Table 11.11. Filtering a red wine on prefilter cartridge and membrane (Serrano, unpublished data)

	Fouling index (IC)	Turbidity (NTU)	Viable yeasts (cells/100 ml)	Viable bacteria (cells/100 ml)
First day				
Filter inlet	27	0.28	500	1 100
Prefilter outlet	18	0.17	5	80
Membrane filter outlet	17	0.16	<1	<1
Second day				
Filter inlet	26	0.29	1200	20 000
Prefilter outlet	21	0.21	2	240
Membrane filter outlet	17	0.18	<1	<1
Third day				
Filter inlet	21	0.40	2400	27 000
Prefilter outlet	18	0.24	4	3 800
Membrane filter outlet	16	0.20	<1	1

about 20 mins. Prefilter cartridges are treated in the same way, except that it is possible to backwash for regeneration and sterilization.

When all the preceding operations are efficiently carried out, simultaneous filtration with prefilter cartridges and membranes gives good results in terms of clarification quality (Table 11.11). Quality remains high for several days with satisfactory flow rates. It has, however, been observed that this technique, combined with the necessary preliminary clarification processes, may have a greater effect on the polysaccharide concentration than flat-sheet filtration (Section 11.10.2). It should therefore be used very carefully. Proper operating conditions are essential and filtration results should be carefully monitored.

Membrane filtration should provide perfect clarification before bottling. Filtration is said to be 'low microbe' if the viable residual population is no more than 1 germ per 100 ml. It is considered 'sterile' if this value is reduced to no more than 1 germ per bottle.

11.9 TANGENTIAL FILTRATION

11.9.1 Principles

Standard filtration techniques are known as 'frontal' or 'transversal', as the liquid circulates perpendicularly to the filter surface. The trapped particles form a 'cake' that may be involved in the clarification mechanism. This 'cake' also causes the filter to become gradually blocked.

In tangential filtration (Guimberteau, 1993; Donèche, 1994), the flow is parallel to the filter surface (Figure 11.14), which consists of a membrane with relatively small pores. This maintains excess pressure in the feed liquid, causing a small quantity to flow through the membrane (3), where it is clarified (4). The solid particles do not accumulate, as they are constantly washed away by the flowing liquid. Clarification efficiency is modulated by adjusting the pressure (2), the flow rate of the liquid to be clarified (1) and the evacuation rate of the filtrate (6). The excess pressure heats the liquid, so refrigeration is required to cool the system (5).

In practice, a certain amount of clogging is inevitable, although this occurs much less often in tangential filtration than in transversal filtration. Clogging may result in a simple accumulation of trapped substances, or these may interact with the membrane surface. As clogging is detrimental to performance, it is minimized by varying the hydrodynamic parameters (circulation rate, temperature, pressure, etc.), the characteristics of the product to be clarified or the type of membranes and their properties. These filters are generally equipped with unclogging systems which operate by reversing the liquid flow.

A distinction is made between tangential ultrafiltration (pore diameter of 0.1–0.001 μm) and

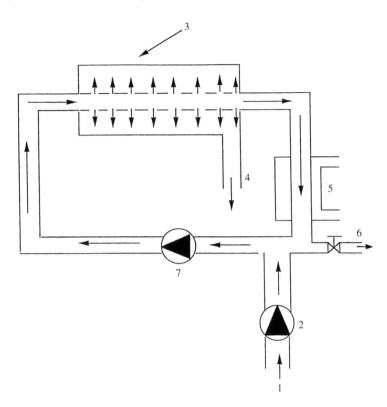

Fig. 11.14. Schematic diagram of tangential filtration: 1, inlet of liquid to be clarified; 2, high-pressure pump; 3, module containing the filtration membrane; 4, clarified liquid outlet; 5, cooling system; 6, adjustable outlet for the concentrate containing the impurities; 7, circulating pump

tangential microfiltration (pore diameter of 10–0.1 μm). In reality, a certain amount of fouling, which tends to reduce the size of the pores, is inevitable. The distinction between these two types of filtration is not, therefore, as clear as it may seem.

The first attempts to apply tangential filtration in winemaking relied on ultrafiltration membranes that were likely to trap not only particles in suspension, but also colloidal macromolecules. In particular, unstable proteins were intended to be eliminated from white wines. However, it very quickly became apparent that the analytical and organoleptic characteristics of white wines were subject to profound modifications under these conditions.

Nevertheless, several enological applications have been developed, using microfiltration membranes with average pore diameters between 0.1 and 1 μm. It is feasible to expect to achieve clarity and microbiological stability in untreated wines, possibly in a single operation, without affecting their composition. Improvements in membrane production techniques and greater diversity in their characteristics have resulted in the availability of equipment suited to a range of different objectives.

11.9.2 Applications in Winemaking

Tangential microfiltration has been used in many wine treatment applications over the past 10 years. Suitable membranes are now available for clarifying must or untreated wines, as well as the final clarification of prefiltered wines. This technique may also provide an alternative to filtration through diatomaceous earth precoats, especially in cases where waste discharges could lead to excessive pollution. Tangential microfiltration is still, however, subject to two major disadvantages: low hourly throughput per m² of filter surface and its high cost, as compared to traditional filtration methods.

Various applications have been suggested:

1. Removing sediment from white grape must: clarification is excellent. It may even be excessive, leading to difficulties during fermentation. The technical and economic advantages of this technique have not been clearly demonstrated.

2. Preparing low-alcohol beverages from grapes—grape juice, sparkling grape juice and partially fermented beverages: tangential microfiltration results in clear, microorganism-free products that may be stored in sterile vats until they are treated for protein (bentonite treatment) and tartrate (cold stabilization) instability.

3. Clarifying wines: it is now possible to integrate tangential microfiltration into the winemaking process, especially for white wines, which achieve better flow rates than reds.

Tangential microfiltration may be used at the end of fermentation to ensure microbiological stabilization or to prepare wines for bottling. However, certain technical aspects of the process make it incompatible with bottling operations. Winemakers must also be aware of the risk of eliminating high molecular weight carbohydrate and protein colloids that are not only an integral part of a wine's composition but also of its organoleptic characteristics. The phenols in red wines tend to clog membranes and reduce flow rates. Furthermore, there is some concern that these phenols may be modified, resulting in a deterioration of the color.

A comparison of the effectiveness of various types of filtration in clarifying white and red wine after fining (Serrano, 1994) highlights the low flow rates of tangential microfiltration as compared to filtration through kieselguhr (Table 11.12). This is particularly true of red wines. However, flow rates are higher than those observed during early tests. Fouling is by no means negligible and explains why it is possible to have a lower flow rate with a 0.4 μm

Table 11.12. Application of various filtration techniques to a white wine after fining with blood albumin (8 g/hl) and to a red wine after fining with gelatin (8 g/hl) (Serrano et al., 1992)

	Control	Precoat and kieselguhr filtration	Tangential filtration		
			Inorganic membrane 0.2 μm	Inorganic membrane 0.2 μm with reverse flow unclogging	Organic membrane 0.4 μm
Flow rate (l/h/m²)					
White wine		1 020	137	245	68
Red wine		950	85	150	57
Turbidity (NTU)					
White wine	7.00	0.32	0.26	0.60	0.28
Red wine	3.00	0.51	0.22	0.21	0.10
Viable yeasts (cells/100 ml)					
White wine	30 000	1 400	<1	10	<1
Red wine	200 000	4 200	16	110	5
Viable bacteria (cells/100 ml)					
White wine	7 200	6 500	130	850	50
Red wine	Uncountable	16 000	8 500	12 500	500

Hydrodynamic conditions: transmembrane differential pressure, 0.7–1.3 bar; tangential flow rate between 2 and 3 m/s; eliminated concentrate, less than 0.2%.

organic membrane than a 0.2 μm inorganic membrane.

All types of tangential microfiltration produce higher quality clarification than those achieved by filtration through a kieselguhr precoat. However, the filtrates are not always sterile, particularly when unclogging by reversed flow has destroyed the polarization layer.

Analysis shows that concentrations of polysaccharides and volatile fermentation products in white wines are reduced by tangential microfiltration, as compared to filtration through a kieselguhr precoat. Anthocyanins and tannins are affected in red wines. However, in view of the natural modifications that occur in wines during aging, these differences tend to become less marked over time. The standard tests used in organoleptic analysis did not identify any significant differences (threshold of 5%) after samples had been kept for 1, 6, and 12 months.

In fact, this type of tangential microfiltration is not a method for achieving final clarification, but rather an alternative to filtration through a kieselguhr precoat to prepare wines (or at least white wines) for final filtration. Flow rates would need to be improved and operating costs reduced for this technique to develop on a wider, industrial scale. It should also be taken into account that this process produces a liquid residue that requires treatment to avoid excessively polluted discharges, although the absence of kieselguhr makes the waste less polluted than earth filtration residues.

4. Clarifying fining lees: tangential microfiltration, using membranes with pore diameters from 0.2 to 0.8 μm, was compared with a rotary vacuum filter (Serrano, 1994). The flow rates were lower (50–100 l/h/m^2 instead of 350–500 l/h/m^2) but clarification was better, both in terms of much lower turbidity as well as the elimination of microorganisms. Wine losses were also lower: 0.2% instead of 4–6% with rotary filters. Modifications in the chemical composition were less marked, especially carbon dioxide and volatile compounds, which were easily eliminated by the vacuum in the rotary filter. Tasting tests did not identify any significant differences (threshold 5%).

11.10 EFFECT OF FILTRATION ON THE COMPOSITION AND ORGANOLEPTIC CHARACTER OF WINE

11.10.1 Various Effects of Filtration

Consumer demand insists on wines that are clear and stable. However, wine quality may be affected by too much, ill-advised treatment. Filtration is known to have potentially harmful effects and is particularly criticized for making wines thinner. Filtration just before bottling is sometimes challenged on these grounds, but this criticism is often unjustified. Properly controlled filtration has positive effects on quality, whereas careless or excessive treatment may have a decidedly negative impact. In filtration, as in all other treatments applied to wine, proper conditions and care are essential. Winemakers are responsible for deciding precisely which operations are necessary.

Several possible consequences of filtration should be considered. Besides any changes in chemical composition (described in the next paragraph), filtration may be responsible for secondary phenomena, due to operating techniques or the use of poor-quality filtration equipment. These problems can, and should, be avoided.

The first important point is that contact with air during filtration should be prevented. Negative effects sometimes attributed to filtration are often simply due to the penetration of air during pumping, which is a necessary part of the process. Wine may be saturated with oxygen when it comes out of the filter, while losing carbon dioxide at the same time. This may cause ferric casse or a loss of aroma, especially in wines with a low free SO_2 content. Wine should be protected from these risks by checking that filtration systems are airtight and purging them to remove air.

It should also be emphasized that poor-quality filter media may transmit an earth, paper or cloth taint to the wine. Generally, only the first few liters of wine are affected, but the defect may be more persistent in certain instances. Off-flavors from cellulose-based filter sheets are the most common. Manufacturers recommend flushing the system with several liters of filtered water per sheet. This operation also cools the filter after sterilization (Section 11.9.4). It is easy to determine whether any off-flavors have been eliminated by tasting. It may be necessary to use 10–20 liters of water per filter sheet to eliminate off-flavors completely.

Cloth, especially cotton, and diatomaceous earth filters can also be responsible for transmitting off-flavors they may have picked up in damp, poorly ventilated storage areas.

11.10.2 Modifications in Wine Composition and their Effect on Flavor

If the preceding precautions are taken, the wine's quality should not be affected as it passes through the filter surface. Filtration, after all, is intended to eliminate turbidity, foreign bodies and impurities that would, in time, form the lees. It would be ridiculous to suggest that these substances make a positive contribution to flavor.

Contrary to a widely held opinion, clear wine always tastes better than the same wine with even slight turbidity. Furthermore, wines made from grapes affected by rot and press wines lose at least part of their bitterness and roughness after filtering, which results in a definite improvement. Filtration through fine filter sheets or sterilizing membranes does not affect flavor, provided that these operations are carefully controlled. The difference is most significant in young wines with a high particle and microorganism content which become more refined and acquire elegance thanks to early filtration through a diatomaceous earth precoat.

The separation capacity of some filter media, however, enables them to eliminate macromolecules that form an integral part of the wine's structure, together with turbidity. These macromolecules contribute to a wine's character, not only by producing an impression of fullness and softness, but also by acting as aroma fixatives. A wine's aromatic character may well be altered if these substances are eliminated.

Serrano and Paetzold (1994) published experimental results on the influence of different types of filtration on chemical composition (polysaccharides, phenols, higher alcohols, fatty acids and esters) and the impact of these modifications on flavor. A number of their conclusions are given below (Tables 11.13 and 11.14):

Table 11.13. Effects of different types of filtration on the chemical composition of a white wine (results in mg/l) (Serrano and Paetzold, 1994)

	Control	Coarse kieselguhr (2.3 Darcy)	Fine kieselguhr (0.35 Darcy)	Clarifying filter sheet, prefiltered through coarse kieselguhr (2.3 Darcy)	Sterilizing filter sheet, prefiltered through coarse kieselguhr (2.3 Darcy)	Membrane: 0.65 μm, prefiltered through fine kieselguhr (0.35 Darcy)
OD 420	0.084	0.087	0.083	0.079	0.080	0.078
Tannins	71	69	68	67	68	66
Total polysaccharides	570	540	517	521	518	454
Higher alcohols (total)	317	312	312	308	309	291
Higher alcohol acetates (total)	3.5	3.5	3.4	3.4	3.2	2.9
Volatile fatty acids (total)	14.3	14	12.8	13.8	13.7	12.3
Ethyl esters of fatty acids (total)	4.3	4.2	4.0	4.4	4.0	3.8

Clarifying Wine by Filtration and Centrifugation

Table 11.14. Effects of different types of filtration on the chemical composition of a red wine (Serrano and Paetzold, 1994)

	Control	Coarse kieselguhr (1.5 Darcy)	Fine kieselguhr (0.06 Darcy)	Clarifying filter sheet, prefiltered through coarse kieselguhr (1.5 Darcy)	Sterilizing filter sheet, prefiltered through coarse kieselguhr (1.5 Darcy)	Membrane: 0.65 μm, prefiltered through fine kieselguhr (0.06 Darcy)
Free polysaccharides (mg/l)	426	420	389	380	385	342
Total polysaccharides (mg/l)	650	630	607	625	620	562
Phenol compound index (D280)	41	40	39	40	39	37
Tannins (g/l)	2.7	2.6	2.4	2.5	2.4	2.3
Total anthocyanins (mg/l)	252	243	225	240	230	208
Color intensity	0.53	0.54	0.62	0.59	0.59	0.57
Hue	0.81	0.79	0.81	0.78	0.80	0.80

1. Filtration through a coarse diatomaceous earth precoat (2.3 and 1.5 Darcy) did not affect chemical composition. The same operation with fine earth (0.35 Darcy) reduced the polysaccharide and condensed tannin content by 10%. No organoleptic effects were identified when the samples were tasted one month after filtration.

2. Neither clarifying nor sterilizing flat-sheet filters caused any more noticeable changes than fine earth filters. A reduction in fermentation esters was noted, although the terpenols in Muscat wines were unaffected. No significant differences were identified when the wines were tasted.

3. It is not advisable, nor is it useful, to filter wines on a fine diatomaceous earth precoat (0.35 Darcy) prior to flat-sheet filtration.

4. Membrane filtration (0.65 μm) caused a more marked reduction in polysaccharides, phenols and esters than flat-sheet filtration. Muscat aromas were not affected. However, no significant differences were found when the wines were left to rest for one month after filtration and then tasted.

5. The first trials of tangential filtration showed that it had a major impact on wine composition, especially the color of red wines. Consequently, a drop in quality was noted. Currently available membranes do not have such a harmful effect on wine composition. It is, however, still true that this technique must be used with great care, and ongoing quality control is essential.

6. It is important not to filter wines too many times as each operation can have a detrimental effect. Each wine must be clarified by a well-defined process, keeping treatment to a minimum.

11.10.3 Comparison of the Effects of Fining and Filtration

One clear advantage of filtration over fining is the speed of clarification. Clarity is immediate, even in a turbid wine, provided, of course, that clogging is not excessive. Fining, however, leads to greater stability as it affects unstable colloids. These may still be in solution after the wine is clarified, but are likely to flocculate later, causing turbidity that will lead to the formation of a deposit. Fining is particularly effective at eliminating colloidal coloring matter from red wine and preventing ferric casse.

In practice, when preparing wine for bottling, these two techniques are by no means mutually exclusive and may, if necessary, be used one after the other. Fining prior to filtration improves filter throughput by flocculating the particles in suspension so that they cause less clogging. Filtration also traps yeasts and bacteria more efficiently when the wine has previously been fined.

It is possible to fine very turbid young wines sooner if they have been filtered, even coarsely. Fining agents are more effective when some of the mucilage and matter in suspension has already been eliminated by filtration.

Of course, fining conditions affect a wine's composition even more than filtration. Fining red wines with protein fining agents or bentonite reduces their color even more than filtration and is more likely to make them seem thinner.

11.10.4 Filtration Prior to Bottling Fine Wines

Fine red wines should not be bottled without filtration unless the necessary precautions are taken. Some wines are still not completely clear after 18–24 months' barrel-aging, especially if they have not been fined. If these wines are bottled without filtration, a sediment of unstable phenolic compounds and, even more importantly, microorganisms, may form on the glass. In some cases, this leads to the development of off-odors. While the presence of acetic and lactic bacteria in the genus *Oenococcus* does not present a real threat to the wine's development, lactic bacteria in the genus *Pediococcus* and yeasts in the genus *Dekkera* (*Brettanomyces*) are much more dangerous (Millet, 2001). Filtration is generally advisable in these cases, depending on residual population levels and the physiological condition of the microorganism cells. If filtration is properly controlled, it should not affect the wine's tasting characteristics (Section 11.10.2)–unsatisfactory results are usually due to poor operating conditions.

It is difficult to envisage bottling great white wines without filtration, as any problem with clarity is immediately obvious. In addition, there is a risk of malolactic fermentation in bottle in wines containing malic acid.

11.11 CENTRIFUGATION

11.11.1 Centrifugal Force

Matter in suspension in wine may be naturally separated out by sedimentation, at a speed proportional to the squared diameter of the particles and the difference between their density and that of the liquid. This speed is also inversely proportional to the viscosity of the medium. Particle sedimentation is also subject to the g factor: acceleration due to the earth's gravitational field.

The aim of centrifugation is to accelerate settling of the sediment by rotating it very fast around an axis. The sediment moves away from the axis due to centrifugal force and, at the same time, the gravitational force is multiplied by a considerable factor, proportional to the speed of rotation squared. The acceleration factor is defined as follows:

$$f = \frac{r\omega^2}{g}$$

where r = particle radius, ω = centrifuge rotation speed, g = acceleration due to gravity (9.81 m/s^2).

A particle revolving in a centrifuge at 4000–5000 rpm is subjected to a force several thousand times greater than g. Particle separation is furthermore accelerated by the small distance the sediment has to fall (a few millimeters), as compared to the large distances (several meters) in other wine containers.

The volume of liquid treated is restricted by the capacity of the system, but this limitation is overcome by using continuous centrifuges. The turbid liquid is fed into the centrifuge and its impurities are removed. The centrifuge is only stopped for removal of the sediment and cleaning when the sludge chamber is full.

Particle sedimentation is subject to forces resulting from rotation and the speed of the liquid to be clarified, i.e. its flow rate. In order to operate with a high throughput, systems must have a large separation surface and small sedimentation height. For this reason, centrifuges are partitioned bowls or plates set a few millimeters apart. Separation operations that would have required several days, or even weeks, by spontaneous sedimentation in tall containers take only a few seconds.

11.11.2 Industrial Centrifuges

Centrifuges used to clarify wine are plate separators (Figure 11.15). Inside the bowl, a pile of truncated cones, known as 'plates', divide the liquid into a large number of thin layers. This decreases the distance over which the solid particles are separated and accelerates clarification. The liquid to be treated is fed into the center of the bowl and directed towards the periphery. The wine then moves upwards through the spaces between the plates, from the outside towards the center of the bowl. The particles are separated out under the influence of centrifugal force and collected on the underside of the upper plate (Figure 11.16). The clarified liquid outlet is at the top of the bowl. The sediment slips along the plates and is collected in the 'sludge chamber' on the outside of the bowl. Sediment may be evacuated continuously through outlets in the bowl. In most systems, sludge is removed at intervals. The feed is cut off and the bowl opened for cleaning by an automated system. This may be controlled in one of three ways: by a solenoid valve connected to an automatic timer operating at fixed intervals, by a nephelometer monitoring clarity at the outlet or by a mechanism that detects clogging in the bowl. The sludge is flushed out with pressurized water or

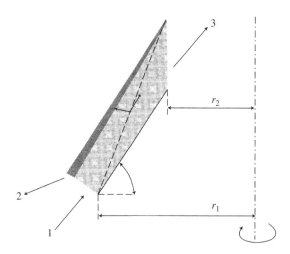

Fig. 11.16. Section diagram of a centrifuge plate: 1, inlet of liquid to be clarified; 2, sediment outlet; 3, clarified liquid outlet

compressed air, which avoids mixing water with the wine and results in a dry by-product that causes less pollution.

Standard centrifuges, with rotation speeds between 5000 and 10 000 rpm, have throughputs between 10 and 200 hl/h (up to 300 hl/h). High-performance centrifuges, with rotation speeds of 15 000–20 000 rpm have a high g factor (14 000–15 000) and are capable of eliminating the lightest particles (bacteria).

11.11.3 Using Centrifugation to Treat Wine

Centrifuges are universal clarification systems that may be used for must and wine at various stages in winemaking. They are mainly installed in large wineries, in view of their major capital investment cost. Most centrifuges used in winemaking are plate separators, with regular, automatic sludge removal.

This technique is particularly efficient when filtration may not be used directly, especially in white winemaking. It is a rapid method for obtaining wines that are clean, stable and ready to drink, without adding excessive amounts of sulfur dioxide. It also minimizes losses of lees wine, which

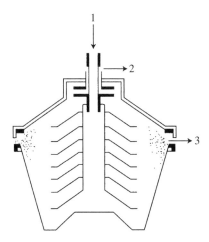

Fig. 11.15. Diagram of a continuous centrifuge with automatically opening bowl for regular removal of the lees; 1, feed; 2, clarified liquid outlet; 3, sediment outlet

is always difficult to process without polluting the environment. Centrifuges must operate under sufficiently air-free conditions to avoid excessive oxidation.

The following are a few applications for centrifuges in white winemaking:

1. Clarifying must after pressing: this is quite effective if the must does not have too high a solid particle content. It may be preferable to centrifuge only the deposits produced by static settling.

2. Clarification during fermentation: by repeating this operation several times, it is possible to stabilize wine permanently through the gradual elimination of yeast and nitrogenated nutrients.

3. Clarification of new white wines at the end of fermentation: this operation is particularly useful for eliminating yeast after brandy has been added to fortified wines. Another desirable effect is a decrease in sulfur dioxide combinations. Furthermore, early centrifugation facilitates later filtration. Centrifugation eliminates yeast as efficiently as filtration, achieving over 99% removal, even at high flow rates. High-performance centrifuges are, however, necessary to achieve a good level of clarity and maximize the elimination of bacteria.

4. Clarifying new red wines just before they are run off into barrels.

5. Clarifying wines after fining: this makes wine perfectly clear in one or two operations, whereas natural sedimentation may take three or four weeks. Fining lees may also be centrifuged.

6. Facilitating tartrate precipitation: simply centrifuging a wine may cause the precipitation of potassium hydrogen tartrate. This may be due to the elimination of protective colloids or the effects of violent agitation. Furthermore, centrifugation has been suggested as a technique for eliminating tartrate crystals after cold stabilization, especially in the contact process, which involves large quantities of small crystals. In view of the abrasiveness of these crystals, it may not be possible to envisage eliminating all of them by centrifugation. The bulk of the crystals are usually removed using a hydrocyclone separator and clarification is completed in a standard centrifuge.

REFERENCES

Donèche B. (1994) *Les Acquisitions Récentes dans les Traitements Physiques du Vin*. Tec. et Doc., Lavoisier, Paris.

Dubourdieu D. (1982) Recherches sur les polysaccharides secrétés par *Botrytis cinerea* dans la baie de raisin. Thèse Doctorat Université de Bordeaux II.

Gaillard M. (1984) *Vigne et Vin*, 362, 22.

Gautier B. (1984) *Aspects Pratiques de la Filtration des Vins*. Bourgogne-Publication, La Chapelle de Guinchay.

Guimberteau G. (1993) La clarification des moûts et des vins. *J. Int. Sci. Vigne et Vin*, hors série.

Lafon-Lafourcade S. and Joyeux A. (1979) *Conn. Vigne Vin*, 13 (4), 295.

Mietton-Peuchot M. (1984) Contribution à l'étude de la microfiltration tangentielle. Application à la filtration des boissons. Thèse Docteur Ingénieur, Institut National Polytechnique, Toulouse.

Millet V. (2001) Dynamique et survie des populations bactériennes dans les vins rouges au cours de l'élevage: interactions et équilibres. Thèse Doctorat, Université Victor Segalen Bordeaux 2.

Millet V. and Lonvaud-Funel A. (2000) *Lett. Appl. Microbiol.*, 30, 136.

Molina R. (1992) *Tecnicas de Filtraction en la Enologia*. A. Madrid Vicente Ediciones, Espagne.

Paetzold M. (1993) La clarification des moûts et des vins (ed. G. Guimberteau). *J. Int. Sci. Vigne Vin*, hors série, Bordeaux.

Ribéreau-Gayon J., Peynaud E., Ribéreau-Gayon P. and Sudraud P. (1977) *Sciences et Technique du Vin*, Vol. IV: *Clarification et Stabilization. Matériels et Installations*. Dunod, Paris.

Serrano M. (1981) Etude théorique de la filtration des vins sur plaques. Thèse Doctorat, Université de Bordeaux II.

Serrano M. (1984) *Conn. Vigne Vin*, 18 (2), 127.
Serrano M. (1993) La clarification des moûts et des vins (ed. G. Guimberteau). *J. Int. Sci. Vigne Vin*, hors série, Bordeaux.
Serrano M. (1994) *Les Acquisitions Récentes dans les Traitements Physiques du Vin* (ed. B. Donèche). Tec. et Doc., Lavoisier, Paris.
Serrano M. and Paetzold M. (1994) *Les Acquisitions Récentes dans les Traitements Physiques du Vin* (ed. B. Donèche). Tec. et Doc., Lavoisier, Paris.
Serrano M. and Ribéreau-Gayon P. (1991) *J. Int. Sci. Vigne Vin*, 25 (4), 229.
Serrano M., Pontens B. and Ribéreau-Gayon P. (1992) *J. Int. Sci. Vigne Vin*, 26 (2), 97.

12

Stabilizing Wine by Physical and Physico-chemical Processes

12.1 Introduction	369
12.2 Heat stabilization	370
12.3 Cold-stabilization treatment	373
12.4 Ion exchangers	376
12.5 Electrodialysis applications in winemaking	382

12.1 INTRODUCTION

Physical processes, mainly heating and cooling, have been used to treat wines for a long time. The oldest technique is certainly the use of heat to destroy microorganisms (pasteurization). Other effects of heat were also discovered many years ago, e.g. the fact that it stabilizes white wines and prevents certain types of colloidal precipitation.

The use of heat to accelerate maceration phenomena is described elsewhere (Volume 1, Section 12.8.2). Cold stabilization is also effective for eliminating insoluble compounds such as tartrates (Sections 1.7.2–1.7.4) or colloidal coloring matter in red wines (Section 6.8). Cooling is also widely used to control fermentation (Volume 1, Section 3.7.1). The practical benefits of both of these techniques were well known before any theoretical studies had been carried out to identify the mechanisms involved and define the optimum conditions for their implementation. This has, however, resulted in a somewhat empirical and approximate approach to their utilization.

In the past (Singleton, 1962), other physical processes have been suggested for treating wines: infrared, ultraviolet and ionizing radiation, as well as various types of electrical currents. In fact, the objective was mainly to combine these treatments with heating and cooling to accelerate aging, or even to improve the wine. The overall results were relatively disappointing, and any improvement in quality was highly debatable. These techniques have not led to any significant developments and have now been abandoned. It is unnecessary to describe them any further.

There is generally, however, a positive attitude toward physical treatments. They are theoretically more natural than chemical treatments and less likely to cause unacceptable modifications to a wine's chemical composition. It is, however, just as true that the effects of cold, and particularly heat, may cause appreciable changes, especially in colloidal structure.

In fact, physical and chemical treatments are complementary. It is true that physical processes are less likely to be harmful than chemical treatments. It is also true that, at least in some cases, chemical treatments are more effective and less expensive. Properly controlled and appropriately used, chemical treatments do not affect quality.

Refrigeration is widely used for stabilization to prevent tartrate precipitation. This treatment alone may be adequate to ensure the stability of red wines. However, if bentonite is not used to treat white wines, prior heating is necessary to prevent protein precipitation. In some countries, equipment capable of applying both heat and cold is used. These processes give satisfactory results in stabilizing white wine with a low iron content, as they have only a limited effect on ferric casse.

This chapter also describes several physico-chemical treatments based on electrical charges in solutions. Ion exchange and electrodialysis are mainly used to prevent tartrate precipitation. These techniques are much more controversial than purely physical methods. If they are not properly used, they may produce unacceptable changes in a wine's chemical composition. This is why they are not legally permitted in all wine growing countries. Their utilization must be carefully controlled by legislation.

Other physical processes consist of concentrating must by eliminating water: vacuum evaporation, reverse osmosis and partial freezing. The first two are used to concentrate must as an alternative to chaptalization (Volume 1, Section 11.5.1). Reverse osmosis, together with ion exchange, have been suggested for eliminating excess acetic acid (Section 12.4.2).

The use of inert gases to protect wines in partly empty containers by preventing oxidation and aerobic microorganism activity may also be considered a physical stabilization process. This technique is described elsewhere (Volume 1, Section 9.6.1) as a complement to the use of sulfur dioxide.

Another physical process currently under study is the use of high pressure to destroy microorganisms. It is certainly possible to sterilize wine by this method, but the difficulty is then to keep it sterile. The same reservations apply to the use of heat (Section 12.2.3).

12.2 HEAT STABILIZATION

12.2.1 Preventing Certain Types of Colloidal Precipitation: Protein Casse and Copper Casse

Heating to high temperatures denatures the unstable proteins in white wines and accelerates flocculation during the cold treatment that follows. The protein deposit is then eliminated by fining. Most white wines will remain stable in the bottle at normal storage temperatures (6–24°C), and even at higher temperatures (30°C) if they are heated to 75°C for 10 min, then fined and filtered.

The same treatment also provides protection from copper casse, by reducing the Cu in the form of colloidal copper sulfide. It is then eliminated by fining and filtration. More intense heating is required, e.g. 75°C for 2 hours, for wines containing 1.5 g/l of copper.

There is certainly a degree of concern that so much heating affects a wine's organoleptic qualities, even if all necessary precautions are taken to avoid oxidation. Young wines stand heating better than older ones. Sweet wines are the most delicate, as sugars, especially fructose, are heat-sensitive and develop caramelized flavors.

General use of bentonite for protein stabilization of white wines has made heat treatment rather rare. It is mainly reserved for those few wines that have such a high protein content that excessive doses of bentonite would be needed to treat them effectively.

12.2.2 Impact of Heating on Physico-chemical Stabilization

One effect of heating is to dissolve crystallization nuclei, which are necessary for crystals to grow and precipitate. New wine is a supersaturated tartrate solution. The precipitation of tartrates, however, requires the presence of submicroscopic nuclei that are the starting point from which molecules build up into crystals (Section 1.5.1). Heating wine, especially when it is already in the bottle, may be sufficient to stabilize it and prevent tartrate precipitation.

The formation of protective colloids is another consequence of heating that deserves further investigation (Ribéreau-Gayon *et al.*, 1977). Red and white wines that have been heated and re-cooled generally have properties similar to those produced by adding a protective colloid, such as gum arabic. These effects include slower sedimentation of particles in suspension, protection from copper casse and flocculation problems when gelatin is used for fining. These properties start to appear at relatively low temperatures (40–45°C) and increase, within certain limits, according to exposure time and temperature. The existence of a colloidal mechanism has been confirmed by comparing the behavior of a white wine heated to 70°C for 30 min, before and after ultrafiltration. When the wine has not been filtered, clarification by fining is difficult after heating (due to the presence of protective colloids formed by heating). Fining is effective if the wine is ultrafiltered before heating, as the protective colloids that would have been formed by heating had already been eliminated.

These factors must be taken into account in the use of heat treatments. A clear wine is more stable after heating, as it is following the addition of gum arabic (Section 9.4.3). If, however, there is any turbidity in the wine, clarification (by settling, fining or filtration) will be more difficult after heating.

Prolonged heat treatment also causes changes in color and flavor similar to certain aspects of aging.

If the wine is heated in the presence of air, there is a 'maderization' effect, with the formation of aldehydes, acetals and other aromatic compounds, giving a 'rancio' character. These are standard practices for certain wines, mainly *vins de liqueur* (sweet fortified wines), including Madeira, Port and French *vins doux naturels*. Heating in these instances may be very intense, e.g. 60 days at 60°C.

It used to be thought that subjecting wines to heating in the absence of air (e.g. a few weeks at 40–45°C) would accelerate aging and improve quality. This practice has almost been abandoned, as the effects did not really correspond to those of aging and there was no obvious improvement in flavor. However, aging bottles of good red wine at 18–20°C for two months causes the wine to develop a balanced flavor that would only be achieved after a year or two at lower temperatures. The positive effect is more marked in wines with good intrinsic quality.

Heating also destroys enzymes and, consequently, inhibits the reactions that they catalyze. This is particularly true of oxidation enzymes (tyrosinase and laccase) that are destroyed by heating to 60–70°C for just a few minutes. This technique is effective for treating must (Volume 1, Section 13.4.3). In new wines, heating would no doubt enhance the effect of sulfuring, but it is rarely used. Inhibition of pectolytic enzymes at high temperatures during fermentation (thermovinification) (Volume 1, Section 12.8.3) should also be taken into account, as it leads to difficulty in clarifying new wines.

12.2.3 Biological Stabilization

Appert was the first to state that it would be possible to preserve wine, like other foods, by applying heat. He observed that, although unheated control wines became spoiled, heated wines were unaffected. Although neither wine spoiled, both were very similar in flavor. Pasteur demonstrated that the effects of heating observed by Appert were due to the destruction of microbes. He introduced the concept of 'pasteurization'.

Initially, pasteurization was used to protect wine from the microbial spoilage caused by acetic and lactic bacteria ('tourne', 'amertume' and 'graisse'). In more recent years, heating has mainly been used to kill yeast, to stabilize sweet wines containing residual sugar. Microbial spoilage can now be avoided by other means, based mainly on careful fermentation management, the use of sulfur dioxide and the reduction of contaminant populations by various clarification processes.

The theory of the destruction of yeast by heating has been described elsewhere (Volume 1, Section 9.4.2). Thanks to the alcohol content and low pH of wine, as well as the presence of sulfur dioxide, it is fairly easy to achieve absolute sterility by relatively limited heating (60°C for 30 s). However, the difficulty of keeping the treated wine under sterile conditions in a normal winery environment is well known. The risk of recontamination after wines have been treated in the vat explains why pasteurization has not been more generally used for sweet wines. The applications that have been reported (Volume 1, Section 9.4.3) require technical conditions that are not always easy to apply in a conventional winery. It is, however, at this stage that heat could make the greatest contribution to stabilizing wine. Of course, wine may be pasteurized in the bottle or just before bottling, but other stabilization techniques, especially sulfuring and sterile filtration prior to bottling, are easier to use.

12.2.4 Practical Implementation of Heat Treatment

The temperature and heating time required depend on the aim of the treatment. Heating is normally used for biological stabilization. The wine's composition (alcohol content, pH and SO_2 content) must be taken into account. As combined forms of SO_2 break down when the temperature rises, its antiseptic properties are enhanced.

Wine may be heated in bulk, before bottling, as recommended by Pasteur. Dimpled-plate heat exchangers may be used to heat, cool, or pasteurize wine. Wine and water circulate in opposite directions on either side of the plates. The space between the plates is too narrow to facilitate heat transfer from the water to the wine. Water at 60–65°C is used to pasteurize wine. The respective throughput rates of water and wine control the treatment time required, i.e. approximately one minute. The pasteurized wine is cooled by passing it through another heat exchanger, where wine being prepared for pasteurization acts as a coolant. Systems using infrared radiation to heat wine have also been designed. Contrary to certain assertions, this radiation does not have any chemical effect, but ensures even penetration of heat into the wine, which circulates through transparent tubes.

Another possibility is flash pasteurization. This consists of heating wine to 90°C for a few seconds and then cooling it rapidly in a high-performance plate heat exchanger. It is considered that this high-speed process is less likely to affect the wine's organoleptic characteristics.

For the reasons described above, other pasteurization techniques have focused on bottled wines, heating them to 60°C for a few minutes. This guarantees that all of the germs are destroyed and prevents later contamination. Furthermore, this process maintains wine quality and allows it to develop properly during aging. However, it is not as widely used for wine as beer. Wine has better natural stability than beer. Corks must be reinforced so that they are not partially expelled from the bottle during heating and a slight ullage develops when the wine cools. There is also a risk of softening the natural cork, which may no longer provide a sufficiently tight seal.

These objections have led to the development of high-temperature bottling. This is now the main heat treatment technique used for wine. It is currently becoming more widely used on an industrial scale. The principle consists of heating wine to the relatively moderate temperature required to destroy yeast (45–50°C, depending on the alcohol content and the possible presence of sugar). The hot wine is transferred directly into bottles, sterilizing both glass and cork as it cools. High-temperature bottling requires a steam generator and a heat exchanger, in which the steam and the wine to be treated circulate in opposite directions. This equipment is easily installed on a standard bottling line. High-temperature bottling is becoming widespread in certain countries, as the effectiveness of this process and the conditions for its use become increasingly well known.

In view of the risk of even slight organoleptic changes, this technique is more suitable for medium-quality wines that are to be drunk young than for fine wines with aging potential. It is particularly useful in stabilizing and protecting sweet white wines from accidental fermentation without using high doses of sulfur dioxide. In red wines, it also prevents the development of mycodermic yeast which may produce 'flor' on the wine's surface if bottles are stored upright. It also stops the development of lactic bacteria and is recommended for bottling red wines containing malic acid.

Of course, this treatment must only be used for wines that have been stabilized in terms of colloidal turbidity, especially protein and copper casse, as these problems would otherwise be likely to be triggered by heating. Precautions must also be taken to avoid excessive oxidation, especially as there is likely to be a variable amount of ullage between the cork and the wine once it has cooled, even if the bottles were initially completely filled. High-temperature bottling may now be replaced by fine filtration processes (sterilizing plates, membranes) that achieve absolute sterility at cool temperatures, provided that perfect hygiene is maintained throughout the bottling system (Section 11.7.4).

12.3 COLD-STABILIZATION TREATMENT

12.3.1 Aim of the Operation

This chapter discusses applications involving chilling wine for the purpose of stabilization and preventing precipitation. The technique of cooling during fermentation is described elsewhere (Volume 1, Section 3.7.1), but the same refrigeration equipment may be used for both purposes.

The positive effect of natural cold on new wines has been known for many years, and winemakers have long taken advantage of low winter temperatures. In order to enhance this effect, wine was then subjected to temperatures below 0°C, close to its freezing point. It was maintained at this temperature for a period of time and then clarified by filtration to eliminate the precipitate. This technique is effective in purifying new wines, as well as stabilizing color and clarity, particularly in red wines and *vins de liqueur* (sweet fortified wines) that are bottled young. Cold stabilization is also used for sparkling wines and brandies.

Any improvement in wines thus treated may not seem so obvious after a few months and, even more so, after a few years of aging. Some wines may also seem to lose a great deal of body, aroma and flavor after this operation. Fortunately, the high cost of purchasing and operating cold stabilization equipment tends to discourage its overuse.

Cold stabilization is mainly used to cause two types of precipitation that help to stabilize wine:

(a) tartrate crystals;
(b) colloidal substances: unstable coloring matter and ferric complexes with phenols in red wines and ferric phosphate and proteins in white wines.

Cold stabilization is mainly used to prevent tartrate precipitation. As there are effective, less expensive treatments are available for other problems.

Cold temperatures are not effective for treating microbial problems. The development of microorganisms slows down at low temperatures, but they become fully active again when the temperature rises.

12.3.2 Preventing Crystal Precipitation

The mechanisms behind this precipitation and ways of forecasting instability in wine are described elsewhere (Sections 1.5 and 1.6). Tartrate solubility is reduced by the presence of ethanol, but precipitation is partially inhibited by colloidal substances that coat the crystal nuclei and prevent them from growing. For this reason, wines, particularly reds, are likely to produce crystal deposits several months after fermentation.

Potassium bitartrate is strongly insolubilized at low temperatures. There is no further risk of precipitation after treatment, provided that the wine is not cooled to a lower temperature than that of the treatment and the colloidal structure is not greatly changed. All the excess calcium tartrate, however, is not always eliminated by this method. In some cases, precipitation of this salt may even be promoted in cold-stabilized wine.

Cold stabilization is not always totally effective and the following additional treatments may be used to ensure complete stability: metatartaric acid (Section 1.7.6), mannoproteins (Section 1.7.7), carboxymethylcellulose (Section 1.7.8), ion exchangers (Section 12.4) and electrodialysis (Section 12.5).

12.3.3 Preventing Colloidal Precipitation

It is well known that some of the coloring matter in red wines is colloidal. It is soluble at normal temperatures but precipitates at low temperatures (0°C), causing turbidity in the wine. This colloidal coloring matter gradually becomes less soluble throughout the winter, due to the drop in temperature. It settles out to form part of the lees of young wines.

The standard technique of fining red wines (using egg albumin, gelatin or bentonite) before bottling is aimed at eliminating this colloidal coloring matter. Cold stabilization has exactly the same effect. It is well established, however, that colloidal coloring matter will form again spontaneously. Stabilization is temporary, but it guarantees clarity for a few months or years. In the long term, a deposit may appear in old wines in the bottle, but is then considered acceptable.

Cold stabilization is also partially effective in preventing other types of colloidal precipitation. It helps to prevent ferric casse by insolubilizing ferric phosphate in white wines and ferric tannate in reds. However, even after aeration to promote the formation of the Fe^{3+} ions involved in these mechanisms, only small quantities of iron are eliminated. Fining at the same time as cold stabilization improves treatment effectiveness but is never sufficient to prevent ferric casse completely.

The situation concerning the flocculation of proteins is similar. They are partially eliminated, but not sufficiently to ensure total stabilization.

The impact of cold stabilization on the elimination of colloids is clearly demonstrated by the improvement in filtration flow rates for certain wines with a high fouling capacity.

12.3.4 Cold Stabilization Procedures

The various processes using cold temperatures to prevent tartrate precipitation have been described elsewhere (Section 1.7.1 to 1.7.5). There are three major procedures:

(a) slow stabilization, without tartrate crystal seeding;

(b) rapid stabilization, involving static contact with seeded crystals;

(c) rapid dynamic contact stabilization.

When cold treatment is used to clarify new wines or prevent colloidal precipitation, the installation in Figure 12.1 is most appropriate. It may also be used for tartrate stabilization without contact (Section 1.7.2). The process involves:

(a) cooling the wine to a temperature close to its freezing point;

(b) keeping it at the same temperature for several days;

(c) filtering the wine at low temperatures.

It is advisable to eliminate at least part of the wine colloids beforehand by centrifugation or filtration, as this enhances precipitation. It has been observed that slow, gradual cooling encourages the formation of large bitartrate crystals. These are easy to remove by filtration, but precipitation is incomplete. Rapid, sudden cooling causes total precipitation, but the tiny crystals are difficult to eliminate and dissolve rapidly if the temperature rises. The quantity of salt precipitated may vary by a factor of two according to the cooling regime. It is recommended to stir the wine throughout the cooling process to facilitate agglomeration of the precipitate.

The time the wine needs to be kept cold depends on the type of wine (precipitation is slower in

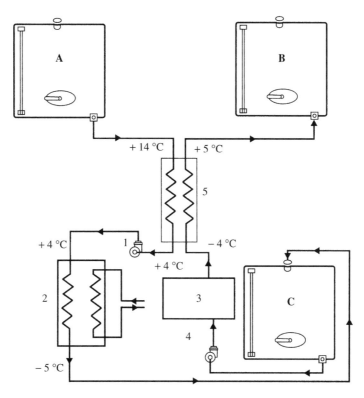

Fig. 12.1. Schematic diagram of a cold-stabilization installation: A, untreated wine (+14°C); B, treated wine (+5°C); C, wine during stabilization (−5°C); 1, untreated wine pump; 2, treating wine at −5°C (refrigeration system and plate heat exchanger); 3, filter at the end of cold treatment; 4, pump for cold-stabilized wine, ready to be filtered 5, heat exchanger for precooling wine to be treated by using it to warm treated wine

red than in white wines) and the purpose of the treatment. In some cases, 7 or 8 days are considered sufficient, but other authors recommend 15 or even 30 days. According to Ribéreau-Gayon *et al.* (1977), 1 or 2 days is enough to eliminate colloidal coloring matter from red wines, but 5 or 6 days are not always sufficient to prevent tartrate precipitation. This takes 10 to 15 days. In any case, if wine is to be kept cold for several days, it should be taken into account that it will inevitably become warmer at some stage and may, therefore, need cooling again.

The time required to achieve stabilization may be reduced by adding 30–40 g/hl of small tartrate crystals and agitating for 36 hours (Section 1.7.2). Stabilization of white wines was obtained in 62 hours by this method, instead of 6 days using the standard process. This is not quite the same as the contact process, which stabilizes wine in just a few hours but requires the addition of 300–400 g/hl of tartrate crystals.

There are several rapid cold stabilization processes for precipitating tartrates (static and dynamic contact processes). These techniques have certain economic advantages, but are not easy to implement. They are described elsewhere (Sections 1.7.3 and 1.7.4).

Furthermore, wine must remain at low temperatures during filtration until the sediment has been removed, to avoid the crystals dissolving again. Heat-insulated filters are used and a heat exchanger at the filter outlet contributes to the cooling process.

Care must be taken to protect wine from oxidation during cold stabilization, as more oxygen can be dissolved at lower temperatures (approximately

11 mg/l at 0°C and 8 mg/l at 15°C). It should also be taken into account that oxidation reactions take place more slowly at low temperatures. The risk of oxidation increases during agitation although this is recommended to maintain the tartrate crystals in suspension.

Refrigerating the environment is another way of utilizing cold to stabilize wine. Winter cold may be used for this purpose, as well as specialized vat room installations where a chilled atmosphere makes it possible to keep wine at a temperature in the vicinity of 0°C for one or two months. The wine must previously have been filtered. Vats may be equipped with individual exchangers to accelerate cooling. This type of installation is highly effective. Furthermore, it minimizes the risk of oxidation, as the wine does not need to be handled. This system may also be used for fermenting white wines, maintaining a fermentation temperature of around 18–20°C, or even lower.

12.4 ION EXCHANGERS

12.4.1 Operating Ion Exchangers

Ion exchange reactions are carried out using insoluble polymer resins, activated with various functional groups. The polymerized material is usually based on a mixture of styrene and vinyl benzene (Figure 12.2). The active radical of cation exchangers is generally sulfonic acid ($-SO_3H$), but carboxylic acid may also be used. The functional radical in anion exchangers consists of a quaternary ammonium or tertiary amine salt.

Figure 12.3 clearly shows both possible types of exchanger reactions: cation exchange, which may be described as acidification if the resin releases H^+ ions, and anion exchange, which may lead to deacidification if the resin releases OH^- ions. Exchanges may also occur between cations and anions other than H^+ and OH^-, in which case they do not alter the pH.

Ion exchange phenomena are stoichiometric, i.e. 37 mg of potassium are exchanged by 23 mg of sodium and 40 mg of calcium by 46 mg of sodium. The ion exchange rate depends on the type of exchanger: grain size, porosity and distensibility.

An exchanger generally has a specific affinity for different ions. This phenomenon is due to many factors, including the polymerized structure of the matrix, the chemical characteristics of the exchanger radicals, the exchange capacity and pH.

In the case of cations, the affinity laws indicate (Ribéreau-Gayon et al., 1977):

1. The ease of exchange increases with the valence of the exchanger ion: $Na^+ < Ca^{2+} < Al^{3+}$. This means that divalent ions in wine, such as calcium and magnesium, are fixed on the resin in preference to monovalent sodium and

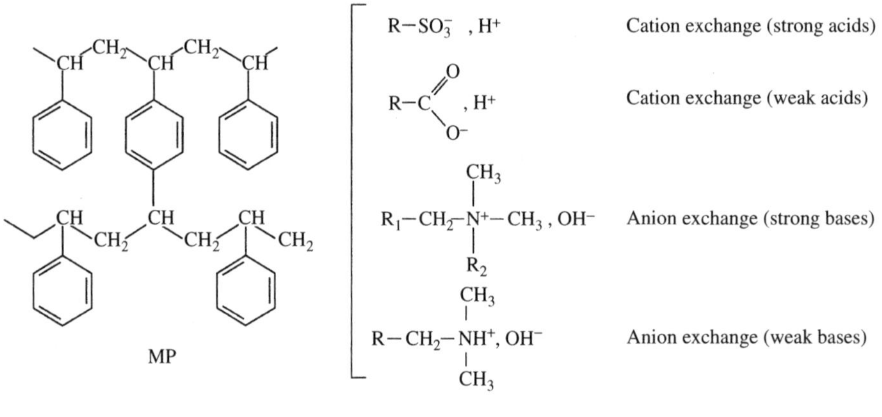

Fig. 12.2. Ion exchange resin composition. Various functional groups are grafted on to a polymerized material (MP, 4 styrene units and 1 vinyl/benzene unit) to facilitate different reactions (Weinand and Dedardel, 1994)

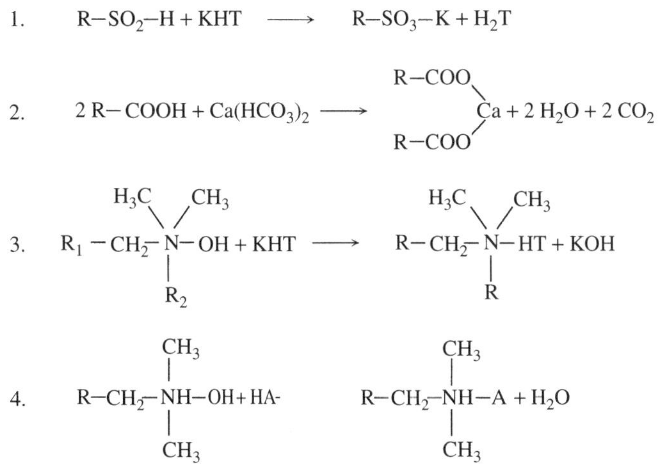

Stabilizing Wine by Physical and Physico-chemical Processes

1. $R-SO_2-H + KHT \longrightarrow R-SO_3-K + H_2T$

2. $2 R-COOH + Ca(HCO_3)_2 \longrightarrow \begin{matrix} R-COO \\ \diagdown \\ Ca + 2 H_2O + 2 CO_2 \\ \diagup \\ R-COO \end{matrix}$

3. $R_1-CH_2-\underset{R_2}{\overset{H_3C\diagdown\diagup CH_3}{N}}-OH + KHT \longrightarrow R-CH_2-\underset{R}{\overset{H_3C\diagdown\diagup CH_3}{N}}-HT + KOH$

4. $R-CH_2-\underset{CH_3}{\overset{CH_3}{NH}}-OH + HA- \longrightarrow R-CH_2-\underset{CH_3}{\overset{CH_3}{NH}}-A + H_2O$

Fig. 12.3. Principles of different ion exchange reactions (Weinand and Dedardel, 1994): 1, cation exchange, e.g. potassium bitartrate (this is an acidification reaction); 2, cation exchange (the reaction does not work with strong acid salts); 3, anion exchange (fixing bitartrate ions is equivalent to deacidification); 4, anion exchange using a tertiary amine (this reaction does not work with strong basic salts)

potassium ions. Ferric iron is fixed before ferrous iron.

2. If two ions have the same valence, the ease of exchange increases with the atomic number. Potassium is fixed in preference to sodium and calcium in preference to magnesium.

3. In the case of heavy metals, present in wine in the form of complexes, the fixing capacity depends on the stability (dissociation constant) of the new complex formed by the heavy metal and the exchanger.

Resins are defined by their exchange capacity, or the total quantity of ions that can be mobilized per unit mass of exchanger. Exchange is expressed in meq/g of cations exchanged or by the weight of the milliequivalent of resin. These calculations are relatively simple in the case of sulfonate cation exchangers, as the exchange reaction is fast. The resin is first regenerated as H^+ and then exposed to excess sodium chloride (NaCl) to neutralize the exchanger's acid functions. The resulting hydrochloric acid is titrated with sodium hydroxide. The cation resins exchange 4–5 meq/g. The weight of the milliequivalent of resin is therefore 200–250 mg.

Resins for use in winemaking must meet several criteria: mechanical strength, total insolubility in wine and the absence of off-flavors. These resins must also be capable of being regenerated many times.

Sulfonic cation exchange resins are totally insoluble, unlike anion exchangers, which may produce organoleptic changes. Anion exchangers may also enhance microbiological stabilization, due to the powerful antiseptic properties of traces of quaternary ammonium salts that are released into the wine. This alone would be sufficient to prohibit anion exchangers from use in treating wine.

In addition to their ion exchange capacity, resins have a microporous structure that gives them absorption properties. This is very useful in the agrifood industry in general, especially in eliminating condensed phenols.

12.4.2 Possible Uses in Winemaking

The first attempts to devise enological applications for ion exchangers date back to the 1950s. These techniques were totally rejected at the time, first by France and then by the EEC, on the recommendation of the OIV (Office International de la Vigne et du Vin).

However, it rapidly became clear that a distinction should be made between:

1. Cation exchangers, which are likely to improve tartrate stability by removing K^+ and Ca^{2+}, acidify wine by adding H^+ and, possibly, prevent ferric casse by reducing Fe^{3+}.

2. Anion exchangers, which make it possible to reduce acidity by adding OH^-, or even to reduce certain specific acids (tartaric or acetic acid). There is, however, a risk of major alterations in flavor and composition.

Some countries, in particular the United States and Australia, have authorized the use of ion exchangers. The recent membership of these two countries in the OIV, as well as the authorization to import treated wines into Europe, has revived this issue.

Cation exchangers are the most widely used in countries where they are permitted, while anion exchangers are also authorized, but very little used. Anion exchangers have mainly been included in double cation/anion exchangers. This technique aims to avoid any significant modification in the pH of the end product, while ensuring tartrate stability by reducing K^+ and Ca^{2+} levels. However, this process is too brutal. It involves excessive, if only temporary, variations in the pH of the wine: the pH drops to 1.8 at the outlet of the cation exchanger and rises to over 7 after it passes through the anion exchanger.

One recently suggested application for anion exchangers is in reducing volatile acidity (Oenovation International Inc., Santa Rosa, California, USA). The wine is treated with reverse osmosis (Volume 1, Section 11.5.1) to remove some of the water, alcohol and acetic acid. The corresponding fraction is passed through an anion exchanger to eliminate the acetic acid and then returned to the remaining wine. This technique is very effective. As only part of the wine passes through the anion exchange resin, no significant organoleptic alterations have been detected. It is, however, highly unlikely that the EEC would ever authorize this process. Not only is the wine subjected to chemical modification during the treatment, but this operation is also capable of bringing a wine that was badly spoiled within legal limits, which is certainly not permitted.

Another application of ion exchangers is in purifying the grape juice used to produce rectified concentrated must.

12.4.3 Using Cation Exchangers to Treat Wine

The main cation resin in winemaking applications is Amberlite IR 120, although IRC 50, Dowex 50 and Duolite C3 are also used. New resins for treating food products and drinking water have recently been developed by ROHM and HAAS, under the IMAC HR brand, while SAC is intended for preparing sweetening products from fruit juice. SAC seems to be beneficial for treating wine to prevent tartrate precipitation and increase acidity. The resins may be directly immersed in the wine, but this technique is only partially effective. It is preferable to use columns (50, 500 and 1000 l) through which the wine percolates.

Hydrogen, sodium and magnesium resins may be used to removed K^+ and prevent tartrate precipitation. These different forms are obtained by the regeneration operation that consists of circulating either an acid or a saline solution (sodium or magnesium chloride) through the column to saturate the sulfonate or acid radicals (Figure 12.2) in the resin with H^+, Na^+ or Mg^{2+} ions.

The hydrogen form replaces almost all the cations in wine with H^+ ions, causing a considerable decrease in pH (from 3.4 to 1.8). Approximately 45 volumes of wine can be treated with one volume of resin in each treatment cycle. The end of the cycle may easily be detected by measuring the pH. This increase in acidity, corresponding to a decrease in pH, may be desirable in certain hot-climate wines with an initial pH above 4. Generally, however, it is preferable to minimize this drop in pH. As explained above (Section 12.4.2), a double cation/anion exchange avoids this alteration in pH, but it is not recommended for other reasons.

In practice, the decrease in pH is limited by applying the treatment to only a fraction of the wine, which is then subjected to a major variation in pH. Mourgues (1993) suggests that the proportion of the total volume to be treated (approximately 20%) should be determined by measuring the decrease in electrical conductivity during a 24 h test at −4°C, seeding the sample with 4 g/l of potassium bitartrate microcrystals. The decrease in conductivity is proportional to the potassium loss. The percentage decrease in conductivity corresponds to the percentage of wine to be treated, in which most of the potassium will be removed.

The figures in Table 12.1 show that, under these conditions of regeneration in an H^+ medium, the decrease in pH is restricted to 0.1 or 0.2 units. The increase in acidity is 0.1–0.4 g/l (H_2SO_4). Sodium levels vary little, with a slight tendency to decrease. These variations may be considered acceptable, and the decrease in potassium (10–20%) is sufficient to ensure stability.

It should be noted that adding tartaric acid to wine, followed by cold stabilization, also reduces the potassium level by insolubilizing the bitartrate. This process causes a less marked decrease in pH.

The use of cation resins such as Na^+ has been envisaged to avoid acidification and the corresponding variations in pH. The potassium in the wine is replaced by sodium from the resin. The exchange process is a little more complex, as the resin has a greater affinity for divalent

Table 12.1. Characteristics of wine treated with hydrogen and sodium cation exchange resins (Mourgues, 1993)

Type of regeneration	Type of wine and percentage of wine treated on resin	pH	Total acidity (g/l of H_2SO_4)	K (meq/l)	Na (meq/l)
H^+	Carignan				
	0	3.60	4.17	1380	30
	10	3.65	4.30	1290	30
	Red wine				
	0	3.70	4.05	1530	25
	15	3.55	4.44	1290	20
	White wine				
	0	3.57	4.05	1210	35
	18	3.36	4.42	1000	30
	Red wine				
	0	3.60	3.05	930	60
	10	3.52	3.14	820	52
	Carignan				
	0	3.68	3.23	1290	26
	20	3.42	3.68	1036	23
Na^+	White wine				
	0	3.31	4.12	875	85
	10	3.32	4.05	800	130
	Red wine 1				
	0	3.48	3.53	1060	25
	15	3.50	3.46	905	175
	Red wine 2				
	0	3.51	3.66	1035	15
	10	3.50	3.59	945	110
	Red wine 3				
	0	3.49	3.85	1135	15
	15	3.48	3.80	970	170

cations (Ca^{2+} and Mg^{2+}), which are eliminated before the potassium. At the end of cycle, when all the sodium in the resin has been exchanged, the potassium fixed on the resin may be replaced by other cations. If the treatment is not stopped at the right time, the wine's potassium content may actually increase. The potassium must be assayed by flame photometry to determine the end of the treatment cycle (Mourgues, 1993).

This technique can only be used for part of the wine to be treated, due to the increase in Na^+ content. Under these conditions, the potassium decreases by only 10–15%, but there is no change in acidity. There is, however, a relatively large increase in the sodium content (Table 12.1). The maximum permitted value in the USA is 200 mg/l, but it is possible to maintain this value below 150 mg/l.

It is possible to regenerate the resin with a hydrogen/sodium mixture, consisting of hydrochloric acid and sodium chloride, provided that the wine requires acidification.

In both of the preceding instances, it is recommended that the cation exchange treatment is applied to approximately 10–20% of the total volume of the wine. The treated wine is then mixed into the rest. It is also advisable to cold-stabilize the wine prior to ion exchange treatment to enhance protection from tartrate precipitation.

In certain countries (such as Australia), ion exchange treatment is applied to grape juice (30–40% of the total volume). A mixed hydrogen/sodium treatment has the advantage of reducing pH.

Applied under the conditions described above, ion exchange treatment does not absorb phenols and nor does it affect color (OD at 520 and 420 nm).

Magnesium regeneration of cation resins has been investigated, with the aim of avoiding significant acidification or an excessive addition of Na^+ (Ribéreau-Gayon *et al.*, 1977). The results were interesting, but this process does not seem to have been incorporated in any practical applications. The advantages of this approach were recently reported in another article (Weinand and Dedardel, 1994). Magnesium exchange is selective. It does not affect the Na^+ content, while it eliminates K^+ and Ca^{2+}. Another advantage is that, compared to sodium, only half of the amount of magnesium is required to remove the same quantity of potassium, due to the equivalent weight (12 for magnesium and 23 for sodium). For example, 78 mg of Mg^{2+} are required to remove 150 mg of K^+ and 50 mg of Ca^{2+} from a liter of wine. It takes 150 mg of Na^+ to obtain the same result, plus the quantity necessary to remove all the natural Mg^{2+} from the wine, as this is exchanged first. It would therefore be necessary to add a total of approximately 150 mg of Na^+. It is obvious that sodium exchange can only be applied to a fraction of the wine.

The figures in Table 12.2 compare the effects of a sodium and a magnesium cycle (Ribéreau-Gayon *et al.*, 1977). The same authors recommend using a long magnesium cycle, i.e. treating a large volume of wine (200 volumes of wine per volume of resin) to ensure proper fixing of K^+ and Ca^{2+} without any appreciable change in the wine's composition. It is thus possible to reduce the K^+ concentration by 10–20% and the Ca^{2+} content by 25–30%. This is generally sufficient to ensure tartrate stability as the sodium content does not increase. If the wine is to be slightly acidified at the same time, mixed hydrogen/magnesium regeneration may be used. The subsequent increase in the wine's magnesium content should not cause any problems. Magnesium is an essential

Table 12.2. Comparison of different ion exchange regeneration methods (Ribéreau-Gayon *et al.*, 1977)

	Control	Sodium cycle	Magnesium cycle
K^+	676	656	598
Na^+	39	225	39
Mg^{2+}	85	30	137
Ca^{2+}	66	20	20
Fe^{2+}	9	4	5

Cation content (mg/l) of a red wine before and after treatment on Amberlite IR 120 regenerated in different ways. The volume of wine treated was 220 times that of the exchanger.

component of the chlorophyll naturally present in wine. Its salts are soluble and stable.

12.4.4 Practical Implementation of Ion Exchange Resins

The first operation consists of washing the column from bottom to top with water. Regeneration is then carried out from top to bottom, with approximately 10 times the volume of the resin, using either:

(a) a 2–4% H_2SO_4 or 2–10% HCl solution (acid cycle),

(b) a 10% NaCl solution (sodium cycle) or

(c) a 2.5% $MgCl_2$ solution (magnesium cycle).

Mixed regeneration is also possible.

Besides their ion exchange capacity, resins may also absorb polyphenols and other polymers that affect their exchange properties. This absorption capacity is sometimes useful, e.g. removing color from fruit juice. Foreign substances fixed on the resin are eliminated by treating it with sodium hypochlorite (3% available chlorine). The column must then be rinsed again, using a volume of water representing 6–12 times the volume of the resin.

The system is then ready to treat wine. The wine flows through the resin column from top to bottom, with a throughput on the order of 8 times the volume of resin per hour. A throughput of 25 volumes per hour has been suggested for magnesium cycles.

The wine or rinsing water may be pumped through the column, while the resins are kept dry by nitrogen back-pressure. It is possible to treat 2000 hl of wine in an 8-hour magnesium cycle (Ribéreau-Gayon *et al.*, 1977) using a 1000 l column (80 cm in diameter and 2 m high). The column must then be regenerated using 100 hl of magnesium chloride solution at a concentration of 25 g/l, i.e. 250 kg of the salt.

Independently of legal problems and government authorization to use this technique, ion exchange treatment raises the issue of recycling and treating residual washing water with a very high salt content due to the concentrations of mineral ions involved.

12.4.5 Conclusion

Of course, it would be useful to harmonize legislation on the use of ion exchangers in the various wine growing countries. Current knowledge indicates that the use of anion exchangers raises serious problems, as they cause excessive changes in wine composition. It would, thus, no longer be possible to use ion exchange to reduce acidity. Similarly, the use of mixed anion/cation exchangers, designed to avoid changes in pH, is also debatable.

It would, however, be possible to envisage improving tartrate stability by eliminating K^+ and Ca^{2+} ions. Unfortunately, acid resins reduce pH (which may be useful under certain conditions) and sodium resins increase the sodium level. Legal limits on these changes should be imposed. It is possible to keep the effect of ion exchange within acceptable limits by varying the type of resin regeneration (hydrogen/sodium) and the percentage of wine treated. This technique is more effective in preventing tartrate precipitation in wines that have been previously cold-stabilized.

The possibility of using magnesium cation resins, already investigated several decades ago, should be reconsidered. This process eliminates K^+ and Ca^{2+} very efficiently. Magnesium resin is suitable for long cycles (a volume of wine representing 200 times the volume of resin may be treated) and there is no significant impact on the wine's composition. The entire volume of wine to be treated may be circulated through the resin and only the excess of undesirable cations is removed.

One of the first objectives in using ion exchangers was to eliminate iron from wine (Ribéreau-Gayon *et al.*, 1977). This technique does not, unfortunately, seem suitable for that purpose. However, the issue of ferric instability is much less acute, since the widespread use of stainless steel has eliminated the problem of excessive iron concentrations in wine.

12.5 ELECTRODIALYSIS APPLICATIONS IN WINEMAKING

12.5.1 Operating Principle

Electrodialysis is a method for separating ions using selective membranes that are permeable to ions according to their charges. An electric field moves the ions in one direction or the other. It is thus possible to extract a large proportion of the charged ions from the solution. The principle of electrodialysis is based on the property of selective membranes to allow only cations or anions to pass through (Escudier *et al.*, 1998). Initial experiments with electrodialysis were carried out as early as 1975, but it took 20 years to develop a system for tartrate stabilization in wine.

Moutounet *et al.* (1994) did a great deal of work on defining the conditions for using electrodialysis to stabilize wine, identifying suitable membranes and process control so that each wine is treated according to its specific level of instability. This has made electrodialysis technology reliable and effective, while avoiding excessive alterations in the wine's chemical composition.

Figure 12.4 shows a simple electrodialysis cell, consisting of two compartments, separated by alternating anion and cation membranes. The difference in potential at the electrode terminals causes the cations to migrate toward the cathode and the anions toward the anode. The cations pass through the cation-permeable membrane and are concentrated in Compartment 2, as the next membrane is only permeable to anions. Similarly, the anions are attracted toward the positive electrode, passing through the anion-permeable membrane and stopping at the next membrane as it only lets cations through. As the process continues, Compartment 1 loses its ions (anions and cations) and its contents are known as the "diluate". The ion-enriched solution in the next compartment is known as the "concentrate".

An electrodialyzer consists of a series of these cells, with up to 700 pairs of membranes, arranged like a filter press. The system is subjected to a potential difference on the order of 1 V/cm and the concentrate gradually builds up in alternating cells, while the solution in the other cells becomes diluted. A separating frame allows a uniform, thin layer (0.3–2 mm) of liquid to flow through each membrane. The diluted solution (treated wine) and the concentrate (saline solution) are collected separately. The solutions may be treated again, either to decrease the ion content of the treated wine, or to increase the ion charge of the concentrate, thus decreasing the volume of waste.

The electrodes at either end of the electrodialyzer are bathed in the electrolyte in a special compartment.

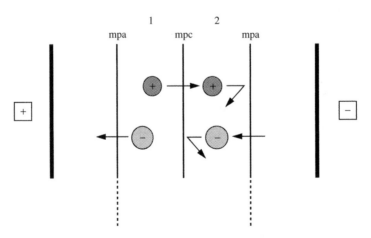

Fig. 12.4. Diagram of a simple electrodialysis cell (Moutounet *et al.*, 1994): m.p.a., membrane permeable to anions; m.p.c., membrane permeable to cations; compartment 1, the ions are diluted; compartment 2, the ions are concentrated

12.5.2 Choice of Membranes

Membranes are films 100–200 μm thick, consisting of ionized function groups grafted on an organic polymer matrix. Sulfonic radicals are used for membranes permeable to cations and quaternary ammonium for membranes permeable to anions.

The ion transfer kinetics of a particular membrane at a given ion strength depend on several parameters, including the dimension and mobility of the solute, which define the speed constant. These variables explain the differences in migration observed when different ions are treated with various membranes (Audinos, 1983).

K^+ is the cation that migrates the most easily, while Na^+ and Ca^{2+} are much less mobile and consequently less reduced. Among anions, tartaric and, possibly, acetic acid are the most reduced.

It has been observed experimentally that various combinations of "cation-permeable" and "anion-permeable" membrane pairs have varying capacities to eliminate different ions. It is possible to enhance potassium elimination by choosing an appropriate pair of membranes, thus achieving tartrate stabilization without greatly modifying the acetic acid content, as this would be unacceptable. Pairs of membranes that eliminate potassium tend to reduce pH, even if the tartaric acid content is reduced. Membranes selected by Moutounet et al. (1994) reduced pH by under 0.2 units and volatile acidity by just a few percent.

Membranes used for electrodialysis in tartrate stabilization must meet regulation standards and specific winemaking criteria. Legal requirements for electrodialysis membranes are specified in the Community Code of Oenological Practices and Processes (EC 1622/2000). Membranes must not excessively modify the physico-chemical composition and sensory characteristics of the wine. They must meet the following requirements:

- They must be manufactured from substances authorized for materials intended to come into contact with foodstuffs.

- They must not release any substances likely to endanger human health or affect wine quality.

- They must not result in the formation of new compounds that were not initially present in the wine.

The stability of fresh electrodialysis membranes is determined using a dilute alcohol solution with an acid pH to simulate the physico-chemical composition of wine and investigate any possible migration of substances from the membranes.

Membranes must also be sufficiently strong and not excessively subject to fouling. To be cost-effective, filter membranes must be capable of operating efficiently for at least 2000 hours and, more usually, for 3000 or 4000 hours. Daily cleaning with acid and alkaline solutions is recommended to maintain membrane performance at its initial level.

Membrane performance must meet the following criteria:

1. A maximum reduction in alcoholic strength of 0.1% vol.

2. A pH reduction of no more than 0.25 pH units.

3. A maximum reduction in volatile acidity of 0.09 g/l (expressed in H_2SO_4)

4. It must not affect the non-ionic constituents of the wine, in particular, polyphenols and polysaccharides.

5. Membranes must be conserved and cleaned using substances authorized for use in the preparation of foodstuffs.

6. Membranes must be marked so that alternation in the stack can be checked.

7. The command and control mechanism must take account of the particular instability of each wine so as to eliminate only the supersaturated fraction of potassium hydrogen tartrate and calcium salts.

12.5.3 Tartrate Stability Test Used to Determine Process Settings

This test consists of analyzing variations in the conductivity of a sample seeded with potassium hydrogen tartrate crystals and kept at negative temperatures ($-4°C$) for 4 hours with continuous stirring (Biau and Siodlak, 1997).

The wine's instability may be affected by various treatments (filtration, fining, etc.) carried out prior to electrodialysis.

Modeling this phenomenon over a 4-hour period makes it possible to assess the theoretical drop in conductivity over an unlimited period. The test results, therefore, indicate the final conductivity value at which the wine no longer presents a risk of tartrate precipitation. The drop in conductivity required for the wine to be stabilized is monitored using an automatic system controlled by a PC (Escudier et al., 1998).

The operational utilization of this process is subject to a prior conductivity test and a properly-regulated automatic control system.

The same system is then used to monitor the operation automatically, using a conductivity set point below which it is useless to continue electrodialysis, as shown by the preliminary test. Treatment is thus adapted to each wine according to its specific instability, under conditions ensuring that there are no excessive alterations in its chemical composition. The decrease in conductivity required depends on the wine, generally varying from 150 to 500 μS. The ion reduction varies from 15 to 20% for young wines and 5 to 15% for older wines. The automatic control system prevents the apparatus from eliminating ions beyond the limit required to ensure stability, so that modifications in wine composition remain within an acceptable range.

The first industrial electrodialysis unit with a 45 m^2 membrane surface went into operation in 1996. Its results, consistent with those of the many earlier pilot tests, showed that electrodialysis can be applied on an industrial scale for the tartrate stabilization of wine. The optimum effectiveness of this technique is subject to choosing a pair of (anion and cation) membranes that best preserve the wine's natural balance, and combining it with an automatic testing system to restrict the treatment to the minimum required to achieve tartrate stability.

12.5.4 Operational Details

Wine is circulated continuously between a vat and the electrodialysis cells until the desired level of treatment is obtained. This is assessed by measuring the wine's electrical conductivity.

A volume of wine is pumped into the treatment vat and then into the 'diluate' circuit of the electrodialysis cells. When conductivity reaches the set point, determined by an instability test, the wine is automatically pumped into a reception vat using a system controlled by solenoid valves. A new volume of wine is then pumped into the system and stabilized under the same conditions. Treatment time and, consequently, the performance of the system depend on the wine's degree of instability. Treatment flow rates vary from 50 to 150 $l/h/m^2$, depending on this parameter. The 'concentrate' circuit consists of a saline solution that collects the ions extracted from the wine. The ion load is adjusted by adding water to avoid the precipitation of bitartrate crystals inside the small, easily blocked cells. This function is also automatically controlled by conductivity measurements.

Electrodialysis is more efficient in treating white than red wines, if the same level of treatment is required. The colloidal structures in red wine increase membrane surface resistance, leading to a decrease in the ion transfer rate. Regular cleaning of the membranes attenuates this effect.

12.5.5 Changes in Wine Composition

Electrodialysis is highly effective in eliminating mineral cations. Potassium ions migrate the most rapidly. There is a linear correlation between the reduction in potassium concentration and the decrease in conductivity. Sodium, iron and copper concentrations decrease slightly, but the calcium content remains almost unchanged. Anions are less affected, so there is a decrease in pH. Tartaric acid is the most strongly affected by electrodialysis. A 20% drop in conductivity corresponds to a decrease of 10–15% in the tartaric acid content. Volatile acidity also drops slightly, as does the alcohol content.

Table 12.3. Changes in analysis parameters of a red wine according to the rate of ion elimination by electrodialysis (Moutounet et al., 1994)

Deionization rate[a]	0%	10%	17%	20%	25%	30%	35%	40%
Alcohol content (% vol at 20°C)	10.70	10.65	10.60	10.60	10.55	10.50	10.40	10.35
Total acidity (g/l of H_2SO_4)	3.10	3.00	2.85	2.80	2.75	2.65	2.55	2.50
pH	3.84	3.79	3.75	3.74	3.72	3.71	3.66	3.64
Volatile acidity (g/l of H_2SO_4)	0.55	0.53	0.54	0.54	0.54	0.53	0.50	0.52
Tartaric acid (g/l)	2.60	2.20	1.80	1.80	1.60	1.40	1.20	1.00
Lactic acid (g/l)	1.40	1.40	1.40	1.40	1.40	1.40	1.30	1.30
K^+ (mg/l)	1690	1440	1280	1190	1100	990	860	780
Ca^{2+} (mg/l)	68	69	67	67	68	67	67	64
Na^+ (mg/l)	21.7	20.0	18.9	18.5	17.9	16.9	15.6	14.7
Abs. 280 nm	40.7	39.7	39.4	39.5	38.9	38.5	37.5	37.5

[a] Deionization rate = $\dfrac{\text{initial conductivity} - \text{conductivity calculated by the instability test} \times 100}{\text{initial conductivity}}$.

The other elements in a wine's chemical composition (polyphenols, polysaccharides, amino acids and volatile compounds) are not greatly affected by electrodialysis. In particular, this treatment has less effect on the colloidal coloring matter of red wines than cold stabilization.

Table 12.3 gives an example of the results obtained by applying treatments of varying intensity, corresponding to increasing ion elimination rates (measured by the drop in conductivity), to the same wine. The level of deionization necessary to obtain tartrate stability in this wine was 17%. According to several authors, the changes in chemical composition that occur under these treatment conditions are acceptable. However, the decrease in tartaric acid is by no means negligible. A treatment aimed at eliminating 35% of the total ions would not be as innocuous. A system that is capable of controlling treatment intensity to a level that just ensures tartrate stability has obvious advantages. In general, the drop in conductivity necessary in red wines varies between 5 and 20% of the initial value. In white and very red young wines, this value is 30%.

From an organoleptic standpoint, wines treated by electrodialysis are considered to be slightly different from those treated by standard cold stabilization. However, the differences are not sufficient to be able to classify the wines in terms of preference.

REFERENCES

Audinos R. (1983) *Les Membranes Artificielles, Collection Que sais-je*, Presses Universitaires, Paris.

Biau G. and Siodlak A. (1997) *Rev. Fran. Œno.*, 162, 18.

Escudier J.L., Moutounet N., Saint-Pierre B. Battle J.L. (1998) Electrodialyse appliquée à la stabilisation tartrique des vins. Vigne et Vin, Publications Internationales, p. 131.

Mourgues J. (1993) *Rev. Œnologues*, 69, 51.

Moutounet M., Escudier J.-L. and Saint-Pierre B. (1994) In *Les Acquisitions Récentes dans les Traitements Physiques du Vin* (ed. B. Donèche), Tec. et Doc., Lavoisier, Paris.

Ribéreau-Gayon J., Peynaud E., Ribéreau-Gayon P. and Sudraud P. (1977) *Sciences et Techniques du Vin*, Vol. IV: *Clarification et Stabilization. Matériels et Installations*. Dunod, Paris.

Singleton V.L. (1962) *Hilgardia*, University of California, 32 (7), 319.

Weinand R. and Dedardel F. (1994) In *Les Acquisitions Récentes dans les Traitements Physiques du Vin* (ed. B. Donèche). Tec. et Doc., Lavoisier, Paris.

13

Aging Red Wines in Vat and Barrel: Phenomena Occurring During Aging

13.1 Oxidation–reduction phenomena	388
13.2 Oxidation–reduction potential	389
13.3 Influence of various factors on the oxidation–reduction potential	393
13.4 Development of the phenolic characteristics of red wines (color and flavor) during aging	397
13.5 Bottle aging red wines	404
13.6 Winemaking practices	409
13.7 Barrel aging red wines	411
13.8 Effect of the type of barrel on the development of red wine	416
13.9 Constraints and risks of barrel aging	424

During the period from the end of the fermentations until bottling, a wine is said to be aging. Aging duration is highly variable according to a wine's origin, type and quality. It must be long enough to stabilize the wine, as well as to prepare great wines for bottle aging. Many changes occur in the composition of the wine during this period, accompanied by the development of color, aroma and flavor. The conditions under which wine is stored and handled, as well as the types of container used, have a very marked effect on these developments, which are closely connected with oxidation–reduction phenomena that take place in the wine.

13.1 OXIDATION–REDUCTION PHENOMENA

13.1.1 Introduction

Varying quantities of oxygen are dissolved in wine during aging, depending on winery practices and the temperature at which the various operations are carried out. Saturation is on the order of 10 mg/l at 5°C and 7 mg/l at 25°C. 'This molecular oxygen fixes directly on certain substances described as auto-oxidizable (Fe^{2+} and Cu^+), forming unstable peroxides that, in turn, oxidize other oxygen-accepting substances. These molecules are not directly oxidized by molecular oxygen, as it is a very weak oxidant' (Ribéreau-Gayon et al., 1976). Peroxides have a greater oxidizing capacity than molecular oxygen. The operations to which wine is subjected are therefore responsible for causing oxidative phenomena. These vary in intensity, according to the composition of the medium. In airtight vats and bottles, wine is deprived of oxygen from the air and is affected by reduction phenomena.

13.1.2 General Reminder of Oxidation–Reduction Concepts

Substances are oxidized when they fix oxygen or lose either hydrogen or one or more electrons. Reduction is the reverse of these reactions. In organic molecules, oxidation produces compounds with a higher oxygen or lower hydrogen content. Both of the reactions (1 and 2) below are examples of oxidation.

In fact, there is always a balance between the two phenomena. When an oxidation reaction occurs, there is always a parallel reduction reaction:

$$Red_1 + Ox_2 \rightleftharpoons Ox_1 + Red_2$$

Reducing agents may be oxidized by the following three mechanisms:

$$Red \rightleftharpoons Ox + ne^-$$
$$Red \rightleftharpoons Ox + nH_2$$
$$Red \rightleftharpoons Ox + n(2H^+) + n(2e^-)$$

This constitutes an oxidation–reduction battery, as a platinum (Pt) filament placed in the ($Red_1 + Ox_2$) solution has a measurable potential as compared to a standard.

Oxidizing–reducing systems are divided into three categories:

1. Directly electroactive substances that react with Pt. These are often pairs of metals: Fe^{2+}/Fe^{3+} and Cu^+/Cu^{2+}.

2. Weakly electroactive substances that do not react with Pt, but are nevertheless active in the presence of these substances. These molecules

(1) CH_3CH_2OH (Ethanol) → CH_3-CHO (Ethanal) → CH_3COOH (Acetic acid)
    H_2 ; $+\frac{1}{2}O_2$

(2) Flavene (colorless) → Red flavylium
    $-\frac{1}{2}H_2, -e^-$

have a conjugated dienol:

$$-\underset{OH}{C}=\underset{OH}{C}-$$

3. Electroactive substances in the presence of dehydrogenases:

$$\text{lactic acid} \rightleftharpoons \text{pyruvic acid}$$
$$\text{ethanol} \rightleftharpoons \text{ethanal}$$
$$\text{butanediol} \rightleftharpoons \text{acetoin}$$

13.1.3 Measuring Dissolved Oxygen

When a wine is in contact with air, the longer and the more vigorously it is agitated, the more oxygen is dissolved. When the wine is no longer in contact with air, this oxygen reacts with the compounds in wine and disappears. This reaction is faster at higher temperatures and in wines with a high concentration of oxidizable molecules. Although the quantity of dissolved oxygen depends on many factors, a wine's oxygen content is still a useful parameter in analyzing its condition.

The first assays (Ribéreau-Gayon, 1931) used a chemical method, based on the oxidation of sodium hydrosulfite into bisulfite by free oxygen, with carmine indigo as the color indicator. The currently preferred method is polarographic analysis, developed by Clark (1960). The apparatus consists of two electrodes, a silver anode and a gold cathode, linked by potassium chloride gel. They are separated from the medium by a membrane selectively permeable to oxygen. The difference in potential established between the two electrodes (on the order of 0.6 to 0.8 volts) is modified by circulating oxygen through the membrane. The following reactions take place:

1. At the cathode: $O_2 + 2H_2 + 4e^- \rightarrow 4OH^-$, the oxygen consumes electrons.

2. At the anode: $Ag + Cl^- \rightarrow AgCl + e^-$, electrons are released.

The intensity of the electrical current, caused by the movement of electrons, is directly proportional to the quantity of dissolved oxygen, expressed in mg/l.

13.2 OXIDATION–REDUCTION POTENTIAL

13.2.1 Measuring the Oxidation–Reduction Potential in a Simple Medium

Many chemical reactions in wine are characterized by electron transfers, leading to the oxidation and reduction phenomena. These reactions occur simultaneously and continue until an oxidation–reduction equilibrium is reached. The oxidation–reduction potential of a wine is an observation of the oxidation and reduction levels of the medium at a certain equilibrium. This value is quite comparable to pH as a measurement of a wine's acidity. Its value is linked to the quantity of dissolved oxygen, just as pH depends on the quantity of (H^+) protons. Furthermore, it is possible to define the normal potential E_0 of a given oxidizing/reducing couple when half the component is oxidized and half is reduced. This characterizes the wine's oxidation capacity in the same way as pK indicates the strength of an acid.

In a simple solution, the ratio between molecules in an oxidized state and those in a reduced state is assessed by the difference in potential between a metal measuring electrode, chemically inert in relation to the solution, and a reference electrode, generally calibrated in relation to the H_2 electrode immersed in the medium under examination. This oxidation–reduction potential E_H, measured in volts (V), is expressed by the Nernst equation:

$$E_H = E_0 + \frac{RT}{nF} \log_n \frac{[\text{oxidized}]}{[\text{reduced}]}$$

where

E_0 = normal potential of the system

R = perfect gas constant = 8.31 J/mole/°K

T = measured temperature (in °K)

n = number of electrons involved

F = Faraday number = 96 500 coulombs

At 25°C, with decimal logarithm:

$$E_H = E_0 + \frac{0.059}{n} \log \frac{[\text{oxidized}]}{[\text{reduced}]}$$

The Nernst equation, as described above, is only strictly valid for mineral oxidation–reduction systems; for example:

$$Fe^{2+} \rightleftharpoons Fe^{3+} + e^-$$

In organic systems involving proton exchanges, the pH must be taken into account:

$$AH_2 \rightleftharpoons A + 2H^+ + 2e^-$$

A combined electrode is used for measurements of oxidation–reduction potentiel in wine. It consists of a platinum measuring electrode and an Ag/AgCl, KCl reference electrode, with a constant potential in relation to the standard hydrogen electrode, on the order of 200 mV at 25°C.

When this combined electrode is immersed in distilled water at 25°C, the positive potential measured is due to the following reactions:

Pt electrode: $4H^+ + O_2 + 4e^- \longrightarrow 2H_2O$

Reference electrode: $Ag + Cl^- \longrightarrow AgCl + e^-$

The potential is then expressed as follows:

$$E_H = E_0 + \frac{0.059}{4} \log \frac{[H^+]^4[O_2]}{[H_2O]^2}$$

and for distilled water at 25°C: $[H_2O] = 55.55$ mole/l and $E_0 = 1.229$ V. Therefore

$$E_H(V) = 1.178 - 0.059\, pH + 0.014 \log[O_2]$$

In an aqueous solution, E_H depends on pH and oxygen content. At a constant oxygen content, any increase in pH leads to a decrease in E_H. At a set pH, any additional dissolved oxygen leads to the opposite phenomenon.

13.2.2 Measuring the Oxidation–Reduction Potential in Wine

Satisfactory results are obtained when the oxidation–reduction potential is determined using a standard electrode in a model medium. However, it is much more difficult to obtain reliable measurements in a complex medium such as wine. It has been observed that the readings do not stabilize and that electrode calibration is disturbed due to pollution (Zamora, 1989).

The electrode is generally calibrated using solutions with a known, constant, oxidation–reduction potential. The equimolar mixture (10 mM) of potassium ferricyanide and ferrocyanide (Michaelis, 1953) used by Deibner (1956) has a potential of 406 ± 5 mV at 20°C and remains stable for approximately two weeks. Its composition is as follows: 0.329 g of $Fe(CN)_6K_3$ + 0.422 g of $Fe(CN)_6K_4$ + 0.149 g of KCl + H_2O qs 1000 ml.

According to the literature, it is necessary to wait for at least 40 minutes, and up to 2 hours, for measurements to reach their limit. In fact, complete stabilization of the electrode is never observed. Furthermore, once the measurement has been made, the electrode is no longer capable of returning to the reference potential. The Deibner protocol (1956) indicates that the electrode must be thoroughly cleaned (H_2O_2 + HNO_3 + HCl) before it returns to its initial value. In view of these difficulties in making measurements, the electrode has been changed to take into account the composition of wine (Vivas et al., 1992).

In the case of the combined standard electrode (Figure 13.1a), the electrons are exchanged by the Pt filament (measuring electrode) and by the diffusion of Cl^- ions (reference electrode):

1. Electrons released by reducing agents in the medium reduce the AgCl and are transmitted by the Pt: $AgCl + e^- \rightarrow Ag + Cl^-$. The Cl^- is diffused in the medium.

2. In the presence of oxidizing agents, the Cl^- ions penetrate the electrode and form AgCl. The

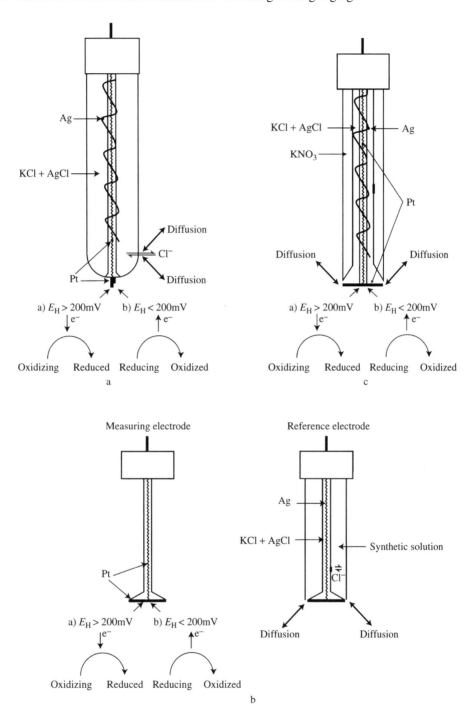

Fig. 13.1. Electrodes for measuring the oxidation–reduction potential and modifications required for their use with wine: (a) standard combined electrode, (b) measuring electrode and modified reference electrode, (c) modified combined electrode (Vivas et al., 1992)

electrons thus released are transmitted towards the oxidizing agents by the Pt: $Ag + Cl^- \rightarrow AgCl + e^-$.

Two types of interference make it difficult to use this system for measurements in wine. On the one hand, certain substances form a deposit on the Pt filament and insulate it. On the other hand, it is difficult to exchange Cl^- ions through the ceramic junction, following changes in the composition of $KCl + AgCl$ due to the diffusion of ethanol and tartaric acid in the wine.

Changes to the measuring and reference electrodes have been envisaged to produce an electrode suitable for use in wine (Figure 13.1b).

1. The contact surface of the platinum electrode may be increased to promote exchanges of electrons and limit the accumulation of deposits.

2. A transition layer (model wine solution) may be introduced to minimize the diffusion of molecules in the wine towards the reference electrode and increase exchanges with the wine. The composition of the transition layer must be adapted to the medium under investigation (*vins doux naturels*, fortified wines, brandy, etc.).

In practice, the two electrodes may be combined into one (Figure 13.1c) that provides satisfactory measurements in wine. The stabilization time is between 5 and 10 min for red wines and somewhat shorter for white wines (2 min). Calibration remains stable for several days.

13.2.3 Correlation Between Dissolved Oxygen and the Oxidation–Reduction Potential

Of course, the oxygen concentration has a major effect on the value of E_H. Differences of 150–250 mV have been observed at oxygen levels ranging from 1 to 6 mg/l. The wine's degree of aeration is, therefore, one of the main factors involved in oxidative phenomena (Table 13.1). However, wine is not merely distilled water

Table 13.1. Effect of oxygen content on the oxidation–reduction potential of a red wine (Vivas *et al.*, 1992)

O_2(mg/l)	E_H(mV)	ΔE_H
0.1	263	
0.8	280	17
2.5	340	77
4.8	424	161
5.0	434	171

(Nernst's law) and oxygen is constantly taken up by oxidation reactions. The dissolved oxygen concentration may vary significantly, depending on the precise time the measurements are made.

Under these conditions, the decrease in oxygen concentration should lead to a decrease in the oxidation–reduction potential $E_H = f(O_2)$ and oxidation of the various oxidizing–reducing systems in the wine should lead to an increase in potential, i.e. ($E_H = f \log [ox]/[red]$). According to Zamora (1989), the equation that integrates all of these phenomena is as follows:

$$E_H = E_0 + A\, pH + B \log[O_2]$$

where A and B are the characteristic coefficients of pH and log $[O_2]$, respectively.

Using the new electrode, Vivas *et al.* (1992) obtained a result very close to that of the Nernst equation on the basis of experimental measurements with distilled water:

Experimental equation

$$E_H(mV) = 1182 - 59.6\, pH + 15.2 \log[O_2]$$

Theoretical equation

$$E_H(mV) = 1178 - 59.1\, pH + 14.8 \log[O_2]$$

It is possible to interpret the potential equation in model wine solution experimentally by studying the impact of factors such as temperature (potential decreases as temperature increases, $\Delta E_H = +40$ mV per 20°C), oxygen (potential increases with the quantity of dissolved oxygen, $\Delta E_H = 250$ mV per 6 mg/l), and, to a lesser extent, pH (variations are very small in the pH range of wine).

A comparison of the effect of pH on E_H in a solution with a low oxygen content (1 µM = 0.032 mg/l) and one near oxygen saturation (235 µM = 7.5 mg/l) produces two straight lines, with the following equations:

$$E_H = 358.5 - 57.83 \text{ pH},$$

with $\quad E_0 + B.\log 10^{-6} = 358.5$

$$E_H = 716.6 - 61.93 \text{ pH},$$

with $\quad E_0 + B.\log 235.10^{-6} = 716.6$

Under these conditions $\begin{pmatrix} B = 151 \\ E_0 = 1264 \end{pmatrix}$

which gives the following equation:

$$E_H(\text{mV}) = 1264 - 59.8 \text{ pH} + 151 \log (O_2)$$

The difference in value of factor B in comparison to distilled water is due to oxidation reactions specific to the model solution and wine.

It is therefore possible to measure the dissolved oxygen in a wine over time, as well as its oxidation–reduction potential. It is also possible to calculate the normal potential of the system, using the preceding equations ($E_H = E_0 + A \text{ pH} + B \log[O_2]$). Finally, the oxygen consumption rate may also be determined.

13.3 INFLUENCE OF VARIOUS FACTORS ON THE OXIDATION–REDUCTION POTENTIAL

13.3.1 Influence of Oxidation–Reduction Agents

The quantity and rate of oxygen consumption are always higher in red wines (Table 13.2) due to their higher concentrations of oxidizable substances (phenolic compounds). Furthermore, iron and copper are oxidation catalysts likely to be oxidized directly by oxygen. They increase the rate of oxygen consumption (Table 13.2) in red and white wines, but the effect is not the same in both cases (Vivas, 1997). When iron and copper are present in red wine, there is

Table 13.2. Influence of iron (8 mg/l of Fe^{II}) and copper (2 mg/l of Cu) on the oxygen consumption and oxidation–reduction potential of a red and white wine (Vivas *et al.*, 1993)

	V_i (mg O_2/l min)	E_{HM} (mV)	dE_H/dt
Red wine Control	0.45	528	−0.70
With added Fe + Cu	0.90	461	−1.42
White wine Control	0.20	574	−0.27
With added Fe + Cu	0.38	530	+0.41

V_i = instantaneous oxygen consumption rate, E_{HM} = maximum potential after oxygen saturation, dE_H/dt = regression line of the potential, 72 h after saturation.

a rapid drop in E_H, reflecting an acceleration in the oxidative phenomena. The reverse occurs in white wine, where these metallic catalysts cause a slower decrease in E_H. This is thought to be due to the formation of peroxides (Chapon and Chapon, 1977). These are consumed more slowly in white wines than in reds, which contain more oxidizable substances. White wine remains in an oxidized state for a longer time; indeed, it may be several weeks before the potential returns to normal levels.

The presence of antioxidants (SO_2 and ascorbic acid) does not produce any significant variation, either in the instantaneous oxygen consumption rate or in the normal potential of red wine.

13.3.2 Influence of Compounds in Wine and Certain External Factors

Varying the composition of a model medium produces the following results (Table 13.3):

1. Ethanol increases the instantaneous oxidation rate and slightly reduces the potential.
2. Tartaric, malic and lactic acids produce only a few minor modifications. Although E_H decreases when pH increases, variations are small, remaining between 3 and 4.

Table 13.3. Influence of the main components of wine on the oxygen consumption and oxidation–reduction potential in a model medium[a] (Vivas and Glories, 1993a)

			V_i (mg O_2/l min)	E_{HM} (mV)	dE_H/dt
Control			0.025	528	−0.01
EtOH	5	%	0.017	560	−0.01
	15	%	0.071	512	−0.02
Glycerol	5	g/l	0.022	521	−0.01
Tartaric acid	3	g/l	0.025	530	−0.01
	7	g/l	0.027	535	−0.02
Malic acid	3	g/l	0.021	526	−0.01
Lactic acid	2.5	g/l	0.023	518	−0.02
Catechin	2	g/l	0.104	506	−0.41
Oligomeric procyanidins	2	g/l	0.101	515	−0.36
Polymeric procyanidins	2	g/l	0.086	517	−0.25
Monoglucoside anthocyanins	200	mg/l	0.112	491	−0.45
Anthocyanin–tannin combinations	2	g/l	0.097	514	+0.4

[a]Model control medium: 12% vol EtOH; 5 g/l tartaric acid, NaOH 1 N qs pH 3.5, distilled water qs 1000 ml.

3. Glycerol has no effect on oxidation mechanisms.

4. Phenols inhibit variations in potential. Anthocyanins, in particular, consume oxygen rapidly, leading to a rapid drop in potential. Catechins and oligomeric procyanidins are more active than polymers. A wine with a high concentration of flavonols and not very highly condensed tannins consumes more oxygen than one that only contains condensed tannins.

5. Furthermore, temperature causes wide variations in the oxidation–reduction potential of wine (100 mV between 0 and 30°C), in proportion to the quantity of dissolved oxygen. Between +5 and +35°C, the amount of oxygen required to saturate wine drops from 10.5 to 5.6 mg/l.

6. Finally, the types of containers used for aging and storing wine have an influence on the oxidative process, depending on their permeability to air (Table 13.4). It is possible to maintain a constant concentration of oxygen and a higher oxidation–reduction potential when wine is aged in oak barrels rather than vats. This feature is attenuated with age, as the pores of the barrels gradually become clogged.

Table 13.4. Influence of the container on the oxidation–reduction condition of a red wine stored for 8 months (Vivas and Glories, 1993a)

	Dissolved oxygen (mg/l)	Mean E_H (mV)
Bordeaux barrel: 2.25 hl		
Age 1 year	0.4	245
Age 2 years	0.2	228
Age 3 years	0.2	218
Stainless steel vat: 70 hl	<0.1	220
Concrete vat: 85 hl	<0.1	215
Plastic vat: 20 hl	<0.1	194

13.3.3 Influence of Various Winemaking Operations

1. Racking (Section 10.3.1) oxidizes wine. The quantity of oxygen dissolved ranges from 2 to 5 mg/l, according to the technique (air pump, mechanical pump, etc.). This oxygen is consumed in 8–10 days. At the same time, the oxidation–reduction potential initially increases from 50 to 100 mV, then decreases sharply until it reaches a minimum value, before returning to its initial level in 15–20 days (Figure 13.2).

2. Air penetration via the bunghole does not depend on the age of the barrel, but rather on the

Aging Red Wines in Vat and Barrel: Phenomena Occurring During Aging

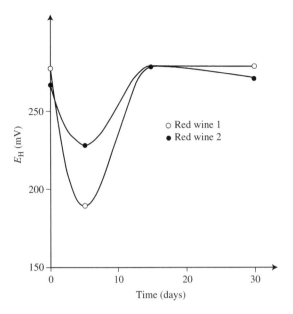

Fig. 13.2. Influence of racking (at time 0) on the development of the mean E_H of two red barrel-aged wines (Vivas and Glories, 1994)

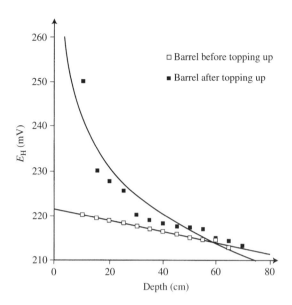

Fig. 13.3. Influence of topping-up operations on the E_H profile of red wines in the barrel according to the level of the wine (Vivas and Glories, 1993a)

type of bung (silicon, wood, cork, glass, etc.). Every year, 0.5 mg of O_2/l may be absorbed into the wine through this orifice. The increase in E_H is mostly noticeable in the 20 cm of wine nearest the bung, where it may reach 20–30 mV.

3. Topping up also causes an increase in E_H in the upper 20–30 cm of wine. This may be on the order of 20 mV, depending on the type of wine (Figure 13.3). Approximately 1 mg/l of oxygen is added, which is capable of initiating surface oxidation reactions.

4. Filtration, centrifugation and pumping may be major oxidation factors if proper precautions are not taken to minimize aeration. These operations may lead to oxygen saturation of the wine and a 50–150 mV increase in the oxidation–reduction potential. Furthermore, the addition of ellagic tannins (major components in oak) in the absence of oxygen leads to a significant increase in the oxidation–reduction potential. This value may reach 30–50 mV in both red and white wines, following the addition of 300–500 mg/l of ellagitannins.

5. It is possible to monitor changes in the oxidation–reduction potential of a wine during fermentation and in the early stages of aging (Vivas and Glories, 1995). It has been observed (Figures 13.4 and 13.5) that prefermentation treatment of the grapes involves rapid oxidation, reinforced by the presence of polyphenoloxidases. Highly reducing media are produced during alcoholic and malolactic fermentation. The oxidation–reduction potential then stabilizes at an average level between 200 and 300 mV. Measuring the oxidation–reduction potential makes it possible to predict the reduction problems that are likely to occur at potentials ≤150 mV.

13.3.4 Impact of Aerating Wine

Tables 13.5 and 13.6 show the overall average quantities of oxygen in red wines during vatting and aging. Experiments show that the quantities of oxygen used in microbubbling are highly variable from one brand of equipment to another, especially during vatting after fermentation. These values are based on the olfactory detection of ethanal (Table 13.7).

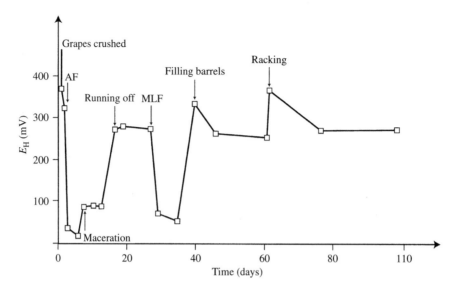

Fig. 13.4. Example showing changes in the oxidation–reduction potential of a red wine during the winemaking process: AF, alcoholic fermentation; MLF, malolactic fermentation (Vivas and Glories, 1995)

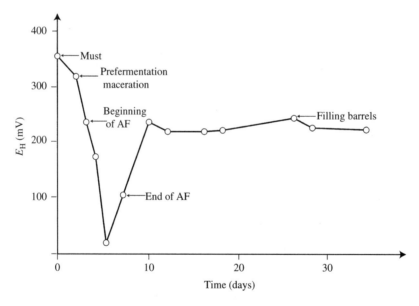

Fig. 13.5. Example showing changes in the oxidation–reduction potential of a white wine during winemaking with stirring of the lees: AF, alcoholic fermentation (Vivas and Glories, 1995)

These values are much higher at the end of vinification than those of wines aged in oak, especially if the wine is racked as well. Ribéreau-Gayon *et al.* (1976) reported values of 3–7 mg/l/year^{-1} during barrel aging, with 10 mg/l/year^{-1} for the top of the barrel before topping up (Section 13.7.2).

Even without racking, this technique maintains the wine's high potential throughout the entire aging period, thus promoting oxidation.

Table 13.5. Estimated quantities of oxygen absorbed during vatting (results are expressed in mg $O_2 . l^{-1}$ wine)

	Maximum	Minimum
Pre-fermentation skin contact	8	5
Alcoholic fermentation (pumping over)	60	30
Post-fermentation vatting	4	1
Running-off	6	4
Total during vinification	78	40

Table 13.6. Overall estimate of the quantities of oxygen dissolved via aeration during the aging of red wines (results are expressed in mg $O_2 . l^{-1}$ wine)

	Barrel	Vat
Aging in new barrel (16 months)*	27–60	0
Aging in used barrel (16 months)*	15–20	
Racking	5–25	10–25
Topping up	3–12	3–12
Pumping	5–10	5–10
Transfers (fining)	7	12
Filtration	4	8
Bottling	3	3
Total during aging	Used barrels: 42–81 New barrels: 54–121	41–70

*Depending on the type and position of the bung (Table 13.14).

Table 13.7. Quantity of oxygen likely to be added by microbubbling during the fermentation and aging of red wines

Phase	Microbubbling time	Quantity of O_2 added	
		mg/l/ month	mg/l
Post-fermentation vatting	15 days	86	43
	2 days	171	11,5
	3 days	100	10
	10 days	86	29
After running-off, before malolactic fermentation	10 days	14	5
Aging	16 months	2	32
		5	80
		7	112

13.4 DEVELOPMENT OF THE PHENOLIC CHARACTERISTICS OF RED WINES (COLOR AND FLAVOR) DURING AGING

13.4.1 Wine Development

The aging of red wine should be characterized by harmonious development of the various components of color, aroma and flavor. The color gradually changes from cherry red to deep red and then brick red. The oldest wines even take on an orange tinge. The flavor also evolves, becoming softer, with less astringency. There is, however, a risk that the wine may become thinner and dry out on the palate as it ages. Furthermore, the rate at which these changes occur is different for each wine, depending on both outside conditions and the wine's specific composition:

1. External conditions include oxidative phenomena (O_2 and SO_2), temperature and time. A great deal of research has focused on the aging of wines prior to bottling (Pontallier et al., 1980; Pontallier, 1981; Ribéreau-Gayon et al., 1983; Glories, 1987; Chatonnet et al., 1990, 1993b; Vivas and Glories, 1993a, 1993b, 1996). There are, however, very few publications on bottle aging (Ribéreau-Gayon, 1931, 1933).

2. The way a wine ages depends on its phenol composition, characterized by the total quantity of phenols (OD 280), the ratio of the various pigments (tannins/anthocyanins) and the type of tannins (seed tannins consisting of procyanidins polymerized to varying degrees and skin tannins with more complex structures) (Section 6.5.2). The presence of polysaccharides of both plant and yeast origin also affect aging potential.

Anthocyanins and tannins extracted from grapes are involved in various reactions that depend to a great extent on external conditions and produce a variety of compounds (Section 6.7). These reactions include degradation, modification, and stabilization of the color, polymerization of tannins and condensation with other components. These reactions are summarized in Figure 13.6.

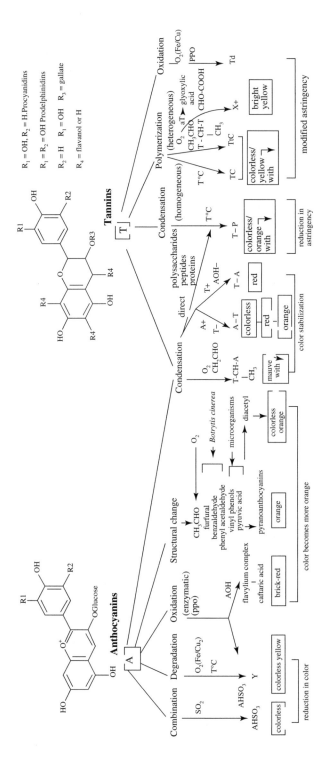

Fig. 13.6. Changes in phenols (A, anthocyanins; T, tannins) in red wine during aging. Impact of these reactions on organoleptic characteristics. (Glories, 2003, unpublished)

↲ = precipitation; Y = anthocyanin degradation products (phenol-acids); TP = tannin-polysaccharide and tannin-protein combinations T-A = tannin-anthocyanin combinations
TC = condensed tannins; TtC = highly condensed tannins; X+ = xanthylium structure; aT = tartaric acid; Td = degraded tannins; ppo = polyphenol oxidase

The main consequences of these reactions involving phenols in red wines are changes in color intensity, a tendency to develop a yellow–orange hue (generally accompanied by loss of color) and various modifications in the tannins, responsible for their gradual softening.

13.4.2 Changes in Color Intensity

In the months that follow malolactic fermentation, the color of red wines evolves, generally becoming more intense, but at varying rates, depending on the conditions. Aeration plays a vital role in these changes, as shown quite clearly by the results in Table 13.8. When wine is not allowed any contact with air, its color intensity remains unchanged and sometimes even decreases. Color intensity is clearly enhanced, however, by aeration, although the results of anthocyanin assays may actually show a decrease. An increase in the PVPP number shows that some anthocyanin molecules have condensed. It has also been demonstrated (Pontallier, 1981) that these changes are increasingly marked as aeration increases. The obvious conclusion is that young red wines require sufficient aeration. They should, however, be protected from excessive oxidation by an appropriate dose of free SO_2. The free SO_2 concentration must not be too high or it will inhibit reactions involving coloring matter.

A more detailed analysis of Pontallier's experiment (1981) makes it possible to distinguish between the free anthocyanin concentration (Al) and that of anthocyanins combined with tannins (T-A) (Section 6.4.2). Multiplying each of these values by the ionization value (Section 6.4.5) indicates the proportion of each type of molecule in colored form. The difference represents the proportion of colorless forms. It has been observed (Table 13.9) that the drop in anthocyanin content following aeration mainly affects the free forms. Concentrations of anthocyanins combined with tannins remain constant, and the proportion of these molecules in colored forms increases regularly with aeration. The overall effect is a decrease in total anthocyanins, but the quantity of molecules in colored forms increases, thus accounting for the intensified color. Furthermore, the development of these combined forms tends to stabilize the color.

The reactions involved in these color changes and the oxidative transformations of phenols in wine mainly involve ethanal. They either result in the formation of an ethyl cross-bond between anthocyanin and tannin molecules (Section 6.3.10), or a cycloaddition to the anthocyanins, producing tannin-pyranoanthocyanins (Atanasova et al., 2002). It has been demonstrated that it is possible to produce a few tens of mg/l of ethanal by oxidizing ethanol in the presence of phenols and Fe^{3+} or Cu^{2+} ions.(Wildenradt and Singleton, 1974; Ribéreau-Gayon et al., 1983). When ethanal is added to wine, it disappears during the oxidative phenomena involved in barrel aging. This rapid reaction initially leads to the development of a more purplish color. Provided that the temperature is not too high, not only does color intensity increase in the presence of oxygen, but also the proportion of blue nuances (OD 620%). However, when large quantities of ethanal are added, the color tends to yellow, as its double bond fixes on the C-4 of the anthocyanin.

As shown in Table 13.8, these reactions occur spontaneously in barrel aged wines, as they are always sufficiently well aerated. Regular aeration during racking of wine aged in the vat may

Table 13.8. Color changes in red wine according to aeration conditions (Ribéreau-Gayon et al., 1983)

Time	Anthocyanins (mg/l)			Color intensity (OD 420 + OD 520)			PVPP index	
	0	10 months	16 months	0	10 months	16 months	10 months	16 months
Non-aerated vat	500	380	340	0.66	0.67	0.63	29	34
Aerated vat	500	300	240	0.66	0.72	0.72	37	45
New oak barrel	500	280	240	0.66	0.83	0.75	42	47

Table 13.9. Different types of anthocyanin formed according to aging conditions (same experiment as Table 13.8 continued for 10 months) (Ribéreau-Gayon et al., 1983)

	Colored	Colorless	Total	Color intensity (OD 420 + OD 520)
Free anthocyanins (mg/l)				
Non-aerated vat	27	243	260	
Aerated vat	18	171	189	
New oak barrel	16	146	162	
Anthocyanins combined with tannins (mg/l)				
Non-aerated vat	49	61	110	
Aerated vat	78	33	111	
New oak barrel	96	22	118	
Total free and combined anthocyanins (mg/l)				
Non-aerated vat	76	304	380	
Aerated vat	96	204	300	
New oak barrel	112	168	280	
Color intensity				
Non-aerated vat				0.67
Aerated vat				0.72
New oak barrel				0.83

Table 13.10. Changes in the phenol content of a Madiran wine after 5 months of oxygen microbubbling at two different doses (Moutounet et al., 1996)

	Color intensity (OD 420 + OD 520 + OD 620)	Hue (OD 420/OD 520)	Anthocyanins (mg/l)	PVPP index	Tannins (g/l)	HCl index
Control	0.82	0.67	612	31	4.9	20
+O_2						
1 ml/l/month	1.07	0.62	566	33	4.4	25
3 ml/l/month	1.67	0.59	417	47	3.8	39

compensate, to a certain extent, for the absence of controlled spontaneous oxidation (Pontallier, 1981). A recently developed system for microbubbling system of oxygen (Section 13.3.4; Table 13.7) may facilitate adjustment of the precise quantity of dissolved oxygen necessary for good development and color stabilization (Table 13.10), as well as flavor enhancement (Moutounet et al., 1996). This process reproduces, in a simplified way, the oxidation–reduction conditions that are an integral part of the traditional barrel aging of great wines. Microbubbling could be accompanied by monitoring of the oxidation–reduction potential.

Other modifications in color compounds lead to intensification and stabilization. Various polymerized pigments are formed and the balance shifts from colorless to colored forms (Section 6.4.5) (Guerra, 1997).

The first phenomenon that should be mentioned is the direct reaction of red anthocyanins, in the form of positive flavylium cations, with flavanol molecules (catechins, procyanidins, etc.). This results in the formation of a colorless complex (flavene) that produces a red pigment when oxidized (Section 6.3.10). This reaction is stimulated by an acid pH (<3.5) and a T/A molar ratio <5. This mechanism causes wine to deepen in color following running-off.

Positive carbocations, formed from procyanidins in an acid medium, may also react with colorless anthocyanins, in the form of carbinol base (Section 6.3.10). The molecule thus produced is colorless, but takes a red color by deshydratation. This color depend on the structure of the tannin involved in the reaction. This condensation reaction does not require oxygen. It is linked to the formation of carbocations, promoted by high temperatures. Condensation reactions are, however, relatively slow, and take place during aging in the bottle as well as in airtight vats.

13.4.3 Development of a Yellow–Orange Color

The above transformations result in a reduced anthocyanin content, contrasting with the increase in color. The new condensed pigments formed are more intensely colored than anthocyanins. Other anthocyanin and tannin breakdown reactions may lead to a loss of color, generally accompanied by a tendency towards yellow–orange hues. This is characteristic of the normal development of bottle-aged red wines. The breakdown of anthocyanins involves a loss of molecular structure in the red coloring matter, possibly accompanied by the appearance of a yellowish hue.

Rapid oxidation (Section 6.3.3) has an effect on all these molecules if they are not protected by a sufficient quantity of tannins. The molar ratio T/A must be at least 2, otherwise wine behaves like a pure anthocyanin solution. There is a much lower risk of breakdown reactions under controlled oxidation conditions as malvidin, the main anthocyanin in wine, is not dihydroxylated and, consequently, is not very sensitive to slow oxidation.

The formation of large quantities of ethanal is also responsible for the development of orange tannin-pyranoanthocyanin complexes.

The presence of furfural and hydroxymethyl-furfural, released from grapes affected by *Botrytis cinerea* as well as from toasted oak in barrels (Table 13.20, Section 13.8.3), leads to the formation of orange-yellow and brick-red compounds. These xanthylium structures are formed by condensation of the aldehydes with catechin and anthocyanins (malvidin and cyanidin) (Es-Safi *et al.*, 2000 and 2002). Oxidation of tartaric acid results in glyoxylic acid, which then condenses with flavanols to produce yellow xanthylium structures.

Temperature has a significant influence on color stability, as it shifts the anthocyanin balance (Section 6.3.2) towards the colorless chalcone form, which is in turn converted into phenol acid. This is more dangerous than the preceding reaction since it mainly affects malvidin (Galvin, 1993). When wines are aged at high temperatures, the color always tends towards orange, i.e. there is an increase in the proportion of yellow color (OD 420%). This property is sometimes used expressly to age wines prematurely, e.g. rosés that take on an 'onion skin' hue. Light, temperature and oxidation are used in making *vins doux naturels* (sweet, fortified wines) with a 'rancio' character, as these factors break down the anthocyanins in Grenache, initially composed of 80% malvidin.

Tannins may also be broken down by oxidation (Section 6.3.6), although less readily than anthocyanins (Laborde, 1987). The formation of quinones following tannin degradation has been demonstrated and is sometimes accompanied by the opening of the heterocycle. These structural modifications cause the color to evolve towards yellowish-brown hues and precipitation may occur. These reactions are characteristic of wines with very high tannin contents and low anthocyanin concentrations.

The factors that inhibit these breakdown reactions are: a good molar balance between anthocyanins and tannins ($T/A \geq 2$), temperature $<20°C$ and controlled oxidation.

13.4.4 Changes in Tannins Produce an Impression of Softness

During the aging of red wines, both before and after bottling, tannins undergo transformations that are not only partially responsible for changes in color but also produce a softening of flavor, accompanied by a reduction in astringency. Several reactions are involved, including oxidative phenomena:

1. In an acid medium, the procyanidins components of tannins produce carbocations. These combine with other flavanols, in a reaction enhanced by high temperatures, to form

Table 13.11. Changes in the phenol composition of a red wine (Merlot) after 6 months, according to aeration and temperature (Glories and Bondet de la Bernardie, 1990)

	$t=0$ control	12°C		25°C	
		N_2	O_2	N_2	O_2
Total phenols (OD 280)	47.5	50	49	52	50
Tannins (g/l)	2.68	2.97	2.76	3.19	2.82
HCl index	14.5	22.5	31	35	47.5
Dialysis index	13	15	20	22	32.5
Gelatine index	45	50	41	45	35
Anthocyanins (mg/l)	556	234	116	60	31
PVPP index	32	58	91	94	100

'homogeneous' polymers (Section 6.3.7). This modifies both the structure and organoleptic characteristics of the tannin molecules. The structure values (HCl and dialysis) increase at high temperatures (Table 13.11) and their reactivity to gelatin decreases, as compared to the control. These changes lead to a certain softening of the wine. Precipitation occurs when the molecules become too bulky.

2. Ethanal is responsible for other polymerization reactions that take place in the presence of oxygen, or in an oxidizing medium, and produce complex structures (Saucier *et al.*, 1997) known as 'heterogeneous' polymers (Section 6.3.7) (Table 13.11). Although the HCl and dialysis values increase more than they would in the absence of air, heterogeneous polymers are less reactive to gelatin than homogeneous molecules. Precipitation may occur, depending on the quantity of ethanal and the type of procyanidins (oligomers or polymers) in the medium.

It is therefore obvious that the way a wine is stored, either in airtight vats or in the presence of oxygen, has a significant effect on the development of its organoleptic characteristics. It has also been observed that bottle aging is not capable of softening wine to any great extent: an initially aggressive wine will always retain that character. There is also a risk of thinning due to precipitation.

3. Tannins may be implicated in other reactions involving plant polymers, proteins (used in fining) and polysaccharides from grapes or microorganisms.

The combinations formed when tannins react with bulky polymers are colloidal. They precipitate during aging when Brownian motion decreases due to low temperatures. This is the origin of colloidal coloring matter in wine.

When smaller polysaccharides are involved, precipitation is less marked, but the stability of the combinations still depends on temperature. Warming facilitates condensation and associations between complexes, i.e. the formation of colloids (Ribéreau-Gayon *et al.*, 1976), which are, in turn, likely to precipitate.

These combined molecules improve a wine's flavor, due to the fact that the tannins become less active in relation to proteins and also make the wine more full-bodied. The technique of aging on the lees commonly used for white wines can also be applied to red wines, to facilitate the release of yeast mannoproteins and peptides. The results are variable, depending on the grape variety, aging period, and vintage. There is a risk that this process may lead to tannin precipitation, which is likely to have a negative effect on quality. Excessively high temperatures must be avoided, as there is also a risk of eliminating too much tannin, thus stripping the wine of its character.

13.4.5 Influence of External Conditions on the Development of Coloring Matter and Tannins

It is apparent from the preceding considerations that two physicochemical factors, oxidation and temperature, have a particularly strong influence on the various reactions responsible for the development of coloring matter and tannins. The following paragraphs summarize all the phenomena involved:

1. Flavenes take on color in an oxidizing medium (due to the presence of oxygen, air or ellagic tannins) at low temperatures (12°C), causing young wines to become more intensely red. Their color also deepens as a result

of the formation of tannin–anthocyanin complexes linked via the intermediary of ethanal (Tables 13.8 and 13.9). OD 620% and IC increase (Table 13.12) at 12°C. The risk of anthocyanin breakdown depends on the wine's composition and is greater at lower concentrations. At this temperature (12°C), a wine's tannic structure is less affected than its anthocyanins, and polymerization is relatively limited, as shown by HCl and dialysis values (Table 13.11).

2. In an oxidizing medium at high temperatures, wine evolves towards an orange color (OD 420% increases and OD 520% decreases) (Table 13.12). Anthocyanins disappear by breaking down (DA % decreases) and combining with tannins. This reaction, either via an ethyl cross-bond from the ethanol or as an orange tannin-pyranoanthocyanin structure, is apparently promoted by higher temperatures. Furthermore, the tannic structure changes faster than at low temperatures (Table 13.11). These phenomena are amplified when increased oxygen is combined with high temperatures.

3. In practice, these various conditions must be taken into account in aging red wines, and modulated according to the type of wine. Aeration is desirable at the beginning, to degas the wine and promote the stable, purplish-colored combinations with ethanal that develop from only slightly polymerized tannins. Aeration should then be reduced to maintain a high oxidation–reduction potential, as this is favorable to the evolution of tannins. Aeration should be modulated according to the phenol content of the wine. If phenol levels are low, there is a greater risk of breakdown reactions, leading to precipitation. If, however, the wine has a high phenol content, the total quantity of pigments acts as a buffer to limit breakdown reactions.

Although sorting has a positive effect, when the grapes are affected by rot the various aldehydes and ketones are partially solubilized and there is a major risk that the color of the resulting wine will turn yellow, even if temperature and oxidation conditions are carefully controlled.

Temperatures above 20°C are dangerous during barrel aging. Excessive heat may cause irreversible color breakdown and the formation of tannin polymers that do not always soften the wine. These changes are even more extreme in wines with a low phenol content and in oxidized media. It is always dangerous to barrel-age wines in cellars without temperature control. Furthermore, higher microbial risks are likely to lead to an increase in volatile acidity.

Low temperatures, on the contrary, do not cause any particular problems, but rather facilitate the precipitation of colloidal coloring matter. Oxygen dissolves more easily, leading to oxidation of the medium, and various reactions take place more slowly. However, it is not advisable to keep wine at low temperatures for too long, as development is inhibited and there is a significant risk of oxidation, but a few weeks' exposure to cold is strongly recommended.

Table 13.12. Effect of aeration and temperature on changes in the color of Merlot wine over a 6-month period (1986 Saint-Emilion) (Glories and Bondet de la Bernardie, 1990)

Merlot	$t = 0$ control	12°C		25°C	
		N_2	O_2	N_2	O_2
Color intensity (OD 420 + OD 520 + OD 620)	0.866	1.058	1.477	0.947	0.891
Hue (OD 420/OD 520)	0.63	0.68	0.64	0.94	0.99
OD 420%	34.5	35	32	42.5	44.2
OD 520%	54.5	51.6	50	45.5	44.4
OD 620%	11	13.4	18	12	11.4
DA %	58.2	53.1	50	40.1	37.4

13.5 BOTTLE AGING RED WINES

13.5.1 Aging Phenomena

During bottle aging, wines develop in a reducing environment, tending towards greater organoleptic quality than they initially possessed. Besides changes in color, this process results in an increase in the complexity and finesse of aroma and flavor. The time necessary to attain this optimum condition varies considerably according to the type of wine—from a few years to several decades. Great wines are generally characterized by their capacity to age for a long time, unlike more modest wines that develop their full potential after a relatively short period in the bottle (Figure 13.7).

The aging process includes three distinct phases (Dubourdieu, 1992). During the first stage, wines mature and there may be some fluctuation in quality (Figure 13.8, A and B). During the second phase,

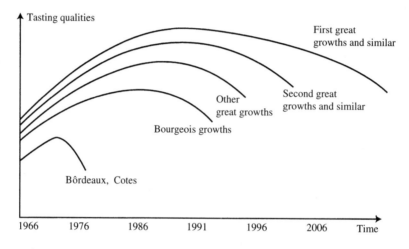

Fig. 13.7. Different aging curves according to *terroir* for the 1966 vintage (Dubourdieu, 1992)

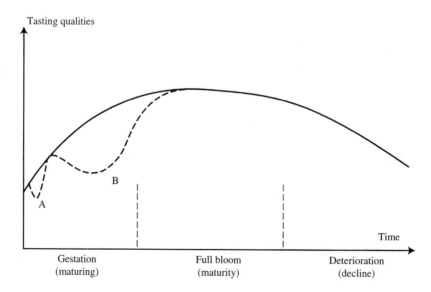

Fig. 13.8. The three phases in the aging of red wine (Dubourdieu, 1992) Tasting qualities

wines reach their peak and are considered to be fully mature. The third stage is one of decline, when wines dry out and become thin. This drop in quality occurs at varying rates and organoleptic changes are accompanied by the gradual stripping of the wine, possibly causing precipitation in the bottle.

Over the same period, the color also evolves toward brick red. The purplish hues disappear completely, making way for more yellowish-orange hues. Under identical storage conditions, the color change occurs at different rates according to the wine's phenol composition. Wines with a high seed-tannin content (e.g. Pinot Noir) age more rapidly than those with a high skin-tanning content (e.g. Cabernet Sauvignon). In comparing wines of the same type, color is generally considered to be a marker for aging. A brick-red color is characteristic of a wine that has aged for some time. From an organoleptic standpoint, a wine's age may be considered in one of three ways: if it has aged too long, a wine is said to be 'over the hill', not long enough and the wine is not fully mature, or it may be just right ('at its peak').

During bottle aging, the previously described reactions make wine particularly sensitive to storage temperatures (Section 13.4.5). Wine may also deteriorate if the cork is no longer airtight. Rapid oxidation, caused by a porous, non-airtight cork, degrades the wine completely in a very short time. Aging phenomena normally take place slowly in a reduced medium, but reactions may be altered by changes in temperature. Wine develops very slowly at 12°C, but much faster at 18°C. This property may be used to prepare medium-quality wines for sale in a short time. Furthermore, variations in temperature between summer and winter must be avoided for wines aging in the bottle, as they modify the volume of the liquid. As the volume shrinks, air may be sucked into the bottle, which is likely to have a negative effect on the wine's development. Wine kept with no added oxygen, in flame-sealed airtight bottles, does not age, at least during the first year. The color changes very little, the bouquet does not develop, the wine remains closed, and seems thinner, with a more marked tannic character (Khan, 2000).

Bottled wines, especially in clear glass bottles, are particularly sensitive to light and should be protected by keeping them in a dark place. Cellars should be sufficiently damp to ensure that corks will remain airtight. The only major drawback to excessive humidity is damage to wine labels. It is also generally considered that wines should be protected from vibrations.

13.5.2 Chemical Explanations

When a wine ages in the bottle, the oxidation–reduction potential decreases regularly until it reaches a minimum value, depending on how well the bottle is sealed. Reactions in bottled wine do not directly involve oxygen. If the cork is no longer airtight, an oxidized character develops. The color of red wines tends towards orange and maderized white wines turn a brownish yellow. However, although they are difficult to measure, microscopic quantities of oxygen have also been shown to play a variable role, depending on the quality of the cork and the position of the bottle (Khan, 2000). It has been observed that changes in color and anthocyanin concentration vary according to temperature.

Several reactions occur during bottle aging:

1. The phenols evolve towards homogeneous polymerization of the tannins, accompanied by condensation of the anthocyanins and tannins, involving the carbocations formed after protonation of the procyanidins (Section 6.3.5). According to Haslam (1980), these types of reaction continue throughout the aging process until the polymers precipitate, which has the effect of softening the wine. The formation of the carbocations required for polymerization implies the breakdown of certain polymers present at the time of bottling. The diversity of the reactions involved is perhaps the cause of the organoleptic variations observed in high-quality wines during the first few years (Figure 13.8).

    These polymerization and condensation processes certainly occur in all red wines, but at different rates and with varying intensity, according to the type of tannins in the wine.

Polymerized procyanidins react more slowly than oligomeric procyanidins, while procyanidins linked with ethyl cross-bonds are even more reactive (Guerra, 1997). This affects the results of the tannin assay, as these combined molecules are converted into anthocyanins on heating in an acid medium (Section 6.4.3). The tannin value decreases during the first few years and then increases again. The values relating to the tannin structure indicate that polymerization occurs, followed by precipitation. The dialysis and HCl values (Section 6.4.4) increase to a maximum of around 30–40 and then decrease. One or more polymerization–precipitation cycles may occur, depending on the wine's phenol content and the length of time it is aged.

2. Color intensity increases while the results of the anthocyanin assay decrease, finally reaching a minimum around of 50 mg/l. The free anthocyanins gradually disappear and color intensity varies little, but the hue evolves towards orange. These modifications are due to transformation of the anthocyanins:

   (a) Over time, anthocyanins are affected by breakdown reactions that lead to the degradation of the flavylium structure (Piffaut *et al.*, 1994). Syringic acid and trihydroxybenzaldehyde may be formed.

   (b) Anthocyanins red and condense directly with flavanols, forming orange complexes (Guerra, 1997) with xanthylium structures (Jurd and Somers, 1970; Santos-Buelga *et al.*, 1995). Other reactions are also involved, producing reddish orange or even yellowish pigments with a new oxygen heterocycle bonded to the flavylium structure in 4 and 5, e.g. following a reaction with a vinyl phenol (Cameira dos Santos *et al.*, 1996; Sarni-Manchado *et al.*, 1996).

At the time of bottling, a large proportion of the anthocyanins (over 50%) is already combined with tannins, in the form of reddish-purple complexes with ethyl cross-bonds, formed during aging in an oxidizing medium (Section 6.3.10). During bottle aging, some of these complexes develop into orange tannin-pyranoanthocyanin structures and the rest may precipitate. The speed of these reactions is highly variable according to the type of combination, but they are much slower than those involving free anthocyanins. The color of a wine with a high level of anthocyanins combined evolves more slowly in the bottle than that of another wine with a low concentration of combined anthocyanins, but more quickly with anthocyanins combined by oxidative process than direct reaction. In a comparison of two wines with the same total anthocyanin content after fermentation, the color of a wine with a high concentration of combined anthocyanins evolved more slowly in bottle than that of another with a low.

3. It has been observed that bottle aging corresponds to a gradual stripping of the wine. It is logical that the length of time a wine will be capable of developing with age depends on its phenol content. 'The more tannins and anthocyanins a wine has, the more likely it is to age well.' It is true that the greatest wines, with good aging potential, have high phenol levels, but the opposite is not always true. The structure of the tannins extracted during fermentation and the type of aging affect all these reactions. The changes in terms of flavor tend to be relatively limited. A wine that tastes hard and astringent at the time of bottling will generally retain that character, even after several years.

   In addition to the molecular structure, the tannin/anthocyanin ratio also affects wine development. Color has been observed to change rapidly due to anthocyanin breakdown reactions when the medium has a low tannin content (molar ratio $T/A \leq 1$) or as a result of tannin polymerization reactions when the tannin concentration is much higher than the anthocyanin content (molar ratio $T/A \leq 4$). If a wine is to develop harmoniously, this ratio should be between 1 and 4, i.e. 500 mg of anthocyanins and 1–3 g of tannins per liter of wine.

4. Polysaccharides affect the speed of the reactions or, more precisely, the stability of the polymers (protective colloids) that have been formed.

They form tannin–polysaccharide complexes by reacting with tannins and thus keep the micelles in solution (Section 9.4.1). This mechanism contributes to the inactivation of tannins and, therefore, the softening of the wine (Augustin, 1986), provided that these complexes are not large enough to produce a colloidal state that would result in their precipitation. Great wines always have a high polysaccharide concentration, compared to that of more modest wines from the same vintage.

A wine's behavior during aging therefore depends on a whole series of factors related to its composition. These include not only phenol content, but also the type of tannins and their structures, as well as the tannin/anthocyanin ratio and the polysaccharide concentration. Furthermore, the wine's initial composition changes during aging, developing complex new structures likely to slow down the process of stripping.

The phenol composition of old wine is relatively simple. The anthocyanins have disappeared, leaving only yellowish-orange complexes and a few procyanidin and flavylium molecules, as well as some xanthylium nuclei. These are present in the form of aggregates, known as condensed tannins (CT and CTt, Figure 13.6), and may also be combined with polysaccharides (TP). The mean degree of polymerization (mDP) can be low (2–5) in wines that have aged too long (dried out, thin, etc.) or high (10–20) when a wine is at its peak (Figure 13.7).

13.5.3 Development of the Bouquet

It has been observed that the bouquet of red and white wines develops after only a short period of bottle aging, generally when all the dissolved oxygen has reacted and the oxidation–reduction potential has reached its lowest value (≤ 200 mV). This varies with the type of wine, its SO_2 concentration and the type of closure (cork, etc.). High temperatures and light stimulate reduction in the medium and modification of the aromatic characteristics.

A wine's bouquet is formed by complex reactions corresponding to the formation of reducing substances and harmonization of the grape aromas developed during the alcoholic and malolactic fermentations, as well as aromas connected with barrel aging (wood, vanilla, etc.). These reactions are inversely proportional to the oxidation–reduction potential, increasing as the latter decreases (Ribéreau-Gayon, 1931, 1933) and continuing as long as the potential remains low. There is, however, apparently, a limit beyond which the bouquet no longer develops (totally airtight containers).

Temperature and light provide favorable conditions for reduction, so they help to accelerate, and even modify, the process. A wine's bouquet is always enhanced after the summer and varies with the temperature of the bottle storage cellar. Burnt overtones are typical of wines stored at temperatures above 25°C. Light is often responsible for off-aromas (Section 8.6.5) linked to homolytic reactions (producing thiols, etc.). Accelerated aging mainly affects the color and rarely results in well-balanced products, as opposed to slow development in a dark place, at temperatures below 20°C.

When wine is aged in glass bottles, the closure is responsible for maintaining an airtight seal. The quality of the cork, the type of capsule, and the position of the bottle determine the wine's reduction state, i.e. consumption of both the oxygen dissolved in the wine and that remaining in the ullage in the bottle. Ribéreau-Gayon et al. (1976) reported that, after four months, the minimum oxidation–reduction potentials of white wines were comparable, whether they had aged in control flasks with glass stoppers (162 mV) or 750 ml bottles with corks (168 mV). The potential may remain as high as 320 mV in a bottle with an ineffective screw cap. When wine is packaged in containers made of insufficiently airtight porous material (PVC, polyethylene, etc.), the SO_2 disappears and large quantities of oxygen are dissolved in the wine. The E_H limit is never reached and the bouquet cannot develop, while the wine runs a major risk of oxidation.

However, even in an ideal situation with a glass bottle and a properly airtight natural cork, differences are observed in the quality of the bouquet depending on storage temperature. These are due to the migration of microscopic quantities of oxygen into the wine. The presence of a capsule and the position of the bottle apparently also have an influence on bouquet (Table 13.13). Color and anthocyanin content vary, and may be considered as markers for the wine's development (Khan, 2000). Each bottle of wine is, quite clearly, a special case, with its own bouquet, due to the heterogeneity of stoppers, corks and storage conditions.

The bouquet of a majority of great wines develops as a result of reduction processes, while on the contrary, flatness, which may even be considered desirable in some cases, is an oxidative phenomenon (Section 8.2.3). Unlike bouquet, flatness is linked to the appearance of oxidizing substances. Both reduction and oxidation are reversible, and disappear under conditions opposite to those that produced them (Ribéreau-Gayon et al., 1976). Flatness corresponds to the disappearance of the bouquet and the appearance of a bitter, chocolate, acrid or burnt character, accompanied by harshness on the palate. This effect, caused by aeration, produces highly oxidized wines. Whites are described as 'maderized' and reds as 'rancio'. The formation of this oxidized character is fostered by high temperatures, so it occurs more easily in summer rather than in winter. Aldehydes responsible for the oxidized character, especially ethanal, are always formed when wine comes into contact with air during handling. They cause this effect if trace amounts remain in a free state, and do not react with phenols or combine with SO_2. The unpleasant, oxidized character disappears rapidly when SO_2 is added.

13.5.4 Accelerated Aging

There is obviously a certain economic advantage in accelerating the transformations that occur during aging, and reducing the length of time wines need to be stored. Standard accelerated aging processes involve strong oxidation with wide variations in temperature, based on the way wine develops during aging. Basically, it has been observed that wine ages mainly in summer, then throws deposits and stabilizes in winter. Aging can, theoretically, be accelerated by reproducing these seasonal effects at shorter intervals. The process consists of saturating wine with air, or oxygen, at low temperatures and then heating it to $20-25°C$. It is subsequently cooled, then oxygenated and heated again, etc. The cycle concludes with low-temperature filtration prior to bottling. After this process, the color is brick red and the aromatic character of the young wine has disappeared, but the characteristic bouquet of old wine has not developed. Wines treated in this way become dried out, and sometimes have burnt or cooked flavors, or even oxidized overtones. Color may be rapidly changed in this way, but the wine does not develop a proper bouquet. The aromas do not develop with the color and changes that occur may even be unpleasant.

A number of other aging techniques have been tested, using various physical processes:

Table 13.13. Changes in the phenol content of a Bordeaux wine (1998) stored in bottle at a temperature of $15-23°C$ under various conditions. Impact of the type of closure on the color of the phenolic compounds (Khan, 2000).

		OD 420%	OD 520%	OD 620%	CI (*)	H (*)	TP (*)	T g/l	A mg/l	PVPP (index)
Control T = 0		40.1	49.3	10.7	0.62	0.81	52	3.2	271	29
Horizontal	} Natural cork	37.9	51.2	10.8	0.62	0.74	45	3	219	38
Horizontal with wax capsule		38	51.2	10.8	0.61	0.74	46	3	220	38
Horizontal with plastic cork		35.7	50.6	13.6	0.94	0.71	46	3	198	52
Vertical	} Natural cork	36.2	52	11.7	0.76	0.70	46.6	3	205	43
Vertical with wax capsule		37.9	51.2	10.9	0.63	0.74	45.4	3	213	38

(*)
CI: color intensity (OD 420 + OD 520 + OD 620)
H: Hue (OD 420/OD 520)
TP: Total phnenol (OD 280)

ultrasound, infrared and ultraviolet radiation, high pressure and even electrolysis. Singleton (1974) also came to this conclusion: accelerated aging leads to results inferior to those obtained by normal aging processes. The effects of all these treatments cannot be controlled.

13.6 WINEMAKING PRACTICES

13.6.1 Winery Hygiene

Wine receives constant care throughout the time it spends in the winery, right up until bottling. Different cellar operations are required, depending on the way the wine is aged, in vat or in barrel. Wine aged in the barrel requires a great deal more attentive care, as it is less well protected and more vulnerable to outside conditions likely to cause extensive, harmful changes.

The most important requirement in cellar work is cleanliness in the winery and hygiene in all containers (Peynaud, 1975). Wine gives the impression of a certain stability, due to the presence of alcohol and relatively high acidity. In fact, it is sensitive, not only to microbial deterioration but also to various types of contamination that may give the wine unpleasant odors and off-flavors that are impossible to eliminate. Cleanliness in all winemaking operations, from fermentation to bottling, is an indispensable prerequisite for quality.

The need for proper cleaning and maintenance must be taken into account from the design stage in all winery buildings used for fermentation or storage in vats. Wineries must be spacious and properly aerated, with artificial ventilation if necessary. They must always be equipped to facilitate washing operations.

Barrel aging cellars must be kept at relatively low temperatures, with no sudden variations and without excessive aeration or ventilation. Small cellars may be easier to regulate. Evaporation from the wine maintains a certain level of humidity. Cellars with mold or saltpeter on the walls should be avoided, as the evaporation of alcohol fosters these growths, which can harbor an undesirable microbe population.

It is preferable to replace beaten-earth floors with concrete or even tiles, ensuring efficient drainage for cleaning water. Walls should be coated with antifungal paint or, better still, tiled.

All equipment and fittings must also be constantly maintained in a perfectly clean condition. Negligence may lead to contamination that would be impossible to eliminate (Sections 8.3, 8.4 and 8.5). It is recommended that installations have built-in systems for cleaning with cold water, hot water and, possibly, steam, with or without detergents.

Another recommendation concerns the elimination of *Drosophila* fruit flies, likely to contaminate must and new wine with acetic bacteria. Systems that constantly release volatile insecticides are easily available. The same type of system may be used in bottle aging cellars to destroy moths and cork worms.

13.6.2 Hygiene Precautions for Wine Containers

All containers must be kept clean to avoid deteriorating wine quality. Wooden vats should be completely dry if they are to be left empty. The slightest trace of humidity rapidly leads to the development of mold. It is sometimes advised not to wash wooden vats after use. Once they have dried, they must be carefully brushed, then treated with burning sulfur. These vats must be properly cleaned and rinsed before reuse. Alternatively, once the tartrates have been removed, the vats may be disinfected with various antiseptic cleansers (Diversey), rinsed with plenty of water, dried with a hot air stream and treated with sulfur.

Concrete vats must, of course, be cleaned regularly, and the protective lining must also be checked. Epoxy resin-based lining is increasingly used and it must be checked frequently for imperfections. Tartrates must be removed as necessary. When the concrete is only conditioned with tartaric acid, care must be taken to avoid accumulating layers of tartrate deposits that may foster microbial spoilage. This is likely to give the wine off-flavors that are impossible to eliminate by simple cleaning. Tartrates must be removed regularly, either by hand, using a welding torch, or by spraying with an alkaline solvent solution.

Easy maintenance is one of the major advantages of stainless steel vats. They still require a minimum of attention, however, especially to remove tartrate deposits.

The sides of empty wooden vats are disinfected by burning sulfur inside. This is not advisable for concrete or stainless steel vats, as the sulfur dioxide attacks the material.

Special care should be taken in maintaining wooden barrels. New barrels must be conditioned to eliminate any bitter, astringent off-flavors (Section 13.9.1), by cleaning with steam or boiling water, or by keeping the barrels full of slightly sulfured cold water for a few days.

During racking, rinsed barrels must be drained for 24 hours and left to dry in a draft before the sulfur wick (5 g) is burnt inside.

The problems raised by the storage of empty barrels prior to reuse are often difficult to solve. Sulfuring is essential, but excessive use may cause the formation of large quantities of sulfuric acid. This accumulates in the wood and acidifies the wine when the barrel is filled. However, if insufficient amounts of sulfur are burned, bacteria may develop. Volatile acidity and ethyl acetate may be produced in the 5 l of liquid (mixture of wine and water) that impregnate the barrel and contaminate the wine when it is filled. Furthermore, once a barrel has been dried, the wine adsorbed in the wood reduces the porosity of the oak. It is advisable to fill barrels that have been stored empty with water for a few days before filling them with wine, to eliminate any possible impurities from the wood. Whenever possible, it is preferable to organize work in the cellar so that there are never any empty barrels. Extreme care must be taken in buying used barrels of unknown origin.

As soon as they have been emptied and the lees removed, the barrels must be carefully cleaned, rinsed with a high-pressure water jet and drained for a few minutes to eliminate excess humidity. Ten or, more generally, 20 g of sulfur are burned inside each barrel, which is then left to drain for 5 or 6 days. These conditions ensure that the oak will dry without any microbial contamination. It has, however, been clearly demonstrated that burning sulfur in damp wood provides much less effective sterilization than in dry barrels. It takes several days to obtain complete sterilization in damp barrels, but only a few hours in dry wood. For this reason, even if sulfur has been burned inside barrels when they are damp, the same quantity of sulfur should be burned in them again once they are dry, as sterilization will be more effective. The barrels are then closed with a bung. In theory, they should keep indefinitely in this condition. In reality, the wood may dry out, so the staves will no longer be airtight and air leaks in, reducing the SO_2 concentration. It is usually considered that empty barrels should not be left for more than two or three months without burning sulfur inside.

13.6.3 Racking

Racking is one of the essential operations in cellar work (Section 10.3) and it has a number of objectives: (a) clarifying the wine by eliminating the lees, (b) homogenizing wine stored in large vats, or blending wines from a number of different barrels, (c) degassing by eliminating excess CO_2 to achieve concentrations suited to the type of wine, (d) introducing oxygen, thus increasing the oxidation–reduction potential (Section 13.3.3) and the controlled oxidation of phenols in red wine, (e) adjusting free SO_2 in order to ensure proper protection from oxidation and, above all, microbial spoilage, and (f) ensuring barrel hygiene (Section 13.6.2).

13.6.4 Topping Up and Wine Loss

Inert gases ($N_2 + CO_2$) may be used to prevent chemical oxidation on the surface of wine and the development of acetic bacteria when it is stored in vats that are not completely full (Volume 1, Section 9.6.1.), although it is generally preferable to keep containers full. Regular topping up, to ensure that containers are always full, is another essential cellar operation.

The wine level in oak barrels drops rapidly after filling. The wood absorbs 5 liters in the first few days and then wine starts to evaporate due to the porosity of the wood and possible slight leakage around the bung. As the volume of wine decreases, the surface area in contact with the air increases, as does the risk of oxidation. Topping up consists of filling barrels regularly to the top, using healthy wine of the same quality set aside for this purpose, either in a demijohn with a floating lid to avoid contact with oxygen or in a vat kept under inert gas ($N_2 + CO_2$).

Evaporation varies according to several factors:

1. Cellar humidity should be between 80 and 90%. Above this level, cellars are too damp and alcohol evaporates rather than water, so the alcohol content decreases. Cellars with humidity below 80% are too dry, water evaporates and large volumes of wine are lost. The average annual evaporation rate is 4–5%.

2. Ventilation plays a major role, especially in air-conditioned wineries where ventilators blow air directly over the barrels. Evaporation varies considerably according to the position of the barrel in relation to the air circulation (on the ground or on upper layers). Furthermore, the effectiveness of the winery's insulation determines how much of the time the ventilators will be in operation.

3. The characteristics of the oak, depending on its origin, affect the intensity of oxidative phenomena and evaporation. Barrel age is also an important factor, as used barrels are less porous than new barrels.

Other factors also cause variations in wine volume in barrels and vats:

1. Fluctuations in atmospheric pressure can cause overflows when pressure drops and ullage when it rises.

2. Temperature variations in wineries without air-conditioning cause expansion and contraction which may also result in a considerable drop in level.

Topping up both vats and barrels is, therefore, an indispensable operation.

In vats, the operation may be automated using a communicating vessel system controlled by solenoid valves and supplied by a storage vat under inert gas. Topping up may be a continuous or weekly operation.

Barrels are topped up manually. Formerly, a can equipped with a long spout was used, but this has now been replaced by a transfer system using nitrogen under pressure. This is especially necessary when the barrels are stored with the bung at the top, particularly if the bung is just sitting in position and there is no airtight seal. Topping up may be necessary twice a week or once every two weeks, according to the desired level of oxidation.

When barrels are stored with the bung on the side or hermetically sealed with a silicon bung, topping up is no longer necessary. The space left by evaporation has a very low oxygen content, as the gas has already been dissolved and consumed by the wine (Vivas and Glories, 1993a). Under these conditions, wine is stored under a nitrogen atmosphere with very little head space, as shown by the air that rushes in when the barrel is opened.

Another system, based on certain ancient practices where pebbles were put inside the wine vats, consists of using a balloon made of food-grade rubber, connected to an air or oxygen source. The balloon is inflated inside the vat to compensate for low wine levels and deflated as required to avoid overflows.

13.7 BARREL AGING RED WINES

13.7.1 Role of Barrel Aging

Traditionally, great red wines are aged in oak barrels from the end of fermentation until bottling. The first motivation in choosing barrels was probably that they were easy for one man to handle and could also be used for shipping. It was not until some time later that their positive effect on wine development, in terms of color, clarity and flavor, came to be appreciated. However, the use of barrels involved a major financial commitment and entailed risks of microbial contamination, as well as the likelihood of communicating organoleptic faults to wine (Section 13.9.2). For these reasons, the practice of aging even high-quality red wines in inert vats became widespread from 1950 until 1960. At that time, the elimination of old barrels, responsible for moldy off-flavors, certainly resulted in improved quality. The red wines were perhaps less complex, but cleaner and fruitier.

Over the past few years, a more favorable economic climate has fostered a new interest in barrel aging. There is greater awareness of the role played by oak in wine development, and a concern to adapt barrel aging to the quality of each wine. Perfect control of the various parameters and techniques has made it possible to fine-tune the use of wood and its influence on wine quality (Guimberteau, 1992). This section deals exclusively with the issues relating to the barrel aging of red wines. White wines must be fermented in the barrel to obtain high-quality results and the techniques are described elsewhere (Volume 1, Section 13.8).

Firstly, clarity is easier to obtain when wine is aged in the barrel rather than in the vat, due to the smaller volume. Clarification is also facilitated by the adsorption phenomena that occur in oak. Furthermore, wine in the barrel is more sensitive to outside temperature, so the precipitation of salts, particles and colloidal coloring matter is much more likely to be triggered by winter cold.

Stabilization reactions affecting color, clarity and colloids, as well as modifications in the phenol structures (softening of the tannins), also occur in wine during aging, while aromas develop. Barrel aging promotes these reactions to a much greater extent than large airtight vats, which, being theoretically inert, are considered not to interact with the wine.

The phenol composition of wine is considerably modified by barrel aging, thanks to controlled oxidation. Color is intensified due to reactions between tannins and anthocyanins, as well as others involving ethanal. The free anthocyanin concentration decreases and the tannin structure evolves, as does its reactivity to gelatin. After ten months of barrel aging, wines have better color than those aged in the vat (Table 13.8; Section 13.4.2) and this color remains more stable during bottle aging. The flavor is also more attractive, characterized by softer tannins.

Wine also acquires aromatic complexity as a result of the odoriferous substances extracted from wood. The oaky aroma must be carefully modulated, to ensure that it blends harmoniously with the wine's overall structure. Even though producers may wish to give their wines an oak character, this must not be overdone. It should never overpower the wine's intrinsic qualities. The barrel's contribution to aroma and flavor may be adjusted by modifying the proportion of wine aged in oak, especially new barrels. Other important factors are the type of oak and the way the barrels are made (degree of toasting), as well as the duration of barrel aging.

Three factors related to this type of aging are responsible for the wine's development: oxidation–reduction reactions, as well as volatile and non-volatile compounds dissolved from the oak.

13.7.2 Oxidation–Reduction

Oxygen in red wines may have various origins. Handling operations, treatments and regular winemaking tasks represent a major proportion (up to 50%), while barrel aging accounts for the remainder. The amount of oxygen absorbed by the wine depends on the origin of the barrels, as well as the type and position of the bung (Table 13.14). It is thought (Vivas, 1997) that oxygen passes through the wood (16%), mainly via gaps between staves (63%). Smaller amounts (21%) are admitted through the bunghole. The position of the bung affects the penetration of oxygen into the wine. Wooden bungs positioned on the side of the barrel and tight silicon bungs produce a vacuum effect on the order of 120 mbar (Moutounet et al., 1994), which increases the quantity of oxygen dissolved in the wine.

It is, however, difficult to determine the precise quantity of oxygen that penetrates into the wine, as measurements, even in model solutions, do not take into account the amounts consumed by ellagitannins in the oak. Dissolved oxygen constantly disappears by oxidizing various components in the wine. Quantities may vary widely, from 100 to 200 mg/l—values significantly higher than those previously reported (Ribéreau-Gayon et al., 1976)—depending on the aging method (in barrel or in vat).

The oxidative phenomena involved in barrel aging are not exclusively due to increases in the wine's oxygen content. Ellagitannins from the oak are dissolved in wine (castalagin, vescalagin, roburins, etc.), and although concentrations

Table 13.14. Oxygen dissolved during various operations involved in the aging of red wines (Vivas, 1997)

Origin	Operation	Dissolved oxygen
Handling	Pumping	2 mg/l
	Transfer from vat to barrel	6 mg/l
	Transfer from vat to vat, filling from bottom	4 mg/l
	Transfer from vat to vat, filling from top	6 mg/l
Treatment	Earth filtration	7 mg/l
	Plate filtration	4 mg/l
	Centrifugation	8 mg/l
	Bottling	3 mg/l
Winery operations	Racking with aeration	5 mg/l
	Racking without aeration	3 mg/l
	Topping up barrels	0.25 mg/l
Wood	New barrels: Limousin	20 mg/l an
	New barrels: Center, Wooden bung hammered in on top	28 mg/l an
	New barrels: Center, Wooden bung on one side	36 mg/l an
	New barrels: Center, Silicon bung on top	45 mg/l an
	Used barrels (5 wines)	10 mg/l an

are difficult to assess accurately, they may be estimated at around a hundred mg/l. They decrease regularly, due to oxidative phenomena catalyzed by these same substances. Even in the absence of oxygen, ellagic tannins are capable of modifying the tannin structure of a wine, as well as combining with anthocyanins and, consequently, stabilizing color (Jourdes et al., 2003).

13.7.3 Non-Volatile Compounds Extracted from Oak Barrels

In addition to ellagitannins, the oak releases a certain number of other compounds (Section 6.2.4), mainly lignins with a high guaiacyl and syringyl content, lignans (lyoniresinol), and triterpenes (Aramon et al., 2003). Coumarins are also present in oak (Section 6.2.1). The concentration in wine depends on the type of wood and the way it is seasoned. These compounds may be dissolved in wine in heteroside (scopoline, esculin) and aglycone (scopoletin, esculetin) form (Salagoïty-Auguste et al., 1987).

Another group of molecules extracted from oak, no doubt produced by the transformation of ellagitannins and possibly lignin, contribute to increasing the phenol acid concentration of wine (Section 6.2.1), mainly gallic acid, by a concentration of about 50 mg/l (Pontallier et al., 1982).

In terms of flavor, studies investigating the organoleptic characteristics of these components produced the following findings (Vivas, 1997):

1. Phenol acids (gallic acid) have an acid taste.

2. Coumarins (aglycones) seem acid and have a harsh character. Their glycosides are very bitter.

3. Ellagitannins are astringent as compared to gallotannins, which give a bitter, acidic impression. The benchmark, procyanidin B3, is both astringent and bitter.

Depending on conditions, oak may also release polysaccharides, mainly consisting of hemicelluloses, that contribute to wine flavor.

It is therefore quite understandable that wines aged in oak barrels have different organoleptic characteristics from those aged in the vat. Oak has two contradictory effects: it strengthens the impression of harshness due to the phenol components it releases, while softening condensed tannins thanks to its heterogeneous polymers. The result depends on the relative intensities of these two phenomena. There is a risk of toughening, depending on the wine's tannic structure and the characteristics of the barrels (origin, type, preparation, technology, etc.).

In any event, even if aging in oak barrels increases the phenol content of wine, it is by no means sure that it increases the overall tannin content, at least in red wines. Analysis has shown that the total phenol index (OD 280) only increases by a few units due to wood tannins, compared to an initial value between 50 and 60. Of course, the increase is proportionally more significant in white wines, with their much lower initial tannin content.

13.7.4 Volatile Compounds Extracted from Oak Barrels

Another fundamental aspect of aging wines in oak concerns the aromatic compounds that are extracted. When these compounds marry perfectly with a wine's intrinsic aromas, they make a significant contribution to the richness and complexity of the bouquet, as well as improving the flavor. Great red wines are almost always aged in oak, as barrel aging enhances their quality and finesse. In order to benefit fully from barrel aging, wine must have a certain aromatic finesse and a sufficiently complex structure to blend well with the organoleptic input from the oak. Ordinary wines cannot be turned into quality wines by exposing them to oak. Attempts at flavoring wines that do not justify this treatment, resulting in a standardized, 'woody' character, should be approached with great caution.

Untreated oak contains a certain number of volatile substances with specific odors (Figure 13.9). As previously described, these lactones, in particular β-methyl-γ-octalactone, with four enantiomers (Figure 13.10), two geometrical isomers and two optical isomers, are produced by the breakdown of complex polymers. The $cis(-)$ isomer has an earthy, rather herbaceous character with

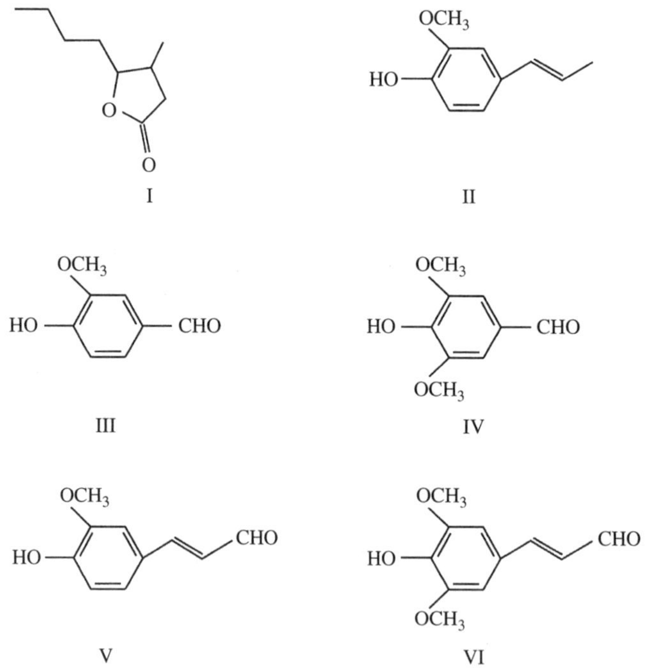

Fig. 13.9. Structure of the main volatile compounds identified in extracts of non-toasted oak wood: I, methyl-octalactone or methyl-4-octanolide, II, eugenol; III, vanillin; IV, syringaldehyde; V, coniferaldehyde; VI, sinapaldehyde (Chatonnet, 1995)

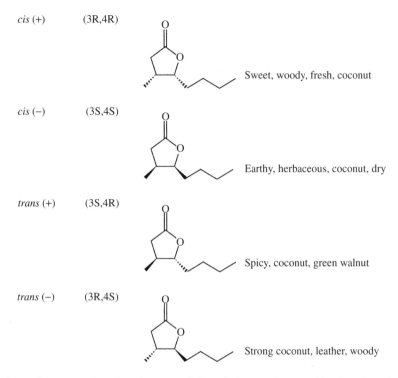

Fig. 13.10. Formulae and aromas of various isomers of β-methyl-γ-octalactone (the first three have been identified in natural oak) (Günther and Mosandl, 1986)

hints of coconut, and is 4–5 times more odoriferous than the *trans* (+) isomer. The latter not only smells of coconut, but is also very spicy (Chatonnet, 1995). Above a certain concentration, excessive amounts of this lactone may have a negative effect on wine aroma, giving it a strong woody or even resinous odor.

Eugenol, with its characteristic odor reminiscent of cloves, is the main volatile phenol (Figure 13.9). Other volatile phenols are present in relatively insignificant quantities.

Phenol aldehydes (Figure 13.9) are also present, but in relatively small quantities. Vanillin and syringaldehyde (benzoic aldehydes) have been identified, as well as coniferaldehyde and sinapaldehyde (cinnamic aldehydes). Vanillin plays an active part in the oaky and vanilla odors that barrels communicate to wine.

Concentrations of *trans*-2-nonenal vary a great deal from one oak sample to another. Together with *trans*-2-octanal and 1-decanal, this molecule is responsible for the odor known as 'plank smell' that wines may acquire during barrel aging. This unpleasant smell is attributed to unseasoned wood (Chatonnet and Dubourdieu, 1997) and may be attenuated by toasting the inside of the barrels more intensely (Section 13.8.3).

Oak may also release norisoprenoid compounds into wine (Section 7.3). The most important of these is β-ionone.

Oak from different origins is odoriferous to varying degrees. Its characteristic odors are mainly revealed during seasoning and barrel manufacture. Heating (Section 13.8.3) forms furanic derivatives, volatile phenols and phenylketones, as well as increasing concentrations of phenol aldehydes and lactones.

The types and concentrations of odoriferous substances released into white wines according to the origin of the wood and degree of toasting of the barrels is described elsewhere (Volume 1, Section 13.8, Table 13.19).

13.8 EFFECT OF THE TYPE OF BARREL ON THE DEVELOPMENT OF RED WINE

Phenomena affecting oxidation and aromas that occur in wine during barrel aging depend on many parameters, such as the type of barrels and the way they are made.

13.8.1 Origins of the Wood

In France, wood for cooperage comes primarily from forests located in four main regions (Limousin, Centre, Bourgogne (Burgundy) and Vosges). Two species are unevenly distributed in these regions:

1. Pedunculate oaks (*Quercus robur* or *Quercus pedunculata*) grow most widely in the Limousin. They are also present in Bourgogne, as well as the south of France. They have a high extractable polyphenol content and relatively low concentrations of odoriferous compounds.

2. Sessile or durmast oaks (*Quercus petraea* or *Quercus sessilis*) are prevalent in the Centre and Vosges regions. They generally have a high aromatic potential and rather low levels of extractable ellagitannins.

Other species of oak grow in France, but are not used in cooperage.

Pedunculate oaks, mainly those from the Limousin, grow in coppices with standards on clay–limestone and rich granite soils. Their annual growth rings are broad and evenly spaced and the wood is coarse-grained (Figure 13.11). Sessile oaks, mainly present in the Centre and Vosges, grow on poorer clay-siliceous soils in high forests. The annual growth rings are narrow and the wood is fine-grained.

The heartwood (duramen) used in cooperage no longer provides any physiological function in the tree. Heartwood is resistant to insects and fungi, and is very hard. Its complex structure consists of three main categories of tissue: fibers (support units), parenchyma and radial cells (reserve tissues), as well as conducting vessels. Vivas (1997) demonstrated a link between the porosity of wood and its ultrastructure, identified under an electron microscope: macroporosity depends on the quantity of large vessels, while the other tissues control microporosity. According to this author, wood from *Quercus robur* grown in the Limousin is less porous than that of *Quercus petraea* from the Vosges and Allier forests (Table 13.15). These findings, in agreement with those of Feuillat *et al.* (1993), were obtained using a different process, and are also quite logical, as they are related to the total quantity of vessels in the wood.

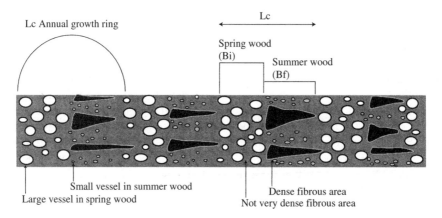

Fig. 13.11. Simplified cross-section diagram showing the structure of heartwood (duramen). Definition of grain (G = Lc (mm)) and texture (T = Bf/Lc): Lc, width of an annual growth ring; Bi, early or spring wood; Bf, late or summer wood (Vivas, 1997)

Table 13.15. Measuring the porosity characteristics of oak samples (porosity estimated by image analysis) (Vivas, 1997)

Origin	Species	nGV.Bi	D (mm)	SU (mm^2)	ST (mm^2)	IP (%)
Limousin	Q. robur	12	320	0.08	0.964	6.5
Vosges	Q. petraea	23	248	0.048	1.11	7.5
Centre	Q. petraea	20	275	0.059	1.187	8
Allier	Q. petraea	27	324	0.082	2.22	15

nGV.Bi = number of spring wood vessels, D = diameter, SU = unit surface area of one vessel, ST = total surface area of the large vessels, IP = porosity value = (ST/total surface area of the image) × 100.

Geographical origin and species have a considerable influence on the aromatic and polyphenol content in oak (Table 13.16). Recent research has shown that the qualities of oak wood vary, not only according to the species of oak but also on the age of the tree, the height of the sample, the direction in which it was facing and the region of production. It is therefore insufficient to classify wood merely by geographical origin. This concept could be replaced by a definition taking into account the type of wood a sample resembles (Allier, Limousin, etc.), even if it is from elsewhere.

Table 13.16. Influence of geographical origin on the composition of French oak, naturally seasoned in the open air[a] (Chatonnet, 1995)

Parameters	Geographical origin			
	Limousin	Centre	Bourgogne	Vosges
Total extractables (mg/g)	140	90	78.5	75
Total polyphenols (OD 280)	30.4	22.4	21.9	21.5
Coloration (OD 420)	0.040	0.024	0.031	0.040
Catechic tannins (mg/g)	0.59	0.30	0.58	0.30
Ellagitannins (mg/g)	15.5	7.8	11.4	10.3
Methyl-octalactone (μg/g)	17	77	10.5	65.5
Eugenol (μg/g)	2	10	1.8	0.6

[a]Mean of 7 samples; compounds extracted in a dilute alcohol medium, under standard conditions.

In Europe, the distribution of sessile and pedunculate oaks varies according to latitude, although pedunculate oaks predominate. It is possible to find wood from forests in Central Europe (Russia and Hungary) from the species *Quercus farnetto* that have certain similarities to French oak. Trials in progress since 1994 (Chatonnet, 1995) have produced some interesting results.

In the USA, the dominant species is American white oak (*Quercus alba*). This species has a low phenol content and a high concentration of aromatic substances, especially methyl-octalactone, which strongly affect the flavor of wine during aging (Table 13.17).

Another characteristic of American white oak is that staves may be sawed thanks to the natural blocking of longitudinal vessels (due to tylosis) which prevents leaks. French oak, however, must be split along the grain, then split again into the staves used to make barrels. This technique is more complex than sawing and results in considerable wastage. This is, however, the only way to avoid leaks due to vessels running through the width of the staves. Barrels made from sawed French oak should be sealed inside to avoid leaks, but wood protected in this way no longer has a beneficial effect on aging.

13.8.2 Influence of Seasoning Conditions

The oak's humidity level should be in equilibrium with that of the surrounding atmosphere, on the order of 14–18% in temperate regions, to ensure the barrel's mechanical strength. Oak wood must

Table 13.17. Variations in volatile and fixed compounds according to the botanical origin of oak wood[a] (Chatonnet, 1995)

	Sessile oak	Pedunculate oak	American white oak
Methyl-octalactone (µg/g)	77	16	158
Eugenol (µg/g)	8	2	4
Vanillin (µg/g)	8	6	11
Total extractables (mg/g)	90	140	57
Extractable polyphenols (OD 280)	22	30	17
Ellagitannins (mg/g)	8	15	6
Catechic tannins (µg/g)	300	600	450

[a]Mean of 10 samples; compounds extracted in dilute alcohol medium, under standard conditions.

therefore be seasoned either by natural or artificial methods prior to use in cooperage.

Natural seasoning is an operation that takes several years, generally 24 months for 21 mm staves and 36 months for 28 mm staves. This length of time is necessary to obtain wood that is properly suited to the aging and improvement of wine (Taransaud, 1976). Seasoning takes place in the open air, in large, level spaces. It has been estimated that oak seasons at a rate of about 10 mm per year. In fact, intense dehydration takes place during the first 10 months. This is followed by a period when the wood actually 'matures', thus improving its physical, aromatic and organoleptic qualities. Seasoning is, however, heterogeneous, depending on the position of the wood in the pile. The outside of the pile is most intensively washed out, while the center is hardly touched by rain, or even sprinkling, and always has a lower humidity level. Enzymic reactions are also involved, caused by enzymes secreted by the fungal microflora that develop on the wood.

According to Vivas and Glories (1993b), *Aureobasidium pullulans*, always present on wood during seasoning, represents 80% of the total microflora. *Trichoderma harzianum* and *Trichoderma komingii* have also been identified, but represent less than 20% of this population. Other, rarer, species account for less than 10%, although if seasoning continues for a very long time, the wood is colonized by a diversity of flora (Larignon et al., 1994; Chatonnet, 1995).

The following observations have been made of phenol composition during natural seasoning:

1. When oak is macerated, the solution becomes less astringent and lighter colored as the wood seasons. The quantities extracted, especially ellagic tannins, also decrease. This decrease mainly affects water-soluble monomers and oligomers, whereas polymerized forms that are insoluble in water only decrease when seasoning lasts more than three years. The reduction in ellagic tannins is due to chemical and enzymic hydrolysis (*Penicillium* and *Trichoderma*), as well as oxidation of any ellagic acid that may be released.

2. Bitter-tasting glycosylated coumarins (esculin and scopoline) are hydrolyzed to form aglycones (esculetin and scopoletin) with relatively neutral or slightly acid flavors. The loss of bitterness and harshness generally observed during the seasoning and aging of wood (Marché and Joseph, 1975; Taransaud, 1976; Vivas, 1993) seems to be linked to the above modifications in phenol composition.

As previously reported, natural seasoning leads to an increase in the concentrations of various aromatic compounds: eugenol, syringic and vanillic aldehydes produced by the breakdown of lignin, as well as both isomers of β-methyl-γ-octalactone, with a higher proportion of the more odoriferous *cis* form. The effect of microorganisms on these odoriferous compounds is not very well known, although they are considered to be responsible to a varying extent for reducing concentrations of these substances.

Artificial seasoning consists of keeping split oak in a ventilated drying oven at 40–60°C for approximately 1 month. This technique considerably reduces seasoning time, without altering the oak's physical properties. It also eliminates the financial investment tied up in a wood seasoning lot. However, green wood must be seasoned gradually, to avoid shrinkage cracks. This entails alternating seasoning times of varying lengths with stabilization periods in a dry, ventilated place.

Nevertheless, this type of seasoning has certain effects on the development of the compounds in the oak. In particular, most of the reactions described in the preceding paragraphs do not occur under these conditions (Table 13.18). Compared to naturally seasoned wood, oven-dried wood has a higher content of astringent tannins and bitter coumarins. It contains less eugenol, vanillin and methyl-octalactone, with a majority of the less odoriferous *trans* isomer.

Natural seasoning is indisputably better for the quality of barrel aged wines. A combined process that has been tested may make it possible to benefit from the main advantages of each technique. Natural seasoning is used for in-depth modifications ('maturing' stage) and artificial seasoning for rapid, homogeneous dehydration. The results are somewhere between those of both techniques, and indicate that it may be possible to implement this system with a certain degree of success.

Another seasoning technique that has also been envisaged involves immersing the wood for a few days in water that is frequently replaced. This enhances the in-depth washing-out of tannins, but hardens the wood and deteriorates its physical properties.

13.8.3 Impact of Barrel Toasting

Once the stave wood is considered to be dry and seasoned, a cooper makes it into staves that are assembled (in groups of 18–25) with metal hoops to form barrels. The oak is then subjected to heating and toasting, both fundamental stages in barrel manufacture. The two stages can be summarized as follows:

1. Heating facilitates the bending of the staves to produce the characteristic barrel shape. It affects the plasticity of the lignin, but has little impact on the glucide polymers (cellulose and hemicellulose), as these compounds are protected by the humidity they absorb. The combination of heat and humidity makes it possible to bend barrels into shape without breaking the staves. Barrels, usually open at both ends, are heated for 20–30 minutes, with regular increases in temperature (<7°C mn), while the staves are gradually bent into shape. At the end of the operation, the inside temperature of the barrel is approximately 200°C, while it is only

Table 13.18. Impact of accelerated artificial seasoning on the aromatic and polyphenol composition of oak[a] (Chatonnet, 1995)

	Limousin		Centre	
	Natural seasoning	Artificial seasoning	Natural seasoning	Artificial seasoning
Dry extract (mg/g)	135	145	90	113
Total polyphenols (OD 280)	30.4	31.2	22.4	27.2
Color (OD 420)	0.040	0.038	0.024	0.030
Catechins (mg/g)	0.59	0.56	0.30	0.60
Ellagitannins (mg/g)	15.5	17.2	7.8	11.9
Methyl-octalactone *cis* (µg/g)	12	0.85	77	25
Methyl-octalactone *trans* (µg/g)	4.5	0.22	10	124
Eugenol (µg/g)	2	0.3	8	4
Vanillin (µg/g)	11	0.5	15	0.3

[a]Mean of 7 different samples; compounds extracted in a dilute alcohol medium, under standard conditions.

50°C on the outside (Chatonnet and Boidron, 1989).

2. The second operation, toasting, gives the barrel its final shape, while at the same time modifying the oak's structure and composition.

Barrel quality depends on successful toasting. This has a major impact on the later development of the wine during aging, as well as the organoleptic characteristics it acquires. However, toasting conditions vary a great deal from one cooperage to another, as well as within the same cooperage. The human element, i.e. the fact that these operations are carried out by true craftsmen, means that these parameters can only be partially controlled. Factors include: the type and intensity of the heat source (wood, gas or electricity), whether the top of the barrel is open or closed, heating homogeneity and final temperature, as well as duration (risk of the wood charring and blistering), how often the wood is moistened and to what extent it changes color.

In view of all these data, an automatic, consistent heating system has been suggested and implemented by Chatonnet et al. (1993a). This has the advantage of achieving reproducible results. The heat source consists of infra red radiation emitted by an electrical resistor. The wood is automatically moistened and the toasting time is programmed.

All of these toasting operations affect the surface and internal structure of the oak (Chatonnet, 1991). There are three levels of toasting:

1. Light toast indicates a toasting time of approximately 5 minutes, with a surface temperature between 120 and 180°C. The inside of the barrel has a spongy appearance, due to modification of the lignins and hemicelluloses, while the cellulose structure remains intact.

2. Medium toast corresponds to a toasting time of approximately 10 minutes, producing a surface temperature of approximately 200°C. The parietal surface components disappear by fusion.

3. Heavy toast corresponds to a toasting time of more than 15 minutes, resulting in a surface temperature of approximately 230°C. The cell structure is considerably disorganized, while the surface is blistered and covered with tiny cracks.

The physical changes are accompanied by modifications in the oak's chemical composition. The parietal polymers (cellulose, hemicellulose and lignin) have different fusion points and give rise to a wide variety of decomposition products.

Analysis of toasted oak extracts shows a breakdown of the ellagitannins, especially after medium toasting (Table 13.19). This is related to the fusion temperatures of a blend of vescalagin and castalagin (163°C) and gallic acid (250°C). Ellagic acid only reacts at higher temperatures ($F > 450°C$), but it is not very soluble in dilute alcohol solutions.

Heating oak also leads to the formation of volatile compounds that may have several origins. Firstly, thermal degradation of polysaccharides produces furanic aldehydes from carbohydrate polymers (mainly hemicelluloses). The resulting compounds include: furfural, methyl-5-furfural (toasted almond aromas) and hydroxymethyl-5-furfural (odorless) (Figure 13.12 and Table 13.20). However, these furanic aldehydes are present in wine at concentrations well below their olfactory perception thresholds, so they have little impact on empyreumatic nuances in barrel-aged wines. Heating also produces enolic compounds with a caramel-toasty character (cyclotene, maltol and isomaltol) (Figure 13.12) derived from

Table 13.19. Impact of toasting intensity on the polyphenols extractable from oak wood[a] (Chatonnet, 1995)

	Toasting intensity			
	Non-toasted	Light	Medium	Heavy
OD 280	17.5	17.2	15.3	13
Ellagitannins (mg/l)[b]	333	267	197	101
Gallic acid (mg/l)	20	103	9.8	2
Ellagic acid (mg/l)	21	18	13.8	13.7

[a]Mean of 3 samples taken at depths of 1.2 and 3 mm; compounds extracted in a dilute alcohol medium, under standard conditions. Results are expressed in mg/l in this solution.
[b]Expressed as hexahydroxy-diphenyl-4-6-glucose.

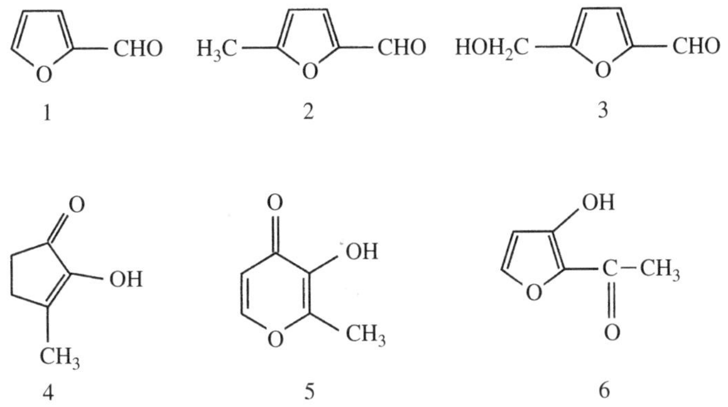

Fig. 13.12. Various compounds likely to develop when oak is heated during barrel-making: 1, furfural; 2, methyl-5-furfural; 3, hydroxymethyl-5-furfural; 4, cyclotene; 5, maltol; 6, isomaltol

Table 13.20. Impact of toasting intensity on the formation of furanic aldehydes[a] (Chatonnet, 1995)

	Toasting intensity			
	Non-toasted	Light	Medium	Heavy
Furfural	0.3	5.2	13.6	12.8
Methyl-5-furfural	0	0.6	1.3	1.5
Hydroxymethyl-5-furfural	0	3.6	6.9	4.8
Σ Furanic aldehydes	0.3	9.4	21.8	19.1

[a]Mean of 3 samples taken at depths of 1.2 and 3 mm; compounds extracted in a dilute alcohol medium, under standard conditions. Results are expressed in mg/l in this solution.

hexoses in the presence of nitrogenated substances. Their olfactory impact is greater than that of furanic aldehydes.

Thermal degradation of lignin and polyols produces volatile phenols and phenol aldehydes. Volatile phenols have smoky, spicy odors. Both monomethyloxylated (gaiacyl G series) and dimethyloxylated (syringyl S series) derivatives are also present (Table 13.21). Methoxyphenols are extractable after toasting and their composition reflects the structure of the lignin (Monties, 1980) as well as the heating temperature. The concentration of syringyl (S) derivatives increases with heating intensity.

Other aromatic compounds present after oak has been toasted (Table 13.22) include: benzoic (vanillin and syringaldehyde) and hydroxycinnamic (coniferaldehyde and sinapaldehyde) aldehydes (Figure 13.9). Maximum quantities are formed when the oak is medium toasted, with a higher quantity of benzoic than cinnamic aldehydes (Puech, 1978).

Toasting barrels also causes the thermal degradation of certain lipids or fatty acids, forming isomers of methyl-octalactone. This reaction increases in proportion to heating intensity (Table 13.23). The more odoriferous cis isomer, which already predominates in non-toasted wood, represents an even higher proportion of the isomers in toasted oak. This compound is heat sensitive and disappears after 15 minutes (heavy toast).

Toasting, or, more precisely, 'hydrothermolysis', leads to the development of new volatile and odoriferous compounds, mainly: (a) furanic aldehydes, (b) phenol aldehydes and volatile phenols, (c) fatty acids, especially acetic acid (from xylanes), and acids with more than six carbon atoms formed by the breakdown of more complex lipids (Chatonnet,

Table 13.21. Impact of toasting on the formation of volatile phenols[a] (Chatonnet, 1995)

Molecules	Toasting intensity			
	Non-toasted	Light	Medium	Heavy
Guaiacol	1	5.2	27.7	30.3
Methyl-4-guaiacol	2	10	38.7	24.7
Ethyl-4-guaiacol	0	0	0	7.7
Propyl-4-guaiacol	0	0	0	6.3
Eugenol	20	17.7	71.7	44.3
Phenol	5	12	11.7	20
Ortho-Cresol	0	0	0	1.7
Meta-Cresol	0	0	0	1.3
Para-Cresol	0	0	0	2
Syringol	0	78.3	310.7	313.3
Methyl-4-syringol	0	17.3	80.7	193.3
Allyl-4-syringol	0	60.3	298.7	204.3

[a]Mean of several analyses, compounds extracted in a dilute alcohol medium, under standard conditions. Results are expressed in µg/l in this solution.

Table 13.22. Impact of toasting intensity on the formation of phenol aldehydes[a] (Chatonnet, 1995)

	Toasting intensity			
	Non-toasted	Light	Medium	Heavy
Vanillin	<0.1	2.1	4.8	3.1
Syringaldehyde	0.2	5.6	12.9	12.2
Coniferaldehyde	tr	3.1	6.2	2.1
Sinapaldehyde	tr	1.9	4.9	2.6
Σ Phenol aldehydes	0.2	12.7	28.8	20

[a]Mean of several analyses; compounds extracted in a dilute alcohol medium, under standard conditions. Results are expressed in mg/l in this solution.

1991), (d) methyl-octalactones and (e) dimethyl pyrazines (cocoa and fresh bread). Significant quantities of phenylketones are also formed during this operation. Heating is furthermore responsible for the breakdown of ellagitannins, present in decreasing concentrations as toasting intensity increases.

The oak aroma becomes more complex as toasting progresses from light to heavy. This aroma is initially characterized by toasty and vanilla overtones from the furanic and phenol aldehydes, as well as smoky, spicy and roasted odors from the volatile phenols. Following heavy toasting, the increase in methyl-octalactones contributes a hint of coconut, but this is generally masked by the overall aromatic complexity. At heavy toast, the intensity of the oak's aroma decreases, and there is an emphasis on smoky and burnt (toasted) odors.

It is therefore possible to modulate the organoleptic impact of oak barrels on wine (not only aromatic character but also overall structure) by choosing a different toasting level. The HCl value and gelatin index decrease with heating intensity, indicating a softening of the tannins (Chatonnet, 1995).

13.8.4 Wine Flavoring Processes

Various processes have been envisaged for giving a wine the required oak character without using barrels in order to avoid the technical and financial constraints involved. Oak has been macerated in wine in the form of staves, splints (2 cm × 2 cm), shavings or chips. Wood extract has even been used, in powder form and in solution. The wine

Table 13.23. Impact of toasting intensity on the formation of β-methyl-γ-octalactone isomers[a] (Chatonnet, 1995)

	Toasting intensity			
	Non-toasted	Light	Medium	Heavy
trans-Methyl-octalactone	0.16	0.11	0.11	0.14
cis-Methyl-octalactone	0.64	0.57	1.38	1.59
Cis + *trans*-Methyl-octalactone	0.8	0.68	1.49	1.73
Ratio of *cis*/*trans*	4	5.3	12.7	11.2

[a]Mean of several analyses; compounds extracted in a dilute alcohol medium, under standard conditions. Results are expressed in mg/l for this solution.

must also come into contact with oxygen. The oxidation reactions that occur during barrel aging are indispensable to bring out the oaky character. Oxygenation may be produced by racking with aeration or oxygen microbubbling.

Although these processes give wine a certain flavor, the character is noticeably less refined and pleasant than that produced by real wood aging. This fault is considerably accentuated by the fact that producers aim for a marked woody character and that wine treated in this way is not generally of very high quality. Maceration with shavings gives less satisfactory results than splints and great care is required in order to obtain reasonable results.

These processes are legally authorized in certain countries. They make it possible to offer reasonably priced wines that attempt to imitate top-quality products. This strategy has, to a certain extent, been successful from a commercial standpoint. These techniques are, however, prohibited in many countries. They are not allowed in France, for example, especially in *appellation contrôlée* wines. Some French producers feel that they are penalized in comparison to their competitors in other countries.

Bertrand *et al.* (1997) tested the effect of various forms of oak in red wine and found a high degree of extraction of the *cis* isomer of β-methyl-γ-octalactone and vanillin (Table 13.24). These substances add a vanilla and coconut character, especially when wines are aged in American oak. Commercial tannins, liquid flavoring and toasted chips lack almost all of the most volatile compounds, with the exception of eugenol and isoeugenol. Other additives, such as toasted wood splints, granules and chips release furfural, methyl-furfural and hydroxymethyl furfural into the wine. Another compound derived from non-toasted European oak is octanal, which has an orange odor (Bertrand *et al.*, 1997). Octanal may contribute to

Table 13.24. Volatile compounds extracted from oak in different forms[a] by two media. Concentrations are expressed in µg/l (Bertrand *et al.*, 1997)

Product used and quantities added (g/l)		*trans*-WL	*cis*-WL	EUG	*cis*-IEU	VAN	4HBE
Model solution							
Toasted splints	10	12	32	16	12	2036	203
Non-toasted splints	10	15	33	13	0	572	57
Liquid wood flavoring	0.3[b]	0	0	0	0	136	142
Chips	2	0	0	9	19	672	138
Toasted granules	2	17	76	6	0	521	128
Non-toasted granules	2	17	62	8	0	430	93
Commercial tannin 1	0.09	0	0	0	0	60	120
Commercial tannin 2	0.09	0	0	0	0	125	114
Shavings	2	12	31	4	0	520	126
Red wine							
Control		0	0	0	0	0	135
Toasted splints	30	35	85	51	52	4930	471
Non-toasted splints	30	49	134	17	0	1103	280
Liquid wood flavoring	0.9[b]	0	0	0	0	395	386
Chips	6	0	0	24	62	2383	407
Toasted granules	6	45	230	6	4	1515	433
Non-toasted granules	6	48	192	27	8	1387	350
Commercial tannin 1	0.27	0	0	0	0	170	428
Commercial tannin 2	0.27	0	0	0	0	304	320
Shavings	6	33	75	8	0	1089	407

[a]*trans*-WL = *trans*-3-methyloctano-4-lactone, *cis*-WL = *cis*-3-methyloctano-4-lactone, EUG = eugenol, *cis*-IEU = *cis*-isoeugenol, VAN = vanillin, 4HBE = ethyl-4-hydroxybenzoate.
[b]ml/l.

the positive impression sometimes given by these products. However, the presence of hexanal and *trans*-2-nonenal, with their unpleasant smells of paper and 'planks' (Chatonnet and Dubourdieu, 1997), can only be perceived as negative. The amount of ellagitannins extracted depends on the type of oak. When present in large quantities, ellagitannins may be accompanied by bitter glycoside coumarins.

It is understandable that the organoleptic consequences of adding these substances are different from those obtained by traditional barrel aging. Typically, these techniques produce a strong woody character, with little finesse or complexity, as well as burnt overtones when the wood fragments have been toasted. Furthermore, extraction is generally rapid and intense due to the large contact surface.

13.9 CONSTRAINTS AND RISKS OF BARREL AGING

13.9.1 Adapting the Type of Oak to Different Wines

Another problem raised by aging wines in oak is that of choosing the right barrels to suit the type of wine. Barrel aging must enable wines to develop their full character, yet remain in balance. If the oak/wine match is not perfect, there is a risk of acquiring a dominant woody character that overpowers the wine and dries it out rapidly.

The variables relating to barrels have been investigated (Section 13.8): origin of the oak, seasoning, degree of toasting and the cooper's expertise. Two other factors that play a major role are the age of the barrel and the way it is prepared before it is filled with wine. These variables make it possible, to a certain extent, to modulate the characteristics of barrel aging and the effects of oak on wine.

It has been observed that oxidative phenomena are more extensive in new barrels and less so in used wood. Furthermore, ever-decreasing quantities of compounds are extracted from oak as it ages and the inner surface of the barrel becomes clogged. One of the major dangers relating to used barrels is the development of off-flavors and unpleasant smells, generally due to mold in the pores and joints between the staves. The length of time barrels may be used for aging wine depends on the care with which they have been maintained (Section 13.6.2).

New barrels may be prepared with cold water, hot water or steam. This operation affects the porosity of the barrel walls. Steaming and prolonged washing with hot water lead to an increase in oxidative phenomena and reduced extraction of phenols. The wines have a more intense color, despite their decreased anthocyanin concentration, and softer tannins (lower gelatin value) than those aged in barrels simply rinsed with cold water. High-temperature cleaning has been shown to affect the fibers on the inside of the barrel wall.

The following list of criteria for selecting barrels is based on the preceding data, taking into account the variable characteristics of wines and barrels:

1. Naturally seasoned oak is generally better for red wines than artificially seasoned wood.

2. Fine-grained oak releases smaller amounts of phenols than coarser-grained oak. This slow, regular release may continue for several years.

3. Toasting eliminates 'green wood' and 'plank' faults, while producing very pleasant vanilla and spicy aromas. Heavy toasting gives a marked burnt, toasty character, making the aroma different, but without any major faults.

4. If a wine has a rich tannic structure, but is lacking in body and roundness, lower density wood (e.g. Vosges and Limousin) tends to reinforce the wine's astringency, but this effect may be partially alleviated by heavy toasting. High-density wood (such as Allier) is better suited to this type of wine, as it releases smaller quantities of phenols. A medium or heavy toast produces aromatic qualities that counteract herbaceous tendencies and attenuate bitterness.

5. The effect of barrel aging is more limited when wines are fine and well-balanced, with good structure and body, as well as powerful, fruity aromas. These wines can handle the phenols

extracted from the oak, but care must be taken to avoid bitterness and 'plank' odors, as well as intense, smoky, toasty overtones that are likely to make the wine seem rather coarse. Under these conditions, too much toasting should be avoided, especially when the barrel has already been closed at the top. Fine-grained wood and a medium toast would be just right.

6. Wines that have a light tannic structure or not very well developed aromas with herbaceous overtones need barrels that can enhance the wine's structure, without any aggressiveness. The right kind of oak contributes aromatic complexity without overpowering the fruit. Heavily toasted barrels are most appropriate in this instance and the density of the oak should be suited to the characteristics of the wine.

Adapting wood to wine is not a matter of following simple rules. Once the selection has been made on the basis of the preceding considerations, testing should be carried out to determine which types of barrels are best suited to each wine, and in what proportions.

Aging in 100% new barrels is only advisable for top-quality wines, with a sufficiently robust structure to resist developing an excessively woody character. It is better to age other wines in batches in different types of containers (new barrels, used barrels and even vats). The final blend is made just before bottling. The wine may be aged in different containers in turn, generally spending a minimum of 6 months in each. This time is necessary for the wine and wood to reach a balance, especially in new barrels, which are likely to give the wine a certain harshness in the first few weeks. A good solution consists of using variable proportions of three types of barrels:

1. New barrels that contribute a classical, strong, oak character.

2. Barrels that have already been used once provide a more discreet, aroma that may give a better impression than new wood at the beginning of barrel aging.

3. Barrels that have been used to age two vintages contribute a more subdued oak character, but there is a risk the wine will dry out.

Older barrels are likely to cause quality problems and should be avoided to minimize the risk of contamination.

It should be emphasized, however, that barrel aging alone cannot be expected to produce high-quality wine. Wood is capable of enhancing a wine's intrinsic qualities and may hide certain faults. Used unwisely, barrel aging may produce disastrous results.

13.9.2 Risks Resulting from the Development of Microorganisms

Increases in volatile acidity, generally 0.1–0.2 g/l (H_2SO_4) and sometimes more (Vivas *et al.*, 1995), have been observed during barrel aging. This phenomenon is partly due to anaerobic lactic bacteria that break down the few remaining milligrams of residual malic acid. However, it is mainly due to aerobic acetic bacteria that may develop around the bunghole. Their metabolism may be sufficient to form a few mg/l of acetic acid after temporary aeration, e.g. during racking (Section 8.3.3). The formation of acetic acid may be accompanied by that of ethyl acetate, which has a very unpleasant smell (Table 13.25). Barrels provide a favorable environment for this type of spoilage, as the wine is in constant contact with oxygen and subjected to variations in temperature. Barrel aging requires special care from late spring until early fall. The greatest risk of microbial spoilage occurs during this period, as cellar temperature rises and the evaporation rate increases.

In addition to acetic bacteria (*Acetobacter* sp.), harmful oxidative and acidifying yeasts (*Candida valida* and *Pichia vini*) have also been identified (Chatonnet *et al.*, 1993b).

It is vital to top barrels up carefully when they are stored unsealed with the bung on top, to avoid an excessive increase in acetic acid. Wines should be maintained at a temperature $<20°C$, with a free SO_2 concentration ≥ 15 mg/l. A relatively low

Table 13.25. Acetic acid, ethyl acetate, and ethyl-phenol content of a red wine aged under various conditions. (After 12 months aging with the bung on the side; blend of 83 barrels for each condition) (Chatonnet et al., 1993b)

Aging conditions	Acetic acid (mg/l)	Ethyl acetate (mg/l)	Σ Ethyl-phenol[a] (mg/l)
Concrete vat	526	44	37
New barrels			
Medium toast	637	46	496
Heavy toast	686	48	550
Used barrels (>5 wines)			
Untreated	526	39	1285
Scraped	563	46	1230

[a]Ethyl-4-guaiacol + ethyl-4-phenol.

temperature combined with the use of antiseptics prevents the development of aerobic germs. It is also recommended to minimize aeration during racking, starting in the spring after the vintage (Section 10.3.3).

The use of new barrels always increases volatile acidity by 0.1 g/l due to the presence of acetic acid formed from acetyl radicals in the wood hemicellulose during toasting (Marsal, 1992). In properly made wines with an initial volatile acidity below 0.40 g/l of H_2SO_4 (or 0.50 g/l of acetic acid), 12–18 months of barrel aging should not result in a volatile acidity noticeably higher than 0.50 g/l of H_2SO_4 at the time of bottling. If this value is as high as 0.60 g/l H_2SO_4 (0.75 g/l of acetic acid), winemaking and/or barrel aging procedures have not been properly controlled.

In addition to bacteria, *Brettanomyces* and *Dekkera* contaminant yeasts, always present in wineries, may develop in new or used barrels during summer, regardless of the presence or absence of air. Their metabolism, accompanied by the production of unpleasant-smelling (Section 8.4.5) ethyl-phenols (Table 13.25), may continue in the bottle.

Sulfur dioxide is used in gas form, generally by burning a tablet or wick, to sulfite wine and disinfect the inside of barrels where microorganisms are most likely to develop (at least 7 g of sulfur must be burnt per barrel) (Section 8.4.6). Adding sulfite solutions directly to wine does not provide totally effective protection.

It is quite clear that barrel aging is a delicate operation, requiring strict hygiene of premises, containers and wine. The risks are known, so they must be minimized by taking the proper precautions.

REFERENCES

Aramon G., Saucier C., Tijou S. and Glories Y. (2003) *L.C.C.*, 21, 910.

Atanasova V., Fulcrand H., Cheynier V. and Moutounet M. (2002) *Anal. Chem. Acta*, 458, 15.

Augustin M. (1986) Etude de l'influence de certains facteurs sur les composés phénoliques du raisin et du vin. Thèse Doctorat Université (ancien régime), Université de Bordeaux II.

Bertrand A., Barbe J.C. and Gazeau O. (1997) *J. Sci. Tech. Tonnellerie*, 3, 37.

Cameira dos Santos P.J., Brillouet J.-M., Cheynier V. and Moutounet M. (1996) *J. Sci. Food Agric.*, 70, 204.

Chapon L. and Chapon S. (1977) Phénomènes d'oxydation catalytique dans les bières. Proceedings of the 16th Congress of the European Brewery Convention, Amsterdam.

Chatonnet P. (1991) Incidences du bois de chêne sur la composition chimique et les qualités organoleptiques des vins. Applications technologiques. Diplôme d'Etudes et de Recherches de l'Université de Bordeaux II.

Chatonnet P. (1995) Influence des procédés de tonnellerie et des conditions d'élevage sur la composition et la qualité des vins élevés en fûts de chêne. Thèse Doctorat, Université de Bordeaux II.

Chatonnet P. and Boidron J.-N. (1989) *Conn. Vigne Vin*, 23, 1.

Chatonnet P. and Dubourdieu D. (1997) *Rev. Œnologues*, 82, 17.

Chatonnet P., Boidron J.-N. and Pons M. (1990) *Sciences des Aliments*, 10, 565.

Chatonnet P., Boidron J.-N. and Dubourdieu D. (1993a) *Rev. Fr. d'Œnologie*, 144, 41.
Chatonnet P., Boidron J.-N. and Dubourdieu D. (1993b) *J. Int. Sci. Vigne et Vin*, 27, 277.
Clark L.C. (1960) *Oxidation–Reduction Potential of Organic Systems*. (ed Baillère) Findall and Co. London.
Deibner L. (1956) *Ann. Technol. Agric.*, 22, 77.
Dubourdieu F. (1992) *Les Grands Bordeaux de 1945 à 1988*. Mollat, Bordeaux.
Es-Safi N.E., Le Guerneve C., Cheynier V. and Moutounet M. (2000), *J. Agric. Food Chem.*, 48 (9), 4233.
Es-Safi N.E., Le Guerneve C., Cheynier V. and Moutounet M. (2002), *Magn. Research Chem.*, 40, 693.
Feuillat F., Huber F. and Keller R. (1993) *Rev. Fr. Œnologie. Cahier Scientifique*, 142, 5.
Galvin C. (1993) Etude de certaines réactions de dégradation des anthocyanes et de leur condensation avec les flavanols. Conséquences sur la couleur du vin. Thèse Doctorat, Université de Bordeaux II.
Glories Y. (1987) *Le bois et la qualité des vins et des eaux-de-vie*. (ed. G. Gimberteau). Spécial *Conn. Vigne Vin*, 81.
Glories Y. (1990) *Rev. Fr. d'Œnologie*, 124, 91.
Glories Y. and Bondet de la Bernardie C. (1990) In *Actualité Œnologiques 89*. Dunod Paris, p. 398.
Guerra C. (1997) Recherches sur les interactions anthocyanes-flavanols: application à l'interprétation chimique de la couleur des vins rouges. Thèse Doctorat, Université de Bordeaux II.
Guimberteau G. (1992) Le bois et la qualité des vins et eaux-de-vie. *J. Int. Sci. Vigne et Vin*, hors série.
Günther C. and Mosandl A. (1986) *Liebigs Ann. Chem.*, 2112.
Haslam E. (1980) *Phytochemistry*, 19, 2577.
Jourdes M., Quideau S., Saucier C. and Glories Y. (2003) *7ème Symposium International d'œnologie*, Lonvaud, de Revel, Darriet éd., Tec-Doc Lavoisier, Paris.
Jurd L. and Somers T.C. (1970) *Phytochemistry*, 9, 419.
Khan N. (2000) Rôle du bouchon de liège sur le vieillissement des vins en bouteille. Etudes de l'apport du liège et des phénomènes oxydatifs. Thèse de Doctorat, Onologie-Ampélologie, Université Victor Segalen Bordeaux 2.
Laborde J.-L. (1987) Contribution à l'étude des phénomènes d'oxydation dans les vins rouges, rôle joué par l'anhydride sulfureux. Thèse Doctorat 3rd Cycle, Université de Bordeaux II.
Larignon P., Roulland C., Vidal J.-P. and Cantagrel R. (1994) etude de la maturation en Charentes du bois de tonnellerie. Bureau National Interprofessionnel du Cognac.
Marché M. and Joseph E. (1975) *Rev. Fr. Œnologie*, 57, 1.
Marsal F. (1992) *Rev. Œnologues*, 63, 22.
Michaelis L. (1953) *Oxidation–Reduction Potential*. Springer, Berlin.
Monties B. (1980) Les lignines, in *Les Polymères Végétaux. Polymères Pariétaux et Alimentaires non Azotés*. (ed. B. Montiès). Bordas, Paris, p. 122.
Moutounet M., Saint-Pierre B., Micaleff J.-P. and Sarris J. (1994) *Rev. Fr. Œnologie*, 74, 34.
Moutounet M., Ducourneau D., Chassin M. and Lemaire T. (1996) 5th International Symposium on *d'Œnologie*, Tec. et Doc., Lavoisier Paris, p. 411.
Peynaud E. (1975) *Connaissance et Travail du Vin*. Dunod, Paris.
Piffaut B., Kader F., Girardin M. and Metche M. (1994) *Food Chem.*, 50, 115.
Pontallier P. (1981) Recherches sur les conditions d'élevage des vins rouges. Rôle des phénomènes oxydatifs. Thèse de Docteur Ingénieur, Université de Bordeaux II.
Pontallier P., Glories Y. and Ribéreau-Gayon P. (1980) *C.R. Acad. Agric.*, 66, 989.
Pontallier P., Salagoïty-Auguste M.-H. and Ribéreau-Gayon P. (1982) *Conn. Vigne Vin*, 16, 45.
Puech J.-L. (1978) Le vieillissement des eaux de vie en fûts de chêne. etude de la lignine et de ses produits de dégradation. Thèse Doctorat, Université Paul Sabatier, Toulouse.
Ribéreau-Gayon J. (1931) Contribution à l'étude des oxydations et réductions dans les vins. Thèse Doctorat ès Sciences Physiques, Bordeaux, 2nd edn 1933.
Ribéreau-Gayon J. (1933) *Bull. Soc. Chim.*, 53, 209.
Ribéreau-Gayon J., Peynaud E., Sudraud P. and Ribéreau-Gayon P. (1976) *Traité d'Œnologie*, Vol. III: *Vinifications, Transformations du Vin*. Dunod, Paris.
Ribéreau-Gayon P., Pontallier P. and Glories Y. (1983) *J. Sci. Food Agric.*, 34, 505.
Salagoïty-Auguste M.-H., Tricard C. and Sudraud P. (1987) *J. Chromatogr.*, 392, 379.
Santos-Buelga S., Bravo-Haro S. and Rivas-Gonzalo J.-C. (1995) *Lebensm Unters Forsch.*, 201, 269.
Sarni-Manchado P., Fulcrand H., Souquet J.-M., Cheynier V. and Moutounet M. (1996) *J. Food Sci.*, 61, 938.
Saucier C., Bourgeois G., Vitry C., Roux D. and Glories Y. (1997) *J. Agric. Food Chem.* (to be published).
Singleton V.L. (1974) In *Chemistry of Wine Making*, (ed. A.D. Webb). American Chemical Society, Washington.
Taransaud J. (1976) *Le Livre de la Tonnellerie*. La Revue à Livres Diffusion, Paris.
Vivas N. (1993) *Rev. Œnologues*, 70, 17.
Vivas N. (1997) Recherches sur la qualité du chêne français de tonnellerie (*Q. petraea* L., *Q. robur* L.) et sur les mécanismes d'oxydoréduction des vins rouges au cours de leur élevage en barriques. Thèse Doctorat, Université de Bordeaux II.

Vivas N. and Glories Y. (1993a) *Rev. Fr. Œnologie*, 142, 33.

Vivas N. and Glories Y. (1993b) *Cryptogenie Mycol.*, 14, 127.

Vivas N. and Glories Y. (1994) *Progr. Agric. Viticult.*, 111, 421.

Vivas N. and Glories Y. (1995) *Rev. Œnologues*, 76, 10.

Vivas N. and Glories Y. (1996) *Am. J. Enol. Viticult.*, 47, 103.

Vivas N., Zamora F. and Glories Y. (1992) *J. Int. Sci. Vigne Vin*, 26, 271.

Vivas N., Zamora F. and Glories Y. (1993) *J. Int. Sci. Vigne Vin*, 27, 23.

Vivas N., Lonvaud-Funel A. and Glories Y. (1995) *J. Sci. Tech. Tonnellerie*, 1, 81.

Zamora F. (1989) Contribution à l'étude des phénomènes d'oxydation du vin. Mémoire de Diplôme d'Etudes Approfondies Œnologie-Ampélologie, Université de Bordeaux II.

Index

Accelerated aging 408–409
Acephate 268
Acephate hydrolysis 268
Acescence 242
Acetals 63
   formation 63
Acetic acid 9, 238, 421, 425
Acetic bacteria, spoilage caused by 241–242
Acetic esters of higher alcohols 59–60
Acetone in degradation of anthocyanins 158
Acetylation 74–75
2-Acetyltetrahydropyridine
   tautomeric forms 281
Acid pectic substances 79–82
Acidification, buffer capacity in 18–21
Acidity, types 8–9
Acrolein 56, 240
   formation 56
Adsorption 295, 342
Affinity laws 376
Aging
   conditions 255
   duration 387
   of red wines 387–426
Aging aroma, defective
   molecules responsible for 274–275
   type of defect 274–275
Aglycones 210
AH in wine 10
Alanine 113, 115, 270
Albumin 318–319
Alcohols 51–64
   biosynthesis 55
   boiling point 56
   esterification balance 59
   higher fermentation 53–55
   originating from plants and yeast 54
Aldehydes 61–63, 209, 270
Aldimine 72, 73, 269

Aldohexose isomers 68
Aldopentoses 69
Aldoses 72
Aliphatic series 58
Alkaline alginates 320–321
Alkaline cupric solutions 66
Allyl gaiacol 143
Alsace grape
   volatile thiol composition 218
Al,TA fraction 180
Aluminum, elimination 105
Amertume 240
Amide nitrogen 113
2-Aminoacetophenone
   formation 274
   formation, impact of aging conditions 276
α-Amino acid
   L configuration 114
   forms 115
Amino acids 110
   assay 115
   chromatograms 116
   concentration 115
      during ripening 118
   presence in must and wine 115–117
   structure 113–115
   trifunctional 114
Amino sugar nitrogen 113
   o-Amino-acetophenone 223
Amino-1-aldose 73
Amino-1-ketose 73
Anderson Hasselbach equation 10, 13
Anion exchangers 377–378
Anthocyanidins
   structure 145
Anthocyanin 3,5-diglucosides 146
Anthocyanin 3-monoglucosides 146
Anthocyanins 145–147, 397
   assay 173–174, 190

Anthocyanins (*continued*)
  bleaching due to pH and sulfur dioxide 156
  breakdown reactions 156–158, 194
  chemical properties 152–171
  combined with tannins 399
  concentration 194, 406
  condensation reactions 168–171
  copigmentation 167
  copigmentation reactions 166–167
  degradation by acetone 158
  development from skins and seeds 187
  direct A-T type condensation 169
  direct T-A type condensation 170
  effect on color 193–195
  equilibrium depending on pH and SO_2 152–156
  evolution as grapes ripen 184–191
  extraction coefficient 189
  extraction during winemaking 191–193
  forms 154, 155, 167
  HPLC chromatogram 175
  in grape skins 184
  molecular structures 187–188
  organoleptic properties 172–183
  oxidative degradation 157–158
  reactions involving 193–195
  reactions with compounds with polarized double bonds 167–168
  thermal degradation 157
  types 400
Apfelsaure (apple acid) 5
Appellation d'origine controlée 9
Arabinans 81, 82
Arabinofuranosyl-β-D-glucopyranoside 210
Arabinogalactan I (AG-I) 79, 80
Arabinogalactan II (AG-II) 79, 80, 82
Arabinogalactan proteins (AGP) 82
  characteristics of fractions 83
  composition of fractions 83
Arabinogalactans 79, 314
  structure 80
Arabitol 56, 57
Arginase 121
Arginine 114, 115, 122, 123
Arginine deiminase 121
Aroma 205–227
  American vine species 222
  development during ripening 223–227
  effect of fining 315
Aroma fractions in grapes 207
Arsenic 105
Artificial cold stabilization 28–37
  monitoring 28–37
    Mextar calculation software 36
Artificial seasoning
  accelerated 419
Asbestos 340, 342

Ascorbic acid 5, 99, 237
  Oxidation–reduction equilibrium 5
Ash 93–94
  alkalinity 94
  preparation 93
Asparagine 113
Aspartic acid 114, 125
Aspergillus niger 246, 247
Aspergillus niger esterases
  effects, in commercial pectinases 248

Bacterial spoilage 238–242
Barrel aging 177, 253–256, 315, 411–415
  adapting to different wines 424–425
  cellars 409
  constraints and risks 424–426
  origins of wood 416–417
  reactions during 197–198
  role 411–412
  seasoning conditions 417–419
Barrel toasting 419–422
Bate–Smith reaction 147
Bentonite 35, 124, 130–132, 134–137, 290, 303, 324–328, 370
  application 326–327
  physicochemical characteristics 325–326
  structure 324–325
  substitution treatments 133–134
  treatment techniques 327–328
  use to eliminate proteins 132–133
Bentotest 130
Benzoic acids 142
Bioamines 113, 121–124
Biochemical phenomena 206
Biological stabilization 371–372
Bitartrate
  instability 25
  stabilization technologies 37–48
Bitter almond flavor 281
Blood by-products 319
Botrytis cinerea 55, 57, 64, 66, 77, 86, 87, 135, 200, 240, 304, 344, 345
Bottle aging 182
  chemical explanations 405–407
  reactions during 197–198
  red wines 404–409
Bottle sickness 237
Bouquet development 407–408
Brettanomyces 253, 255, 281, 305
Brettanomyces bruxellensis 251
Brettanomyces intermedius 253
Brownian motion 314, 328
Bubble degassing model 24
Bubble formation 25
Buffer capacity 11–18
  acidobasic 18
  calculation of 15

Index 431

definition   13
in acidification and deacidification of wine   18–21
oxidation   236–237
Buffer solutions   10
Buffer zones   11
Butanedioic acid   7
2,3-Butanediol   57
   Oxidation–reduction balances   57
Butyric acid   117
γ-Butyrolactone, formation   63
τ-Butyrolactone, formation   63

Cadmium   105
Caffeic acid   142, 200
Caftaric acid   194, 200
Calciphos   101
Calcium carbonate   12, 40
Calcium cation   95
Calcium gluconate   5
Calcium phytate   101–102
Calcium tartrate   12, 18
   problems   39–40
   solubility in water   39
Calcium tartromalate   16, 21, 22
Candida mycoderma   242
Capillary electrophoresis (CE)   125, 127, 135, 136
Carbamic acid   119
Carbocations   400
   formation from procyanidins   159–162
Carbohydrate colloids   66
Carbohydrates   65–89
   characteristics   65
Carbon dioxide diffusion   24
Carbonic acid   19
   amino derivatives   119
Carbonylated compounds   61–63
Carboxymethylcellulose (CMC)   46–48
Carotenoids
   breakdown   223
   oxidative degradation   211, 212
Casein as fining agent   319
Castalagin   148
Castalin   149
Castavinols   147
   structure   147
Catechins   150
   reaction with malvidin 3-glucoside   171
Cation exchangers   377–378
   operation   378–381
   to treat wine   378–381
Cellulose   339
   formula for etherification   47
Cellulose-based filter sheets   351–355
Cellulose ester membranes   341
Cellulose flat-sheet filter   352
Centrifugal force   364

Centrifugation   364–366
   applications   365
   to treat wine   365
Centrifuges   365
   applications   365
   industrial   365
Chair conformations   67
Champagne grape varieties, ripening in   117
Charcoal   281
Chardonnay wines
   composition of, after tartaric stabilization   20
Chloroanisoles   258, 260
   assay of   260
Chlorophenols   258, 260
   assay of   260
Cinnamate decarboxylase (CD)   245, 251
Cinnamate esterase   246
Cinnamic acids   142, 200
   derivatives   143
Cinnamyl esterase (CE)   246
Citramalic acid   7
Citric acid   5, 9, 10, 99
Citronellol   207
Citrulline   114, 117
Clarification   303
   miscellaneous treatments   328–330
   products used   323
   quality assessment   336–338
   quality monitoring   348
   stages   353
   treatments   301–330
   see also Centrifugation; Filtration
Clarity   301
   and stability   285–300
   observing   286–287
   problems related to   285–286
Clogging   342, 356
Coenzyme A   59
Cold stabilization   373–376
   aim   373
   installation   375
   procedures   374–376
   rapid processes   375
Cold-stabilized wine   38
Colloid reactivity   290–295
Colloid stability   290–292
Colloidal particles
   Brownian motion   289
   double layer   291
   electrical charges   290
   protection by polysaccharides   296
Colloidal precipitation, prevention   369–371, 374
Colloidal solutions   288, 289
Colloidal state   287–290
Colloidal sulfur   104
Colloidal tannins   293

Colloidal transformations, diagram   288
Colloids
   mutual flocculation   294–295
   properties   289
   types   288–289
Colony forming units (CFU)   338
Color   178–179
   changes in red wine   399
   composition   178
   development during aging   397–403
   effect of anthocyanins   193–195
   intensity   178, 399–401
   of Merlot wine   403
   stability   198–199
   of white wines   179, 199–201
   yellow–orange development   401
Coloring matter
   development   402–403
   precipitation
      in old wines   199
      in young wines   198–199
Condensation   397
   reactions   196
Conductivity   26, 27, 30, 40
   meter cell   28
   variation   29
Contaminant populations   234
Continuous centrifuge
   diagram   365
Continuous cold stabilization system   39
Continuous treatment   25
Copper   95, 102–104
   elimination   103
   presence and state in wine   102
Copper casse   102–104, 124, 271, 370
   mechanisms   102–103
   preventing   103–104
Copper sulfate   102
Copper sulfide   102
Copper turnings   267
Cork taint   256–261
   compounds responsible for   257
Coumaric acid   5, 142, 200, 245
Coumarins   143, 157
Coumaryl tartaric acid   5, 143
Coutaric acid   200
C, P fraction   180
Cream of tartar   25, 29, 33
Crystal precipitation prevention   373–374
Crystallization   24, 26
   inhibitors   26
   kinetics   40
   rate   26
CT fraction   180
Cyanidin   147
Cysteine   117, 119
   precursors of thiols derived from   219–222

Cysteinylated thiol precursors
   distribution in Sauvignon Blanc grapes   221

β-Damascenone
   concentrations   212, 213
   formation   213–214
Da Millipore membrane   35
Deacidification, buffer capacity in   18–21
Delphinidin   154, 194
Denatured proteins   292–294
Depletion   296, 297
Deposits, formation   304
Depsidase   246
Destemming   130
Dialysis index   177
Diammonium phosphate   95
Diatomaceous earths   336, 339–340, 346–351
Diatomite   339
DICALCIC process   21
Diethoxyethane
   formation   53
Diffusion speed   26
Dihydroquercetin   144
Dihydroxyacetone   70
Dihydroxyacetone-1-phosphate   71
Dimethyldisulfide (DMDS)   273
Disaccharides   71–72
Dispersed systems, classification   287–288
Dissolved oxygen
   and oxidation–reduction potential correlation   392–393
   measuring   389
Disulfide formation   266, 269
Dithiocarbamates   267
Double layer
   and electrostatic potential   309
   charge distribution   290
   colloidal particles   291
Dry extract   91–107
   calculation   92, 93
   types   92
Dynamic continuous contact process   38–39

Earth filtration   346–351
Earthy-smelling defect   277–279
Effervescence   24
Egg albumin   318
Egg white   318–319
Electrodialysis   26
   applications   382–385
   changes in wine composition   384–385
   operating principle   382
   operational details   384
Electrodialysis cell   382, 384
Electrophoresis   124
Electrostatic phenomena   309

Index

Electrostatic potential 309
Ellagic acid 149
Ellagitannins 148
  structure of 148
Ene-diols 270
Enological Codex 299, 317, 321
Enzyme mechanisms 245
Epifluorescence 338
Erythritol 56, 57
Esculetin 144
Esculin 144
Esters 59–61
  chemical origin 60–61
Ethanal 61, 170, 171, 235, 242, 402
  acetalization 53
Ethanol 242, 245, 247, 393
  chemical properties 52
  solvent properties 52
  structure 52
  test 131
  see also Ethyl alcohol
Ethanolamine 117
Ethyl acetates 59, 241
  of fatty acids 59–60
Ethyl alcohol 51–53
  fermentation 52
Ethyl carbamate 119–121
Ethyl decanoate 60
Ethyl gaiacol 142
Ethyl hexanoate 60
Ethyl lactate 7, 61
Ethyl octanoate 60
Ethyl-3-mercaptoproprionate 216
Ethyl-phenol 142, 243, 245, 280, 426
  acetic acid content 426
  concentration during barrel aging 250
  concentration in red wines 253–256
  content in red wines 244
  defects in red wines 249–251
  production mechanism 251
4-α-Ethylthioflavan-3-ol 160
  structure 162
Exocellular polysaccharides
  from yeast 83–86
  influence of temperature 85
Exponential hypersolubility curve 24

Fatty acids 58
  biosynthesis 60
  ethyl acetates of 59–60
Fatty degeneration 240
Fehling's solution 74
Fermentable sugar 66
Fermentation 132–133, 236, 249, 276
  ethyl alcohol 52
  metabolisms 206
  organic acids from 6–8

  reactions occurring after 206
Ferric casse 94, 96–102
Ferric colloids 98
Ferric ferrocyanide 290
Ferric hydroxide 10
Ferric iron 96
  forms 98
Ferrocyanide treatment 295
Ferulic acid 143, 200
Filter press
  adjuvants 349, 350
  diagram 349
  operation 349–350
Filtration
  adjuvants 338–342
  cellulose-based filter sheets 351–355
  characteristics determination 352
  comparison with fining 363
  deep-bed 335–336
  earth 346–351
  effects on chemical composition 361–364
  effects on organoleptic character 361–364
  equipment 338–342, 346–347
  filter layers function 342–346
  flat-sheet 339–341, 351, 355
  with gradual clogging of pores 335
  with intermediate clogging of pores 336
  laboratory tests 346
  laws 334–336
  lenticular module 355
  mechanisms 342–343
  precoat 346, 347
  preparing filter layers and operating filters 347–349
  principles 333–334
  prior to bottling 364
  properties of filter medium 338–339
  quality, impact of filter sterilization 355
  rotary vacuum filter 350–351
  selection of parameters 353–354
  sterilizing equipment 354–355
  with sudden clogging of pores 335
  techniques 360
  types used in winemaking 334
  see also Filter press; and under specific methods
Fining 302
  applications 311–312
  background research 307–308
  charges of particles involved in 308–310
  comparison with filtration 363
  effect on aroma 315
  effect on phenolic compounds 315
  modeling 311, 312
  preliminary trials 322
  procedures 322–324
  products used 316–322
  protein agents 316–317

Fining (*continued*)
   techniques   322–324
   theory   307–312
Fischer projection   67, 70
   of aldoses   68
Fixed acidity   9
Flash pasteurization   372
Flatness   237–238
Flavanols   401
   colloidal properties   294
   oligomeric   181
   polymerization   166
   polymerized   181
   small   181
Flavanones   144
Flavanonols   144
Flavenes   402
Flavones   144, 145
Flavonoids   144–145
Flavonols   144
Flavor
   balance   8
   development during aging   397–403
   effect of composition modifications   362–363
   effect of tannins   195–197
Flavoring processes   422–424
Flocculation   128, 290–294, 296, 303, 307, 329
   mechanism   308
Flor   242
Folin–Ciocalteu value   172
Fouling index   335, 343, 356
Freezing temperature   37
French paradox   141
Fructose
   chemical structure   67–68
   presence in grapes and wine   66–67
Fructose-1, 6-diphosphate   71
Fumaric acid   7
Furaneol   223
Furanic aldehydes
   formation   421

Gaiacol   143
Galactaric acid   5
Galaturonic acid   77
Gallic acid   148, 149
Gallotannins   148
Gas-phase chromatography   74–75
   coupled with mass spectrometry (GCMS)   257
Gel   288
Gelatin   313, 317–318
   influence on tannin elimination from wine   314
Gelatin index   177–178, 181
Gentisic acid   142
Geosmin   277–279
   content in red wines   278
   content in white wines   278

Geraniol   207
Geranium odor   279–280
Glass fiber membranes   342
Glucanase   87, 344
Glucane   86–89
   structural unit   344
Glucanex   86, 135
Glucofuranose   68, 70
Glucomannoprotein complexes   85
Gluconic acid   5, 77
Glucopyranose   67, 71
   conformation equilibrium   69
   epimerization equilibrium   70
Glucopyranoside   210
Glucose   66
   chemical structure   67–68
   epimerization   67
   presence in grapes and wine   66–67
Glucose/fructose ratio   66–67
Glucosyl-*p*-coumaric acid   143
Glutamic acid   63, 114, 115
Glutamine   113, 115
Glutathion   119
   structure   118
Glyceraldehyde   70
Glyceraldehyde-3-phosphate   71
Glycerol   56–57, 240, 394
   conversion into acrolein   240
Glycine   114
Glycosides   75–77, 209–211
Glycosylated aromatic potential of grapes   211
Graisse   240
   polysaccharides in   88–89
Grape metabolism   206
Grape skins   184, 210
   anthocyanins in   184
Grapes
   amino acids   111
   Botrytis cinerea affected
      polysaccharide content   87
   fractionation   180
   glycosylated aromatic potential of   211
   organoleptic defects   277–279
   phenolic maturity   188–189
   polysaccharide content   87
Guaiacol   257
Guanidine   119
Gum arabic   99
   use to stabilize clarity   298–300
Gums   77–78

HCl index   177
Heat-generated volatile sulfur compounds   268–271
Heat stabilization   370–373
Heat treatment, implementation   372–373
Heavy metals, definition   104–105
Hemiacetalization   67, 69

Index

Hexadecyltrimethylammonium bromide 84
Hexahydroxydiphenic acid 149
Hexametaphosphate 40
Hexyl acetate 60
High-temperature bottling 372
Histidine 114, 122
Homogalacturane 77
Ho-trienol 207
Hue 178
Hydrogen sulfide 53, 262, 267
Hydrophilic colloids
    flocculation 293
Hydrophilic colloids, stability 292–294
Hydrophobic effects 312
Hydroxybenzoic acid 142
γ-Hydroxybutyric acid 63
Hydroxycinnamic acids 246
Hydroxy-3-proline 82
Hygiene 409
Hyperoxygenation of must 133
Hypersolubility exponential curve 22–24, 33

In-depth prefilters 356
Initial streaming potential (PEI) 310
Inorganic acids, state of salification 7
Inorganic anions 95–96
Inorganic cations 91
Inorganic ceramic membranes 342
meso-Inositol 58
Interfacial surface energy 24
Ion exchange
    reactions 377
    regeneration methods 378, 380
    resin composition 376
Ion exchange resins, implementation 381
Ion exchangers 376–381
    operating 376–377
    possible uses in winemaking 377–378
Ionization value 178
β-Ionone, concentrations 213
Iron 95–102
    elimination 98
    presence and state in wine 96
Isinglass 318
Isoamyl acetate 60
Isobutyl acetate 60
Isobutylmethoxypyrazine (IBMP) 224, 225
    distribution in Cabernet Sauvignon grape 216
Isoleucine 55, 117

Jelly 288

Kaempferol 144
Kaolin 326
Ketimine 72, 269
2-keto D-gluconic acid 5

Ketones 61–63
Kieselguhr 338–340, 346–351, 360
Klebosol 328
Krebs cycle 7, 110
KTH, *See* Potassium bitartrate (KTH)

Laccase 201
Lactic acid 6, 19, 393
Lactic bacteria, spoilage caused by 239–240
Lactic fermentation 239
Lactobacillus 122, 239, 281
Lactobacillus plantarum 120, 252
Lactones 63–64
Lactose 71
Lannate 268
Lannate hydrolysis 269
Lead 105–107
    contamination 106
    elimination 105
    evolution 106
    human exposure 106
    pathological effects 106
Lead arsenate 106
Lead capsules 107
Lead–tin capsules 107
Lenticular filter 355
Lenticular module filtration 355
Lenticular modules 340–341
Leucine 117
Leuconostoc 239
Leuconostoc dextranicum 86
Leuconostoc oenos 251
Linalol 207
Liquid chromatography 125
Low-temperature crystallization 25
Lysine 114, 137

Maceration 191
Macromolecular colloids 289, 292
Maderization effect 371
Magnesium 95
    regeneration of cation resins 380
Maillard reaction 268–271
Malic acid 5, 11, 12, 22, 425
    buffer capacity, variations in 17
Malolactic fermentation 395
Maltose 71
Malvidin 146
Malvidin monoglucoside 146
Malvidin-3-*p*-coumarylglucoside 168
Malvidin-3-glucoside 172
    reaction with catechin 171
Manganese elimination 105
Mannitol 57, 58
Mannoproteins 41, 84–86, 135, 136
    crystallization-inhibiting effect 43–46

Mannoproteins (*continued*)
  HPLC analysis   44, 45
Mannostab   45–46
Mean degree of polymerization (mDP)   407
Megastigmanes   211
Melibiose   71
Membrane filtration   356
  preparing wines   356–357
  selecting parameter   357–358
  tests   356–357
Membranes   341–342
  choice of   383
Mercaptans   104
2-Mercaptoethanol   265
3-Mercaptohexanol
  cysteine conjugate form   220
Mercaptopentanone   63
Mercury   105
Merlot wine, color of   403
Metatartaric acid   26, 40–43
  analysis   41
  effectiveness   42
  esterification number   41, 42
  hydrolysis rate   42, 43
  impurities   42
  instability   43
  polyesterification reaction   41
Methanethiol   262
Methanol, *See* Methyl alcohol
Methionine   114, 117, 262
Methionol   265, 270
  formation   266
2-Methoxy-3-*sec*-butylpyrazine   216
4-Methoxy-2,5-dimethyl-3-furanone   223
Methoxy-1-glucopyranoses   75
2-Methoxy-3-isobutylpyrazine   215–216
Methoxypyrazines   206, 214–216, 224
  olfactory perception thresholds   215
Methyl alcohol   53
Methyl anthranilate   222
Methyl D-glucopyranoside   74
Methyl gaiacol   143
Methyl D-glucopyranoside
  formation   75
Methylation   74–75
α-Methylmalic acid   7
4-Methyl-4-mercaptopentanone   261
β-Methyl-γ-octalactone   64, 415, 422, 423
Methyl syringol   143
Micellar colloids   288
Microbial contamination   234
Microbiological analyses   337–338
Microfiltration membranes   341
Microorganisms development   425–426
Milk as fining agent   319
Minerals   91–107

Mini-contact test   28–29
  limitations   30
Moldy flavor   257
Monoterpenes
  characteristics   207
  and derivatives   207–209
Monoterpenols and derivatives   209
Montmorillonites   132, 324
  flake structure   325
Mousiness   280–281
MP32   137
Mucic acid   5, 6
Must
  hyperoxygenation   133
  protein concentration   128–130
  soluble polysaccharide content   79
Mutual flocculation   300
Mycoderma aceti   242
Mycoderma vini   242
Mycodermic yeast contamination   242
Myricetin   144

Natural protective colloids   297–298
Nephelometer   336
Nephelometric turbidity units (NTU)   336
Nernst equation   390, 392
Nerol   210
Neutral pectic substances   82
Nitrogen   109
  concentration   109
  forms   109–113, 119–124
  mineral   110
  nucleic   113
  organic   110–113
  total   109–110
Nitrogen compounds   109–138
Non-colloids   188, 288
Non-volatile compounds
  extracted from oak barrels   413–414
Norisoprenoid derivatives   211–214, 223
  main families   212
  precursors   213–214
Norisoprenoids   206
Nucleation   24, 26, 28

OD 280 value   172
Odoriferous C_{13} norisoprenoid derivatives   211–214
Odoriferous compounds   206
Odoriferous terpenes   206–209
Odoriferous volatile thiols   216–219
Odors
  phenol   243
  reduction   261–273
OIV (Office International de la Vigne et du Vin)   52
Olfactory defects   242–244, 266, 282
  volatile phenols responsible for   242–244

Index

Oligopeptides 113, 117–119
Onion skin hue 401
Oomycetes fungi 87
Ordinary solutions 287
Organic acids 3–48, 91
  from fermentation 6–8
  in grapes 4–6
  state of salification 7
  steric configuration 3
Organoleptic characteristics of red wine 398, 402
Organoleptic defects 233–282
  eliminating 281–282
Organoleptic effect 281
Organoleptic quality 315–316, 370
Ornithine 114, 122
Osazone formation 74
Overfining 315–316
Oxaloacetic acid 6, 41
Oxidation
  buffer capacity 236–237
  role of 235–236
Oxidation–reduction agents 393
Oxidation–reduction concepts 388–389
Oxidation–reduction phenomena 388–389, 412–413
Oxidation–reduction potential 236, 389
  changes in 396
  and dissolved oxygen correlation 392–393
  impact of aerating wine 395–396
  influence of compounds in wine and external factors 393–394
  influence of oxidation–reduction agents 393
  influence of racking 394
  influence of topping up operations 395
  influence of wine components 394
  influence of winemaking operations 394–395
  measuring 391
  measuring in simple medium 389–390
  measuring in wine 390–392
  red wine
    effect of oxygen content 392
  white wine 396
Oxidative defects 235–238

Pall membranes 339
  characteristics 357
Particle charge detector 309
Particle counts 337
Pasteurization 371
Pectic substances 77–83
  acid 79–82
  impact on wine character 82–83
  molecular structures 79–82
  monomer composition 77–78
  neutral 82
  terminology 77–78
Pectinases 248
Pectins 77, 314

Pediococcus 66, 122
Penicillium frequentans 278
2,3,4,6-Pentachloroanisole (PCA) 260
2,3,4,6-Pentachlorophenol (PCP) 260
Pentathionic acid 104
Pentoses 66
Peonidin 154
PEP 4 gene 134
Peptidases, proteins resistance to 134
Perception threshold 205
Perlite 340
Permeability 338
Petunidin 194
pH
  and anthocyanin equilibrium 152–156
  applications 9–21
  concept 9–21
  decrease in 18
  definition 9–10
  differential equation 13
  expression in wine 10–11
  values of wines 10
pH meter 8, 10, 13
Phenol acids 5
  decarboxylation 246
Phenol aldehydes
  formation, impact of toasting 422
Phenol character 242–244
Phenol characteristics of red wine during aging 397–403
Phenol content
  during aging 198
  of red wine 172–173, 402, 403
  of white wines 172–173, 200
Phenol odors 243
Phenolic acids 142
  derivatives 142–144
  in grapes and wine 142
  structure of 148
Phenolic alcohols 144
Phenolic compounds 141–201
  fining effect on 315
  fractionation in grapes and wine 179–181
  location in grapes 184–186
  maturity 189–191
  maturity measuring 189–191
  organoleptic properties in red wines 181–183
  with polarized, double bonds, reactions with anthocyanins 167–168
  types of substances 142–152
  in white wines 199–200
Phenolic maturity 188–189
Phenols
  oxidation 118, 119
  properties 152
  see also Volatile phenols
Phenyl-2-acetate 60

Phenyl alanine 117
Phenyl esterase 246
Phenylhydrazine addition 74
Phloroglucinol 168
Phosphoenol pyruvate 6
Phosphomolybdic acid 130
Phosphophenolpyruvic acid 6
Phosphorus 94
Physicochemical parameters, influence of pretreatment 34
Physicochemical stabilization, impact of heating 371
Phytic acid 101
Pigments, extraction during vatting 191–192
pK 11
Polyacrylamide gel 124
Polyamide/polyimide membranes 342
PolyDADMAC 310, 311
Polygalacturonic acid 314
Polymer composition 176
Polymerization 397
   homogeneous 405
   reactions 195
Polymethylsiloxanes 259
Polyols 55–58
   concentrations in wines 56
Polypeptides 113
Polyphenols 252
   and protein interaction 158, 159
   extracted from oak
      impact of toasting intensity 420
   oxidation 163
   protein precipitation by 160
Polypropylene membranes 342
Polysaccharide colloids 344
Polysaccharides 66, 77, 296–298, 402, 406
   from *Botrytis cinerea* 86–88
   in Graisse 88–89
   negative charges attributable to 311
   reactions involving tannins with 158
   variations in must during ripening 78–79
Polytetrafluoroethylene (PTFE) 309
   membranes 342
Polyvinyl polypyrrolidone (PVPP) 173, 303, 329–330, 399
Polyvinylidene fluoride membranes 342
Pore sizes 341
Porosity 338
Porosity characteristics
   of oak, measurement 417
Potassium bicarbonate 19
Potassium bitartrate (KTH) 18–21, 23, 26, 27, 30, 373
   crystallization kinetics 36
   crystals 30, 31, 37
   precipitation inhibition 42
   solubility in water 22
   solubilization 30
Potassium calcium tartrate 22
Potassium cation 95
Potassium concentration 22
Potassium crystallization 45
Potassium ferrocyanide 43, 98–101
Potassium hydroxide 11
Potassium tartrate 21
Preference threshold 206
Prefilter cartridges 356, 358
Prelog rules 4
Proanthocyanidins 151
Procyanidins 149, 159
   breakdown by acid catalysis 161
   concentrations 186
   direct T-A type condensation 170
   flavanol precursors 149
   formation of carbocations from 159–162
   heterogeneous polymerization 165, 171
   oxidation reactions 162–164
   polymerization reactions 164–166
   structure 151
   type-A 150
   type-B 150
   type-C 151
   type-D 151
Proline 115, 117
Propyl gaiacol 143
Protective colloids 24, 30, 33, 43, 296–300, 303, 371
   composition and properties 296–297
   natural 297–298
Protein casse 124–131, 370
   mechanism 128
   prevention 132–138
Protein fining, *See* Fining
Protein stability
   tests 130–131
   white wines aged in the lees 134–138
Proteins 113, 124–131
   assay 125
   concentration in must 128–130
   and polyphenols interaction 158, 159
   precipitation by tannins 294
   reactions involving tannins with 158
   resistance to peptidases 134
   separation 124–128
      by capillary electrophoresis (CE) 127
      by chromatofocusing 125
      by electrophoresis 125
      by liquid chromatography 126
Protocatechuic acid 200
Pyrazines 113, 224
   structure 114
Pyrocatechol 152
Pyrogallol 152
Pyruvic acid 6

Index

Quercetin 200
Quinones 118

Racking 395, 410
  frequency 305
  influence on wine aeration 306
  role 304–307
  techniques 305–307
Raffinose 71
Rapid cold stabilization 38–39
Recognition threshold 205
Red wines
  acetic acid content 426
  aging 387–426
  aging phenomena 404–405
  barrel aging 411–415
  bottle aging 404–409
  color changes 399
  development, effect of type of barrel 416–424
  dual contamination 261
  ethyl acetate content 426
  ethyl-phenol content 253–256
  ethyl-phenol defects 249–251
  fractionation 180
  organoleptic characteristics of 402
  phases in aging 404
  phenol characteristics during aging 397–403
  phenol composition 172–173, 403
  phenolic compounds in 181
Redox potential 236
Redox system 5
Reducing sugar 66
Reduction odors 261–273
Reduction ratio (RR) 339
Refrigerator test 28
Resistivity 27, 28
Resveratrol 144
Retention efficiency 356
Reverse osmosis 341
Rhamnogalacturonan I (RG-I) 79
Rhamnogalacturonan II (RG-II) 81, 82
Rhamnosylquercetin 144, 145
Riboflavin 271
  photochemical reduction 272
Ribofuranose 70
Ripeness index 117
Ripening in Champagne grape varieties 117
Rot
  defects associated with 277
Rotary vacuum filter 350–351

Saccharomyces cerevisiae 83, 134, 137, 234, 242, 245, 248, 251, 277, 279
  amino acids sequence 137
Saccharomycodes ludwigii 234

Saccharose 71–72
  formation 71
Salicylic acid 142
Saturation temperature
  calculation 30–32
  concept 29–32
  linear correlation 32
  and stabilization temperature 33–35
    in full-scale production 35–36
Sauvignon Blanc wines
  glutathion content 276
  proteins, heat stability of 135
Sclerotium rolfsii 87
Scopoletin 144
Scopoline 144
Screening 342
Sedimentation 303–304
  conditions 303–304
Serine 115, 125
Siliceous 328
Siliceous earths 303
  properties 328
  use in winemaking 328
Silicone oil
  non-reactive 259
  reactive 259
Silicones 259
Sinapic acid 251
Slow cold-stabilization 34, 37–38
Slow stabilization 25
Sodium alginate 320–321
Sodium bentonite 326
Sodium cation 95
Sodium hydroxide 8, 14
Sol 288, 328
Solid content determination 337
Solubility exponential curve 22–24
Sorbic acid 279–280
  transformation 280
Sorbitol 56, 58
Sotolon 275
Sotolon, formation 64
  impact of aging conditions 276
Sparkling wine 24
Specific adsorption surface 325
Specific retention 356
Specific swelling volume 326
Spontaneous clarification 303
Spontaneous crystallization 24, 46
Spraying 24
Stability and clarity 285–300
Stability tests 28–37
Stabilization
  by physical and physicochemical processes 369–385
  treatments 301–330
  see also specific methods

Stabilization temperature
    and saturation temperature   33–35
        in full-scale production   35–36
Stabilization treatments
    principle   132
Stainless-steel equipment   234
Standard suspensions   287
Static contact process   38
Stereoisomerism   4
Sterilizing equipment   354–355
Stilbenes   144
Stokes' law   303
Storage conditions   241
Storage premises contamination   260–261
Streaming potential (PE)   309, 311
Strecker reaction   73
Succinic acid   7, 15
Sugar derivatives   75–77
Sugars   65, 68–72
    chemical properties   72–75
    oxidation products   77
Sulfates, reduction   264
Sulfur amino acids   262, 272
Sulfur compounds   206, 216–222
    volatile   262–273
Sulfur derivatives   61, 261–273
    light and heavy   263
Sulfur dioxide   5, 77, 173, 234, 236, 237, 246, 249,
    253–256, 273, 302, 426
    and anthocyanin equilibrium   152–156
Sulfuric acid   8
Sunlight flavor   272, 273
Supersaturation   24, 26, 27, 29, 31, 33
Surface charge density   310, 312
    of grape seed phenolic fractions   313
Surface prefilters   356
Swelling number   325
Synthetic membranes   341
Syringic acid   406
Syringol   143
S-3-(Hexan-1-ol)-Glutathion   222

Tangential filtration   358–361
    principles   358–359
    schematic diagram   359
Tangential microfiltration   359, 360
    applications   359–361
Tangential ultrafiltration   358
Tannic strength   183
Tannin-protein complexation   128, 133, 313
Tannin-protein interactions   312–314
    influence of medium   314
Tannins   130, 132, 147–152, 292, 297
    anthocyanins combined with   399
    assay   174–176, 180
    changes in   401–402
    characteristics   176–178

    chemical properties   152–171
    condensation reactions   168–171
    condensed   149, 159
    development   402–403
    development from skins and seeds   187
    direct A-T type condensation   169
    effects on flavor   195–197
    enological   321–322
    evolution as grapes ripen   184–191
    extraction during winemaking   191–193
    flavanol precursors   149
    hydrolyzable   148
    negative charges attributable to   311
    organoleptic properties   172–183
    polymerization   164
    properties   149
    reactions involving   195–197
    reactions with protein and polysaccharides   158
    types   184
    use in fining   321–322
Tartaric acid   4, 8, 9, 18, 41, 200, 240, 401
    buffer capacity, variations in   17
    derivatives   143
    solubility in water   22
Tartrate crystal seeding   37
Tartrate crystallization and precipitation   26–27
Tartrate instability   33
Tartrate precipitation   21–28, 138
    monitoring   27
    prevention   37–48
    principle   21–26
Tartrate stability test   383–384
Tartrate stabilization   25
Tartrates   5
Tartrazine   5
Taxifolin   145
Teinturiers   184
Temperature gradient method   32
Terpene compounds   206–211
Terpene glucosides   210
    forms   210
Terpenols   209, 210, 223
    glycosylated forms   209–211
α-Terpineol   207
2,3,4,6-Tetrachloroanisole (TeCA)   260
2,3,4,6-Tetrachlorophenol (TeCP)   260
Thiamin pyrophosphate (TPP)   6, 55
Thiocarbamic acids   267
    formation   267
Thiol/disulfide system, Oxidation–reduction balance
    53
Thiols   216–222, 271
    precursors   219–222
THK aggregates   26
Threonine   117, 125
Thresholds, concept of   205
Titration   310

Titration curves 13
Topping up 410–411
Total acidity 8–9
Tourne 240
TP fraction 180
Treatment effectiveness, monitoring 38
Treatment temperature 37
Trehalose 71
Trichloroacetic acid test 131
2,4,6-Trichloroanisole (TCA) 258, 260
  biosynthesis 259
2,4,6-Trichlorophenol (TCP) 260
Trichoderma 88, 344
Trihydroxy-3.5,4'-stilben 144
Trisulfide formation 269
Tryptophan 66
Turbidity 128, 286, 299, 301, 336
  appearance 336
  measurement 336
Tyndall effect 286, 301
Tyrosine 114, 117
Tyrosol 143, 199

Urea 119–120

Valine 117
Van der Waals attraction 291, 312
Vanillic acid 200
Vanillin 257
  bacterial transformation into guaiacol 258
Varietal aroma 205–227
  general concept 205–206
  use of term 206
Vescalagin 148
Vescalin 149
Vine sprays 267–268
Vinegar mother 241
Vinyl gaiacol 142
Vinyl-4-guaiacol 243
Vinyl-4-phenol 243
Vinyl-phenol 142, 167, 168, 243, 245
  concentration in white wines 244
  concentrations of white wines 245–249
  formation mechanism 247
Vinylpyrrolidone
  polymerization 329
Vitamin B1 6
Vitamin C 273
Vitis labrusca 222, 223
Vitis riparia 146
Vitis rotundifolia 222, 223
Vitis rupestris 146
Vitis vinifera 144, 145, 147, 206, 223
Volatile acidity 9–10
  formation by bacteria 238

Volatile compounds 51–64, 205
  extracted from oak 414–415, 423
    structure 414
Volatile phenols 244
  formation, impact of toasting 422
  in wine 143
  microbiological origin and properties 242–256
Volatile sulfur compounds 262–273
  heat-generated 268–271
  photochemical origin 271–273
Volatile thiol concentration
  impact of barrel aging 275
Volatile thiols
  assay in Sauvignon Blanc wines 218
  formation from cysteinylated precursors 221
  organoleptic impact 217

Weinsaure (wine acid) 5
White casse 94, 124
White wine aroma
  premature aging 274–277
White wines
  color of 199–201
  components contributing to color 200–201
  oxidation–reduction potential 396
  phenolic compounds in 199–200
  vinyl-phenol concentrations 245–249
Wine
  amino acids 111
  classification 245
  contamination due to corks 256–259
  treated with hydrogen and sodium cation exchange resins 379
Wine composition modifications 362–363
Wine loss 410–411
Winemaking practices 409–411
Winemaking, adapting to various factors 192–193
Wines, aerated
  iron reactions 97
Wurdig test 29–32

Xylopyranose 70

Yeast 66, 133, 234, 235, 238, 242, 245, 337
  cell walls 135
  exocelluar polysaccharides from 83–86
  mannoproteins 26, 43–46
  metabolism 262–267
Yellow–orange color development 401

Zeta potential 308
Zinc 107
  elimination 105
Zygosaccharomyces 234
Zymaflore VL1 248